THE DYNAMICAL PROCESSES OF BIODIVERSITY – CASE STUDIES OF EVOLUTION AND SPATIAL DISTRIBUTION

Edited by **Oscar Grillo**
and **Gianfranco Venora**

The Dynamical Processes of Biodiversity – Case Studies of Evolution and Spatial Distribution
Edited by Oscar Grillo and Gianfranco Venora

Published by InTech

Janeza Trdine 9, 51000 Rijeka, Croatia

Publishing Process Manager Dragana Manestar

Technical Editor Teodora Smiljanic

Cover Designer InTech Design Team

Additional hard copies can be obtained from orders@intechopen.com

The Dynamical Processes of Biodiversity – Case Studies of Evolution and Spatial Distribution, Edited by Oscar Grillo and Gianfranco Venora
p. cm.
ISBN 978-953-307-772-7

For EU product safety concerns:
IN TECH d.o.o., Prolaz Marije Krucifikse Kozulić 3, 51000 Rijeka, Croatia,
info@intechopen.com or visit our website at intechopen.com.

Meet the editors

Dr. Oscar Grillo was born in Catania (Sicily) in 1977, he is a food technologist with an international PhD in applied and environmental botany. Since 2003 he has been working as researcher at the Stazione Sperimentale di Granicoltura per la Sicilia, a governmental institute of agronomic research, mainly working with computer vision applied to food matrices and plant structures, above all seeds, and in particular studying wheat and the related leguminous. Currently, he is working also at the Sardinian Germplasm Bank of the Biodiversity Conservation Centre of the University of Cagliari on projects devoted to seed characterization and identification by image analysis. Results of his work have been published in many peer-reviewed journals papers and international conference papers. Referee for a few peer-reviewed journals, many times he has been invited as teacher/lecturer/speaker by some universities and research centres in Spain and Italy. He has trained many MSc and PhD students, who have made their own contributions to the agronomical and botanical research.

Dr. Gianfranco Venora is a biologist, born in Caltagirone (Sicily) in 1958, where he lives and works. He took his University degree in 1981, and since 1982 he has been working as researcher at the Stazione Sperimentale di Granicoltura per la Sicilia. His working expertise is mainly about durum wheat and leguminous breeding. At the beginning of 1990, after some years of experience on karyotyping cropped and wild species of agronomical importance, he was fascinated by computer vision applied to food matrices and plant structures, above all seeds, studying wheat and the related leguminous. He was recently nominated as professor to the research doctorate on Applied and Environmental Botany of the University of Cagliari. Many peer-reviewed journals published his papers and he was invited as speaker in many international conferences, and as teacher/lecturer by some universities and research centres in Germany, Netherlands, Bulgaria, Spain, Czech Republic and Italy. Gianfranco Venora is currently referee for about 15 peer-reviewed journals and tutor of many MSc and PhD students.

Contents

Preface

Discoveries of new species have always represented a demanding challenge for mankind, derived from the human wish to improve the quality of his own life. Each new specie has always been considered as a potential new food or medicine, as well as a possible source of fuel or clothes. But today, exploring new animal and plant species mainly derives from men's effort to try to understand the life on Earth in order to tackle some of the problems caused by his own species.

The current world's biodiversity consists of an innumerable amount of dynamic species in constant pursuit of the best solutions to react and survive the natural and anthropic environmental changes, suggesting us innovative strategies to overcome human limits and live better.

> *"Blind metaphysical necessity, which is certainly the same always and every where, could produce no variety of things. All that diversity of natural things which we find suited to different times and places could arise from nothing but the ideas and will of a Being necessarily existing."*
>
> (from *Philosophiae naturalis principia matematica*, Isaac Newton, 1687)

Divided into 15 chapters written by internationally renowned contributors, this book offers a few case studies about the diversity of many life forms. It includes systematic overviews, biogeographic and phylogenic backgrounds, species composition and spatial distribution in more or less restricted areas of the world, offering to the reader an overall view of the present condition in which our planet is.

Oscar Grillo
Stazione Sperimentale di Granicoltura per la Sicilia, Caltagirone
Biodiversity Conservation Centre, University of Cagliari
Italy

Gianfranco Venora
Stazione Sperimentale di Granicoltura per la Sicilia, Caltagirone,
Italy

1

Biodiversity and Evolution in the *Vanilla* Genus

Gigant Rodolphe[1,2], Bory Séverine[1,2], Grisoni Michel[2] and Besse Pascale[1]
[1]*University of La Reunion, UMR PVBMT*
[2]*CIRAD, UMR PVBMT,*
France

1. Introduction

Since the publication of the first vanilla book by Bouriquet (1954c) and the more recent review on vanilla biodiversity (Bory et al., 2008b), there has been a world regain of interest for this genus, as witnessed by the recently published vanilla books (Cameron, 2011a; Havkin-Frenkel & Belanger, 2011; Odoux & Grisoni, 2010). A large amount of new data regarding the genus biodiversity and its evolution has also been obtained. These will be reviewed in the present paper and new data will also be presented.

2. Biogeography, taxonomy and phylogeny

2.1 Distribution and phylogeography

Vanilla Plum. ex Miller is an ancient genus in the Orchidaceae family, Vanilloideae sub-family, Vanilleae tribe and Vanillinae sub-tribe (Cameron, 2004, 2005).

Vanilla species are distributed throughout the tropics between the 27th north and south parallels, but are absent in Australia. The genus is most diverse in tropical America (52 species), and can also be found in Africa (14 species) and the Indian ocean islands (10 species), South-East Asia and New Guinea (31 species) and Pacific islands (3 species) (Portères, 1954). From floral morphological observations, Portères (1954) suggested a primary diversification centre of the *Vanilla* genus in Indo-Malaysia, followed by dispersion on one hand from Asia to Pacific and then America, and on the other hand from Madagascar to Africa. This hypothesis was rejected following the first phylogenetic studies of the genus (Cameron, 1999, 2000) which suggested a different scenario with an American origin of the genus (160 to 120 Mya) and a transcontinental migration of the *Vanilla* genus before the break-up of Gondwana (Cameron, 2000, 2003, 2005; Cameron et al., 1999). The genetic differentiation between New World and Old World species observed would therefore be a consequence of the further separation of the continents. Our recent molecular phylogeny using chloroplastic *psa*B, *psb*B, *psb*C, and *rbc*L regions (Bouetard et al., 2010) supported the hypothesis of an American origin of the genus (figure 1). However, the recent discovery of a fossilized orchid pollinaria (20 Mya) (Ramirez et al., 2007) allowed the dating of Vanilloidae sub family at 72 Mya, well after the separation of Gondwana which questions the hypothesis of a vicariate evolution of the *Vanilla* genus (Bouetard et al., 2010).
Transoceanic dispersion appears more credible and would have been implied at least three times in the evolution of the *Vanilla* genus (figure 1). This was demonstrated by dating a *Vanilla* molecular phylogeny, testing these two extreme evolutionary scenarios (vicariate

versus transoceanic dispersion) (Bouetard et al., 2010) (figure 1). The Gondwanan dispersion scenario used 95 Mya as prior on the NW/OW node (the minimum age assumption for the break-up of Gondwana), whereas the NW/OW transoceanic dispersion scenario used 71 Mya as prior on the Vanilloidae node (a date estimated from fossil orchid pollinaria dating (Ramirez et al., 2007)) (figure 1). This provided evidence for at least three transoceanic dispersion events whatever the original scenario retained for the differentiation of NW versus OW species: from Africa to Asia, from Africa to the South West Indian Ocean Islands, and from Africa back to America (Carribean region) (Bouetard et al., 2010) (figure 1).

2.2 Taxonomy and phylogeny

Taxonomic classification is based on morphological variations in vegetative and floral characters. Ephemeral flowers and their scarce availability in herbarium specimens associated with the fact that vegetative characters show important intra-specific variations are responsible for the difficulties in providing a clear taxonomic classification in *Vanilla* (Bory et al., 2010).
The first classification (Rolfe, 1896) distinguished two sections in the genus: section Foliosae, and section Aphyllae with leafy or leafless species, respectively. Portères (1954) then divided section Foliosae in three sub-sections: Papillosae, with thick leaves and a labellum with fleshy hairs, Lamellosae with thick leaves and a labellum with scaly lamellae, and Membranacae with thin membranous leaves.
The *Vanilla* genus taxonomy has recently greatly beneficiated from molecular phylogenetics. The sequences used were chloroplastic *rcb*L (Cameron et al., 1999; Soto Arenas & Cameron, 2003), *psa*B (Cameron, 2004), *psb*B and *psb*C (Cameron & Molina, 2006), and the results obtained showed that Rolfe's sections and Portères' sub-sections classically used for taxonomy in *Vanilla* did not have a phylogenetic value. A recent study (Bouetard et al., 2010), based on these four markers combined, revealed three major clades in the genus, called groups α, β, et γ (figure 1). Group α is represented by *V. mexicana* and is ancestral. Separation between group β (composed of New World/American Foliosae species) and group γ (composed of Old World/African and Asian Foliosae and American, Asian and African Aphyllae species) is more recent. This study confirmed an American origin of the genus, and also showed that the sections Foliosae and Aphyllae are not monophyletic (figure 1), a statement that questions the classical taxonomic treatment of the genus proposed by Rolfe (1896) and Portères (1954).
Recently, based on phylogenetic data of 106 species, (Soto Arenas & Cribb, 2010) proposed a new taxonomic classification, differentiating two sub-genera in the *Vanilla* genus. A group contains species previously classified as sub-section Membranaceae: *V. angustipetala, V. martinezii, V. inodora, V. mexicana, V. parviflora, V. edwalii* and the monospecific genus *Dictyophyllaria dietschiana* now *V. dietschiana* (Bouetard et al., 2010; Cameron, 2010; Pansarin, 2010a2010b; Soto Arenas & Cameron, 2003). It was named genus *Vanilla* sub-genus *Vanilla* as it contains the typus species for the genus (*V. mexicana*). It corresponds to the ancestral phylogenetic group α (figure 1). The remaining *Vanilla* species are included in genus *Vanilla* sub-genus *Xanata,* which is further divided in two sections: section *Xanata* (corresponding to phylogenetic group β) and section *Tethya* (group γ) (figure 1). Within section *Xanata,* an early diverging group is noteworthy (figure 1) containing *V. palmarum, V. lindmaniana* and *V. bicolor* (Bouetard et al., 2010; Cameron, 2010; Soto Arenas & Cameron, 2003). This preliminary revised classification is a major step towards a needed complete revision of the genus based on molecular analyses.

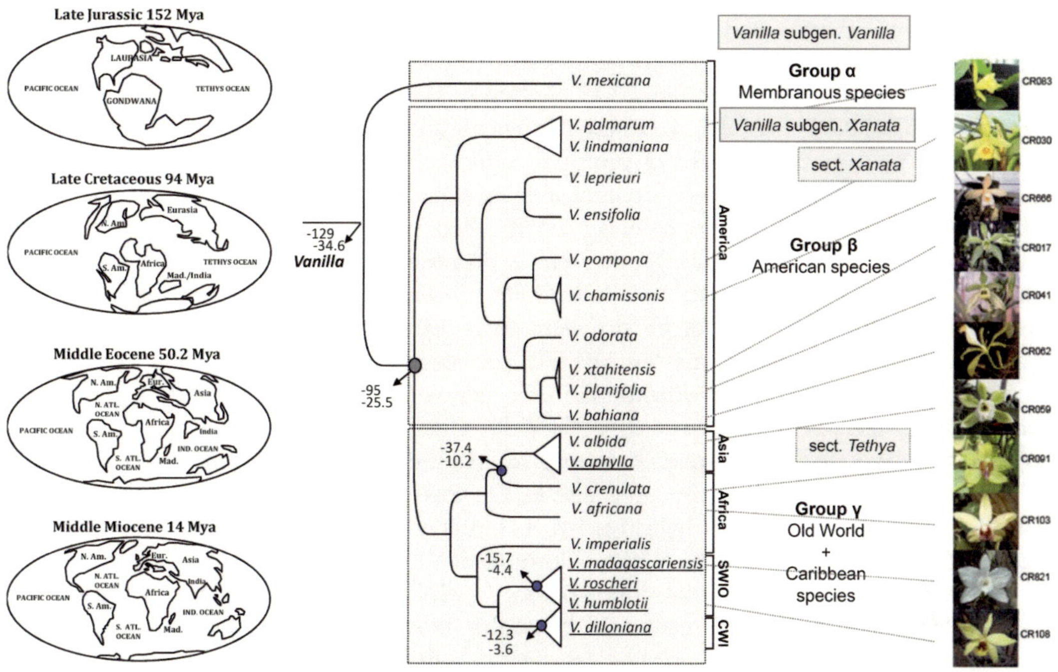

Fig. 1. Schematic representation of the molecular phylogeny of the *Vanilla* genus based on *rbc*L, *psa*B, *psb*B and *psb*C (Bouetard et al., 2010), distinguishing clades α, β and γ. The geographical origin of the species is indicated. Species underlined are from sect. Aphyllae, others are from sect. Foliosae (as per Rolfe's classification). Taxonomic classification as per Soto Arenas & Dressler (2010) is indicated. Flowers of representative species and their voucher number (CR) in the BRC Vatel collection are presented (photographs: M Grisoni). Estimated divergence times (in Mya) derived from Bayesian relaxed clock analyses (uncorrelated exponential relaxed molecular clock model) (Bouetard et al., 2010) are indicated for key nodes: (i) origin of *Vanilla*, (ii) separation between New and Old World *Vanilla* species; (iii) separation between African and Asian species; origin of Aphyllae species (iv) in the South West Indian Ocean area and (v) in the Caribbean-West Indies area. Upper values correspond to the Gondwanan dispersion scenario and lower values correspond to the transoceanic dispersion scenario. Blue dots on clade nodes indicate transoceanic dispersion whatever the scenario tested. World maps at different geological times are provided.

In the first thorough taxonomic treatment of the genus published, Portères (Portères, 1954) described 110 species in the *Vanilla* genus. This number was reduced by different authors (Cameron et al., 1999; Soto Arenas, 1999, 2006; Soto Arenas & Dressler, 2010), but some species were not included (Hoehne, 1945) and new species have since been described (Z.J. Liu et al., 2007; Pignal, 1994; Soto Arenas, 2006, 2010; Soto Arenas & Cameron, 2003; Szlachetko & Veyret, 1995). There are to date more than 200 *Vanilla* species described (Bory et al., 2008b; Cameron, 2011b), but numerous synonymies remain and there is therefore an urgent need to thoroughly revise the taxonomic classification of the *Vanilla* species. We recently reviewed (Bory et al., 2010) the complexity of the processes involved in the evolution and diversification of the *Vanilla* genus and concluded that *Vanilla* must be considered as a TCG, a "Taxonomic Complex Group" (Ennos et al., 2005). Indeed, it exhibits (i) an uniparental reproduction mode (vegetative growth) (Portères, 1954) (ii) interspecific hybridization in sympatric areas (Bory et al., 2010; Bory et al., 2008c; Nielsen, 2000; Nielsen & Siegismund, 1999) and (iii) polyploidy (Bory et al., 2010; Bory et al., 2008a; Lepers-Andrzejewski et al., 2011a; Lepers-Andrzejewski et al., 2011b). These mechanisms have profound effects on the organization of the biological diversity and have been described as responsible for the difficulty to define discrete, stable and coherent taxa in such TCGs (Ennos et al., 2005). *Vanilla* is a typical example of a genus for which the barcoding protocols (*mat*K and *rbc*L) as proposed by the CBOL (M.L. Hollingsworth et al., 2009; P.M. Hollingsworth & CBOL Plant Working Group, 2009 ; Ratnasingham & Hebert, 2007), will therefore not be sufficient to revise the species taxonomy. The lack of genetic incompatibility between most *Vanilla* species (Bory et al., 2010) and the proven occurrence of inter-specific hybridizations in the genus (Bory et al., 2010; Bory et al., 2008c; Nielsen, 2000; Nielsen & Siegismund, 1999) will necessitate the obligate survey of nuclear regions in addition to cpDNA markers to resolve introgression patterns and correctly identify *Vanilla* species (Rubinoff, 2006). As an example, the species *V. ×tahitensis* was recently shown to be a *V. planifolia* x *V. odorata* hybrid using a combined ITS and chloroplastic phylogenetic analysis (Lubinsky et al., 2008b), when chloroplastic DNA alone repeatedly identified this species as identical to its maternal donor parent *V. planifolia* (figure 1). Moreover molecular genetic diagnostics can only be useful for barcoding biodiversity when species delimitations are either subtle or cryptic but nonetheless clear-cut. In a TCG, taxon limits are themselves diffuse, therefore genetic analysis alone might fail in the identification of discrete species (Ennos et al., 2005). A typical example of expected difficulties will be within the *V. pompona* species complex which was recently described as containing subspecies *pompona, pittieri,* and *grandiflora* based on ITS data, although the latter two are rather paraphyletic (Soto Arenas & Cribb, 2010) . In *Vanilla,* taxonomic revision of species will therefore have to use a combination of taxonomic, morphological, ecological, reproductive biology, cytogenetic (polyploidy estimates) and genetic (nuclear and chloroplastic) assessments.

3. *Vanilla* biodiversity in the wild

Most *Vanilla* species are hemiepiphytic vines climbing up to 30 meters high (*V. insignis*) (Soto Arenas & Dressler, 2010) and growing in tropical wet forests between 0-1000m (Portères, 1954). Only a few species are adapted to drier conditions (*V. calycullata,* (Soto Arenas & Dressler, 2010)), although extreme xeric adaptation is observed in the 18 leafless species of the genus (Portères, 1954). Vegetative reproduction (by natural stem cuttings) is the predominant reproduction mode adopted by most *Vanilla* species to develop settlements, such as *V. bahiana, V. chamissonis, V. madagascariensis, V. dilloniana, V. barbellata, V. claviculata* (reviewed in (Bory et al., 2010)). Some vines can grow up to 100 meters long (*V. insignis* (Soto Arenas & Dressler, 2010)) and in *V. planifolia* the same individual can cover up to 0.2ha (Soto Arenas, 1999). However a few species might be strictly sexually reproducing, such as *V. bicolor* and *V. palmarum* which are described as epiphytic on palm trees (Householder et al., 2010; Pignal, 1994), and *V. mexicana* (Bory et al., 2010; Cameron, 2010). Another notable exception is the species *V. dietschiana* which is non lianescent and 40 cm high, and has long been classified for these reasons as a different genus *Dictyophyllaria* (Pansarin, 2010a, 2010b; Portères, 1954).

In natural conditions, vanilla plant density can be extremely variable from being very high in certain areas (*V. trigonocarpa* (Soto Arenas & Dressler, 2010), *V. pompona* (Householder et al., 2010)) from very low as reported for wild *V. planifolia* in Mexico with less than one plant found per square kilometre (Soto Arenas, 1999). Some species are known to flower very frequently (*V. chamissonis,* (Macedo Reis, 2000)) to very un-frequently (*V. planifolia, V. hartii,* (Schlüter, 2002; Soto Arenas & Dressler, 2010)). A single flower per inflorescence generally opens in *Vanilla*, except 2-3 in some species (*V. odorata, V. martinezii, V. insignis*) and flowers are ephemeral (one day) except for some rare species such as *V. inodora* (2-3 days) (Soto Arenas & Dressler, 2010) or *V. imperialis* for which the flowers can be fertilized 4-5 days after opening (unpublished data). Seedlings can be found very frequently for species such as *V. bicolor* and *V. palmarum* (Householder et al., 2010) or be extremely rare as in *V. pompona* in Madre de Dios (Householder et al., 2010) or *V. planifolia* in Mexico (Schlüter, 2002). All these natural history traits will have deep effects on the levels of *Vanilla* species biodiversity that can be found in the wild. Particularly, the relative balance between vegetative and sexual reproduction and their relative efficiency will be of major importance in shaping populations genetic diversity. Exploring *Vanilla* species reproductive systems is therefore essential in this context.

3.1 *Vanilla* pollination

Vanilla species, like other orchids, are characterized by the presence of a rostellum membrane separating female and male reproductive systems, therefore limiting self-pollination. The diverse floral morphology observed in *Vanilla* species (figure 1) suggests that they have evolved to adapt to different pollinators (Soto Arenas & Cameron, 2003).

3.1.1 Self-pollinating species

A few *Vanilla* species are described as spontaneously self-pollinating (Householder et al., 2010; Soto Arenas & Cameron, 2003; Soto Arenas & Dressler, 2010; Van Dam et al., 2010), as suggested by their abnormally high fruit set (table 1). This is consistent with general data in orchids showing that autogamous species display a much higher fruit set (77%) than cross pollinating species for which the majority show fruit set <20% (Tremblay et al., 2005). Based on high fruit set, these suggested autogamous species are *V. palmarum, V. savannarum, V. bicolor* (American species of the *V. palmarum* group), *V. guianensis, V. martinezii* (American species of the *V. mexicana* group) and *V. griffithii* (an Asian species). Possible self-pollination for *V. inodora* is also reported (Soto Arenas & Dressler, 2010), due to the large fruit set observed in some populations, although others have a fruit set as low as 2.5%.

Species	Natural fruit set (self-pollination)	Reference
V guianensis	78%	(Householder et al., 2010)
V. palmarum	76%	(Householder et al., 2010)
V bicolor	71%	(Householder et al., 2010)
V. bicolor	42.5% per raceme	(Van Dam et al., 2010)
V. martinezii	53% in a clone	(Soto Arenas & Dressler, 2010)

Table 1. Suggested self-pollinating *Vanilla* species and recorded natural fruit sets.

More precise observations are available for some of these species. *V. guianensis* is supposedly self-pollinated at early anthesis, as it was observed that the stigma and the anther grew to contact one another; and no pollinators were observed despite the high fruit set recorded in Peru (Householder et al., 2010). The lack of observed local pollinators and the high fruit set also suggested that *V. bicolor* and *V. palmarum* were self-pollinating species in Peru (Householder et al., 2010).

Two mechanisms were proposed to account for self-pollination in *Vanilla* species (Van Dam et al., 2010): true self-pollination occurring by either stigmatic leak and/or the presence of a dehydrated or reduced rostellum, or agamospermy. In *V. bicolor*, pollen removal experiments showed that agamospermy was not the mechanism in play (Van Dam et al., 2010). Also all fertilized flowers showed fully developed rostellum. This suggested that a stigmatic leak, where stigma lobes release a fluid that contacts the pollen and induces germination of the pollen tubes (Van Der Pijl & Dodson, 1966) was the more likely explanation for self-pollination in this species (Van Dam et al., 2010). The observation of the occurrence of a thick rostellum in *V. palmarum* led to the suggestion of an identical mechanism (Householder et al., 2010). Our own observations on *V. palmarum* reveal self-pollination most likely due to a rostellum reduced in width, allowing pollinaria to get in contact with the stigmata on both sides of the rostellum (figure 2). A similar situation is found for the self-fertile species *V. lindmaniana* (data not shown).

Fig. 2. Detailed structure of the pollinaria, rostellum and stigmata in the species *V. palmarum:* (a) and (b) accession CR0891, (c) accession CR0083, maintained in BRC Vatel (Reunion Island).

Spontaneous self-pollination is sometimes described even in classically outcrossing species. In Oaxaca plantations, cases of *V. planifolia* self-pollination are reported (Soto Arenas & Cameron, 2003) with rates reaching 6% of covered flowers giving fruit. Similar rates (6.06%) were reported for bagged *V. chamissonis* flowers in Sao Paulo (Macedo Reis, 2000). Nothing is known about the mechanisms involved in such exceptional cases.

3.1.2 Outcrossing species and pollinators

For the majority of *Vanilla* species, self-pollination does not occur due to an efficient rostellum and sexual reproduction therefore relies on the intervention of pollinators. Consequently, relatively low natural fruit sets are observed in natural conditions ((Bory et al., 2008b), table 2), consistent with the 17% median natural fruit set reported for tropical orchids (Tremblay et al., 2005). Reproductive success in orchids is pollination - rather than by resource - limited and could depend on pollinator effectiveness, abundance and diversity, and pollen quantity and quality (self *versus* allopollen) (Tremblay et al., 2005). This was demonstrated by crossing experiments in temperate and tropical orchids showing that cross hand-pollination shows significantly greater success (80%) than natural open pollination (26.6%) (Tremblay et al., 2005). Further studies are needed in *Vanilla* to determine the highest fruit sets achievable, but results on *V. barbellata, V. claviculata, V.dilloniana,* and *V. poitaei* have showed up to 100% fruit set under hand pollination experiments (Tremblay et al., 2005), and 75.76% in *V. chamissonis* (Macedo Reis, 2000), much higher values than what can be observed in natural conditions (table 2).

Species	Natural fruit set (open pollination)	Reference
V. barbellata	18.2 %	(Tremblay et al., 2005)
V. chamissonis	15%	(Macedo Reis, 2000)
V. claviculata	17.9 %	(Tremblay et al., 2005)
V. crenulata	0%	Johansson 1974, as cited in (Soto Arenas & Cameron, 2003)
V. cristato-callosa	6.6%	(Householder et al., 2010)
V. dilloniana	14.5 %	(Tremblay et al., 2005)
V. planifolia	1% to 1‰	(Soto Arenas, 1999)
V. planifolia	1%	(Childers & Cibes, 1948)
V. planifolia	1%	(Tremblay et al., 2005)
V. planifolia	1 à 3%	(Weiss, 2002)
V. poitaei	6.4 %	(Tremblay et al., 2005)
V. pompona subsp. grandiflora	0.9%	(Householder et al., 2010)
V. riberoi	1.1%	(Householder et al., 2010)

Table 2. *Vanilla* out-crossing species and natural fruit sets recorded.

If the pollinator of *V. planifolia* was long been considered as a social bee from the *Melipona* genus, as reported by Deltiel (as cited in (Rolfe, 1896)) and then mentioned in (Bouriquet, 1954a, 1954b; Stehlé, 1954), these records are now admitted as doubtful (Soto Arenas & Cameron, 2003; Van Der Cingel, 2001) as the bee is too small to perform the necessary pollination steps (Lubinsky et al., 2006; Soto Arenas & Cameron, 2003). Lubinsky (2006), during observations of *V. planifolia* in Oaxaca (Mexico) and *V. pompona subsp. grandiflora* in Peru, indeed noticed *Melipona* visits, but no pollen movement was recorded. In tropical America (Guadeloupe (Stehlé, 1952) and Mexico (Stehlé, 1954)), authors have also reported the intervention of *Trigona* bees for *Vanilla* pollination, but this has never been confirmed. In Puerto Rico, leafless *Vanilla* species might be pollinated by *Centris* bees (Soto Arenas & Cameron, 2003). Hummingbirds are considered as vanilla pollinators in tropical America (Bouriquet, 1954a1954b; Stehlé, 1954). Lubinsky (2006) did indeed observe occasional *V. planifolia* visits by hummingbirds in Oaxaca, but with no pollen movement. Finally some authors (Dobat & Peikert-Holle, 1985; Geiselman et al., 2004) have suggested that the species *V. chamissonis* could be pollinated by two species of bats, although this fact was recently questioned (Fleming et al., 2009).

It is much more likely that in the American tropics, *Vanilla* is pollinated by large euglossine bees, as suggested by Dressler (1981) and demonstrated by such bees caught with *Vanilla* species pollinaria (Ackerman, 1983; Roubik & Ackerman, 1987). The principal reward offered by orchid flowers is nectar (Dressler, 1993), the most common reward for pollination (Van Der Pijl & Dodson, 1966). No *Vanilla* species has been described as producing floral nectar to our knowledge. However, the pollinators that visit orchid flowers can also obtain a variety of rewards (Singer, 2003; Tremblay et al., 2005) including oil, floral fragrances and, occasionally, pollen or stigmatic exudates (Bembe, 2004).

From years of observations in Mexico, Soto Arenas (Soto Arenas, 1999; Soto Arenas & Cameron, 2003) suggested the existence of three pollination systems for American *Vanilla* species (Bory et al., 2008b).

The first system relies on fragrance collection on flowers by male bees of the *Euglossa* genus, and has been suggested to concern the species of the *V. pompona* group as well as *V. hameri*, *V. cribbiana*, and *V. dressleri* (Soto Arenas, 1999; Soto Arenas & Cameron, 2003; Soto Arenas & Dressler, 2010). In this 'male euglossine syndrome' (Williams & Whiten, 1983) also referred to as 'perfume flower syndrome' (Bembe, 2004), now well known in many non nectar producing orchid species, male bees are attracted solely by the flower fragrance, and rub the surface of the flower with special tarsal brushes to collect fragrance materials, and subsequently store them in swollen glandular tibiae of the rear legs (Dodson et al., 1969). This fragrant orchid- male euglossine bee relationship is often highly specific (Dodson et al., 1969; Williams & Whiten, 1983). Bees then supposedly use these fragrance compounds as precursors for their own sex pheromones (Williams & Whiten, 1983) or in a "spraying" (of the fluid substances from their mid tibial tufts by vibrating action of their hind wings) behaviour as part of their courtship displays (Bembe, 2004). No study has so far been conducted to analyze *Vanilla* species flower fragrance compounds diversity and their relationship with pollinator specificity. This could give great insights on *Vanilla* evolution and diversity. On the other hand no direct evidence has been provided with regards to this male euglossine scent collection behaviour in any *Vanilla* flowers so far. Pollination of *V. trigonocarpa* by male *Euglossa asarophora* in Panama was reported (Soto Arenas & Dressler, 2010), with no information regarding scent collection behaviour. Male *Eulaema meriana* was identified as a possible pollinator for the species *V. pompona* subsp. *grandiflora* in Peru following observations of visits accompanied by pollen movement, but no scent collection behaviour was observed (Lubinsky et al., 2006). Similarly, some particularly fragrant flowers of this species were shown to attract two species of euglossine bees, *Eul. meriana* and *Eug. imperialis* (Householder et al., 2010). Only *Eul. meriana* was observed pollinating flowers on two occasions, but no floral fragrance collection was recorded (Householder et al., 2010). This does not so far therefore confirm the suggested male euglossine syndrome within the *V. pompona* group. Most species seem to be pollinated under a deceptive system, as also suggested for *V. planifolia, V. odorata, V. insignis* and *V. hartii,* with flower visits by either male or female bees and an absence of reward (Soto Arenas, 1999; Soto Arenas & Cameron, 2003). This particular pollination system, using different strategies to lure pollinators, is mainly encountered in orchids with a third of the species in this family supposedly using this pollination system (Jersakova et al., 2006; Schiestl, 2005; Singer, 2003; Tremblay et al., 2005), particularly low density species (Ackerman, 1986), as it is the case for *V. planifolia* (Bory et al., 2008b; Soto Arenas, 1999). Soto Arenas considers the bee *Eugl. viridissima,* and maybe bees from the *Eulaema* genus, to be the real pollinators of *V. planifolia* (Bory et al., 2008b; Soto Arenas & Dressler, 2010). These species (as well as *Exeretes*) were recorded as occasional visitors of *V. planifolia* in Oaxaca (Mexico) without pollen movement (Lubinsky et al., 2006). *V. cribbiana* is reported to be pollinated by an unidentified *Eulaema* bee, *V. hartii* flowers are visited by female *Euglossa* bees and *V. insignis* flowers by male bees of *Eul. polychroma* (Soto Arenas & Dressler, 2010). The true pollinators of *V. planifolia* and most allied species therefore remain to be elucidated.

The last system might imply strong and large carpenter bees (*Xylocopa* species) and would concern the species *V. inodora.* This was suggested based on the peculiar floral structure of this species and allied Membranaceae (Soto Arenas & Cameron, 2003) characterized by a frontally closed labellum (the column apex lying on the lip) which is similar to that of other orchid species pollinated by carpenter bees (Soto Arenas & Cameron, 2003). These bees were observed visiting *V. inodora* but no proof of true pollination has been provided so far (Soto Arenas & Cameron, 2003; Soto Arenas & Dressler, 2010). The only data available on *Vanilla* potential pollinators, although partial, is therefore from America. There is a considerable lack of knowledge of potential *Vanilla* pollinators in other geographical areas. In Africa, euglossine bees do not occur, but other large bees may be pollinators there (Van Der Cingel, 2001). Despite three years of observation of the species *V. crenulata* in Africa, no pollinator visit was recorded (Johansson, 1974, as cited in (Soto Arenas & Cameron, 2003)). Observations in Madagascar of occasional natural fruit set in the introduced species *V. planifolia,* were attributed locally to sunbirds of the *Cynniris* genus (so called 'Sohimanga") (Bouriquet, 1954a). Similarly, in Reunion Island, rare natural pollination events of the introduced *V. planifolia* may be linked to noticed visits by the bird *Zosterops* (Zosteropidae) (Bory et al., 2008b), an Angraecoid orchid pollinator there (Micheneau et al., 2006). These hypotheses have not been confirmed, and remain unlikely as flower structure in *Vanilla* is indicative more of a bee pollination system (Dressler, 1981). Finally, a large bee of the *Aegilopa* genus was recorded pollinating *V. cf. kaniensis* in Papua New Guinea (Soto Arenas & Cameron, 2003). Although fruits of *V. albida* and *V. aphylla* from Java were described and illustrated in 1832, the introduced species *V. planifolia* did not naturally set fruit there, showing the need for different pollinators (Arditti et al., 2009). No other information is available regarding *Vanilla* pollinators in Asia (Van Der Cingel, 2001). It will be important to assess whether *Vanilla* species with higher fruit set (table 2) are characterized by reward pollination mechanisms as it was demonstrated that rewarding orchids show significantly higher fruit set than deceptive ones (twice as much) (Tremblay et al., 2005). Reproductive success might also be related to the fragrance attractiveness of flowers, even in a deceptive system. Further insights on this matter could be obtained by characterising *Vanilla* species floral fragrance and colour as well as identifying their respective pollinators and behaviour.

Partial information is available (Soto Arenas & Dressler, 2010) for *V. planifolia* stating the presence of 1-2-dimethyl-cyclopentane, ethyl acetate,1-8-cineol and ocimene-trans, and for *V. insignis* possessing the same principal constituents although ocimene-trans is notoriously absent. 1-8-cineol is especially well known to be a strong attractant for euglossine bees (Soto Arenas & Dressler, 2010). Our own observations (unpublished data) show that the species *V. chamissonis* displays particularly strongly fragrant flowers (more than *V. planifolia*), this could explain why its fruit set is amongst the highest.

3.2 Myrmecology

An obvious interaction exists between *Vanilla* and ants, as also demonstrated for other orchid species (Peakall, 1994). Extrafloral nectar is produced in immature bud abscission layer in many *Vanilla* species such as *V. pompona, V. cristato-callosa* in Peru (Householder et al., 2010) and *V. planifolia* in Panama (Peakall, 1994) and ants were observed in these species feeding on sugary exudates. Ants were also reported visiting *V. planifolia* flowers in Oaxaca (Lubinsky et al., 2006), without pollination. *V. planifolia* also occasionally inhabits ant nests, and was also observed to support ant nests in its root mass (Peakall, 1994).

The benefit of the association is obvious for the ant (food and shelter), but the benefit (if any) for the *Vanilla* plant remains to be elucidated. In some orchid species, ants visiting extrafloral nectaries have been shown in some cases to protect them against herbivory or to be attractors to bird pollinators (Peakall, 1994). Close association between ant nests and orchids have also suggested a role of ants in seed dispersion particularly in orchids with oily seeds (Peakall, 1994). In fragrant *Vanilla* fruits, seeds are held in an oily matrix (Householder et al., 2010). Ants have been reported in vanilla crop to be important for humus disintegration (Stehlé, 1954). On the other hand, the presence of ants could simply be indicative of the presence of mealybugs, softscales or aphids rather than an indication of a mutualistic interaction (Chuo et al., 1994). In *V. planifolia*, associations between scale and the black ant *Technomyrmex albipes* in Seychelles, as well as between ants and the aphid *Cerataphis lataniae* have been reported (Risbec, 1954).

3.3 Fragrance and bees and fruit dispersion

Seed dispersal mechanism(s) of *Vanilla* remains enigmatic. Fruits reaching maturity in many *Vanilla* species show dehiscence (Bouriquet, 1954c). This character favours seed dispersal, although it is noticeably not interesting in fruit crop production. In aromatic fruits, *Vanilla* seeds are easily rubbed off and are extremely sticky due to a thin covering of oil, which may favour epizoochorous seed dispersal by any visitor, insect or vertebrate (Householder et al., 2010). Soto Arenas and Cameron (2003) mentioned that *Vanilla* species producing fragrant fruits are restricted to tropical America and proposed the designation of group β (figure 1) as the 'American fragrant species' group, but this should not include species from the *V. palmarum* group as these were described as non-fragrant ((Householder et al., 2010), see below). Fruit fragrance was described as a pleisiomorphic character in orchids as it is present in *Vanilla* and in three other primitive groups (*Cyrtosia, Neuwiedia, Selenipedium*) (Lubinsky et al., 2006).

It has been demonstrated that euglossine bees are attracted by fragrant *Vanilla* fruits and act as seed collectors and potential dispersers. Van Dam et al. (2010) have photographed male *Eul. cingulata* with a typical scent collection behaviour on *V. pompona* subsp.*grandiflora* fruits in Peru. Householder et al. (2010) also reported strong attractiveness of fruit of this species to *Eul. meriana* and *Eug. imperiali* which may stay on the same fruit for 15 minutes displaying typical scent collection behaviour. They also observed a similar behaviour by a metallic green *Euglossa* sp. on old and dehiscent *V. cristato-callosa* fruits. This confirmed previous observations of euglossine bees brushing on *Vanilla* fruits (Madison, 1981) and demonstrated the particular attractiveness of these bees to fragrant *Vanilla* flowers as well as to fragrant fruits, an important evolutionary step in the orchid/orchid-bee relationship in *Vanilla*. As discussed by Lubinsky et al. (2006), this demonstrates that the orchid/orchid-bee relationship has evolved in *Vanilla* as a mode of flower pollination as well as fruit dispersion *Trigona* bees were observed in Peru transporting sticky *V. pompona* seed packets on their hind tibia and often dropping them (Householder et al., 2010). These bees are not typical scent collectors and could just be interested in the nutritional value of the oils (Householder et al., 2010). One species of carpenter bee (*Xylocopa* sp) is also mentioned visiting *V. pompona* fruits (Householder et al., 2010).

Fruit dispersal by bats was suggested for *V. insignis* and observed for *V. pompona* (Soto Arenas & Dressler, 2010). Occasional total or partial herbivory of the fruit was also noticed for *V. pompona* in Peru, possibly attributed to bats or marsupials (Householder et al., 2010).

Bird dispersal is expected in some Asian species, as *V. abundiflora* and *V. griffithii*, as in the closely related Vanilloideae *Cyrtosia* genus (Soto Arenas & Dressler, 2010). However *Cyrtosia* has fleshy fruits like *Vanilla* but these are bright red presumably acting as an attractor to birds or mammals (Cameron, 2011b)

For some other *Vanilla* species however, fruits are non fragrant and seeds are not held in a particularly oily matrix. This is the case for *V. bicolor* and *V. palmarum* (Householder et al., 2010). Dehiscence of the fruits and canopy habitat suggested a different mechanism of seed dispersal in such species, by a combination of wind turbulence and gravity (Householder et al., 2010).

3.4 Conclusions

Many *Vanilla* species are threatened in the wild. This is particularly the case for *V. planifolia* in Mexico, its centre of origin. Proper conservation strategies need to be developed, but this will require gaining a better knowledge on the reproductive strategies and the derived levels of genetic diversity in these *Vanilla* species. This will include assessing the relative contribution of vegetative *vs* sexual reproduction, self-compatibility (auto *vs* allo fecundation success), pollination syndromes (pollinators, reward/deceit) and seed dispersion systems.

There is a considerable lack of genetic studies of *Vanilla* species biodiversity in the wild. The only published data concern the aphyllous species *V. barbellata*, *V. dilloniana* and *V. claviculata* on the island of Puerto Rico (Nielsen, 2000; Nielsen & Siegismund, 1999) using isozyme markers. Genotypic frequencies were in accordance with Hardy-Weinberg proportions for all species, which could suggest random crosspollination. High differentiation among populations was detected, supposedly attributed to limited seed dispersal by bees. Genetic drift was also demonstrated in some isolated populations (Nielsen & Siegismund, 1999). Soto Arenas also conducted *V. planifolia* population genetic studies in Mexico using isozymes (Soto Arenas, 1999), surprisingly demonstrating homozygous excess corresponding to preferential autogamous reproduction for this species.

Development of suitable approaches to the analysis of genetic diversity in a spatial context, where factors such as pollination, seed dispersal, breeding system, habitat heterogeneity and human influence are appropriately integrated in combination with molecular population genetic estimates, will be essential (Escuderoa et al., 2003) to provide new insights in the understanding of the mechanisms of maintenance and dynamics of *Vanilla* populations and to provide guidelines for their preservation.

4. Vanilla biodiversity in cultivated conditions

Vanilla is the only orchid with a significant economic importance in food industry. It is cultivated for its aromatic fruit, a character restricted to some species from the American continent (Soto Arenas & Cameron, 2003). Only two species are grown to produce commercial vanilla: *V. planifolia* and *V.* ×*tahitensis*; with *V. planifolia* providing 95% of the world production, mainly originating from Madagascar, Indonesia, Comoros, Uganda and India (Roux-Cuvelier & Grisoni, 2010). Biodiversity in cultivated conditions depends on the level of diversity originally introduced and on cultivation practices used in different countries during domestication. Vanilla crops are established from stem cuttings of 8–12 nodes, collected from healthy and vigorous vines (Bory et al., 2008b; Bouriquet, 1954a; Purseglove et al., 1981; Soto Arenas & Cameron, 2003; Stehlé, 1952). As natural pollinators are absent in the areas of vanilla production, pollination is performed by hand following a simple method discovered by the slave Edmond Albius in Reunion Island in 1841 (Kahane et al., 2008). Given these cultivation practices, low levels of genetic diversity are expected in cultivation areas. However, for both species, different varieties, showing recognized but poorly defined morphological, agronomical and aromatic properties, are often cultivated by growers (Duval et al., 2006). Given the vegetative mode of propagation and the absence of pollinators, five hypotheses have been proposed to explain these variations (Bory et al., 2008b): (i) multiple introduction events, (ii) somatic mutations, (iii) sexual reproduction, (iv) polyploidy and (v) epigenetic modifications. In recent years, these hypotheses were explored, giving new insights on the processes involved during the dispersion and domestication of the two main cultivated *Vanilla* species. These results also give important clues to the understanding of *Vanilla* evolutionary processes in natural conditions.

4.1 *V. planifolia* in Reunion Island

The species *V. planifolia* originated in Mesoamerica (Portères, 1954). Some of the history of vanilla follows the history of chocolate because vanilla was gathered from the wild for use in flavoring chocolate beverages in the pre-Columbian Maya and Aztec cultures of southeastern Mexico and Central America. However, the Totonac people of Papantla in north-central Veracruz (Mexico) were probably the first group to cultivate *V. planifolia* (Lubinsky et al., 2011). The species *V. planifolia* has an interesting history of dispersal to other tropical regions between 27° N and 27° S latitudes (Lubinsky et al., 2008a). After the discovery of the Americas by C. Colombus, the whole history of *V. planifolia* dissemination, following the discoveries of manual pollination by the slave Edmond Albius in 1841 and curing process by E. Loupy and D. De Floris is intimately linked to Reunion Island (Kahane et al., 2008). From then, *V. planifolia* was renowned as 'Bourbon Vanilla' since it was produced originally from Reunion Island (from 1848) and later from a cartel of Indian Ocean Island producers (Madagascar, Reunion, Comoros and Seychelles).

The true origin of cultivated vanilla outside of Mexico was unclear until AFLP and microsatellite markers were used to elucidate the patterns of introduction of *V. planifolia.* These studies showed that most of the accessions cultivated today in the islands of the Indian Ocean and worldwide (Reunion Island, Madagascar, French Polynesia, French West

Indies, Mexico) and of different morphotypes (from Reunion 'Classique', 'Mexique', 'Sterile', 'Grosse Vanille' (table 3) and from Mexico 'Mansa', 'Acamaya', 'Mestiza') (Bory et al., 2008c; Lubinsky et al., 2008a) derive from a single introduced genotype. It could correspond to the lectotype that was introduced, early in the nineteenth century, by the Marquis of Blandford into the collection of Charles Greville at Paddington (UK) (Portères, 1954). Cuttings were sent to the botanical gardens of Paris (France) and Antwerp (Belgium) from where these specimens were disseminated to Reunion Island (by the ordinance officer of Bourbon, Marchant) and then worldwide (Bory et al., 2008b; Kahane et al., 2008).

Consequently, cultivated accessions in Reunion Island exhibit extremely low levels of genetic diversity and have evolved by the accumulation of point mutations through vegetative multiplication (Bory et al., 2008c) (table 3). Maximum genetic distance (Dmax) was 0.106 and the majority of the polymorphic AFLP bands revealed had frequencies in the extreme (0-10% and 90-100%) ranges, therefore corresponding to rare AFLP alleles (presence or absence) a pattern typical of point mutations (Bory et al., 2008c). One peculiar and rare phenotype 'Aiguille' found in Reunion Island was shown to result from sexual reproduction (selfing) (Bory et al., 2008c) (table 3) as its AFLP pattern fell within a group of selfed progeny with Dmax=0.140 and showed a strong pattern of segregation bands. The hypothesis was that it resulted from manual self-pollination and subsequent seed germination from a forgotten pod (Bory et al., 2008c). Flow cytometry, microdensitometry, chromosome counts and stomatal lengths showed that polyploidization has been actively involved in the diversification of *V. planifolia* in Reunion Island (Bory et al., 2008a). Three ploidy levels (2x, 3x, 4x) were revealed that allowed to explain the features of the 'Sterile' type which is auto-triploid and of the 'Grosse Vanille' type, auto-tetraploid (Bory et al., 2008a). It was suggested that these resulted from the production of non-reduced gametes during the course of manual self-pollination performed by growers (Bory et al., 2010; Bory et al., 2008a).

As the particular phenotype 'Mexique' encountered in Reunion could not be explained by genetic or cytogenetic variations, we tested whether it could have resulted from epigenetic modifications as some studies showed that morphological variations in clonal populations could be explained by a combination of genetic and epigenetic factors (Imazio et al., 2002). Epigenetics corresponds to reversible but heritable modifications of gene expression without changes in the nucleotidic sequence (Mathieu et al., 2007; Wu & Morris, 2001), such as DNA methylation (Finnegan et al., 1998). Epigenetic modifications are heritable (Akimoto et al., 2007; Finnegan et al., 1996; Grant-Downton & Dickinson, 2006; Martienssen & Colot, 2001) and transmitted as well as by asexual propagation (Peraza-Echevarria et al., 2001).

Sometimes, a phenotypic reversion correlated with demethylation of the epi-mutated gene can occur and its expression is restored (Jaligot et al., 2004). These epigenetic mutations have important phenotypic as well as evolutionary consequences, this representing a current field of investigation (Finnegan, 2001; Kalisz & Purugganan, 2004; B. Liu & Wendel, 2003). DNA methylation proceeds by the addition in a newly replicated DNA of a methyl group by a DNA methlytransferase (Finnegan et al., 1998; Martienssen & Colot, 2001). Cytosine is the most frequently methlylated base, resulting in 5-methylcytosine formation (^{5m}C) (Martienssen & Colot, 2001). Plant methylation is restricted to the nuclear genome and is concentrated in repeated sequence regions (Finnegan et al., 1998). Methylation is implied in many biological processes such as 'gene silencing', mobile DNA elements control, DNA replication duration, chromosome structure determination, and mutation frequency increase (Finnegan et al., 1998; Paszkowski & Whitham, 2001). Many spontaneous or induced epimutations are known in maize, Arabidopsis and other plant species and are responsible

Morphotypes	Characteristics	Diversity/genetics	Origin
'Classique'	The most cultivated type	Point mutations Dmax = 0.106	Mexico then Antwerp Botanical Gardens
'Aiguille'	Slender leaves and thin pods	As self progenies Dmax=0.140	Selfing of 'Classique'
'Sterile'	'Classique', but self-sterile	Same AFLP profile as 'Classique', auto-triploid (3x)	Selfing of 'Classique', unreduced gamete (2n x n)
'Grosse Vanille'	Bigger leaves, stems, flowers and fruits than 'Classique'	Same AFLP profile as 'Classique', auto-tetraploid (4x)	Selfing of 'Classique', unreduced gametes (2n x 2n)
'Mexique'	Darker bluish leaves with central gutter and curved sides, cylindrical pods	Same AFLP and MSAP profile as 'Classique'	Epigenetic or genetic single dominant mutation with pleiotropic effects

Table 3. *V. planifolia* morphotypes encountered in Reunion Island and their description.

Accession	Morphotype	Collection	Accession	Morphotype	Collection
CR0217	'Classique'	Provanille 3A11	CR0493	'Mexique'	Provanille 15A8
CR0218		Provanille 3A11	CR0494		Provanille 15A8
CR0219		Provanille 3A11	CR0495		Provanille 15A8
CR0343	'Classique'	Provanille 6A8	CR0334	'Mexique'	Provanille 6A5
CR0344		Provanille 6A8	CR0335		Provanille 6A5
CR0345		Provanille 6A8	CR0336		Provanille 6A5
CR0457	'Classique'	Provanille 15A6	CR0337	'Mexique'	Provanille 6A6
CR0458		Provanille 15A6	CR0338		Provanille 6A6
CR0459		Provanille 15A6	CR0339		Provanille 6A6
CR0563	'Classique'	Provanille 16B2	CR0001	'Mexique'	BRC Vatel
CR0564		Provanille 16B2	CR0002	'Mexique'	BRC Vatel
CR0565		Provanille 16B2	CR0627	'Mexique'	BRC Vatel StP
CR0340	'Classique'	Provanille 6A7	CR0649	'Mexique'	BRC Vatel StP
CR0341		Provanille 6A7	CR0632	'Mexique'	BRC Vatel StP
CR0342		Provanille 6A7	CR0711	'Classique'	BRC Vatel SteR
CR0647	'Classique'	BRC Vatel StP	CR0714	'Classique'	BRC Vatel SteR
CR0650	'Classique'	BRC Vatel StP			

Table 4. *V. planifolia* Reunion Island accessions surveyed in the MSAP analysis (StP: Saint Philippe; SteR: Ste Rose).

for the generation of mutant phenotypes (Finnegan et al., 1996; Martienssen & Colot, 2001). To assess whether 'Mexique' morphotypes might have resulted from epigenetic modifications, we selected the MSAP (Methylation-sensitive amplified polymorphism) method (Reyna-López et al., 1997), an AFLP-derived methodology which allows the visualization of a large number of markers revealing cytosine methylation state at each digestion site, without any *a priori* knowledge of genomic sequences. MSAP analyses were performed on a sample of 'Classique' and 'Mexique' accessions (table 4). Twenty-four accessions were collected in the collection of Provanille in Bras-Panon (Reunion Island), corresponding to 8 varieties with three cuttings. This was to verify if genetic or methylation polymorphism, if existing, is transmitted through vegetative multiplication. Others were collected in vanilla plantations in Reunion Island (St-Philippe or Ste-Rose) and are maintained in the BRC Vatel collection.

We used the restriction enzyme *Eco*RI as well as *Msp*I and *Hpa*II, isochizomers that cut the same restriction site CCGG but show different sensitivity to methylation (table 5). The MSAP methodology used was as described in (Reyna-López et al., 1997). *Hpa*II digests were repeated twice. The adaptators used are presented in table 6 and 8 *Eco/Hpa* primer combinations were used for selective amplification.

	*Eco*RI/*Hpa*II	*Eco*RI/*Msp*I	CCGG methylation
Case number 1	1	1	CCGG
Case number 2	1	0	^{5hm}CCGG
Case number 3	0	1	C^{5m}CGG
Case number 4	0	0	^{5m}C^{5m}CGG or ^{5m}CCGG

Table 5. Methylation sensitivity of *Hpa*II and *Msp*I (^{m} : methylation; ^{hm} : hemimethylation).

The comparison of the profiles from the amplification after DNA digestion with *Eco*RI/*Hpa*II and *Eco*RI/*Msp*I gives informations on the methylation status of the internal cytosine in sequence CCGG (table 5). For example a band present in the *Msp*I profile and absent in *Hpa*II indicates a methylation of the internal cytosine, whereas the opposite situation indicates an hemimethylation of the external cytosine. A methylation event was considered as polymorphic when at least one accession differed from the others in its profile.

Name	Sequence (5′-3′)
Double strand adaptators	
Ad EcoRI1	CTC GTA GAC TGC GTA CC
Ad EcoRI2	AAT TGG TAC GCA GTC
Ad HpaII1	GAT CAT GAG TCC TGC T
Ad HpaII2	CGA GCA GGA CTC ATG A
Pre-amplification primers	
Eco-A	GAC TGC GTA CCA ATT CA
Hpa-A	TCA TGA GTC CTG CTC GGA
Selective amplification primers	
Eco-AC	GAC TGC GTA CCA ATT CAC
Eco-AG	GAC TGC GTA CCA ATT CAG
Hpa-ATT	ATC ATG AGT CCT GT CGG ATT
Hpa-ATG	ATC ATG AGT CCT GT CGG ATG
Hpa-AAC	ATC ATG AGT CCT GT CGG AAC
Hpa-AAG	ATC ATG AGT CCT GT CGG AAG

Table 6. Adaptator and primer sequences used in MSAP analysis.

Between 48 and 70 fragments were revealed by primer combination. On the 483 CCGG sites observed, 188 were non methylated (38.9%), 36 were methylated (7.45%), with 5 sites only presenting methylation polymorphisms (1.03%) in 4 accessions. Accessions CR0340 and CR0341 were hypomethylated, they showed bands in both their *Hpa*II and *Msp*I profiles whereas the other accessions only presented these bands with *Msp*I. CR0340 was hypomethylated at locus Eco-AG/Hpa-AAC/98bp and CR0341 at locus Eco-AC/Hpa-ATT/426bp. Accessions CR0632 and CR0711 were hypermethylated, they presented some bands in their *Msp*I profiles whereas the other accessions presented these bands in both their *Hpa*II and *Msp*I profiles. Accession CR0632 was hypermethylated at locus *Eco*-AG/*Hpa*-ATG/205bp and at locus *Eco*-AG/*Hpa*-AAC/382bp. Finally, accession CR0711 was hypermethylated at locus *Eco*-AG/*Hpa*-AAG/393bp.

These results showed that methylation is present in *V. planifolia* genome, with 7.45% of the fragments revealed being methylated. This value is in accordance with methylation rates reported in banana (7,5%, (Noyer et al., 2005)), but less than what is revealed in other conventionally propagated plant species such as rice (16.3%, (Xiong et al., 1999)), other bananas (18.4%, (Peraza-Echeverria et al., 2001)), apple (25%, (Xu et al., 2000)) and cotton (32%, (Keyte et al., 2006)).

A limited amount of methylation polymorphism (1%) was detected among 'Classique' and 'Mexique' accessions but the methylation patterns revealed were accession specific. Even for CR0340/0341/0342 which are three clones of the same accession, two different methylation polymorphisms were revealed in CR0340 and CR0341 and none in CR0342, showing that these methylation patterns are either not transmitted trough asexual propagation, or have appeared after clonal propagation. In all cases, no methylation marker was identified which could allow to specifically distinguish the 'Classique' and 'Mexique' morphotypes. A similar conclusion was obtained in studies performed on vegetatively propagated plants such as banana (Baurens et al., 2003; Noyer et al., 2005). Methylation polymorphisms were revealed but could not be correlated to morphological variations.

We therefore conclude that the 'Mexique' morphotype showing no detectable AFLP or MSAP polymorphism is most probably the result of a limited genetic or epigenetic dominant mutation event with pleiotropic effects.

4.2 *V. ×tahitensis* in French Polynesia

The mysterious history of the origin of *V. ×tahitensis*, the so called Tahitian vanilla, has partly been solved. As opposed to its allied species (*V. planifolia*) it cannot be found wild in tropical American forests (Moore, 1993; Portères, 1954; Soto Arenas & Cameron, 2003) but was described from cultivated material found in the Island of Raiatea (Lubinsky et al., 2008b), where it had been introduced via the botanical garden of Papeete from the Philipines in 1848 (Soto Arenas & Dressler, 2010). Molecular sequencing (ITS and cpDNA) have recently shown that *V. ×tahitensis* would be a hybrid, intentional or not between *V. planifolia* and *V. odorata* dating from vanilla exploitation by Mayas in Mesoamerica between years 1359-1500 (Lubinsky et al., 2008b).

As much as 18 different morphotypes are described in *V. ×tahitensis* in French Polynesia beside the most widely cultivated type 'Tahiti' (Lepers-Andrzejewski et al., 2011a). These include: 'Haapape' (the second most cultivated type because of its bigger fruits), 'Tahiti Court', 'Tiarei', 'Ofe Ofe', 'Oviri', 'Parahurahu' and 'Sterile'. A study of 16 different accessions using AFLP markers revealed a Dmax value of 0.150, a slightly higher value than what was revealed in *V. planifolia*. All accessions had patterns related to that of 'Tahiti' (either identical or showing missing bands) which led the authors to conclude of a single introduction event in French Polynesia of a 'Tahiti' vine, consistently with the fact that this accession is the oldest one recorded in Polynesia (Lepers-Andrzejewski et al., 2011a). Ten accessions showed more or less the AFLP profile of 'Tahiti'. These included 'Haapape' and 'Tiarei', which were shown to be autotetraploids based on flow cytometry and chromosome counts (Lepers-Andrzejewski et al., 2011b). Similarly as in *V. planifolia* (Bory et al., 2008a), 'Sterile' morphotypes in *V. ×tahitensis* were also related to autotriploidy (Lepers-Andrzejewski et al., 2011b). It was hypothesized that they originated from a cross between the two most cultivated morphotypes 'Tahiti' (2x) and 'Haapape' (4x) (Lepers-Andrzejewski et al., 2011b). The remaining accessions showed a pattern related to 'Tahiti' but with 15 to 30 missing bands (Lepers-Andrzejewski et al., 2011a), a pattern consistent with segregation, as shown in *V. planifolia* for the 'Aiguille' morphotype or selfed progenies (Bory et al., 2010). For these accessions, graphical genotypes were constructed based an AFLP 'Tahiti' map and showed that morphotypes such as 'Parahurahu', 'Rearea', 'Oviri' and 'Tahiti court' displayed patterns consistent with an origin via self-pollination of 'Tahiti' (one single recombination event per bivalent) whereas others such as 'Popoti' and 'Paraauti' most probably resulted from a second generation of self-pollination (two recombinations events in the same bivalent) (Lepers-Andrzejewski et al., 2011a).

4.3 Conclusions

These results therefore highlight two different domestication models. In both cases, the genetic base of the cultivated material is very narrow with obviously a single genotype introduced ('Classique' *V. planifolia* in Reunion Island and other cultivation areas; 'Tahiti' *V. ×tahitensis* in French Polynesia). Genetic variation revealed is however slightly higher in *V. ×tahitensis* than in *V. planifolia* because most of *V. ×tahitensis* morphotypes have resulted from selfing of the original 'Tahiti' (with sometimes more than one generation involved) (Lepers-Andrzejewski et al., 2011a). Only one rare case of self-pollination ('Aiguille') was detected in Reunion (Bory et al., 2010). This shows that deliberate or inadvertent seed germination has been strongly involved in the domestication of *V. ×tahitensis* in French Polynesia. In Reunion Island, the limited amount of variation revealed is more related to vegetative propagation and the consecutive accumulation of point mutations.

In both cases however, a noticeable diversification was achieved through polyploidy. Autotetraploidy generated varieties with bigger leaves and fruits, and autotriploidy generated self-sterile individuals. It is noteworthy that self-sterile *V. planifolia* varieties were also described in Mexico ('Oreja de Burro') (Castillo Martinez & Engleman, 1993; Soto Arenas & Dressler, 2010). It is most likely that these have resulted as well from autotriploidy. These results, as well as those that surveyed genome sizes in a wide range of *Vanilla* species (Bory et al., 2010) provide converging evidences for the importance of polyploidy and genome rearrangements during *Vanilla* evolution. Polyploidy can be of major importance in cultivation as well as in natural populations as triploidy and to a certain extent tetraploidy can be responsible for dramatic loss in fruit set. Further work is therefore needed to assess polyploidization consequences on *Vanilla* reproductive biology.

5. *Vanilla* genome dynamics

Concordant data obtained on *V. planifolia* as well as *V. ×tahitensis* demonstrated an abnormal mitotic behaviour in the *Vanilla* genus, with a combination of somatic aneuploidy and partial endoreplication (Bory et al., 2008a; Lepers-Andrzejewski et al., 2011b).

5.1 Somatic aneuploidy

Most data in the literature give a basic number n=16 for *V. planifolia* with 2n= 32 (Chardard, 1963; Heim, 1954; Hoffmann, 1929, 1930; Martin, 1963). Hurel-Py (1938) was the first to show the existence of a variable number of chromosomes in differentiated cells (13 to 32 chromosomes). Similarly, Nair & Ravindran (1994) described an important variation in chromosome numbers, from 20 to 32 with 28 being the most encountered. Recent analyses confirmed the existence of such somatic hypo-aneuploidy (i.e. chromosome number is always below an exact multiple of the usually haploid number) in root tip cells of *V. planifolia* (Bory et al., 2008a), *V. ×tahitensis* (Lepers-Andrzejewski et al., 2011b) as well as other *Vanilla* species (Bory, 2007). This aneuploidy could be explained by somatic associations of chromosomes (Nair & Ravindran, 1994) but as well by chromatin elimination (Lepers-Andrzejewski et al., 2011b). Interestingly, it was recently demonstrated that somatic aneuploidy is regulated between somatic and gametic cells in *V. ×tahitensis*, with the full genome complement present in germ cells (Lepers-Andrzejewski et al., 2011b). This suggests that a regulatory mechanism functions during meiosis to stabilize the genome and chromosome number

5.2 Progressively partial endoreplication

Flow cytometry genome size estimates and chromosome counts have been successfully used to demonstrate the occurrence of diploid, triploid and tetraploid accessions of *V. planifolia* in Reunion Island (Bory et al., 2008a) and *V. ×tahitensis* in French Polynesia (Lepers-Andrzejewski et al., 2011b). Genome size variations were also demonstrated in some other species of the *Vanilla* genus (Bory et al., 2010). Flow cytometry revealed endoreplication in somatic cells of *V. planifolia* and *V. ×tahitensis*. In *V. planifolia* the marginal replication ratio, which is the ratio between each peak position, was irregular with 1.43, 1.63, 1.76, 1.82 instead of 2.00 (Bory et al., 2008a). In *V. ×tahitensis* it was 1.38, 1.65, 1.77, 1.79 and 1.81 (Lepers-Andrzejewski et al., 2011b). The almost perfect linearity found between DNA content and the number of endoreplication cycles suggested that the same genome part (or chromosome batch) (P, figure 3) is amplified at each cycle. A matrix of only 43.73% and 38% of the holoploid nucleus is replicated at each cycle in *V. planifolia* and *V. ×tahitensis*, respectively (Bory et al., 2008a; Lepers-Andrzejewski et al., 2011b).
More importantly, this phenomenon is apparently present in all the *Vanilla* species surveyed so far. Flow cytometry genome size estimates for 38 accessions representing 17 different *Vanilla* species and 3 artificial inter-specific hybrids revealed, for each accession, fluorescence histograms with five endoreplicated peaks and the marginal replication ratio was still irregular (from 1.5 to 1.8 instead of 2) (Bory et al., 2010). Nothing is known concerning the mechanisms in play, whether it results from partial replication of the DNA or excision of DNA (possibly chromatin elimination) following whole genome replication, but they occur in many orchids (Bory et al., 2008a). It will be important in the near future to gain knowledge on this developmentally regulated "progressively partial endoreplication" phenomenon unique to orchids. Available data already show that it is vegetatively, as well as sexually transmitted as demonstrated by surveying interspecific hybrids, such as the natural hybrid *V. ×tahitensis* (Lepers-Andrzejewski et al., 2011b) and artificial hybrids (*V. planifolia × V. planifolia, V. planifolia × V. ×tahitensis, V. planifolia × V. phaeantha)* (Bory et al., 2010). This phenomenon is technically important as the first peak (2C) is often very small, and this was shown to be responsible for considerable errors in the genome size estimates that have been published in the literature for *Vanilla* species (Bory et al., 2008a; Lepers-Andrzejewski et al., 2011b). This phenomenon is also evolutionary important as it was shown to be a source of polyploidization in many plant species. However it cannot itself explain the origin of autotetraploid types in *V. planifolia* and *V. ×tahitensis* as these have

exactly double the amount of DNA than their diploids counterparts, unless endoreplication in meristematic cells is regulated (Lepers-Andrzejewski et al., 2011b).

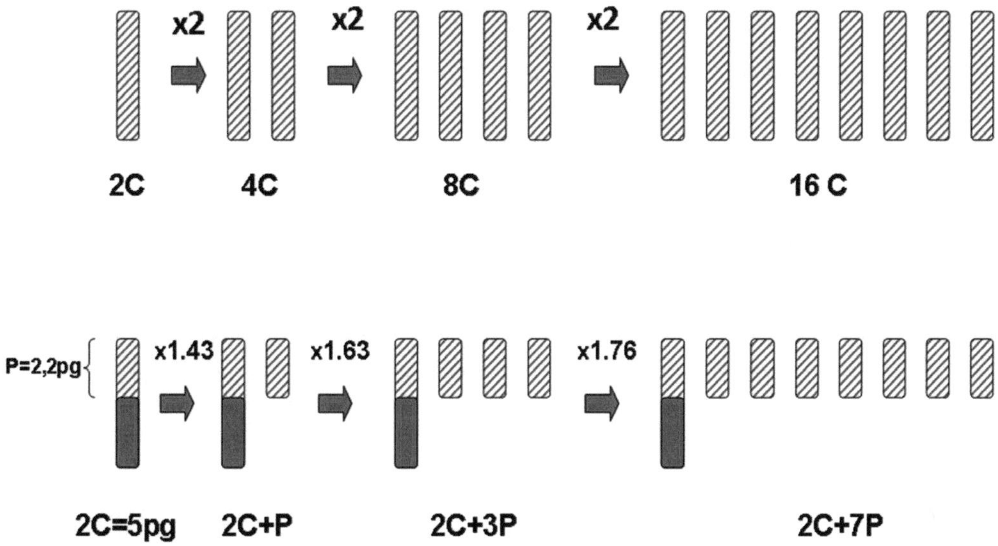

Fig. 3. Partial progressive endoreplication in *V. planifolia* (below) as compared to normal endoreplication (above). The replicated part (P) of the *V. planifolia* genome is indicated (hatched).

6. Conclusion

Although considerable progress has been made in recent years in the precision of the taxonomy and the discovery of evolutionary processes in the *Vanilla* genus (reproduction, genetic diversity, polyploidy, hybridization), many questions remain unanswered. These include elucidating the complex processes involved in genome dynamics and its possible implications on the genus diversification. Evolutionary pathways of important traits in the genus such as self-pollination ability and aromatic compounds accumulation in fruits, which are major targets for vanilla breeding, will need to be surveyed. Self-pollination appears as an ancestral character in the genus, shared by species from group α and early diverging species from group β. Furthermore, although allied genera possess aromatic fruit, this character is found in *Vanilla* within American group β, but not in ancestral American nor in more recent species from Africa and Asia. The aromatic character of both flowers and fruit in *Vanilla* has evolved in a specialized relationship with euglossine bees involved in both flower pollination and fruit dispersion. This represents an exciting further area of investigation. Molecular and cytogenetic studies will have to be combined with morphological, history traits and ecological assessments to provide a thorough revision of the genus taxonomy. In particular, more data is needed to fully characterize the reproductive biology of *Vanilla* species and its implication on the levels of genetic diversity in natural populations. This will be essential to provide conservation guidelines for the many endangered species of the genus.

7. References

Ackerman, J.D. (1983). Specificity and mutual dependency of the orchid-euglossine interaction. *Biol. J. Linn. Soc.*, Vol.20, pp.301-314.

Ackerman, J.D. (1986). Mechanisms and evolution of food-deceptive pollination systems in orchids. *Lindleyana*, Vol.1, pp.108-113.

Akimoto, K., Katakami, H., Kim, H.J., Ogawa, E., Sano, C.M., Wada, Y.& Sano, H. (2007). Epigenetic Inheritance in Rice Plants. *Annals of Botany* Vol.100, pp.205-217.

Arditti, J., Rao, A.N.& Nair, H. (2009). Hand-polination of Vanilla: how many discoverers? In: *Orchid Biology: Reviews and Perspectives XI*, J. Arditti, Wong, (Eds.), pp. 233-249. Springer

Baurens, F.C., Bonnot, F., Bienvenu, D., Causse, S.& Legavre, T. (2003). Using SD-AFLP and MSAP to Assess CCGG Methylation in the Banana Genome. *Plant Mol. Biol. Rep.*, Vol.21, pp.339_348.

Bembe, B. (2004). Functional morphology in male euglossine bees and their ability to spray fragrances (Hymenoptera, Apidae,Euglossini). *Apidologie*, Vol.35 pp.283-291.

Bory, S. (2007). *Diversité de Vanilla planifolia dans l'Ocean Indien et de ses espèces apparentées: aspects génétiques, cytogénétiques et épigénétiques*, Ph D thesis Reunion Island University (France).

Bory, S., Browm, S., Duval, M.F.& Besse, P. (2010). Evolutionary Processes and Diversification in the Genus *Vanilla*. In: *Vanilla*, E. Odoux, M. Grisoni, (Eds.), pp. 15-28. Taylor and Francis Group

Bory, S., Catrice, O., Brown, S.C., Leitch, I.J., Gigant, R., Chiroleu, F., Grisoni, M., Duval, M.-F.& Besse, P. (2008a). Natural polyploidy in *Vanilla planifolia* (Orchidaceae) *Genome*, Vol.51 pp.816-826.

Bory, S., Grisoni, M., Duval, M.-F.& Besse, P. (2008b). Biodiversity and preservation of vanilla: present state of knowledge. *Genetic Resources and Crop Evolution*, Vol.55, pp.551-571.

Bory, S., Lubinsky, P., Risterucci, A.M., Noyer, J.L., Grisoni, M., Duval, M.-F.& Besse, P. (2008c). Patterns of introduction and diversification of *Vanilla planifolia* (Orchidaceae) in Reunion island (Indian ocean). *American Journal of Botany*, Vol.95, pp.805-815.

Bouetard, A., Lefeuvre, P., Gigant, R., Bory, S., Pignal, M., Besse, P.& Grisoni, M. (2010). Evidence of transoceanic dispersion of the genus *Vanilla* based on plastid DNA phylogenetic analysis. *Molecular Phylogenetics and Evolution*, Vol.55, pp.621-630.

Bouriquet, G. (1954a). Culture. In: *Le vanillier et la vanille dans le monde*, G. Bouriquet, (Ed.), pp. 430-458. Paul Lechevalier, Paris

Bouriquet, G. (1954b). Historique. In: *Le vanillier et la vanille dans le monde*, G. Bouriquet, (Ed.), pp. 9-26. Paul Lechevalier, Paris

Bouriquet, G. (1954c). *Le vanillier et la vanille dans le monde. Encyclopédie Biologique XLVI.*, Paul Lechevalier, Paris VI.

Cameron, K.M. (1999). Biogeography of Vanilloideae (Orchidaceae), 749.

Cameron, K.M. (2000). Gondwanan biogeography of vanilloid orchids, 25-26.

Cameron, K.M. (2003). Vanilloideae. In: *Genera orchidacearum: Orchidoideae*, A.M. Pridgeon, P.J. Cribb, M.W. Chase, F.N. Rasmussen, (Eds.), pp. 281-285. Oxford University Press, USA

Cameron, K.M. (2004). Utility of plastid *psa*B gene sequences for investigating intrafamilial relationships within Orchidaceae. *Mol. Phylogenet. Evol.*, Vol.31, pp. 1157-1180.

Cameron, K.M. (2005). Recent Advances in the Systematic Biology of *Vanilla* and Related Orchids (Orchidaceae: subfamily Vanilloideae), 89-93.

Cameron, K.M. (2010). Vanilloid orchids. In: *Vanilla*, E. Odoux, M. Grisoni, (Eds.). CRC Press Taylor and Francis Group

Cameron, K.M. (2011a). Vanilla Orchids: Natural History and Cultivation (Ed.)^(Eds.), p. 212. Timber Press

Cameron, K.M. (2011b). *Vanilla* phylogeny and classification. In: *Handbook of vanilla science and technology*, D. Havkin-Frenkel, F.C. Belanger, (Eds.), pp. 243-255. Wiley-Blackwell, Chichester (UK)

Cameron, K.M., Chase, M.W., Whitten, W.M., Kores, P.J., Jarrell, D.C., Albert, V.A., Yukawa, T., Hills, H.G.& Goldman, D.H. (1999). A phylogenetic analysis of the Orchidaceae: evidence from *rbc*L nucleotide sequences. *Am. J. Bot.*, Vol.86, pp.208-224.

Cameron, K.M.& Molina, M.C. (2006). Photosystem II gene sequences of *psb*B and *psb*C clarify the phylogenetic position of *Vanilla* (Vanilloideae, Orchidaceae). *Cladistics*, Vol.22, pp.239-248.

Castillo Martinez, R.& Engleman, E.M. (1993). Caracterizacion de dos tipos de *Vanilla planifolia*. *Acta Botánica Mexicana*, Vol.25, pp.49-59.

Chardard, R. (1963). Contribution à l'étude cyto-taxinomique des orchidées. *Revue de Cytologie et de Biologie Végétales*, Vol.26, pp.1-58.

Childers, N.F.& Cibes, H.R. (1948). *Vanilla culture in Puerto Rico*, Circular N°28, Federal Experiment Station in Puerto Rico of the United States Department of Agriculture, Washington DC.

Chuo, S.W., Ernst, R., Arditti, J.& Hew, C.S. (1994). Orchid pests - A compendium. In: *Orchid Biology - Reviews and perspective VI*, J. Arditti, (Ed.). John Wiley & Sons, New York

Dobat, K.& Peikert-Holle, T. (1985). *Blüten und fledermäuse.* , Waldemar Kramer.

Dodson, C.H., Dressler, R.L., Hills, H.G., Adams, R.M.& Williams, N.H. (1969). Biologically Active Compounds in Orchid Fragrances. *Science*, Vol.164, pp.1243-1249.

Dressler, R.L. (1981). *The Orchids: Natural History and Classification*, Harvard University Press, Cambridge, MA.

Dressler, R.L. (1993). *Phylogeny and Classification of the Orchid Family.*, Dioscorides Press, Portland, OR.

Duval, M.-F., Bory, S., Andrzejewski, S., Grisoni, M., Besse, P., Causse, S., Charon, C., Dron, M., Odoux, E.& Wong, M. (2006). Diversité génétique des vanilliers dans leurs zones de dispersion secondaire. *Les Actes du BRG*, Vol.6, pp.181-196.

Ennos, R.A., French, G.C.& Hollingsworth, P.M. (2005). Conserving taxonomic complexity. *TRENDS in Ecology and Evolution*, Vol.20 pp.164-168.

Escuderoa, A., Iriondob, J.M.& Torresb, M.E. (2003). Spatial analysis of genetic diversity as a tool for plant conservation. *Biological Conservation*, Vol.113 pp.351-365.

Finnegan, E.J. (2001). Epialleles - a source of random variation in times of stress. *Current Opinion in Plant Biology*, Vol.5, pp.101-106.

Finnegan, E.J., Genger, R.K., Peacock, W.J.& Dennis, E.S. (1998). DNA methylation in plants. : . *Annual Review of Plant Physiology and Plant Molecular Biology*, Vol.49, pp.223-247.

Finnegan, E.J., Peacock, W.J.& Dennis, E.S. (1996). Reduced DNA methylation in Arabidopsis thaliana results in abnormal plant development. *Proc. Natl. Acad. Sci. U.S.A.*, Vol.93, pp.8449-8454.

Fleming, T.H., Geiselman, C.& Kress, W.J. (2009). The evolution of bat pollination: a phylogenetic perspective. *Annals of Botany*, Vol.104, pp.1017-1043.

Geiselman, C.K., Mori, S.A.& Blanchard, F. (2004). Database of neotropical bat/plant interactions. http://www.nybg.org/botany/tlobova/mori/batsplants/database/dbase_frameset.htm

Grant-Downton, R.T.& Dickinson, H.G. (2006). Epigenetics and its Implications for Plant Biology 2. The 'Epigenetic Epiphany': Epigenetics, Evolution and Beyond. *Annals of Botany*, Vol.97, pp.11-27.

Havkin-Frenkel, D.& Belanger, F.C. (2011). Handbook of vanilla science and technology (Ed.)^(Eds.), p. 360. Wiley-Blackwell, Chichester (UK)

Heim, P. (1954). Le noyau dans le genre Vanilla. In: *Le vanillier et la vanille dans le monde*, G. Bouriquet, (Ed.), pp. 27-43. Paul Lechevalier, Paris

Hoehne, F.C. (1945). Orchidaceas Fasc. 8. In: *Flora Brasilica*, pp. 13-43. Secretara da Agricultura, Industria e Comerao, Sao Paulo, Brazil

Hoffmann, K.M. (1929). Zytologische Studien der Orchidaceen (Vorläufige Mitteilung.). *Ber. deutschen Bot. Gesell.*, Vol.47, pp.321-326.

Hoffmann, K.M. (1930). Beiträge zur Cytologie des Orchidaceen. *Planta*, Vol.10, pp.523-595.

Hollingsworth, M.L., Clark, A.A., Forrest, L.L., Richardson, J., Pennington, R.T., Long, D.G., Cowan, R., Chase, M.W., Gaudeul, M.& Hollingsworth, P.M. (2009). Selecting barcoding loci for plants: evaluation of seven candidate loci with species level sampling in three divergent groups of land plants. *Molecular Ecology Resources*, Vol.9, pp.439–457.

Hollingsworth, P.M.& CBOL Plant Working Group (2009). A DNA barcode for land plants. *Proc. Natl. Acad. Sci. U.S.A.*, Vol.106 pp.12794-12797.

Householder, E., Janovec, J., Balarezo Mozambite, A., Huinga Maceda, J., Wells, J.& Valega, R. (2010). Diversity, natural history, and conservation of *Vanilla* (Orchidaceae) in amazonian wetlands of Madre De Dios, Peru. *J. Bot. Res. Inst. Texas*, Vol. 4 pp.227 - 243.

Hurel-Py, G. (1938). Etude des noyaux végétatifs de *Vanilla planifolia. Revue de Cytologie et de Cytophysiologie Végétales*, Vol.3, pp.129-133.

Imazio, S., Labra, M., Grassi, F., Winfield, M., Bardini, M.& Scienza, A. (2002). Molecular tools for clone identification: the case of the grapevine cultivar 'Traminer'. *Plant Breeding*, Vol.121, pp.531-535.

Jaligot, E., Beulé, T., Baurens, F.-C., Billotte, N.& Rival, A. (2004). Search for methylation-sensitive amplification polymorphisms associated with the "mantled" variant phenotype in oil palm (*Elaeis guineensis* Jacq.). *Genome Note*, Vol.47, pp.224-228.

Jersakova, J., Johnson, S.D.& Kindlmann, P. (2006). Mechanisms and evolution of deceptive pollination in orchids. *Biol. Rev.*, Vol.81, pp.219-235.

Kahane, R., Besse, P., Grisoni, M., Le Bellec, F.& Odoux, E. (2008). Bourbon Vanilla: Natural Flavour with a Future. *Chronica Horticulturae*, Vol.48, pp.23-28.

Kalisz, S.& Purugganan, M.D. (2004). Epialleles via DNA methylation: consequences for plant evolution. *TRENDS in Ecology and Evolution*, Vol.19, pp.309-314.

Keyte, A.L., Percifield, R., Liu, B.& Wendel, J.F. (2006). Infraspecific DNA Methylation Polymorphism in Cotton (*Gossypium hirsutum* L.). *Journal of Heredity* Vol.97, pp.444-450.

Lepers-Andrzejewski, S., Causse, M., Caromel, B., Wong, M.& Dron, M. (2011a). Genetic linkage map and diversity analysis of tahitian vanilla (*Vanilla* x *tahitensis*, Orchidaceae). *Crop Science*, Vol.In press.

Lepers-Andrzejewski, S., Siljak-Yakovlev, S., Browm, S., Wong, M.& Dron, M. (2011b). Diversity and dynamics of the Tahitian Vanilla (*Vanilla* x *tahitensis*, Orchidaceae) genome: cytogenetic approaches. *American Journal of Botany*, 98(6): in press

Liu, B.& Wendel, J.F. (2003). Epigenetic phenomena and the evolution of plant allopolyploids. *Molecular Phylogenetics and Evolution*, Vol.29, pp.365–379.

Liu, Z.J., Chen, S.C.& Ru, Z.Z. (2007). *Vanilla shenzhenica* Z. J. Liu & S. C. Chen, the first new species of Orchidaceae in Shenzhen, South China. *Acta Phytotaxonomica Sinica*, Vol.45, pp.301-303.

Lubinsky, P., Bory, S., Hernandez, J.H., Kim, S.-C.& Gomez-Pompa, A. (2008a). Origins and Dispersal of Cultivated Vanilla (*Vanilla planifolia* Jacks. [Orchidaceae]). *Economic Botany*, Vol.62, pp.127-138.

Lubinsky, P., Cameron, K.M., Molina, M.C., Wong, M., Lepers-Andrzejewski, S., Gómez-Pompa, A.& Kim, S.-C. (2008b). Neotropical roots of a polynesian spice: the hybrid origin of tahitian vanilla, *Vanilla tahitensis* (Orchidaceae). *American Journal of Botany*, Vol.95, pp.1–8.

Lubinsky, P., Romero-Gonzalez, G.A., Heredia, S.M.& Zabel, S. (2011). Origins and patterns of vanilla cultivation in tropical america (1500-1900): no support for an independant domestication of vanilla in south america. In: *Handbook of vanilla science and technology*, D. Havkin-Frenkel, F.C. Belanger, (Eds.), pp. 117-138. Wiley-Blackwell, Chichester (UK)

Lubinsky, P., Van Dam, M.& Van Dam, A. (2006). Pollination of Vanilla and evolution in Orchidaceae. *Orchids*, Vol.75, pp.926-929.

Macedo Reis, C.A. (2000). *Biologia reprodutiva e propagacao vegetativa de Vanilla chamissonis Klotzsch: subsidios para manejo sustentado*, Escola Superior de Agric Luiz de Queiroz, Piracicaba, Sao Paulo, Brasil.

Madison, M. (1981). Vanilla beans and bees. *Bull. Marie Selby Bot. Gard.* , Vol.8, pp.8.

Martienssen, R.A.& Colot, V. (2001). DNA methylation and epigenetic inheritance in plants and filamentous fungi. *Science* Vol.293, pp.1070-1074.

Martin, F.W. (1963). Chromosome number and behavior in a *Vanilla* hybrid and several *Vanilla* species. *Bulletin of the Torrey Botanical Club*, Vol.90, pp.416-417.

Mathieu, O., Reinders, J., Caikovski, M., Smathajitt, C.& Paszkowski, J. (2007). Transgenerational stability of the *Arabidopsis* epigenome is coordinated by CG methylation. *Cell. Mol. Life Sci.*, Vol.130, pp.851-862.

Micheneau, C., Fournel, J.& Pailler, T. (2006). Bird Pollination in an Angraecoid Orchid on Reunion Island (Mascarene Archipelago, Indian Ocean). *Annals of Botany*, Vol.97, pp.965–974.

Moore, J.W. (1993). New and critical plants from Raiatea. *Bulletin du Bernice P. Bishop Museum*, Vol.102, pp.25.

Nair, R.R.& Ravindran, P.N. (1994). Somatic association of chromosomes and other mitotic abnormalities in *Vanilla planifolia* (Andrews). *Caryologia*, Vol.47, pp.65-73.

Nielsen, R.L. (2000). Natural hybridization between *Vanilla claviculata* (W.Wright) Sw. and *V. barbellata* Rchb.f. (Orchidaceae): genetic, morphological, and pollination experimental data. *Bot. J. Linn. Soc.*, Vol.133, pp.285-302.

Nielsen, R.L.& Siegismund, H.R. (1999). Interspecific differentiation and hybridization in *Vanilla* species (Orchidaceae). *Heredity*, Vol.83, pp.560-567.

Noyer, J.L., Causse, S., Tomekpe, K., Bouet, A.& Baurens, F.C. (2005). A new image of plantain diversity assessed by SSR, AFLP and MSAP markers. *Genetica*, Vol.124, pp.61–69.

Odoux, E.& Grisoni, M. (2010). Vanilla. In: *Medicinal and Aromatic Plants: Industrial Profiles* (Ed.)^(Eds.), p. 387. Taylor and Francis Group

Pansarin, E.R. (2010a). Taxonomic notes on Vanilleae (Orchidaceae: Vanilloideae) *Vanilla dietschiana,* a rare south american taxon transferred from *Dictyophyllaria. Selbyana,* Vol.30, pp.198-202.

Pansarin, E.R. (2010b). *Vanilla diestchiana*, an endangered species native to Brazil. *Orchids*.

Paszkowski, J.& Whitham, S.A. (2001). Gene silencing and DNA methylation processes. *Current Opinion in Plant Biology*, Vol.4, pp.123-129.

Peakall, R. (1994). Interactions between orchids and ants. In: *Orchid Biology - Reviews and perspective VI*, J. Arditti, (Ed.). John Wiley & Sons, New York

Peraza-Echevarria, S., Herrera-Valencia, V.A.& James-Kay, A. (2001). Detection of DNA methylation changes in micropropagated banana plants using methylation-sensitive amplification polymorphism (MSAP). *Plant Science*, Vol.161, pp.359-367.

Peraza-Echeverria, S., Herrera-Valencia, V.A.& James-Kay, A. (2001). Detection of DNA methylation changes in micropropagated banana plants using methylation-sensitive amplification polymorphism (MSAP). *Plant Sci.*, Vol.161, pp.359-367.

Pignal, M. (1994). Deux vanilles du Brésil : *Vanilla palmarum* Lindley et *Vanilla bahiana* Hoehne. *L'Orchidophile*, Vol.110, pp.23-25.

Portères, R. (1954). Le genre *Vanilla* et ses espèces. In: *Le vanillier et la vanille dans le monde*, G. Bouriquet, (Ed.), pp. 94-290. Paul Lechevalier, Paris

Purseglove, J.W., Brown, E.G., Green, C.L.& Robbins, S.R.J. (1981). Chapter 11: Vanilla. In: *Spices: Tropical Agricultural Series*, Longman, (Ed.), pp. 644-735, London

Ramirez, S.R., Gravendeel, R.B., Singer, R.B., Marshall, C.R.& Pierce, N.E. (2007). Dating the origin of the Orchidaceae from a fossil orchid with its pollinator. *Nature* Vol.448, pp.1042-1045.

Ratnasingham, S.& Hebert, P.D.N. (2007). BOLD: The Barcode of Life Data System (www.barcodinglife.org). *Molecular Ecology Notes*, Vol.7, pp.355-364.

Reyna-López, G.E., Simpson, J.& Ruiz-Herrera, J. (1997). Differences in DNA methylation pattern are detectable during the dimorphic transition of fungi by amplification of restriction polymorphisms. *Mol Gen Genet*, Vol.253, pp.703-710.

Risbec, J. (1954). Insectes et animaux nuisibles au vanillier. In: *Le vanillier et la vanille dans le monde*, G. Bouriquet, (Ed.), pp. 492-512. Paul Lechevalier, Paris

Rolfe, R.A. (1896). A revision of the genus *Vanilla. Journal of the Linnean Society*, Vol.32, pp.439-478.

Roubik, D.W.& Ackerman, J.D. (1987). Long term ecology of euglossine orchid-bees (Apidae: Euglossini) in Panama. *Oecologia (Berlin)*, Vol.73, pp.321-333.

Roux-Cuvelier, M.& Grisoni, M. (2010). Conservation and Movement of Vanilla Germplasm. In: *Vanilla*, E. Odoux, M. Grisoni, (Eds.). CRC Press Taylor and Francis Group

Rubinoff, D. (2006). Utility of Mitochondrial DNA Barcodes in Species Conservation. *Conservation Biology*, Vol.20, pp.1026-1033.

Schiestl, F.P. (2005). On the success of a swindle: pollination by deception in orchids. *Naturwissenschaften*, Vol.92, pp.255-264.

Schlüter, P.M. (2002). *RAPD variation in Vanilla planifolia Jackson (Orchidaceae) and assessment of the putative hybrid Vanilla tahitensis Moore,* MBiochem Thesis, University of Oxford.

Singer, R.B. (2003). Orchid pollination: recent developments from Brazil. *Lankesteriana*, Vol.7, pp.111-114.

Soto Arenas, M.A. (1999). *Filogeografia y recursos genéticos de las vainillas de México*, Herbario de la Asociación Mexicana de Orquideología, México, 102 p.

Soto Arenas, M.A. (2006). La vainilla: retos y perspectivas de su cultivo. *Biodiversitas*, Vol.66, pp.2-9.

Soto Arenas, M.A. (2010). A new species of *Vanilla* from south America. *Lankesteriana*, Vol.9, pp.281-284.

Soto Arenas, M.A.& Cameron, K.M. (2003). *Vanilla*. In: *Genera orchidacearum: Orchidoideae*, A.M. Pridgeon, P.J. Cribb, M.W. Chase, F.N. Rasmussen, (Eds.), pp. 321-334. Oxford University Press, USA

Soto Arenas, M.A.& Cribb, P. (2010). A new infrageneric classification and synopsis of the genus *Vanilla* Plum. ex Mill. (Orchidaceae: Vanillinae). *Lankesteriana*, Vol.9, pp.355-398.

Soto Arenas, M.A.& Dressler, R.L. (2010). A revision of the mexican and central american species of *Vanilla* Plumier ex. Miller with a acharacterization of their ITS region of the nulcear ribosomal DNA. *Lankesteriana*, Vol.9, pp.285-354.

Stehlé, H. (1952). Le vanillier et sa culture. *Fruits*, Vol.7, pp.50-56, 99-112, 253-260.

Stehlé, H. (1954). Ecologie. In: *Le vanillier et la vanille dans le monde*, G. Bouriquet, (Ed.), pp. 292-334. Paul Lechevalier, Paris

Szlachetko, D.L.& Veyret, Y. (1995). Deux espèces nouvelles de *Vanilla* (Orchidaceae) de Guyane Française. *Bulletin du Muséum National d'Histoire Naturelle*, Vol.16 section B, pp.219-223.

Tremblay, R.L., Ackerman, J.D., Zimmerman, J.K.& Calvo, R.N. (2005). Variation in sexual reproduction in orchids and its evolutionary consequences: a spasmodic journey to diversification. *Biol. J. Linn. Soc.*, Vol.84, pp.1–54.

Van Dam, A.R., Householder, J.E.& Lubinsky, P. (2010). *Vanilla bicolor* Lindl. (Orchidaceae) from the Peruvian Amazon: auto-fertilization in *Vanilla* and notes on floral phenology. *Genet Resour Crop Evol*, Vol.57, pp.473-480.

Van Der Cingel, N.A. (2001). *An atlas of orchid pollination - America, Africa, Asia & Australia*, Balkema A.A., Rotterdam, Brookfield

Van Der Pijl, L.& Dodson, C.H. (1966). *Orchid Flowers, Their Pollination and Evolution*, University of Miami Press, Coral Cables, FL.

Weiss, E.A. (2002). Chapter 7: Orchidaceae. In: *Spice Crops*, pp. 136-154. CABI publishing, Wallington, UK

Williams, N.H.& Whiten, W.M. (1983). Orchid floral fragrances and male euglossine bees:Methods and advances in the last sesquidecade. *BioL Bull.* , Vol.164, pp.355-395.

Wu, C.& Morris, J.R. (2001). Genes, genetics, and epigenetics: a correspondence. *Science* Vol.293, pp.1103-1105.

Xiong, L.Z., Xu, C.G., Saghai Maroof, M.A.& Zhang, Q. (1999). Patterns of cytosine methylation in an elite rice hybrid and its parental lines, detected by a methylation-sensitive amplification polymorphism technique. *Mol. Gen. Genet.*, Vol.261, pp.439-446.

Xu, M., Li, X.& Korban, S.S. (2000). AFLP-based detection of DNA methylation. *Plant Mol. Biol. Rep.*, Vol.18, pp.361-368.

2

The Origin of Diversity in *Begonia*: Genome Dynamism, Population Processes and Phylogenetic Patterns

A. Dewitte[1], A.D. Twyford[2,3], D.C. Thomas[2,4], C.A. Kidner[2,3] and J. Van Huylenbroeck[5]
[1]*KATHO Catholic University College of Southwest Flanders, Department of Health Care and Biotechnology*
[2]*Royal Botanic Garden Edinburgh, 20A Inverleith Row, Edinburgh*
[3]*Institute of Molecular Plant Sciences, University of Edinburgh, King's Buildings, Edinburgh*
[4]*University of Hong Kong, School of Biological Sciences, Pokfulam, Hong Kong,*
[5]*Institute for Agricultural and Fisheries Research (ILVO), Plant Sciences Unit,*
[1,5]*Belgium*
[2,3]*United Kingdom*
[4]*PR China*

1. Introduction

Species diversity is unequally distributed across the globe, with more species found in the tropics than any other ecosystem in the world. This latitudinal gradient of species richness illustrates the complex evolutionary history of global biodiversity, and many studies have placed it in the context of geological history and rates of speciation and extinction (Mittelbach et al., 2007). Historical biogeographic studies, using molecular phylogenies calibrated with a relative dimension of time, indicate that the accumulation of this diversity is both ancient ("museum" model) and recent ("cradle" model) within groups (Bermingham & Dick, 2001; McKenna & Farrell, 2006). An additional layer of complexity that makes it difficult to untangle the evolutionary processes driving tropical speciation are biotic interactions, such as plant competition and parasite interactions (Berenbaum & Zangerl, 2006). Much of our understanding of the processes underlying speciation comes from mathematical models or studies of model organisms. However, some of the classical questions of evolutionary biology, such as what factors are driving speciation in species rich biomes, can only be understood by detailed evolutionary and ecological studies of specious groups.

Begonia is a genus of about 1550 described species, placing it in the top ten most speciose angiosperm genera (Frodin, 2004; Hughes, 2008). This makes it an ideal model for studying the processes and patterns underlying the generation of diversity (Forrest et al.,

2005; Neale et al., 2006). The distribution of *Begonia* diversity is uneven throughout tropical regions, with the greatest diversity in America and Asia (>600 species each), whilst being relatively species poor in Africa (160 species) and absent in Australia (Goodall-Copestake et al., 2010). The genus is thought to have originated in Africa, while South American and South East Asian species are the results of parallel radiations over the last 20 - 50 million years (Goodall Copestake et al., 2010; Plana et al., 2004; Thomas et al., 2011). Long distance dispersal is rare, for example *Begonia* species have failed to cross the Torres strait from Papua New Guinea to Australia.

Begoniaceae are easily recognizable by diagnostic characters such as asymmetrical leaves, unisexual monoecious flowers, twisted-, papillose stigmas, and dry-, three-winged capsules (Doorenbos, 1998). However, there are numerous deviations from these typical character states. Within the genus *Begonia* there is a large range of morphological diversity, particularly in vegetative form, and this is linked to adaptation to a variety of ecological conditions. Vegetative adaptations such as the evolution of perennating rhizomes, leaf micromorphology optimised for low, scattered light; or stomatal clustering may underlie their ability to thrive in diverse niches. Phenotypic polymorphism within populations occurs, most frequently in the anthocyanin patterns on the leaves, which although striking have not been shown to have measurable effects on light capture (Hughes et al., 2008).

The genetic and morphological diversity of the genus *Begonia* has been exploited through cultivation to produce over 10,000 cultivars. These are horticulturally divided into 5 classes: a) the tuberous begonias (*B.* x *tuberhybrida*), a complex group derived from crosses between species such as *B. boliviensis* or *B. pearcei*, b) Elatior begonias (*B.* x *hiemalis*), a cross between tuberous begonias and *B. socotrana,* c) Lorraine begonias (*B.* x *cheimanta*), a cross between tuberous hybrids and *B. dregei,* d) semperflorens begonias (*B. semperflorens*-cultorum), with *B. cucullata* and *B. schmidtiana* as important ancestors, and e) begonias grown for their ornamental foliage (*B. rex*-cultorum), Asiatic in origin (Haegeman, 1979; Hvoslef-Eide et al., 2007). Commercial interest in this group has promoted research into a variety of topics, including hybridisation and polyploidy.

Phylogenetic and cytological research in the last decade has significantly increased our knowledge of diversity within the genus *Begonia*. In this review, the current classification and the evolution of species diversity is discussed with reference to recent progress in: a) population genetic and phylogenetic techniques using genetic markers in association with morphological characters and b) cytological techniques such as mitotic or meiotic chromosome visualisation, linked to genome size studies. We also present new data on barriers to hybridisation between *Begonia* species. We focus on how genetic, cytological and local ecological effects may contribute to diversity in this genus, particularly the evolution of species diversity.

2. *Begonia* classification

Morphological and molecular data firmly place the Begoniaceae within the order Cucurbitales, which also includes the large and economically important family Cucurbitaceae (950-980 species), and the small families Anisophylleaceae (29-34 species), Apodanthaceae (19 species), Coriariaceae (15-20 species), Corynocarpaceae (5-6 species), Datiscaceae (2 species) and Tetramelaceae (2 species) (APG, 2009; Matthews & Endress, 2004; Schaefer & Renner, 2011; Zhang et al., 2006). Analyses of DNA sequence data from 14 nuclear, mitochondrial and plastid markers strongly support a close relationship of Begoniaceae with Datiscaceae and Tetramelaceae (Schaefer & Renner, 2011).

Only two genera are currently recognized in the Begoniaceae: the monotypic genus *Hillebrandia*, and the species-rich and morphologically diverse genus *Begonia* (Doorenbos et al., 1998; Forrest & Hollingsworth, 2003). A third genus, *Symbegonia*, was previously included in Begoniaceae, and separated from *Begonia* by floral characters (syntepalous perianth and a monadelphous androecium). Based on molecular data, the genus *Symbegonia* has been shown to be nested within *Begonia* section *Petermannia* (Forrest & Hollingsworth, 2003). *Hillebrandia sandwicensis*, which is endemic to Hawaii, can be differentiated from *Begonia* by a suite of morphological characters. These include more differentiated segments of the perianth, semi-inferior ovaries (inferior in *Begonia*), and fruit dehiscence between the styles in contrast to the usually loculicidal dehiscence in *Begonia* (Clement et al., 2004; Forrest & Hollingsworth, 2003; Forrest et al., 2005).

A reliable infrageneric classification and subdivision of large genera such as *Begonia* is crucial in order to inform taxonomic monographs, biogeographic and evolutionary studies. A revision of circumscriptions of *Begonia* sections by Doorenbos et al. (1998) provides a foundation for the subdivision of the genus. In this revision 63 sections were recognized, and another three sections have been subsequently proposed (de Wilde & Plana, 2003; Forrest & Hollingsworth, 2003; Shui et al., 2002). The distributions of all but one of the currently accepted *Begonia* sections are limited to single continental regions, i.e. Africa, Asia, or America. Only section *Tetraphila* can be found in multiple continents and a single, recently discovered, and still to be named species in this predominantly African section is indigenous to continental Southeast Asia (de Wilde, 2011).

DNA sequence data from non-coding regions plays an important role in plant classification and barcoding (CBOL Plant working group, 2009), and has widely been used to resolve relationships at the species and sectional level in *Begonia*. A framework phylogeny of *Begonia* based on analyses of c. 13000 bases of plastid and mitochondrial DNA of 30 *Begonia* species (Goodall-Copestake et al., 2010; Fig. 1) indicates that African taxa form the earliest divergent clades in the *Begonia* phylogeny and that both Asian and American *Begonia* lineages are derived from African ancestors. The phylogenetic relationships within the relatively small group of African *Begonia*, which comprises around 160 species subdivided into 17 sections (de Wilde & Plana, 2003; Doorenbos et al., 1998), are relatively well understood. African *Begonia* species are not retrieved as monophyletic, but South African species placed in section *Augustia* were shown to be closely related to a clade of American taxa, and Socotran *Begonia* species (section *Peltaugustia*) were shown to form a monophyletic clade with Asian taxa (Forrest et al., 2005; Goodall-Copestake et al., 2010; Plana et al., 2004; Thomas et al., 2011). Revisions exist for the majority of the African sections (see references in Plana, 2003), and the intersectional relationships of African *Begonia* species have been studied using molecular systematic approaches and have been discussed in some detail in Plana (2003) and Plana et al. (2004). Most African sections are well circumscribed and seem to represent monophyletic taxa, but section *Mezierea* is polyphyletic (Forrest et al., 2005; Plana, 2003; Plana et al., 2004); and the lack of resolution or support in phylogenies makes the assessment of the monophyly of some sections problematic. Apart from the grade of continental African taxa, a major Madagascan radiation can be detected. Only one of the c. 50 Madagascan (incl. Comores and Mascarenes) *Begonia* species, *Begonia oxyloba*, is also widespread on the African continent (Keraudren-Aymonin, 1983). The other Madagascan species seem to be the result of a single dispersal event from continental Africa and a subsequent radiation (Plana, 2003; Plana et al., 2004).

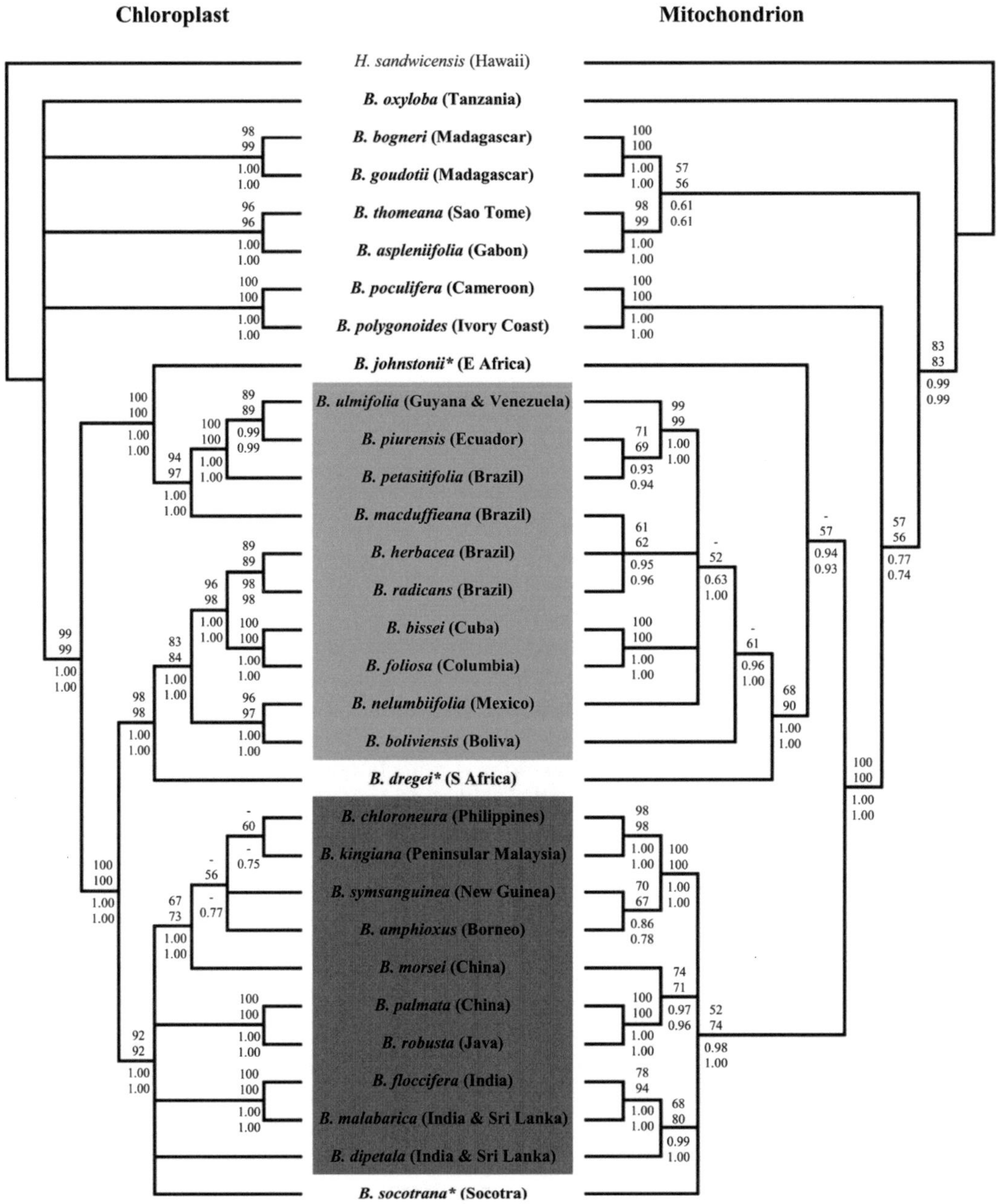

Fig. 1. Cladograms showing relationships supported by ≥ 50% parsimony bootstrap and ≥ 50% Bayesian posterior probability support after analysis of the genome-level DNA sequence datasets with indel data (Goodall-Copestake et al., 2010). The geographic origin of taxon samples is given in brackets and the African, American, and Asian continental species assemblages are indicated by pale, mid, and dark grey shading, respectively. Seasonally-adapted African *Begonia* species are indicated by asterisks.

Asian and Socotran *Begonia* species form a well supported clade in recent analyses of nuclear, mitochondrial and plastid DNA sequence data (Goodall-Copestake et al., 2010; Thomas, 2010; Thomas et al., 201)). The c. 750 Asian *Begonia* species have been divided into 20 sections (Doorenbos et al., 1998; Forrest & Hollingsworth, 2003; Shui et al., 2002; Thomas et al., 2011). However, the circumscriptions of several of these sections are questionable and phylogenetic analyses of sequence data provide evidence for the non-monophyly of most species-rich Asian sections (Forrest & Hollingsworth, 2003; Tebbitt et al., 2006; Thomas, 2010; Thomas et al., 2011). Moreover, several small or monotypic sections were shown to be nested within clades of species assigned to the larger, species-rich sections. For example, section *Baryandra* is nested within Philippine *Diploclinum* (Rubite, 2010; Thomas, 2010); sections *Alicida* and *Putzeysia* fall into a grade of continental Asian species predominantly assigned to section *Diploclinium* (Rajbhandary, 2010; Thomas, 2010; Thomas et al., 2011); and section *Monopteron* is nested within section *Platycentrum* (Forrest, 2001; Rajbhandary, 2010; Thomas, 2010).

Within Asian *Begonia* two well supported major clades can be differentiated based on molecular data. The first is dominated by continental Asian taxa, and mainly includes species of section *Parvibegonia,* continental Asian species of the polyphyletic section *Diploclinium,* and species in sections *Platycentrum* and *Sphenanthera*. The second major clade contains the predominantly Chinese section *Coelocentrum* that forms the sister clade to a clade of Malesian species placed in sections *Ridleyella, Bracteibegonia, Petermannia,* and Malesian species placed in the polyphyletic sections *Diploclinium* and *Reichenheimia* (Thomas, 2010; Thomas et al., 2011).

American *Begonia* comprise more than 600 species divided into 29 sections (Burt-Utley, 1985; Doorenbos et al., 1998). American *Begonia* species are retrieved as monophyletic in analyses of nuclear ribosomal and mitochondrial DNA sequence data, but form two distinct clades with African sister groups in analyses of plastid DNA sequence data (Forrest & Hollingsworth, 2003; Goodall-Copestake et al., 2010). This phylogenetic incongruence may indicate an ancient hybridisation event in this group (Goodall-Copestake et al., 2010). In contrast to African and Asian *Begonia*, the phylogenetics of American *Begonia* are still poorly understood, and the circumscriptions of several neotropical sections are highly problematic. Doorenbos et al. (1998: 181) emphasized that several neotropical sections "shade off into each other," i.e. sectional delimitations often lack autapomorphic characters and morphologically transitional species are present. Low-density taxon samples of 9-11 American species were included in the analyses by Clement et al. (2004), Forrest & Hollingsworth (2003), Forrest et al. (2005) and Goodall-Copestake et al. (2010), but further studies including a much wider and geographically robust sampling of American *Begonia* species are needed to identify major subdivisions within American *Begonia,* clarify sectional delimitations and investigate intersectional relationships.

3. Historical biogeography

Molecular divergence age estimates based on DNA sequence data and the calibration of molecular changes with a dimension of time, allow inference of: i) the time of origin of monophyletic groups, and ii) the timing of dispersal to a geographic area (Renner, 2005). However, molecular divergence age estimates for clades of Begoniaceae is problematic, as the family has a poor fossil record, and suitable fossils for direct calibration are lacking

(Stults & Axsmith, 2011). Previous studies have addressed this problem by using island emergence dates as calibration points (Plana et al., 2004) or by putting Begoniaceae in a wider phylogenetic context using suitable fossils calibrations in related taxa (Clement et al., 2004; Goodall-Copestake, 2005; Goodall-Copestake et al., 2009; Thomas, 2010).

Most mean age estimates indicate an Eocene origin (c. 40-46 Ma) for the Begoniaceae crown group diversification (Clement et al., 2004; Goodall-Copestake, 2005; Plana et al., 2004; Thomas, 2010). However, there are considerable confidence intervals depending on the methods and calibration points used (Fig. 2). The geographic origin of Begoniaceae remains enigmatic, with the geographically isolated *Hillebrandia* in Hawaii and the disjunct distribution of the two species of the putative sister family Datiscaceae, in southwestern North America and southwestern Asia. Clement et al. (2004) hypothesized a widespread most recent common ancestor of Begoniaceae in Eurasia.

The mean age estimates for the *Begonia* crown group diversification indicate an Oligocene origin (c. 23-34 Ma), although large confidence ranges have to be considered (Fig. 2) (Goodall-Copestake et al., 2009; Thomas, 2010). The earliest divergent clades in the *Begonia* phylogeny are African, suggesting an early diversification on the African continent long after the Gondwanan break-up (Goodall-Copestake et al., 2010; Plana et al., 2004; Thomas, 2010). From Africa, *Begonia* dispersed independently to both America and Asia, and the early Miocene has been inferred as the most likely timeframe for these dispersal events (Goodall-Copestake, 2005; Goodall-Copestake et al., 2010). African *Begonia* species (sections *Rostrobegonia, Sexalaria, Augustia* and *Peltaugustia*), which are phylogenetically closely related to Asian and American *Begonia* lineages, show adaptations to seasonally drier climate. Goodall-Copestake et al. (2010) hypothesised that these drought adaptations made the ancestors of the American and Asian lineages more resilient to intercontinental dispersal than the vast majority of African species, which require moist and shaded habitats. Intercontinental dispersal likely occurred by long-distance dispersal, or alternatively, dispersal to Asia may have occurred by an overland dispersal route via the Arabian Peninsula during a more mesic period than at present (Goodall-Copestake, 2005; Goodall-Copestake et al., 2010). The hypothesis of dispersal from Africa to western Asia is consistent with phylogenetic analyses and ancestral area reconstructions of Asian *Begonia*, which indicate an initial diversification of Asian-Socotran *Begonia* in South India-Sri Lanka, Socotra and continental Asia. This was followed by multiple dispersal events into Malesia, and a predominantly west to east colonisation of the Malesian archipelago (Thomas, 2010).

The relatively low species diversity of African *Begonia* (c. 160 species) in comparison to the species diversity in the New World (>600 species) and Asia (c. 750 species) may partly be explained by extinction of African *Begonia* species. This may be due to large-scale aridification during cooler and dry periods in Africa, especially during the pronounced climate oscillation in the Pleistocene. In addition to this, rapid diversifications in the Asian tropics and the Neotropics may explain the uneven distribution of *Begonia* diversity (de Wilde, 2011; Forrest et al., 2005; Thomas, 2010). Weakly developed mechanisms to maintain species cohesion in fragmented habitats (Hughes & Hollingsworth, 2008), in association with the formation of topographical heterogeneity caused by mountain uplift from the early Pliocene onwards, may have been major drivers of rapid *Begonia* diversification in Southeast Asia (Thomas, 2010). Moreover, the diversification of Southeast Asian *Begonia* may have been accelerated by cycles of range fragmentation and amalgamation caused by Pleistocene climate and sea-level changes (Thomas, 2010). Cyclic vicariance due to climate oscillations and Pleistocene diversification have also been hypothesised for some species-rich African taxa such sections *Loasibegonia* and *Scutobegonia* (Plana et al., 2004; Sosef, 1994), but Plana et al. (2004) emphasised that a significant proportion of *Begonia* species diversity is of pre-Pleistocene origin.

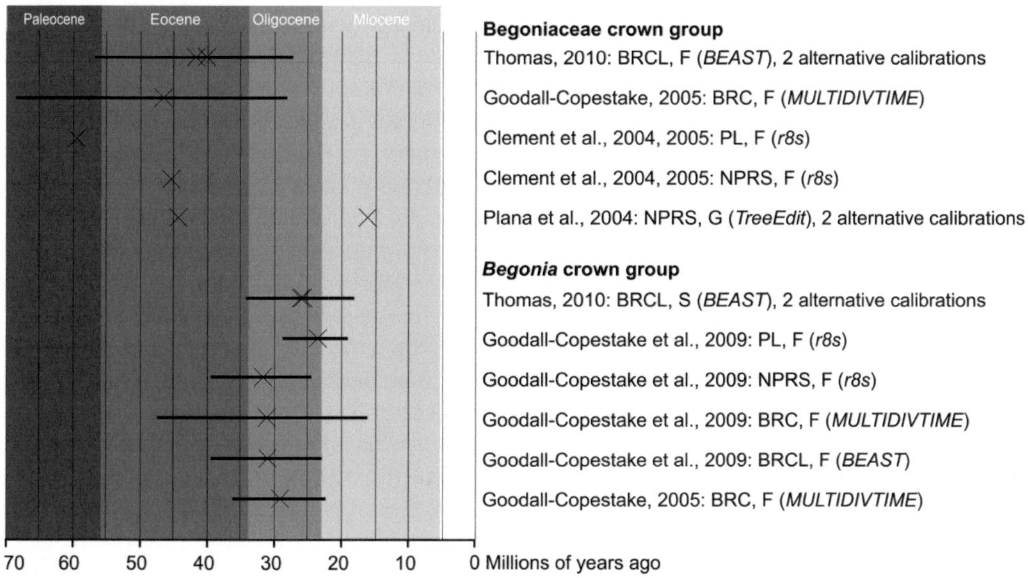

Fig. 2. Molecular divergence age estimates for the Begoniaceae and *Begonia* crown groups. Crosses indicate mean estimates, and bars indicate 95% highest posterior density age ranges and 95% confidence ranges. BRC: Bayesian auto-correlated rates relaxed clock method; BRCL: Bayesian uncorrelated rates relaxed molecular clock method assuming a lognormal distribution of rates; F: fossil calibration; G: geological calibration; NPRS: Non-Parametric Rate Smoothing method; PL: Penalized Likelihood method; S: secondary calibration. Software packages used are indicated in italics in brackets. For details of calibration regimes and methods see: Clement et al., 2004; Goodall-Copestake, 2005; Goodall-Copestake et al., 2009; Plana et al., 2004; Thomas, 2010.

4. Population processes

4.1 Restricted gene flow

Whilst the genus *Begonia* has a very broad distribution, differentiation occurs over very local scales (Hughes, 2008). Three studies have used molecular markers to study local patterns of differentiation. Matolweni et al. (2000) investigated allozyme variation in 12 populations of *B. dregei* and 7 populations of the closely related *B. homonyma* from isolated forest patches in South Africa. This data showed little, if any, gene flow among populations of either *B. dregei* or *B. homonyma*, even between populations that are located only a few kilometers apart within the same forest. Allelic variation and heterozygosity were low, alleles were frequently unique to individual populations, and population differentiation (Fst values) for each locus were very high. This suggests strong and longterm isolation between populations generating extensive genetic divergence and high potential for speciation.

A comparable study was performed by Hughes & Hollingsworth (2008). Seven populations of the South African *B. sutherlandii* were sampled throughout the mist belt forests of

Kwazulu-Natal, and population structure assessed using microsatellite markers. A similar population structure was observed to *B. dregei* and *B. homonyma*. Levels of population differentiation were high and there was significant differentiation between populations, even within subpopulations at a small spatial scale. The genetically isolated nature of *B. sutherlandii* populations suggest effective interpopulation dispersal is rare. There was no significant link between genetic and geographic distance suggesting that this differentiation is caused by genetic drift rather than through long-term isolation.

A third study of population structure in *Begonia socotrana* and *B. samhaensis* at the Socotra archipelago also shows the same pattern of strongly isolated populations (Hughes et al., 2003). The low intraspecific gene flow could be due to both poor seed dispersal in the sheltered conditions of the forest floor and to limited pollen flow. *Begonia* flowers do not attract specialist pollinators but practice deceit pollination by generalist pollinators such as small bees and flies. Analysis of this pollination mechanism in the wild has confirmed that it does results in a low seed set (Ågren & Schemske, 1993; de Lange & Bouman, 1999).

Isolation-by distance may contribute to speciation in the genus. As described by Hughes & Hollingsworth (2008), the population-level data are congruent with the macro-evolutionary patterns observed in the genus. Molecular phylogenies confirm that *Begonia* is characterized by geographically constrained monophyly, species with narrow geographical ranges, very few widespread species, and high levels of morphological differentiation between populations of the few widespread species (Hughes & Hollingsworth, 2008).

4.2 Refuge *Begonia*

The limited dispersal of many *Begonia* makes them a useful group to search for the biogeographic signature of refugia (Sosef, 1994). Refugia are regions where vulnerable lineages could survive periods of dramatic climate change, such as ice ages. When the climate becomes favorable again, species disperse outwards from the refugia. This generates a distinctive pattern in the geographic distribution of genetic diversity, with former refugia having greater genetic diversity. The poor dispersal of *Begonia* means regions with high numbers of *Begonia* species are possible former refugia. Species of the African sections *Loasibegonia* and *Scutobegonia* are restricted to humid and shady locations and are proposed to have been especially sensitive to the period of Pleistocene climate oscillations, and dispersed slowly afterwards (Sosef, 1994). Centers of diversity for these sections are situated in West and Central Africa. Plana et al. (2004) also used *Begonia* diversity to identify the island of Sao Tome as a pre-Pleistocene refuge, and suggested different mainland areas in West Africa as refuges. It is quite possible that a number of different refugia existed for lineages with different ecological tolerances.

5. Cytological investigations and genome size comparisons

Begonia exhibit a highly dynamic genome, with large variation in chromosome number, genome size and mean chromosome size as well as divergent chromosome structure, even between closely related species (Dewitte et al., 2009a). Traditionally, investigations at the chromosome level have been hindered by the small size of *Begonia* chromosomes, the difficulty in visualising centromeres, and few reliable karyograms. However, recent investigations using horticultural hybrids between divergent species, as well as cytological studies in the context of phylogenetic relationships, have greatly improved our understanding of this cytologically interesting genus.

5.1 Chromosome number

The occurrence of particular chromosome numbers in a given group is important for predicting reproductive barriers between species and the potential fertility of the hybrids, and can be indicative of a close evolutionary relationship. Among *Begonia* species, chromosome numbers range from 2n = 16 for *B. rex* to 2n = 156 for *B. acutifolia*. Between these extremes, a wide range of chromosome numbers have been described (Doorenbos et al., 1998; Legro and Doorenbos, 1969; Legro and Doorenbos, 1971; Legro and Doorenbos, 1973). Many species or cultivars exhibit chromosome numbers of 2n = 26 or 28 (x = 13 or 14) or a multiple of this number. Within the horticultural tuberous begonia group, derived from interspecific crosses between American *Begonia*, chromosome numbers of 2n=27,28 (diploid), 41,42 (triploid) and between 52 and 56 (tetraploid) are most common (Legro and Haegeman, 1971; Haegeman, 1979), but variation outside this sequence exists. In Asian *Begonia*, 2n = 22 (x = 11) is the most frequently observed chromosome number. A phylogeny of non-coding cpDNA also indicates a base chromosome number of x = 15 may be ancestral within Asian *Begonia*, with chromosome counts of 30 or 44 as diploid and triploid derivatives (Thomas, 2010; Thomas et al., in press).

The search for a basic chromosome number is complex as there is no common number observed in the group, even taking into account the prevalence of polyploidy in the horticutural varieties assayed. Some authors (Matsuura & Okuno, 1936; Matsuura & Okuno, 1943; Okuna & Nagai, 1953; Okuna & Nagai, 1954) have suggested x = 6, x = 7 and x = 13 as the basic chromosome number, where x=13 may be of secondary origin. By using genomic *in situ* hybridisation (GISH), Marasek-Ciolakowska (2010) concluded that x = 7 may be the basic chromosome number of *B. socotrana*. They based this conclusion on the presence of 7 *B. socotrana* chromosomes and 56 chromosomes derived from tuberous *Begonia* in Elatior hybrids. An alternative explanation of the genomic composition of these Elatior hybrids is selective chromosome elimination of *B. socotrana* chromosomes after hybridisation. Selective chromosome elimination is a genome stabilisation process, and cytological investigation by Arends (1970) supports a role for it in the breeding of Elatior *Begonia*, observing 9 or 12 *B. socotrana* chromosomes in some Elatior hybrids.

The inferred African origin for *Begonia* may suggest a basic chromosome number will be found in these taxa, especially early branching lineages. However it is uncertain how the genomic composition, particularly in terms of chromosome number, has changed in extant African species relative to their ancestors. Most of the described chromosome numbers in African taxa vary between 36 and 38, but counts of 22, 26 and 28 have also been made. Chromosome numbers of 22, 26 and 28 appear to be prevalent in the East-African seasonally adapted *Begonia* from the sections *Rostrobegonia* and *Sexalaria,* which diverged very early during *Begonia* evolution, and from the sections *Augustia* and *Peltaugustia.* These sections show a closer relationship to American and Asian sections than to other African sections.

The closely related *Hillebrandia sandwichensis* has a chromosome number of 2n = 48 (Kapoor, 1966), probably the result of a polyploidisation of a 'diploid' with 2n = 24. Within the most related family Datiscaceae, 2n = 22 is reported for *Datisca cannabina* (Gupta et al., 2009). The chromosome numbers within the family Cucurbitaceae are very diverse, but many species posses chromosome numbers between 2n = 20 and 26 or multiples of these numbers, and 2n = 14, 16 and 18 are also widely reported. In the Coriariaceae, multiples of 20 were observed (2n = 20, 40 and 60) while in Corynocarpus, 2n = 44 or 46 is reported. An exact list of chromosome numbers and references in the abovementioned families is available at the TROPICOS® database (www.tropicos.org). These numbers indicate that basic chromosome numbers within these families are situated between x = 10 and 13. However, within the basally branching family Anisophylleaceae, chromosome numbers of x = 7 and x = 8 are reported (Tobe & Raven, 1987).

The above data suggest a diploid chromosome number for the genus *Begonia* between 20 and 26; chromosome numbers of 2n = 36 or 38 were probably established after polyploidisation and genome stabilisation early in the evolution of this genus. Given that the 'older' sections within the genus have a chromosome number of 2n = 26, we suggest that x = 13 is the original basic chromosome number, a chromosome number observed among sections across the world. However, it cannot be excluded that the basic chromosome number of x = 13 is of secondary origin and arose from a fusion between cells with x = 6 and 7 (as suggested by meiotic studies on *B. evansiana* by Okuna & Nagai (1953)), although karyomorphological studies in *Begonia* do not support x = 6 or 7 as basic chromosome number. If this was the case, the hybrid B276 (2n = 50; Dewitte et al., 2009a) would be a near octoploid and 8 homoeologous chromosomes should be identified, instead of the near tetraploid spread observed (x = 14). Further evidence from Oginuma & Peng (2002) show *B. palmata* and *B. aptera* (both 2n = 22) have 2 secondary constriction chromosomes, which supports a basal number of x = 11 in Taiwanese *Begonia.*

Subsequently, chromosome numbers have diverged resulting in some 'new' basic chromosome numbers for some subgroups (eg x = 11 or x = 14), through selective chromosome elimination and polyploidy. Other derived numbers may have resulted from interspecific hybridisation, and Legro & Doorenbos (1969) suggested that 2n = 22 possibly originated out of a cross between 2n = 16 and 2n = 28.

5.2 Chromosome structure

Few studies have looked into changes in chromosome structure during *Begonia* evolution. Although karyomorphological data in *Begonia* are limited, many secondary constriction (SC) chromosomes have been detected. A SC chromosome consists of a satellite connected to the main body by the secondary constriction, a thin strip of the chromosome (Fig. 3). Oginuma & Peng (2002) showed that up to 63% of the chromosomes in a cell can be secondary or tertiary constriction chromosomes. Moreover, all 14 species of Taiwanese *Begonia* (including *B. palmata* and *B. aptera* with 2n = 22) surveyed possessed SC, except for *B. fenicis* (2n = 26), where the exact position of centromeres could not be determined for some chromosomes. Dewitte et al. (2009a) observed SC chromosomes in 6 out of 11 genotypes investigated. During prophase, the satellites were so loosely associated to their main body that they could easily be misidentified as separate chromosomes. Additional evidence for SC chromosomes in *Begonia* was presented by Legro & Doorenbos (1969).

These data suggest that after polyploidisation, chromosome translocations occur, which are followed by a decrease in chromosome number and genome stabilisation (Oginuma & Peng, 2002).

5.3 Chromosome size

Flow cytometric measurements of genome size, combined with chromosome counts, revealed a 12-fold difference in the mean chromosome size between the species *B. dietrichiana* and *B. pearcei* (Dewitte et al., 2009a). Moreover, as only 15 of the 66 sections of the genus *Begonia* were involved in this study it is likely that greater differences will be found when other sections are surveyed. Differences in chromosome size have also been observed in horticultural hybrids, including Elatior hybrids that are characterized by small *B. socotrana* and large *B. x tuberhybrida* chromosomes.

Apart from *B. pearcei* and *B. boliviensis*, South American species possess smaller chromosomes than Asian, African and Middle American species (Dewitte et al., 2009a). Different mechanisms, such as transposon activity, multiple deletions or other genetic rearrangements may have played an important role in the generation of this chromosome size variation (Devos et al., 2002; Kubis et al., 1998; Sanmiguel & Bennetzen, 1998). Bennetzen (2002) showed that more than 60% of some plant genomes consist of transposable elements and their defunct remnants.

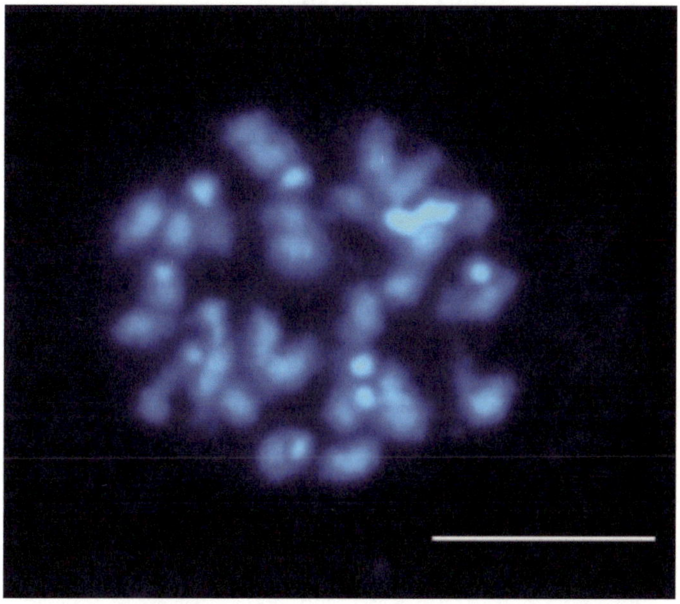

Fig. 3. Prometaphase mitotic chromosome spread of *B.* 'Tamo' (2n = 23), stained with DAPI. Intense fluorescence signal indicates satellites (Dewitte et al., 2009a). Bar = 5 µm.

5.4 Polyploidisation

Polyploidisation within *Begonia* may play a role in the diversification of lineages, the formation of new species, and morphological innovation. Evidence for the importance of polyploidy in *Begonia* evolution comes from the wide range of chromosome number, even in related species. Polyploids can arise by means of somatic mutations in meristematic cells or unreduced (2n) gametes (Bretagnolle & Thompson, 1995; Otto & Whitton, 2000). Harlan & De Wet (1975) showed that many plant species produce 2n gametes, and it is now generally accepted that 2n gametes are the driving force behind the formation of polyploids in nature (Bretagnolle & Thompson, 1995). The occurrence of 2n gametes in a *Begonia* collection has been described by Dewitte et al. (2009b), where 10 of the 70 investigated genotypes (collections of both species and cultivars) produced 2n pollen. The unreduced pollen grains were able to germinate and could be used to produce progeny. The occurrence of 2n egg cells was not investigated, but ploidy analysis of the progeny proved that 2n egg cells do occur (Dewitte et al., 2010a). Unreduced egg cells have also been observed in the hybrid *B. cucullata* x *B. schmidtiana* during breeding of semperflorens *Begonia* (Dewitte et al., 2010a; Horn, 2004).

Unreduced gametes have been shown to occur in species as well as hybrids (Dewitte et al., 2010a; Dewitte et al., 2009b). This suggests natural polyploid lineages may occur at a low frequency in the wild. However, hybrids may produce 2n gametes at c. 50 times higher frequency than species (Ramsey & Schemske, 1998). In horticulture, interspecific hybrids are often a source of 2n gametes. These 2n gametes can transfer high levels of heterozygosity (Bretagnolle & Thompson, 1995; Ramanna & Jacobsen, 2003), promoting polyploid establishment, and in some cases 2n gametes may be the only viable gametes in hybrids (Barba-Gonzalez et al., 2004; Dewitte et al., 2010b).

5.5 Genome size variation

The combined effect of polyploidy and genome stabilisation is a highly variable genome size across the genus. By using flow cytometry, Dewitte et al. (2009a) observed variation in genome size between 1C = 0.23 pg and 1.46 pg, but genome size did not positively correlate with chromosome number. The largest known discrepancies between genome size and chromosome number were observed in *B. pearcei* and *B. boliviensis,* which have the largest genome size measured in *Begonia* to date (1C = 1.46 pg), although they contain the lowest chromosome number of all genotypes analysed (2n = 26 and 28, respectively). The lack of correlation between genome size and chromosome number can be explained by the strong differences in chromosome size. Dewitte et al. (2009a) showed that the total chromosome volume in a cell was positively correlated to the genome size. Smaller chromosomes contribute less to the genome size than large chromosome do, even if the number of chromosomes is equal.

5.6 Meiosis

Chromosome pairing at metaphase I can indicate whether plants are diploids or secondary polyploids (autopolyploid or allopolyploid), but few cytological studies have focused on meiosis in *Begonia*. In *B. evansiana* (Okuna & Nagai, 1953) and *B. x chungii* (Peng & Ku, 2009), several loosely associated bivalents (paired chromosomes), which almost resemble univalents (unpaired chromosomes), were observed during chromosome pairing. Furthermore, many secondary associations between bivalents were observed in *B. evansiana* (Okuna & Nagai, 1953), consistent with observations in *B. semperflorens* and *B. rex* (Matsuura & Okuno, 1943). These observations suggested *B. evansiana* is a secondary polyploid, derived from two species with X=6 and 7 as their basic chromosome numbers.

In horticultural crosses, the interspecific hybrid *B.* 'Tamo' (2n = 23) showed irregular chromosome pairing with multivalents of up to 7 chromosomes (Dewitte et al., 2010c). These associations were often observed between chromosomes with large differences in size, which indicated associations between non-homologous chromosomes and possible recombination between these non-homologous DNA segments (illegitimate recombination). Illegitimate recombination is the driving force behind genome size decrease in *Arabidopsis* (Bennetzen et al., 2005; Devos et al., 2002), but can also be an important mechanism for exon shuffling, a major process for generating new genes (Long, 2001; Long et al., 2003).

In some hybrids, tight associations between chromosomes resulted in chromosome bridges and fragments during anaphase I (Dewitte et al., 2010c). This might be indirect evidence for the presence of paracentric inversions during crossing-over (Newman, 1966). Paracentric inversion loops are crossing over configurations that result in abnormal meiotic end products such as dicentric chromosomes (with 2 centromeres) and acentric (without centromere) fragments. Consequently, new chromosome constitutions may arise during meiosis. Furthermore, chromosome inversions may affect rates of adaptation and speciation because it promotes reproductive isolation of species (Noor et al., 2001; Hoffmann & Rieseberg, 2008). However, the importance of inversions in adaptive evolution has rarely been addressed and therefore its role in adaptive shifts is not yet clear.

Univalent formation was frequently observed in *Begonia* (Dewitte et al., 2010; Okuna & Nagai, 1953). The presence of these univalents resulted in lagging chromosomes: chromosomes with a delayed movement to the poles. In some cases (dependent on the hybrid), univalents did not migrate to the poles but formed micronuclei. Although this usually leads to unbalanced chromosome segregation and sterility (Bretagnolle & Thompson, 1995), most of the studied hybrids were fertile. It is uncertain to what degree *Begonia* are 'buffered' against unbalanced chromosome segregation. If the resulting aneuploid gametes are fertile, this process may influence the chromosome number transmitted through the progeny and the genome stabilisation process. However, more detailed studies on this topic are required.

In general, cytological studies (mainly in artificial hybrids) have shown that chromosome behaviour during meiosis is very dynamic which may have important consequences for chromosome evolution within the genus. However, further clarification with molecular evidence is required.

6. Hybridisation

Hybridisation and polyploidisation have played a major role in the evolution of plant species, and the stabilisation of hybrid lineages may contribute to species diversity (Mallet, 2007; Paun et al., 2009; Rieseberg & Carney, 1998). However, estimating the frequency of hybrid speciation is difficult, as evidence that hybridisation has let to speciation is hard to obtain, particularly in homoploid hybrids (hybrids without a change in ploidy level) (Mallet, 2007). The role that hybridisation has played in contributing new lineages in diverse tropical biomes has yet to be addressed in detail (with some exceptions such as neotropical orchids, eg Pinheiro et al., 2010), however hybrid speciation may be one of the factors contributing to species diversity in large genera such as *Begonia*. Support for hybridisation and introgression being a mechanism for the generation of diversity would require an understanding of the strength of reproductive isolation between parental species, including factors such as the fertility of hybrids, and few studies have focused on this in *Begonia*.

Several natural *Begonia* hybrids have been described including: *B. x breviscapa*, *B. x chungii* and *B. x taipeiensis* (Peng & Ku, 2009; Peng & Sue, 2000; Peng et al., 2010). This illustrates weak reproductive barriers between *Begonia* species that co-occur in the wild, which is typical of a genus that has been widely exploited in the development of horticultural varieties (Tebbitt, 2005). However, in most cases the number of species that co-occur at a single locality in *Begonia* is low. The ease of hybridisation under experimental conditions, combined with the low frequency of hybrid occurrence in the wild, indicates that habitat specialisation and non-overlapping geographic distributions may be important in maintaining the distinct identities of *Begonia* species.

Natural hybrids of recent origin can often be recognised by low pollen fertility, however fertile *Begonia* hybrids have also been observed (*B. decora* x *B. venusta*, Kiew et al., 2003; Teo & Kiew, 1999). The fertility of *Begonia* hybrids varies, and low hybrid fertility is likely to be due to genome divergence and incompatibilities between the progenitors. However, if F1 hybrids retain a low level of fertility then there is an opportunity for backcross progeny to be formed. Even if natural hybrids can be observed, further evidence is required to support the establishment of new hybrid species over time. For hybrid species to become established, they must become reproductively isolated from the parental species. Recent hybrids are likely to be swamped by well adapted parental species, however novel genetic combinations and the resulting heterosis (hybrid vigour) may explain their establishment (Lippman & Zamir, 2007; Paun et al., 2009)

Allopolyploid speciation, where hybridisation is combined with polyploidisation, offers a likely path to speciation, because the hybrid will have a high degree of post-zygotic reproductive isolation from their progenitors (Paun et al., 2009), due to the low fertility of triploid plants. Polyploid plants have a number of notable changes relative to their progenitors, including increased cell size, gene dosage effects, increased allelic diversity (level of heterozygosity), gene silencing and genetic or epigenetic interactions (Leitch & Bennett, 1997; Levin, 1983; Lewis, 1980; Osborn et al., 2003). The occurrence of viable unreduced gametes in *Begonia* hybrids has been described earlier, which is an important precursor for allopolyploid formation. Once polyploids are formed and established, they may enter an evolutionary trajectory of diploidization, a gradual conversion to diploidy through genetic changes that differentiate duplicated loci (Comai, 2005; Levy & Feldman, 2002).

To understand the potential role of hybridisation in *Begonia* evolution, we performed a series of cross-fertilization experiments. Firstly, the effects of parental ploidy level, genome size and geographic origin on seedling viability and fertility was investigated. This experiment allowed us to compare the effects of hybridizing species with highly differentiated genomes, and to understand the role of polyploidy and genome size changes in reproductive isolation in *Begonia*. Secondly, we investigated the cross compatibility of more closely related species (from the Central American Section *Gireoudia*), in order to assess the likelihood that hybridisation may occur between species with less differentiated genomes that are more likely to co-occur in the wild.

In the first experiment, we cross hybridised 19 mainly South-American *Begonia* species (Table 1). From the 156 cross combinations performed, pollinating 5 flowers per cross, 27 combinations generated viable hybrids (Table 2). Successful crosses were observed between species with different chromosome numbers and genome sizes (e.g. *B. cucullata* 2n = 34, 1C = 30 x *B. odorata* 2n = 52, 1C = 0.56 pg), and even between species from different continents (e.g. *B. dregei* x *B. coccinea; B. pearcei* x *B. grandis*).

The likelihood that F1 hybrids were obtained increased when the species used for crossing came from the same section (*Gaerdtia, Begonia*), although this was not the case for the section *Pritzelia*. For example, within the section *Begonia, B. cucullata, B. subvillosa* var. *leptotricha* and *B. schmidtiana* had equal chromosome numbers (2n = 34), while *B. odorata* contained a higher chromosome number (2n = 52). The genome sizes of *B. cucullata* and *B. subvillosa* var. *leptotricha* were similar (about 0.30 pg), but lower compared to those of *B. schmidtiana* (0.38 pg) and *B. odorata* (0.56 pg). Another species from the section *Begonia, B. venosa* (2n = 30; 1C = 0.25 pg), did not cross with these species.

	origin	section	Genome size 1C (pg DNA)	Chromosome number (2n)
B. listada	America	*Pritzelia*	0.31	56
B. echinosepala var. *elongatifolia*	America	*Pritzelia*	0.32	56
B. coccinea	America	*Pritzelia*	0.56	56
B. corallina	America	*Gaerdtia*	0.78	56
B. albo-picta	America	*Gaerdtia*	0.58	56
B. solananthera	America	*Solananthera*	0.57	56
B. luxurians var. *ziesenhenne*	America	*Scheidweileria*	0.32	56
B. odorata	America	*Begonia*	0.56	52
B. subvillosa var. *leptotricha*	America	*Begonia*	0.29	34
B. cucullata	America	*Begonia*	0.30	34
B. schmidtiana	America	*Begonia*	0.38	34
B. venosa	America	*Begonia*	0.25	30
B. ulmifolia	America	*Donaldia*	0.25	30
B. heracleifolia	America	*Gireoudia*	0.75	28
B. boliviensis	America	*Barya*	1.46	28
B. pearcei	America	*Eupetalum*	1.46	26
B. dregei	Africa	*Augustia*	0.66	26
B. grandis var. *evansiana*	Asia	*Diploclinium*	0.68	26
B. diadema	Asia	*Platycentrum*	0.58	22

Table 1. Origin, classification, genome size and chromosome numbers of *Begonia* species used in the first experiment (classification from Dewitte et al., 2009).

These results suggest post-pollination barriers to hybrid formation in *Begonia* are complex, as no strong trend can be seen between parental genome size, chromosome number and the area of origin in the likelihood that F1 hybrids can be obtained. This is in agreement with the many horticultural interspecific crosses described (Tebbitt, 2005).

To assess the fertility of the hybrids, 5 to 10 of the F1 hybrid progeny were scored for pollen morphology and germination according to Dewitte et al. (2009b). The majority of hybrids examined were male sterile and either dropped their male flowers before opening or produced inviable pollengrains. Only a few combinations, mainly those between *B. albo-picta, B. corallina* and *B. coccinea,* were male fertile (Table 3). They are all South American species with 2n = 56. *B. albo-picta* and *B. corallina* belong to the section *Gaerdtia* while *B. coccinea* belongs to the section *Pritzelia.* However, *B. coccinea* is morphologically very similar to *B. corallina* and their sectional affinities may require reappraisal.

In the second experiment, 12 species from the Central American section *Gireoudia* were crossed over a four year period (2005-2009) to investigate the cross compatibility of more closely related species. Of the 144 potential combinations between the parents, 92 (64%) have been attempted - the other combinations did not overlap in their flowering time. Of the 92 combinations attempted, 89 (97%) produced plump seeds, with little or no variability in success between repeats of the same cross. Only a single cross has failed to produce F1 seeds in both directions (*B. theimei x B. heracleifolia)* and two crosses have failed in one direction only (*B. peltata x B. serioceneura, B. nelumbiifolia x B. heracleifolia*). However, these are based on a single parental accession and few replicates, and this requires further crosses to confirm this result. Overall, the ability to form F1 hybrids between related *Begonia* species with the same chromosome number (2n=28; the common chromosome number in this section), further supports weak postzygotic reproductive isolating barriers, as shown in Figure 4.

♀ \ ♂	*B. listada*	*B. echinosepala*	*B. coccinea*	*B. corallina*	*B. albo-picta*	*B. solananthera*	*B. luxurians*	*B. odorata*	*B. subvillosa*	*B. cucullata*	*B. schmidtiana*	*B. venosa*	*B. ulmifolia*	*B. heracleifolia*	*B. boliviensis*	*B. pearcei*	*B. dregei*	*B. grandis*	*B. diadema*
B. listada		x	x	x	x	x	x	x	x	x		x	x		x	x	x	x	x
B. echinosepala			x	x	x	x	o	x	x	x		o	x		x				x
B. coccinea		x		o	o	o	x	x	x	o		x	x					x	
B. corallina		x	o		o	o	x	x	x	x		x	x		x	x	o	x	x
B. albo-picta		x	o	o		o	x	x	x	x		x	x		x				x
B. solananthera		x			o		x	x	x			x	x						
B. luxurians		x	x	x		x		x	x	x		x	x		x				x
B. odorata		x	x	x	x	x	x		o	o		x	x		x		o	x	x
B. subvillosa		x	x	x	x		x	o		o		x	o						
B. cucullata		x	x	x	x		x	o	o			x			x		x		x
B. schmidtiana										o									
B. venosa			x																
B. ulmifolia		x	x	x	x	x	x	x	x			x							
B. heracleifolia		x	x	x	x	x	x	x	x	x		x	x						
B. boliviensis			o					x		x									
B. pearcei								x				x						o	
B. dregei			o	o			x		x	x		x						x	x
B. grandis										x									
B. diadema												x							

Table 2. Outcomes of interspecific crosses. Combinations resulting in viable hybrid are marked with 'o', combinations that did not result in seedlings with an 'x'.

We then germinated the seeds from each seed capsule, to test for evidence of seedling mortality and the fertility of the hybrids. No evidence of seedling mortality was observed, that would indicate incompatible genome interactions which affect fitness at early growth stages (hybrid dysgenesis). Successful crosses were grown to maturity, and the 19 F1 hybrids that flowered in the Spring 2009 were scored for pollen viability. For each accession 100 pollen grains were stained with fluorescein diacetate (FDA) in 5% sucrose solution, and percentage of well stained pollen scored. The mean pollen viability was 3.5% (±1.9%), and 73% of hybrids were pollen sterile.

cross		% bad, shrunken pollen grains	% germination
♀	♂		
B. albo-picta	*B. corallina*	7,6 ± 2,9	48,4 ± 21,0
B. albo-picta	*B. coccinea*	4,5 ± 3,7	39,6 ±28,5
B. coccinea	*B. albo-picta*	8,0 ± 5,2	45,3 ± 20,0
B. coccinea	*B. corallina*	15,0 ± 6,5	41,5 ± 17,6
B. corallina	*B. coccinea*	13,0 ± 6,9	50,0 ± 36,7
B. corallina	*B. albo-picta*	18,8 ± 6,5	38,5 ± 15,4
B. echinosepala	*B. luxurians*	98,6 ± 0,6	0,4 ± 0,3

Table 3. Pollen fertility of fertile F1 hybrid combinations (means ± SD; n = 5). Only combinations producing male fertile progeny are included.

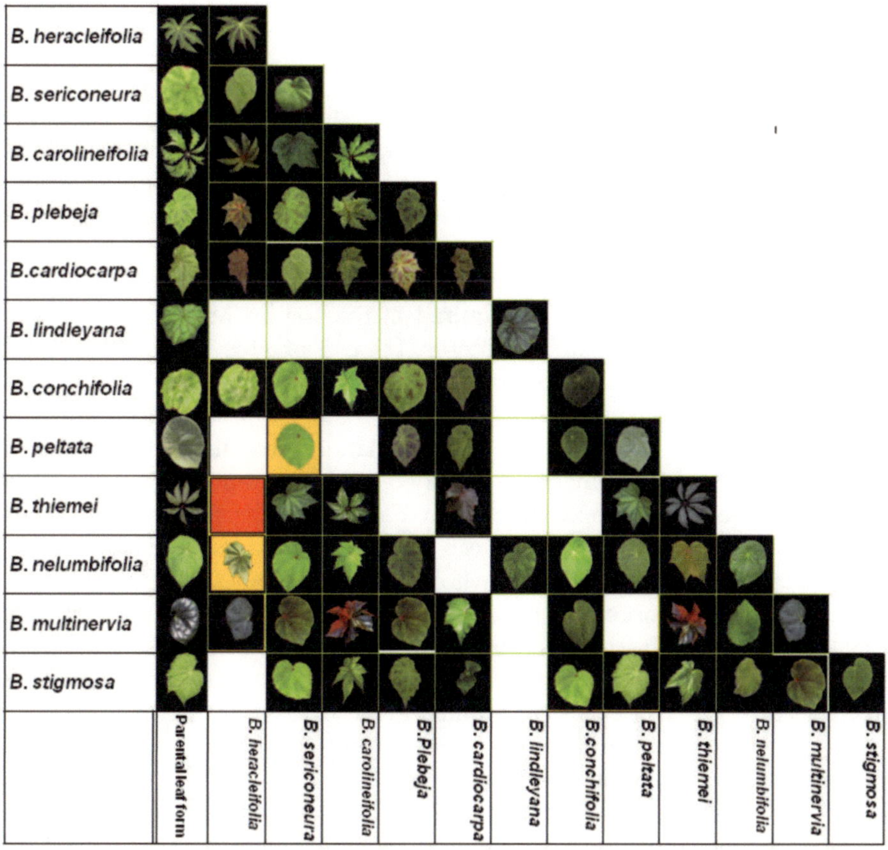

Fig. 4. Interspecific crossing barriers in the section *Gireoudia*. Leaves on a dark background indicate a successful F1 cross. A red square indicates an unsuccessful cross and a yellow background a cross that has worked in one direction only (Twyford & Kidner, previously unpublished data).

These results highlight that interspecific *Begonia* crosses have a significantly reduced pollen fertility. There are a number of notable exceptions, particularly in F1 crosses in the section *Gireoudia*, where pollen stainability is above 30% (Matthews, unpublished). This includes crosses between divergent species found in different geographical areas, such as *B. conchifolia* and *B. plebeja*. Hybrids with good pollen fertility were also observed in section *Gaerdtia* (see above). Overall, this suggests that hybrid fertility is constraining in the formation of later generation hybrids in the wild. However, male fertility is not necessarily linked to female fertility as different genes may underlie male and female meiosis. Some of the male sterile hybrids have been successfully used as a seed parent in crosses (Dewitte et al., 2010a; Twyford & Kidner unpublished data).

The role of hybrid speciation in *Begonia* should be investigated further, but several preconditions for hybrid speciation are fulfilled in this genus: weak crossing barriers, a very labile genome, the presence of natural interspecific hybrids and the occurrence of 2n gametes. The fertility of artificial hybrids is very variable and depends on the cross combination. If a hybrid retains some fertility and becomes isolated from its parent plant, hybrid speciation may be a possible outcome.

7. Future perspectives and conclusions

Begonia is a remarkable genus both for its size and diversity. As a pantropical genus, it provides the opportunity to compare biogeographic patterns in different regions using like to like. It also offers the chance to examine the reasons behind the high levels of tropical diversity through comparisons with speciation patterns in correspondingly widespread temperate genera.

Major progress has been made clarifying the evolutionary relationships in the genus *Begonia* at the sectional level. Further work is required at the species level, to assess lower level relationships, and reveal biogeographic patterns. The lack of a fossil record for the genus has been a stumbling block to the generation of detailed evolutionary scenarios (Goodall-Copestake et al., 2009). Moreover, more informative genetic markers are needed to obtain well-resolved phylogenies of the recent rapid radiations that have occurred in different regions.. However, the development of barcoded next generation sequencing techniques holds out the promise that phylogenies of many hundreds of accessions could be produced en-mass, transforming our ability to detect speciation patterns. The decreasing cost of sequencing means that transcriptomes, whole plastid genomes and even draft nuclear genomes can be produced for reasonable amounts of money (Ng & Kirkness, 2010).

The prevalence of geographic isolation between *Begonia* species may be one of the factors explaining the limited postzygotic reproductive isolating barriers between species. The lack of interspecific crossing barriers has contributed to its success as a horticultural subject, and provides us with the opportunity to use classical and quantitative genetic techniques to study the genes that vary between species.

On one level such analysis can be done by transcriptome or genome comparisons and the screening for genes that show the signature of diversifying or purifying selection (Biswas & Akey, 2006). However, these techniques can provide only statistical support for the importance of any one locus, and are not informative about which traits are affected. In a non-model group the link between locus and function can only be based on comparisons with model species which may be misleading.

A second approach works in the other direction, from the trait down to the locus. Once a genetic map has been constructed, quantitative trait locus (QTL) analysis can reveal the genetic architecture of species level differences: the number of loci that affect a trait, along with their sizes and interactions (Zeng, 2005). The traits most divergent between species may be related to the selective forces that drove speciation and can be linked to plant fitness as measured by seed production or to relative growth rate (Taylor et al., 2009). An ideal experiment would combine both approaches, mapping highly divergent genes relative to QTLs.

One of the great advantages of *Begonia* for this work is the parallel evolution that is common in the genus on many levels. Given sufficient genetic resources, comparisons can be made between the genetic architecture of traits independently evolved within and between sections and continents. This allows hypothesises about trait evolution to be tested using multiple replicates, providing greater robustness than is usually possible for evolutionary studies.

One topic that has yet to be studied in depth and deserves further attention is natural hybridisation and hybrid speciation. Hybrid speciation seems very likely within the genus, and interspecific processes may be the cause of the hard incongruence of phylogenetic trees derived from plastid and nuclear ribosomal data within some species-rich sections (*Platycentrum, Petermannia*) (Goodall-Copestake (2010), Thomas (2010). However, homoploid hybrids are hard to detect; and only a handful of species in evolutionary model systems have proven to be hybrids without a change in ploidy level (e.g. *Helianthus, Argyranthemum, Ceanotus, Pinus* (listed in Gross & Rieseberg, 2005). Investigation of the frequency of natural hybridisation, together with studies of genome stabilisation after polyploidisation and more karyomorphological data may be useful to understand chromosome evolution within the genus. The use of molecular techniques, such as comparative transcriptional profiling or targeted genome resequencing of species and their hybrids (see Twyford & Ennos, 2011), or cytological techniques such as GISH, may shed light on the role of natural hybridisation.

The genus *Begonia* has been studied for many reasons, including: horticulture, taxonomy, as an indicator for biogeographic variation, to understand the development of their distinctive leaf forms and their aberrations, population genetics of endemic species, hybridisation and genome dynamics. As the distinction between model plants with extensive genetic resources, and non-models without these resources becomes less well defined, we expect that *Begonia* will continue to provide insights into the nature and origin of tropical plant diversity.

8. References

Ågren, J. & Schemske, DW. (1993). Outcrossing rate and inbreeding depression in two annual monoecious herbs, *Begonia hirsuta* and *B. semiovata*. *Evolution*, Vol.47, No.1, (February 1993), pp. 125–135, ISSN 1558-5646

APG (2009). An update of the Angiosperm Phylogeny Group classification for the orders and families of flowering plants: APG III. *Botanical Journal of the Linnean Society*, Vol.161, No.2, (October 2009), pp. 105-121, ISSN 0024-4074

Arends, JC. (1970). Somatic chromosome numbers in 'Elatior'-Begonias. *Mededelingen Landbouwhogeschool Wageningen*, Vol.70, No.20, pp. 1-18, ISSN 0369-0598

Barba-Gonzalez, R.; Lokker, AC.; Lim, K-B.; Ramanna, MS. & Van Tuyl, JM. (2004). Use of 2n gametes for the production of sexual polyploids from sterile Oriental x Asiatic hybrids of lilies (*Lilium*). *Theoretical and Applied Genetics*, Vol.109, No.6, (October 2004), pp. 1125-1132, ISSN 0040-5752

Bennetzen, JL. (2002). Mechanisms and rates of genome expansion and contraction in flowering plants. *Genetica*, Vol.115, No.1, (May 2002), pp. 29-36, ISSN 0016-6707

Bennetzen, JL.; Ma, J. & Devos, KM. (2005). Mechanisms of recent genome variation in flowering plants. *Annals of Botany*, Vol.95, No.1, (january 2002), pp. 127-132, ISSN 0305-7364

Berenbaum, R. & Zangerl, AR. (2006). Parnsnip webworms and host plants at home and abroad: trophic complexity in geographic mosaic. *Ecology*, Vol.87, No.12, (December 2006), pp. 3070-3081, ISSN 0012-9658

Bermingham, E. & Dick, C. (2001). The *Inga*--Newcomer or Museum Antiquity? *Science*, Vol.293, No.5538, (September 2001), pp. 2214-2216, ISSN 0036-8075

Biswas, S. & Akey, JM. (2006). Genomic insights into positive selection. *Trends in Genetics*, Vol.22, No.8, (August 2006), pp. 437-446, ISSN 0168-9525

Bouman, F. & de Lange, A. (1983). Structure, micromorphology of *Begonia* seeds. The *Begonian*, Vol.50, (May 1983), pp. 70-78, 91

Bretagnolle, F. & Thompson, JD. (1995). Gametes with the somatic chromosome number: mechanisms of their formation and role in the evolution of autoploid plants. *New Phytologist*, Vol.129, No.1, (Januari 2005), pp. 1-22, ISSN 0028-646X

Burt-Utley, K. (1985). A revision of the Central-American species of *Begonia* section *Gireoudia* (Begoniaceae). *Tulane Studies in Zoology and Botany*, Vol.25, No.1, pp. 1–131, ISSN 0082-6782

CBOL Plant working group (2009). A DNA barcode for land plants. *Proceedings of the National Academy of Sciences*, Vol.106, No.31, (August 2009), pp. 12794-12797, ISSN 0027-8424

Chen, ZJ. (2007). Genetic and epigenetic mechanisms for gene expression and phenotypic variation in plant polyploids. *Annual Review of Plant Biology*, Vol.58, (June 2007), pp. 377-406, ISSN 1543-5008

Clement, WL.; Tebbitt, MC.; Forrest, LL.; Blair, JE.; Brouillet, L.; Eriksson, T. & Swensen, SM. (2004). Phylogenetic position and biogeography of *Hillebrandia sandwicensis* (Begoniaceae): a rare Hawaiian relict. *American Journal of Botany*, Vol.91, No.6, (June 2004), pp. 905-917, ISSN 0002-9122

Comai, L. (2005). The advantages and disadvantages of being polyploid. *Nature Reviews Genetics*, Vol.6, No.11, (November 2005), pp. 836-846, ISSN 1471-0056

de Lange, A. & Bouman, F. (1999). Seed micromorphology of neotropical *Begonias*. *Smithsonian Contributions to Botany*, Vol. 90, (September 1999), pp. 1-49, ISSN 0081-024X

de Wilde, JJFE. & Plana, V. (2003). A new section of *Begonia* (Begoniaceae) from west central Africa. *Edinburgh Journal of Botany*, Vol.60, No.2, (October 2003), pp. 121-130, ISSN 0960-4286

de Wilde, JJFE. (2011). Begoniaceae. In: *Familes and Genera of Vascular Plants (Vol 10; Flowering Plants), Eeudicots*, Kubitzki, K.., (ed.), 56-71, Springer, ISBN 9783642143960, Berlin, Germany

Devos, KM.; Brown, JKM.; Bennetzen, JL. (2002). Genome size reduction through illegitimate recombination counteracts genome expansion in *Arabidopsis*. *Genome Research*, Vol.12, No.7, (July 2007), pp. 1075-1079, ISSN 1088-9051

Dewitte, A.; Eeckhaut, T.; Van Huylenbroeck, J. & Van Bockstaele, E. (2010a). Induction of unreduced pollen by trifluralin and N_2O treatments. Euphytica, Vol.171, No.2, (October 2010), pp. 283-293, ISSN 0014-2336

Dewitte, A.; Eeckhaut, T.; Van Huylenbroeck, J. & Van Bockstaele, E. (2010c). Meiotic aberrations during 2n pollen formation in *Begonia*. *Heredity*, Vol.104, No.2, (February 2010), pp. 215-223, ISSN 0018-067X

Dewitte, A.; Eeckhaut, T.; Van Huylenbroeck, J. & Van Bockstaele, E. (2009b). Occurrence of viable unreduced pollen in a *Begonia* collection. *Euphytica*, Vol.168, No.1, (February 2009), pp. 81-94, ISSN 0014-2336

Dewitte, A.; Van Laere, K..; Van Huylenbroeck, J. & Van Bockstaele, E. (2010b). Inheritance of 2n pollen formation in an F1 and F2 population of *Begonia* hybrids. Proceedings of the 23th international Eucarpia symposium, section colourful breeding and genetics Part II, ISBN 978-90-66051102, Leiden, The Netherlands, September 2010

Dewitte, A.; Leus, L.; Eeckhaut, T.; Vanstechelman, I.; Van Huylenbroeck, J. & Van Bockstaele, E. (2009a). Genome size variation in *Begonia*. *Genome*, Vol.52, No.10, (October 2010), 829-838, ISSN 0831-2796

Doorenbos, J.; Sosef, M.; de Wilde, J. (1998). *The sections of Begonia (Studies in Begoniaceae VI)*, Backhuys Publishers, ISBN 9057820072, Leiden, The Netherlands

Noor, MA.; Grams, KL.; Bertucci, LA. & Reiland, J. (2001). Chromosomal inversions and the reproductive isolation of species. *Proceedings of the National Academy of Sciences*, Vol.98, No. 21 (October 2001), pp. 12084-12088, ISSN 0027-8424

Forest, LL.; Hughes, M. & Hollingsworth, PM. (2005). A phylogeny from *Begonia* using nuclear ribosomal sequence data and morphological characters. *Systematic Botany*, Vol.30, No.3, (July 2005), pp. 671-682, ISSN 0363-6445

Forrest, LL. & Hollingsworth, PM. (2003). A recircumscription of *Begonia* based on nuclear ribosomal sequences. *Plant systematic and evolution.*, Vol.241, No.3-4, (November 2003), pp. 193-211, ISSN 0378-2697

Forrest, LL. (2001). *Phylogeny of Begoniaceae, PhD theses*. University of Glasgow, Glasgow.

Frodin, DG. (2004). History and concepts of big plant genera. *Taxon*, Vol.53, No.3, (August 2004), pp. 753-776, ISSN 0040-0262

Goodall-Copestake, W. (2005). *Framework phylogenies for the Begoniaceae, PhD theses*. University of Glasgow, Glasgow.

Goodall-Copestake, W.; Harris, DJ. & Hollingsworth, PM. (2009). The origin of a mega-diverse genus: dating *Begonia* (Begoniaceae) using alternative datasets, calibrations and relaxed clock methods. *Botanical journal of the linnean society*, Vol.159, No.3, (March 2009), pp. 363-380, ISSN 0024-4074

Goodall-Copestake, W.; Pérez-Espona, S.; Harris, DJ. & Hollingsworth, PM. (2010). The early evolution of the mega-diverse genus *Begonia* (Begoniaceae) inferred from organelle DNA phylogenies. *Biological journal of the linnean society*, Vol.101, No.2, (October 2010), pp. 243-250, ISSN 0024-4074

Gross, BL. & Rieseberg, LH. (2005). The ecological genetics of homoploid hybrid speciation. *Journal of heredity*, Vol.96, No.3, (May 2005), pp. 241-252, ISSN 0022-1503

Gupta, RC.; Kumar, HP. & Dhaliwal, RS. (2009). Cytological studies in some plants from cold desserts of India, Lahau and Spiti (Himachal Pradesh). *Chromosome botany*, Vol.4, No.1, (July 2009), pp. 5-11, ISSN 1881-5936

Haegeman, J. (1979). *Tuberous Begonias. Origin and development.* J. Cramer, ISBN 376821219X, Vaduz, Liechtenstein.

Harlan, J. & De Wet, J. (1975). On Ö. Winge and a prayer: the origins of polyploidy. *The botanical review*, Vol.41, No.4, (October 2004), pp. 361-390, ISSN 0006-8101

Hoffmann, AA. & Rieseberg, LH. (2008). Revisiting the impact of inversions in evolution: from population genetic markers to drivers of adaptive shifts and speciation. *Annual review of ecology, evolution and systematics*, Vol.39, (December 2008), pp. 21–42, ISSN 1543-592X

Horn, W. (2004). The patterns of evolution and ornamental plant breeding. Proceedings of the 21st international Eucarpia symposium on classical versus molecular breeding of ornamentals Part II, ISBN 9789066056879, Munchen, Germany, August 2003

Hughes, M. & Hollingsworth, PM. (2008). Population genetic divergence corresponds with species level biodiversity patterns in the large genus Begonia. *Molecular ecology*, Vol.17, No.11, (June 2008), pp. 2643-2651, ISSN 0962-1083

Hughes, M. (2008). *An annotated checklist of Southeast Asian Begonia.* Royal Botanic Garden Edinburgh, ISBN 9781906129149, Edinburgh, UK.

Hughes, M.; Hollingsworth, PM. & Miller, AG. (2003). Population genetic structure in the endemic *Begonia* of the Socotra archipelago. *Biological Conservation*, Vol.113, No.2, (October 2003), pp. 277–284, ISSN 0006-3207

Hughes, NM.; Vogelmann, TC. & Smith, WK. (2008). Optical effects of abaxial anthocyanin on absorption of red wavelengths by understorey species: revisiting the back-scatter hypothesis. *Journal of Experimental Botany*, Vol.59, No.12, (September 2003), pp. 3435-3442, ISSN 0022-0957

Hvoslef-Eide, AK. & Munster, C. (2007). *Begonia.* History and breeding. In: *Flower Breeding and Genetics, chapter 9*, Anderson, NO., (ed.), 241-275, Springer, ISBN 9781402044274, Dordrecht, The Netherlands

Kapoor, BM. (1966). IOPB chromosome number reports VIII. Taxon, Vol.15, No.7, (September 1966), pp. 279–284, ISSN 0040-0262

Keraudren-Aymonin, M. (1983). Begoniacees. In: *Flore de Madagascar et des Comores*, Leroy, J-F., (ed.), 7-108, Museum national d'histoire naturelle, ISBN 2856541658, Paris, France

Kiew, R.; Teo, LL. & Gan, YY. (2003). Assessment of the hybrid status of some Malesian plants using amplified fragment length polymorphism. *Telopea*, Vol.10, No.1, pp. 225-233, ISSN 0312-9764

Kubis, S.; Schmidt, T. & Heslop-Harrison, JS. (1998). Repetitive DNA elements as a major component of plant genomes. *Annals of Botany*, Vol.82, Supplement I, (December 1998), pp. 45-55, ISSN 0305-7364

Legro, RAH. & Doorenbos, J. (1969). Chromosome numbers in *Begonia* 1. *Netherlands Journal of Agricultural Sciences*, Vol.17, pp. 189-202, ISSN 0028-2928

Legro, RAH. & Doorenbos, J. (1971). Chromosome numbers in *Begonia* 2. *Netherlands Journal of Agricultural Sciences*, Vol.19, pp. 176-183, ISSN 0028-2928

Legro, RAH. & Doorenbos, J. (1973). Chromosome numbers in *Begonia* 3. *Netherlands Journal of Agricultural Sciences*, Vol.21, pp. 167-170, ISSN 0028-2928

Legro, RAH. & Haegeman, JFV. (1971). Chromosome numbers of tuberous begonias. *Euphytica*, Vol.20, No.1, (February 1971), pp. 1-13, ISSN 0014-2336

Leitch, I. & Benett, M. (1997). Polyploidy in angiosperms. *Trends in Plant Science*, Vol. 2, No.12, (December 1997), pp. 470-476, ISSN 1360-1385

Levin, D. (1983). Polyploidy and novelty in flowering plants. *American Naturalist*, Vol.122, No.1, (July 1983), pp. 1-25, ISSN 0003-0147

Levy, AA. & Feldman, M. (2002). The impact of polyploidy on grass genome evolution. *Plant Physiology*, Vol.130, No.4, (December 2002), pp. 1587-1593, ISSN 0032-0889

Lewis, W. (1980). Polyploidy in species populations. In: *Polyploidy: biological relevance*, Lewis, W., (ed.), 103-144, Plenum Press, ISBN 0306403587, New York

Lippman, ZB. & Zamir, D. (2007). Heterosis: revisiting the magic. *Trends in Genetics*, Vol.23, No.2, (February 2007), pp. 60-66, ISSN 0168-9525

Long, M. (2001). Evolution of novel genes. *Current opinion in genetics and development*, Vol.11, No.6, (December 2001), pp. 673-680, ISSN 0959-437X

Long, M.; Betran, E.; Thornton, K. & Wang, W. (2003). The origin of new genes: glimpses from the young and old. *Genetics*, Vol.4, (November 2003), pp. 865-875, ISSN 0016-6731

Mallet, J. (2007). Hybrid speciation. *Nature Reviews*, Vol.446, (March 2007), pp. 279-286, ISSN 1471-0056

Marasek-Ciolakowska; A., Ramanna, MS.; ter Laak, WA. & Van Tuyl, JM. (2010). Genome composition of 'Elatior'-begonias hybrids analysed by genomic in situ hybridisation. *Euphytica*, Vol.171, No.2, (January 2010), pp. 273-282, ISSN 0014-2336

Matolweni, LO.; Balkwill, K. & McLellan, T. (2000). Genetic diversity and gene flow in the morphologically variable, rare endemics *Begonia dregei* and *Begonia Homonyma* (Begoniaceae). *American Journal of Botany*, Vol.87, No.3, (March 2000), pp. 431-439, ISSN 0002-9122

Matsuura, H. & Okuno, S. (1936). Cytogenetical studies on *Begonia*. *The Japanese Journal of Genetics*, Vol.12, No.1, pp. 42-43, ISSN 0021-504X

Matsuura, H. & Okuno, S. (1943). Cytogenetical study in *Begonia* (Preliminary survey). *Cytologia*, Vol.13, No.1, (November 1943), pp. 1-18, ISSN 1348-7019

Matthews, ML. & Endress, PK. (2004). Comparative floral structure and systematics in cucurbitales (Corynocarpaceae, Coriariaceae, Tetramelaceae, Datiscaceae, Begoniaceae, Cucurbitaceae, Anisophylleaceae). *Botanical Journal of the Linnean Society*, Vol.145, No.2, (June 2004), pp. 129-185, ISSN 0024-4074

McKenna, DD. & Farrell, BD. (2006). Tropical forests are both evolutionary cradles and museums of leaf beetle diversity. *Proceedings of the National Academy of Science*, Vol.103, No.29, (July 2006), pp. 10947-10951, ISSN 0027-8424

Neale, S.; Goodall-Copestake, W. & Kidner, CA. (2006). The evolution of diversity in *Begonia*. In: *Floriculture, Ornamental and Plant Biotechnology*, Teixeira da Silva, JA. (ed.), Global Science books, ISBN 4903313093, Middlesex, UK

Newman, LJ. (1966). Bridge and fragment aberrations in *Podophyllum Peltatum*. *Genetics*, Vol.53, No.1, (January 1966), pp. 55-63, ISSN 0016-6731

Ng, PC. & Kirkness EF. (2010). Whole Genome Sequencing. In: *Genetic variation: methods and protocols 628 (Methods in Molecular Biology)*, Barnes, R. & Breen, G., (Eds.), 215-226, Humana press, ISBN 1603273662, Totowa, NJ, US

Oginuma, K. & Peng, CI. (2002). Karyomorphology of Taiwanese Begonia: taxonomic implications. *Journal of Plant Research*, Vol.115, No.3, (June 2002), pp. 225-235, ISSN 0918-9440

Okuno, S. & Nagai, S. (1953). Cytological studies on *Begonia evansiana* Andr. with special reference to its meiotic chromosomes. *The Japanese Journal of Genetics*, Vol.28, No.4, pp. 132-136, ISSN 0021-504X

Okuno, S. & Nagai, S. (1954). Karyotypic polymorphism in *Begonia tuberhybrida*. *The Japanese Journal of Genetics*, Vol.29, No.5-6, pp. 185-196, ISSN 0021-504X

Osborn, T.; Pires, J.; Birchler, J.; Auger, D.; Chen, Z.; Lee, H.; Comai, L.; Madlung, A.; Doerge, R.; Colot, V. & Martienssen, R. (2003). Understanding mechanisms of novel gene expression in polyploids. *Trends in Genetics*, Vol.19, No.3, (March 2003), pp. 141-147, ISSN 0168-9525

Otto, SP. & Whitton, J. (2000). Polyploid incidence and evolution. *Annual Review of Genetics*, Vol.34, pp. 401-437, ISSN 0066-4197

Paun, O.; Fay, MF.; Soltis, DE. & Chase, MW. (2007). Genetic and epigenetic alterations after hybridization and genome doubling. *Taxon*, Vol.56, No.3, (August 2007), pp. 649–656, ISSN 0040-0262

Paun, O.; Forest, F.; Fay, MF. & Chase, MW. (2009). Hybrid speciation in angiosperms: parental divergence drives ploidy. *New Phytologist*, Vol.182, No.2, (February 2009), pp. 507-518, ISSN 0028-646X

Peng, C-I. & KU, S-M. (2009). *Begonia ×chungii* (Begoniaceae), a new natural hybrid in Taiwan. *Botanical Studies*, Vol.50, No.2, (April 2009), pp. 241-250, ISSN 1817-406X

Peng, C-I. & Sue, C-Y. (2000). *Begonia x taipeiensis* (Begoniaceae), a new natural hybrid in Taiwan. *Botanical Bulletin of the Academy Sinica*, Vol.41, No.2, (April 2000), pp. 151-158, ISSN 0006-8063

Peng, C-I.; Liu, Y.; Ku, S-M.; Kono Y. & Chung, K-F. (2010). *Begonia × breviscapa* (Begoniaceae), a new intersectional natural hybrid from limestone areas in Guangxi, China. *Botanical Studies*, Vol.51, No.1, (January 2010), pp. 107-117, ISSN 1817-406X

Pinheiro, F.; De Barros, F.; Palma-Silva, C.; Meyer, D.; Fay, MF.; Suzuki , RM.; Lexer, C. & Cozzolino, S. (2010). Hybridization and introgression across different ploidy levels in the Neotropical orchids *Epidendrum fulgens* and *E. puniceoluteum* (Orchidaceae). *Molecular Ecology*, Vol.19, No.18, (September 2010), pp. 3981-3894, ISSN 0962-1083

Plana, V. (2003). Phylogenetic relationships of the Afro-Malagasy members of the large genus *Begonia* inferred from *trnL* Intron sequences. *Systematic botany*, Vol.28, No.4, (October 2003), pp. 693-704, ISSN 0363-6445

Plana, V.; Gascoine, A.; Forest, LL.; Harris, D. & Pennington, RT. (2004). Pleistocene and pre-pleistocene *Begonia* speciation in Africa. *Molecular Phylogenetics and Evolution*, Vol.31, No.2 (May 2004), pp. 449-461, ISSN 1055-7903

Rajbhandary, S. (2010). *Systematic revision of the genus Begonia L. (Begoniaceae) in the Himalayas, PhD theses*. Tribhuvan University, Kathmandu, Nepal.

Ramanna, MS. & Jacobsen, E. (2003). Relevance of sexual polyploidization for crop improvement-a review. *Euphytica*, Vol.133, No.1, (July 2003), pp. 3-18, ISSN 0014-2336

Ramsey, J & Schemske, DW. (1998). Pathways, mechanisms and rates of polyploidy formation in the flowering plants. *Annual Review of Ecology and Systematics*, Vol.29, (November 1998), pp. 267-501, ISSN 0066-4162

Renner, SS. (2005). Relaxed molecular clocks for dating historical plant dispersal events. *Trends in Plant Science*, Vol.10, No.11, (November 2005), pp. 550-558, ISSN 1360-1385

Rieseberg, H. & Carney, S. (1998). Tansley review 102: Plant hybridization. *New Phytologist*, Vol.140, No.4, (December 1998), pp. 599-624, ISSN 0028-646X

Rubite, RR. (2010). *Systematic studies on Philippine Begonia L. section Diploclinium (Lindl.) A. D.C. (Begoniaceae), PhD theses*. De La Salle University, Manila

Sanmiguel, P. & Bennetzen, JL. (1998). Evidence that a recent increase in maize genome size was caused by the massive amplification of intergene retrotransposons. *Annals of Botany*, Vol.82, Supplement 1, (December 1998), pp. 37-44, ISSN 0305-7364

Schaefer, H. & Renner, SS. (2011). Phylogenetic relationships in the order Cucurbitales and a new classification of the gourd family (Cucurbitaceae). *Taxon*, Vol.60, No.1, (February 2011), pp. 122-138, ISSN 0040-0262

Shui, YM.; Peng, CI. & Wu, CY. (2002). Synopsis of the Chinese species of Begonia (Begoniaceae), with a reappraisal of sectional delimitation. *Botanical Bulletin of Academia Sinica*, Vol.43, No.4, (October, 2002), pp. 313-327, ISSN 0006-8063

Sosef, MSM. (1994). *Refuge begonias. Taxonomy, phylogeny and historical biogeography of Begonia sect. Loasibegonia and sect. Scutobegonia in relation to glacial rain forest refuges in Africa, PhD theses*. Wageningen Agricultural University, ISBN 9054852437, Wageningen, The Netherlands

Stults, DZ. & Axsmith, BJ. (2011). First macrofossil record of *Begonia* (Begoniaceae). *American Journal of Botany*, Vol.98, No.1, (January 2011), pp. 150-153, ISSN 0002-9122

Taylor, SJ.; Arnold, M. & Martin, NH. (2009). The genetic architecture of reproductive isolation in Louisiana irises: hybrid fitness in nature. *Evolution*, Vol.63, No.10, (October 2009), pp. 2581-2594, ISSN 1558-5646

Tebbitt, MC. (2005). *Begonias: cultivation, identification and natural history*. Timberpress, ISBN 0881927333, Portland, USA.

Tebbitt, MC.; Lowe-Forrest, LL.; Santoriello, A.; Clement, WL. & Swensen, SM. (2006). Phylogenetic relationships of Asian Begonia, with an emphasis on the evolution of rain-ballist and animal dispersal mechanisms in sections *Platycentrum, Sphenanthera* and *Leprosae. Systematic botany*, Vol.31, No.2, (April 2006), pp. 327-336, ISSN 0363-6445

Teo, L-L. & Kiew, R. (1999). First record of a natural *Begonia* hybrid in Malaysia. *Garden Bulletin of Singapore*, Vol.51, No.1, pp. 103-118, ISSN 0374-7859

Thomas, DC. 2010. *Phylogenetics and historical biogeography of Southeast Asian Begonia L. (Begoniaceae), PhD theses*. University of Glasgow, Glasgow.

Thomas, DC.; Hughes, M.; Phutthai, T.; Rajbhandary, S.; Rubite, R.; Ardi, W.H. & Richardson, J.E. (2011). A non-coding plastid DNA phylogeny of Asian *Begonia* (Begoniaceae): Evidence for morphological homoplasy and sectional polyphyly.

Molecular Phylogenetics and Evolution, Vol. 60, No. 3, (September 2011), pp. 428-444, ISSN 1055-7903

Tobe, H. & Raven, PH. (1987). Systematic embryology of the Anisophylleaceae. *Annals of the Missouri botanical garden*, Vol.74, No.1, pp. 1-26, ISSN 0026-6493

Twyford, AD. & Ennos, RA. (2011). Next generation hybridization and introgression. *Heredity*, doi: 10.1038/hdy.2011.68., ISSN 0018-067X

Zeng Z-B. (2005). QTL mapping and the genetic basis of adaptation: recent developments. *Genetica*, Vol.123, No.1-2, (February 2005), pp. 25-37, ISSN 0016-6707

Zhang, L-B.; Simmons MP.; ocyan, A. & Renner SS. (2006). Phylogeny of the Cucurbitales based on DNA sequences of nine loci from three genomes: Implications for morphological and sexual system evolution. *Molecular Phylogenetics and evolution*, Vol.39, No.2, (May 2006), pp. 305-322, ISSN 1055-7903

3

Olive (*Olea Europaea* L.): Southern-Italian Biodiversity Assessment and Traceability of Processed Products by Means of Molecular Markers

V. Alba[1], W. Sabetta[1], C. Summo[2], F. Caponio[2], R. Simeone[1], A. Blanco[1], A. Pasqualone[2] and C. Montemurro[1]
[1]*Department of Agro-forestry and Environmental Biology and Chemistry,*
[2]*Section of Genetics and Breeding and Section of Food Science and Technology,*
University of Bari,
Italy

1. Introduction

Southern Italy has geographical, climatic, historical and traditional peculiarities summarized in a landscape to which olive trees are the backdrop. Olive (*Olea europaea* L. subsp. *europaea*) is cultivated since the third millennium B.C. in the Eastern region of the Mediterranean sea (Loukas & Krimbas, 1983) and spread later around the basin following land and maritime routes to Italy, Spain, North Africa and France (Angiolillo et al., 1999).
The plant longevity, the cross-pollinating nature of the species and its secular history contributed to determine a wide germplasm biodiversity with more than 1200 cultivars, 9.992 thousand harvested hectares and a total production of 18 million tons (Tab.1; FAO 2009) mostly concentrated in the main olive oil producing countries (Alba et al., 2009a). Europe alone produces the 69% of the total worldwide production (Fig. 1). Italy is one of the main producers and has a rich assortment of olive cultivars, with a total number of 395 registered entries in the national list of *O. europaea* L. Italian cultivars (Official Journal of the Italian Republic, 1994), covering 1,18 million hectares and a total production of 3.6 million tons (ISTAT, 2010). The 80% of the cultivation is mostly concentrated in Southern Italy, where the production touches the 88% of the whole Italian olive production. In particular, Apulia region alone produces 1/3 of the total (about 1,7 milion tons) (Tab. 2).
Nowadays olive-growing has a crucial commercial role in many Italian regions with a richness of biodiversity spread from North to South in terms of cultivars grown, agronomic practices, processes of transformation of the raw material. Therefore, the preservation and protection of Italian olive biodiversity depends upon the use of genetic research tools like molecular markers for a correct cultivar identification. In countries that belong to the "Union Internationale pour la Protection des Obtentions Végetales" (UPOV) the registration, multiplication, certification, and commercialization of olive varieties rely on the evaluation of morpho-agronomic descriptors. Recently, morphological, biochemical and agronomic traits have been complemented with the large array of DNA molecular marker types

available, that provide an accurate and unambiguous tool for a correct identification of cultivated olive germplasm. Corrado et al. (2009) report that morpho-agronomic analyses on quantitative data in olive often are in contrast to DNA molecular analyses. Despite this, the authors consider the use of morpho-agronomical descriptors irreplaceable for the description of local or neglected olive cultivars since they are simple to be recorded and focus on important agronomic traits, while an accurate and unambiguous olive cultivars identification cannot be performed without the use of molecular markers.

In these last years the Department of Agro-forestry and Environmental Biology and Chemistry of the University of Bari (DiBCA) employed molecular markers both for olive cultivar identification and olive oil traceability. Amplified fragment Polymorphism (AFLP), in absence of sequence information about the olive genome, have been largely employed for simultaneous screening of a large number of loci with high efficiency in revealing inter and intra-cultivar variability (Angiolillo et al., 1999; Belaj et al., 2003; Grati-Kamoun et al., 2006; Montemurro et al., 2005; Montemurro et al., 2008; Sensi et al., 2003). Also Inter Simple Sequence Repeats (ISSR) have been efficiently employed for cultivar identification of olive (Pasqualone et al., 2001). As sequences of genomic DNA in olive became available, Simple Sequence Repeats (SSR) have become the markers of choice for varietal identification studies as they are transferable, hypervariable, highly polymorphic, multiallelic, polymerase chain reaction (PCR)-based co-dominant markers. They are relatively simple to interpret and show a high information content (Ganino et al., 2006). In fact, SSR result less laborious and cheaper than AFLP and do not require special equipment and manipulation of radiolabelled substances, making possible to transfer SSR protocols for routine analyses. Recently, several olive SSRs have been published (Carriero et al., 2002; Cipriani et al., 2002; De la Rosa et al., 2002; Diaz et al., 2006; Sefc et al., 2000) and screened on a huge number of cultivars. The most part of these SSR have been evaluated by Baldoni et al. (2009) on a set of 77 cultivars from different geographical origins. The authors provided details on reproducibility, power of discrimination and number of amplified loci/alleles and precise band size alleles, indicating the most suitable SSR markers for olive cultivar identification. The allele sequencing makes it possible to have a reference point when screening olive cultivars with SSR, since the right size dimension of SSR alleles represents a bottleneck when phenomena of stuttering or genotyping errors occurr. Further information on olive germplasm, cultivar names, homonyms and synonims, SSR data and Expressed Sequence Tag (EST) are available at Oleadb database (www.Oleadb.it).

The advantages enfisized for SSR are much more justified by their applicability on processed material since significant amounts of DNA, suitable for PCR analysis, are present in residues from olive oil mechanic extraction (Alba et al., 2009b; Breton et al., 2004; Consolandi et al., 2008; de la Torre et al., 2004, Montemurro et al., 2008; Pasqualone et al., 2004a; 2004b).

Extra-virgin olive oil represent the pride of the Italian olive production and is only obtained by mechanical or physical extraction with any chemical change. It must fufil a series of physic-chemical quality parameters , such as a free acidity below 0.8 g/100 g expressed in free oleic acid content (Commission Regulation EU No 61/2011). For its fatty acid composition, with a high monounsaturated - to - polyunsaturated fatty acid ratio and the presence of minor compounds having anti-oxidant activity, extra-virgin olive oil is especially appreciated for its high resistance to oxidative deterioration, playng a crucial role in the so called Mediterranean diet (Bendini et al., 2007; Harwood and Yaqoob, 2002; Visioli and Galli, 2002).

	Area Harvested (thousand Ha)	Total Production (million tons)
Albania	40,0	0,05
Algeria	288,4	0,48
Croatia	15,3	0,03
Cyprus	12,0	0,01
Egypt	110,0	0,50
France	19,0	0,03
Greece*	800,0	2,31
Israel	16,0	0,03
Italy	1.159,0	3,60
Jordan	60,7	0,14
Libya*	0,2	0,18
Morocco	550,0	0,77
Portugal	380,7	0,36
Spain	2.500,0	6,20
Syrian	635,7	0,89
Tunisia	2.300,0	0,75
Turkey	727,5	1,29
World	**9.922,8**	**18,24**
Europe	4.923,2	12,61
Northern Africa	3.248,6	2,68
Western Asia	1.604,0	2,54
Others	146,9	0,42

Table 1. Area harvested and total olive production in the main producing countries of the Mediterranean Bacin and in the main worldwide geographical productive areas. (FAO, 2009). * Data are referred to 2008.

	Italian total area allocated to olive		Total olive production (table, oil, double purpose)	
	thousand ha	% on the total area	million tons	% on the total production
Northern Italy	28,4	2,4	0,49	1,4
Central Itali	205,2	17,3	3,88	10,7
Southern Italy	949,8	80,3	31,75	87,9
Total	**1183,4**		**36,11**	
Apulia region	377,6	39,8	11,72	36,9
Other Southern regions	572,2	60,2	20,03	63,1

Table 2. Consistency of the Italian olive cultivation (ISTAT, 2010)

Unique sensorial, aromatic, gustatory, nutraceutical and healthy properties of the different Italian extra-virgin olive oil justify the attention devoted to the respect of standard parameters of quality all along the productive chain.
The skill developed by olive growers after centuries of cultivation has led Italy to become one of the first producing countries of olive oil considering the ratio between quality and quantity. In order to guarantee, preserve and promote the both the olive genetic variability and the technical expertise in cultivating, collecting and transforming processes of olives, Italian producers consortiated and obtained marks of protected designation of origin (PDO) at European level according to EC Regulation no. 510/06. An official production protocol of any PDO extra-virgin olive oil must meet a specific varietal requirement alone or in combination with other varieties eventually allowed by the protocol at well defined percentages. Therefore, it is clear that traceability, of raw and processed materials, specially by means of molecular markers, assumes a crucial role in the certification of products for the protection of consumers and for fraud prevention.

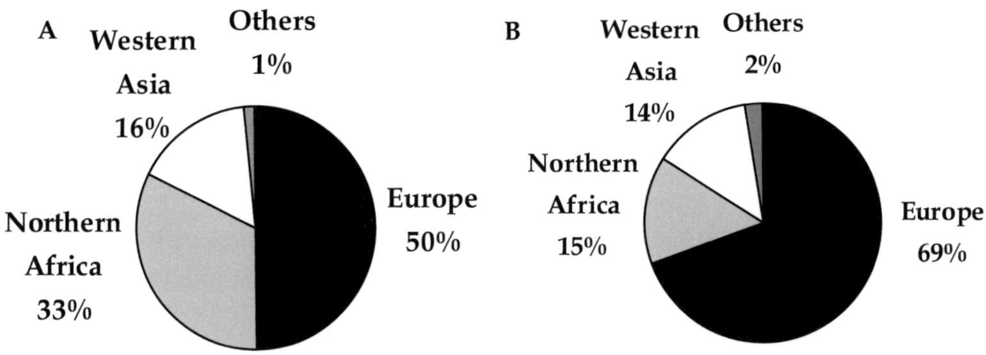

Fig. 1. Percentage distribution of harvested area (A) and total olive production (B) in the main worldwide geographical productive areas. (FAO, 2010).

The aim of this paper is to review the research activity conducted in the last decade by the DiBCA in the fields of Southern Italian olive cultivar identification and monovarietal olive oil traceability (tab. 3, tab. 4). To this purpose, only certified material from the Experimental Station "P. Martucci" of the Plant Production Department (D.P.V.) and the DiBCA, from the Nursery "Vivai Giannoccaro" in Bari-Italy, from the Olive Pre-multiplication Centre "Concadoro" in Palagiano - Italy and from the Department of Olive Research, Idleb, Syria was analysed.

2. Molecular markers for olive cultivar identification and monovarietal olive oil traceability

2.1 ISSR markers

ISSR are DNA based markers that involve amplification via polymerase chain reaction (PCR) of DNA regions situated between adjacent and inversely oriented 16–18 bp long simple sequence repeats, already employed successfully for cultivar identification purposes. ISSR, together with physico-chemical analyses, were used for olive cultivar identification in 2001 by Pasqualone et al.. Here we focus on the molecular aspect of that paper. In particular, 10 olive cultivars were screened with 10 ISSR markers. DNA was extracted from fresh leaves

Molecular marker class	Matrix for DNA extraction	N° Markers tested	N° Olive cultivars	Minimum n° of markers distinguishing cultivars	Source
ISSR	Leaves Drupes Virgin oil	10	10	2	Pasqualone et al., 2001
SSR	Leaves Virgin oil	10	5	2	Pasqualone et al., 2003
SSR	Leaves Drupes Virgin oil	7	10	3	Pasqualone et al., 2004a
SSR	Leaves Virgin oil	4	2	1	Pasqualone et al., 2004b
AFLP SSR	Leaves	3 27	60	-	Montemurro et al., 2005
SSR	Virgin oil	7	7	-	Pasqualone et al., 2007a
SSR	Leaves Virgin oil	1	75	-	Pasqualone et al., 2007b
SSR	Leaves	4	9	-	Alba et al., 2008
AFLP	Leaves Virgin oil	6	10	1	Montemurro et al., 2008a
SSR AFLP	Leaves	27 3	37	-	Montemurro et al., 2008b
SSR	Leaves	16	30	2	Alba et al., 2009a
SSR	Leaves Virgin oil	10	7	1	Alba et al., 2009b

Table 3. Cultivar identification and olive oil traceability research activities conducted at the Department of Agro-forestry and Environmental Biology and Chemistry, University of Bari, Italy.

and ripe drupes following the protocol proposed by Sharp et al. (1988) with some modifications. Monovarietal olive oil was obtained by mechanical extraction with the method proposed by Catalano & Caponio (1996) and, after centrifugation of 250 ml oil, the cell residue sediment was recovered and used for the DNA extraction with the same protocol of leaves and drupes. DNA extracted from leaves and drupes was rich in PCR interfering substances such as phenolic compounds and lipids, suggesting a further purification achieved by chromatography on a commercial device. On the contrary, the DNA extraction protocol revealed to be inadequate when processing cell residue sediments, yielding degraded or unsuitable for PCR experiments (i.e. all 10 ISSR primers failed to amplify). Three ISSR completely failed to amplify both on leaf and drupe DNA, while 7 ISSR gave distinguishable electrophoretic patterns with appreciable polymorphic content, so that a set-up of an identification key with the lowest number of molecular markers that distinguish among the 10 cultivars was obtained by means of only 2 ISSR. The paper

Olive cultivar	Molecular marker class	Olive cultivar	Molecular marker class
Aglandau	AFLP, SSR	Manzanilla	AFLP, SSR
Ascolana Tenera	AFLP, SSR	Maurino	AFLP, SSR
Atena	AFLP, SSR	Mele	AFLP, SSR
Bella di Cerignola	AFLP, SSR	Mora	AFLP, SSR
Bella di Spagna	AFLP, SSR	Moraiolo	AFLP, ISSR, SSR
Bouteillan	AFLP, SSR	Nocellara del Belice	AFLP, SSR
Canino	AFLP, SSR	Nocellara Etnea	AFLP, SSR
Carolea	AFLP, SSR	Nocellara Messinese	AFLP, SSR
Cassanese	ISSR	Nociara	AFLP, ISSR, SSR
Cazzinicchio	AFLP, SSR	Nolca	AFLP, SSR
Cellina di Nardò	AFLP, ISSR, SSR	Ogliarola barese	AFLP, SSR
Cerasella	AFLP, SSR	Ogliarola garganica	SSR
Cima di Bitonto	AFLP, ISSR, SSR	Ogliarola salentina	ISSR, SSR
Cima di Mola	SSR	Oliastro	AFLP, SSR
Cima di Melfi	AFLP, SSR	Oliva rossa	SSR
Cipressino	SSR	Pasola	AFLP, SSR
Conservolia	AFLP, SSR	Pasola di Andria	AFLP, SSR
Coratina	AFLP, ISSR, SSR	Pendolino	AFLP, SSR
Corniola	SSR	Peppino	AFLP, SSR
Dolce di Cassano	AFLP, SSR	Peranzana	AFLP, SSR
Dolce di Montescaglioso	AFLP, SSR	Picholine	AFLP, SSR
Donna Francesca	SSR	Picual	AFLP, SSR
Donna Giulietta	SSR	Primicea	AFLP, SSR
Frantoio	AFLP, SSR	S. Agostino	AFLP, SSR
Gentile di Larino	ISSR	S. Caterina	AFLP, SSR
Giarraffa	AFLP, SSR	S. Francesco	AFLP, SSR
Gioconda	AFLP, SSR	Simone	AFLP, SSR
Grignan	AFLP, SSR	Selvatic entry	AFLP, SSR
Grossane	AFLP, SSR	Sigoise	AFLP, SSR
Itrana	AFLP, ISSR, SSR	Tanche	AFLP, SSR
Kalamata	AFLP, SSR	Termite di Bitetto	AFLP, SSR
Leccino	AFLP, ISSR, SSR	Tonda Iblea	AFLP, SSR
Leccio del Corno	AFLP, SSR	Toscanina	AFLP, SSR
Leo	AFLP, SSR	Verdale	AFLP, SSR
Maiatica	AFLP, SSR	Vergiola	AFLP, SSR

Table 4. List of the olive cultivars screened by means of AFLP, ISSR and SSR at the DiBCA

concluded that ISSR represent a powerful tool to distinguish among olive cultivars but revealed an important aspect in relation to the establishment of more accurate protocols to be used for DNA extraction from cell residue sediments of olive oil. A robust and efficient method for extracting DNA from olive oil is a prerequisite for developing a platform of marker-based assessment of olive oil cultivar composition (Consolandi et al, 2008). Today, many commercial kits to extract DNA from processed food are available and help researchers to overcome the problem.

2.2 AFLP markers

AFLP markers (Vos et al., 1995) allow multi-locus screening of a genome in absence of preliminary sequence knowledge. The use of AFLP for olive cultivar identification and oil traceability has been described before. In particular, Montemurro et al. (2005) tested 3 AFLP primer combinations and 27 SSR (discussed later) on a set of 60 olive cultivars, some of them present in replicates and coming from different germplasm collections, for a total of 112 samples, from the Mediterranean basin. DNA was extracted from lyophilized leaves with the commercial Gene Elute Plant kit and a biodiversity study was conducted. A dendrogram of genetic similarity based on Jaccard index (1901) was calculated with NTSYS-PC (Rohlf, 1992) using AFLP and SSR data. All the cultivars resulted distinguished and clustered according to their type of use: oil, table, and dual purpose cultivars. Differences between the two techniques were detected. The ratio of polymorphic bands/total bands ranged from a maximum of 100% for SSRs to 42% for AFLP, in line with other authors (Belaj et al., 2003; Powell et al., 1996; Russell et al., 1997) that confirmed the higher efficiency of SSR to distinguish among olive cultivars, but revealed the ability of AFLP to distinguish intra-varietal clones respect to SSR. Therefore, AFLP represent the best choice for studies into genetic relationship or for accurate evaluation of intra-cultivar variability.
Based on this experience, the authors hypothesized and wanted to verify the effectiveness of AFLP in distinguishing among oils from different cultivars. To this purpose, Montemurro et al. (2008) tested the efficiency of AFLP on 10 different virgin oils. Oils were obtained with a laboratory-scale pilot plant consisting of a SK1 hammer-crusher (Retsch, Haan, Germany) and 200 ml were centrifuged to recover residues. In this case, in order to overcome problems of DNA of low quality encountered in the past, the commercial Gene Elute Plant kit (Sigma) to extract DNA from cell residue sediments and lyophilized leaves was adopted. All the 6 AFLP primer combination yielded polymorphic electrophoretic patterns for all the 10 cultivars both for leaf and oil DNA. The level of degradation, the low DNA concentration extractable from oil and the template needed by the restriction enzymes for the initial digestion (100 ng) in the AFLP method suggested a standard protocol implementation in case of DNA from oil (Montemurro et al., 2005). Specifically, no dilution of both restricted/ligated and preamplified products were performed before selective amplifications, on the contrary 30 µl of restricted/ligated product, much higher than usual, was used for ensuring good amounts of DNA template. Differences about the band intensity emerged for the amplicons of DNA from oil, reducing the number of scorable bands after electrophoresis, but showing sufficient information useful for olive oil traceability.

2.3 SSR markers

As discussed before, SSR efficiency compared with other DNA-based markers in olive was demonstrated and nowadays they are preferred by several authors for biodiversity studies, germplasm characterization and varietal fingerprinting. Moreover, the recent introduction of fluorescent labeled primers that avoid radioactive labeled ones for PCR-based markers let the use of automatic sequencers, less time consuming and with minor health implications for operators. Besides, automatic sequencers are more effective, because capillary electrophoresis allows to distinguish alleles with very small differences in molecular weight, and has a higher resolutive power than classical methods. The sensibility of the method allows to evidence also weak signals, such as those that could come from olive oil DNA, known to be partially degraded (Alba et al., 2009b).

Baside AFLP analysis, Montemurro et al. (2005) started to screen 65 SSR on a sub-set ot an entire collection of 112 olive accessions. The goal was to select those SSR most efficiently employable in further studies of biodiversity and traceability in olive. Based on simplicity and reproducibility of electrophoretic patterns, 27 were chosen and analysed further on the entire set of olive accessions. Dendrograms of genetic similairity based on Jaccard index (1901) were calculated by means of NTSYS-PC (Rohlf, 1992) using AFLP and SSR data separately and combined. The analysis revealed the ability of AFLP to distinguish accessions belonging to the same cultivars. Conversly, SSR were not able to distinguish all the accessions, making necessary the combination with AFLP data for a complete distinction of all the accessions.

As described, the first attempts to amplify DNA from oil at the DiBCA by ISSR failed, suggesting a different molecular approach to the problem. The analysis of SSR at the DiBCA started in 2004 by Pasqualone et al.. DNA from leaves, drupes and virgin oil of 10 olive cultivars usually grown in Southern Italy were screened with 7 SSR. Virgin olive oil was obtained mechanically as reported before and the commercial Gene Elute Plant kit (Sigma) was used to extract DNA from all matrices. As expected, all the SSRs employed, resolved on agarose gel and ethidium bromide staining, yielded clear and reproducible electrophoretic patterns for leaves, drupes and virgin oil. Full correspondence of the alleles amplified was verified when comparing the electrophoretic profiles from the three different matrices of each olive cultivar. Three SSRs were used as a minimum number of molecular markers able to distinguish among the 10 olive cultivars.

These encouraging results and the standardization of DNA extraction and SSR amplification conditions led the DiBCA to verify the suitability of SSR for the traceability of a PDO extra-virgin olive oil like "Collina di Brindisi" (Pasqualone et al., 2007a). Collina di Brindisi PDO is a typical Apulian olive oil (Tab. 5) prepared from Ogliarola (minimum 70%), and other olive cultivars diffused in the production area alone or together, accounting for a maximum of 30%, with no other indications of their single ratios (Official Journal Euopean. Community, 1996; Official Journal of the Italian Republic. 1998). The research consisted on the screening of different virgin oils, filtered and unfiltered, obtained by different oil mills and derived from the olive cultivars included in the disciplinary protocol. The screening was conducted both on monovarietal and binary blends. DNA was extracted by Gene Elute Plant kit (Sigma) and 7 SSR were employed to amplify DNA from different matrices. The suitability of the SSR technique in olive oil traceability was confirmed by the good amplification levels obtained both from cloudy unfiltered and from clear filtered oils. Sufficient cellular sediments for DNA extraction was obtained from filtered oils by augmenting starting oil volumes, centrifuging at higher speeds and for longer times respect to unfiltered oils, and improving the efficiency of the commercial kit used for DNA extraction by using higher amounts of the extracting reagents. The interesting aspect of this paper was represented not only by the fact that SSR confirmed the composition of a certain olive oil, but GAPU89 discovered a wrong designation attributed by one of the mills that provided the experimental material. In particular, an oil supposed to be composed by Leccino was verified to be composed by Ogliarola. Moreover, the paper tried to shed light on the cultivar "Ogliarola", registered in the national list of O. europaea Italian cultivars

(Official Journal of the Italian Republic. 1998). In Apulia region at least 3 different type of Ogliarola, namely barese, salentina, and garganica, derive their qualifications from the three different geographical districts of diffusion. The SSR GAPU 89 confirmed the assumption, due to geographical aspects, of the identity of the generic Ogliarola with Ogliarola salentina, since its intense cultivation with respect to the other two in the production area of the "Collina di Brindisi" PDO oil. Infact, Ogliarola garganica e Ogliarola barese showed different electrophoretic pattern between them and with respect to Ogliarola salentina, which showed monomorphic pattern when compared to Ogliarola profile (Tab. 6). Beside this, the allelic profile of "Ogliarola" was detectable also in the binary blends, providing a qualitative result of a product. Its prevailing in Collina di Brindisi PDO oil in amounts higher than 70% could be used as a positive control, since the absence of the Ogliarola salentina pattern in a commercial Collina di Brindisi PDO oil could be revealed to be a fraud.

In Apulia region, Southern Italy, five extra-virgin olive oil bear the PDO mark of quality (Tab. 5). Besides the "Collina di Brindisi", "Terra di Bari" represents one of the most produced PDO oils in Apulia and can be subdivided into three distinct sub-denominations: "Bitonto", "Castel del Monte" and "Murgia dei Trulli e delle Grotte". In particular, the sub-denomination "Bitonto" was focused by the DiBCA for traceability purposes by means of SSR (Alba et al., 2008a). The disciplinary of production of "Terra di Bari", sub-denomination "Bitonto" provides that this oil is obtained from the following varieties, present for at least 80%, alone or in combination: Cima di Bitonto, Ogliarola barese or Coratina. These 3 cultivars were screened for allele biodiversity at 4 SSR loci together with other 6 cultivars diffused in Southern Italy and a dendrogram of genetic similarity based on Jaccard index was calculated (Fig. 2). Despite the low number of SSRs tested, all the cultivars were clearly distinguished, revealing a Jaccard value of at least 0.50 when comparing all the cultivars each other, except Cima di Mola e Ogliarola barese (Jaccard = 0.68). More, table 7 shows the allelic profile at 3 SSR loci investigated on the cultivars comprised in the disciplinary of production of "Terra di Bari" and it is evident that all the SSR were singularly able to discriminate cultivars. The automation of the electrophoresis, conducted on a sequencer ABI PRISM 3100 Avant Genetic Analyzer, instead of agarose or PAGE gels, allowed a more precise read of the amplicons, revealing minimal differences between alleles differing by 2 bp. Fig. 3 provides a comparison between SSR amplification resolved on agarose gel and on automatic sequencer. As an example, samples amplified with DCA03 and run on sequencer revealed an allele at 237 bp in Cima di Bitonto and at 239 bp in Coratina, undistinguishable when resolved on agarose gel (data not shown).

Once established the efficiency of SSR in olive biodiversity assessment and in olive oil traceability, they have been largely used in the last years supplanting other more laborious and time-consuming molecular marker classes. Pasqualone et al. (2007b) presented a paper at the 3rd Commission Internationale du Genie Rural (CIGR) International Symposium held in Naples (September 24th - 26th) in which a set of 75 olive cultivars native from the Mediterranean basin and coming from different germplasm collections was subjected to SSR analysis with a DCA04 SSR locus, in order to provide a DNA-bank database. The choice of the SSR DCA04 (Sefc et al., 2000) employed in this paper was based on the simple repeat motif [$(GA)_{16}$]. Genotyping with di-nucleotide SSRs is difficult because separation from neighbourhood alleles requires very precise and reliable protocols for allele separation and identification to avoid allele misidentification (Baldoni et al., 2009). However, despite its di-nucleotide nature, the SSR DCA04 revealed to be reliable in our previous SSR screening, in

Italian PDO and IGP extra-virgin olive oils	O.J.I.R.* legislative reference	Italian Region	O.J.E.C.** legislative reference
Colline Teatine	O.J. 7.10.1998, n° 234	Abruzzo	n° 1065/97
Aputino Pescarese	O.J. 20.8.1998, n° 193	Abruzzo	n° 1263/96
Pretuziano delle Colline Tramane		Abruzzo	n° 1491/03
Dauno	O.J. 20.8.1998, n° 193	Apulia	n°2325/97
Terra d'Otranto	O.J. 20.8.1998, n° 193	Apulia	n° 1065/97
Terra di Bari	O.J. 29.9.1998, n° 227	Apulia	n° 2325/97
Collina di Brindisi	O.J. 21.12.1998, n° 297	Apulia	n° 1263/96
Terre Tarentine		Apulia	n° 189/04
Bruzio	O.J. 28.10.1998, n° 252	Calabria	n° 1065/97
Lametia	O.J. 11.11.1999, n° 265	Calabria	n° 2107/99
Alto Crotonese	O.J. 21.8.2003, n° 193	Calabria	n° 1257/03
Irpinia - Colline dell'Ufita		Campania	n° 203/2010
Cilento	O.J. 20.8.1998, n° 193	Campania	n° 1065/97
Penisola Sorrentina	O.J. 20.8.1998, n° 193	Campania	n° 1065/97
Colline Salernitane	G.U 11.9.2003, n° 211	Campania	n° 1065/97
Brisighella	O.J. 11.11.1998, n° 264	Emilia Romagna	n° 1263/96
Colline di Romagna	O.J. 11.9.2003, n° 211	Emilia Romagna	n° 149/03
Canino	O.J. 11.11.1998, n° 264	Lazio	n° 1263/96
Sabina	O.J. 20.6.1995, n° 142	Lazio	n° 1263/96
Riviera Ligure	O.J. 20.8.1998, n° 193	Liguria	n° 123/97
Molise	G.U 21.8.2003, n° 193	Molise	n° 1257/03
Cartoceto		Marche	n° 1897/04
Laghi Lombardi	O.J. 7.10.1998, n° 234	Lombardy	n° 2325/97
Garda	O.J. 7.10.1998, n° 234	Lombardy	n° 2325/97
Terra di Siena	O.J. 15.1.2001, n° 11	Tuscany	n° 3446/00
Chianti Classico	O.J. 17.1.2001, n° 13	Tuscany	n° 3446/00
Lucca		Tuscany	n° 1845/04
Toscano		Tuscany	n° 644/98
Tuscia		Tuscany	n° 1623/05
Umbria		Umbria	n° 2325/97
Veneto	O.J. 15.11.2001, n° 266	Veneto	n° 2036/01
Val di Mazara	O.J. 28.3.2001, n° 73	Sicily	n° 138/01
Valli Trapanesi	O.J. 26.10.1998, n° 250	Sicily	n° 2325/97
Monti Iblei	O.J. 24.10.1998, n° 249	Sicily	n° 2325/97
Valdemone		Sicily	n° 205/05
Monte Etna	O.J. 21.8.2003, n° 193	Sicily	n° 1491/03
Sardegna		Sardinia	n° 148/07

Table 5. Main Italian PDO and IGP extra-virgin olive oils from different regions and legislative references at Italian and European level. (*Official Journal of Italian Republic; **Official Journal of European Community) Source: http://www.frantoionline.it/dop-e-igp/olio-dop-zone-di-produzione.html

	Allele size (bp)
Ogliarola Barese	180 - 200
Ogliarola Garganica	180 - 190
Ogliarola Salentina	180 - 210
Ogliarola	**180 - 210**

Table 6. Allelic profile for the SSR GAPU89 in the "Ogliarola" cultivars

SSR marker	Cima di Bitonto		Ogliarola barese		Coratina	
DCA03	237	243	243	256	239	243
DCA04	130	132	140	190	130	165
DCA18	177	179	177	177	177	181

Table 7. Allelic profile for 3 SSR markers of the 3 olive cultivars comprised in the disciplinary of production of "Terra di Bari", sub-denomination "Bitonto",

line with Baldoni et al. (2009) which reported the SSR DCA04 to have low peak stuttering and strong peak intensity during electrophoresis. The main drawbacks for DCA04 were represented by the n° of loci amplified (2) and by a high frequency of null alleles (> 0.05). DCA4 was used to amplify DNA from both the leaves and the oils of 5 cultivars: Cellina di Nardò, Coratina, Frantoio, Leccino, Ogliarola barese. Materials and methods on how to extract monovarietal virgin olive oil is reported in this and other here revised papers produced by the DiBCA. In this research, in particular, the authors begun to question the possibility that the full DNA match between the leaves and oil can be undermined by the emergence of different alleles of paternal origin. In fact, if leaf DNA is practically identical to oil DNA since oils derived mainly from olive flesh, that is derived from the evolution of the ovary (e.i. from mother plant), on the other side the contribution of the pollen from the pollinator plant resides in the seed whose oily fraction represents only a minor percentage of the final oil, and whose solid phase is almost totally eliminated during oil processing. It is evident that this problem occurs for self-incompatible cultivars. Thus, before treating as a fraud the incoming of different alleles when a DNA from oil is amplified with SSR and compared to DNA from leaves, one should ask the pollinating nature of the cultivar employed to produce oil, the percentage of allogamy, the presence in the same cultivations areas of other olive cultivars genetically compatible.

Therefore, given the efficiency and speed with which SSR analyses can be conducted on DNA extracted from both leaves and oil, a study of correspondence between genetic profiles obtained from leaves and oil was undertaken at the DiBCA by considering 7 olive cultivars, and 10 SSR (Alba et al., 2009b). The cultivars chosen are all used in different disciplinaries of production of PDO oils. Three kg of drupes from certified clones were collected and milled to obtain monovarietal oils. DNA was extracted as previously reported, yielding as expected a mean concentration of 100 ng/µl for DNA from leaves and 5 ng/µl from DNA from oil. Six out of 10 SSR gave clear and unequivocal electrophoretic patterns, while 4 SSRs failed to amplify on at least 2 of the 7 cultivars tested, being the Coratina the most difficult to amplify. The research revealed identical patterns between leaves and oil DNA for 90% of the experiments, even in cases where SSR revealed a multilocus nature. Fig. 4 shows a comparison between oil and leaves electrophoretic patterns of cultivar Leccino obtained with SSR DCA03 and it is evident how peaks for alleles amplified from oil were different in

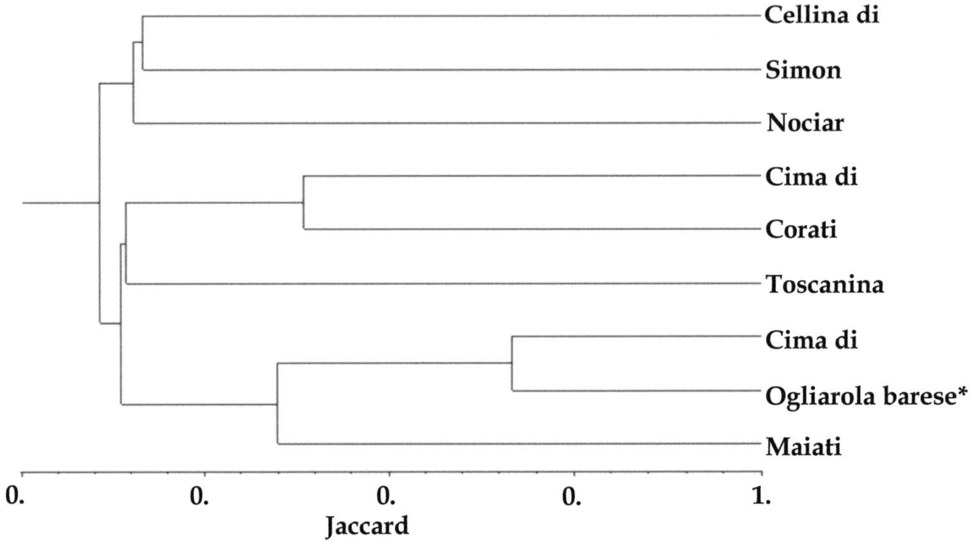

Fig. 2. Dendrogram of genetic similarity based on Jaccard Index of 3 Italian olive cultivars included in the disciplinary of production of "Terra di Bari", sub-denomination "Bitonto", indicated with *, compared to other 6 Italian olive cultivars.

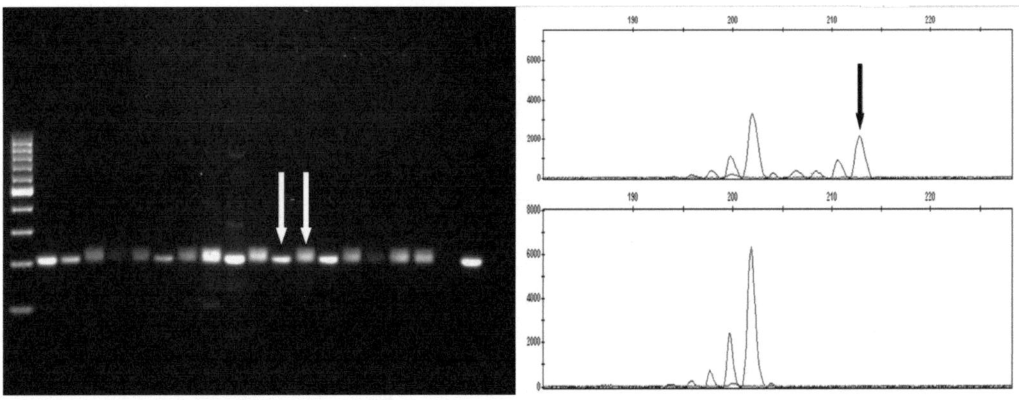

Fig. 3. Comparison of SSR electrophoretic pattern on agarose gel 2% (left) and on ABI PRISM 3100 Avant Genetic Analyzer (right). Arrows indicate polymorphisms between olive cultivars. The polymorphism revealed by the sequencer is more evident than that revealed by agarose gel.

shape and height but not in molecular weight with respect to alleles from leaves. The reduced height of oil peaks was determined by the lower quality and concentration of template with respect to leaves. In our previous researches cases of a non-complete concordances between leaves and oil DNA did not emerged, probably due to the low resolution of the methods employed to resolve amplifications (agarose, PAGE). In this research the use of a sequencer let to identify some miss-matches between leaves and oil DNA due to the loss or gain of alleles in the oil DNA. The loss of some peaks in oil profiles was hypothesized to be related to the degraded DNA for the process of olive oil extraction, to a low signal of the amplicons, or to troublesome working conditions for the polymerase. On the other hand, a wrong denominations of the cultivars or the occurrence of somaclonal mutations as causes of the gain of additional alleles in oil profiles with respect to leaf DNA could be excluded since the certified origin of cultivars and the belonging of leaves and drupes to the same clone. An accredited hypothesis is represented by a paternal contribution of embryos in oil samples extracted from entire drupes, in line with Doveri et al. (2006) that recorded diverse profiles in oil and plant tissues of the cultivar "Leccino", suggesting a nonmaternal origin of additional alleles in oil profiles arisen from out-crossing. A confirmation of the hypothesis of a paternal contribution is offered by the presence of all the scored additional alleles in the electrophoretic profiles of the other cultivars of the research, suggesting, in fact, that out-crosses could be occurred between clones grown in the same field.

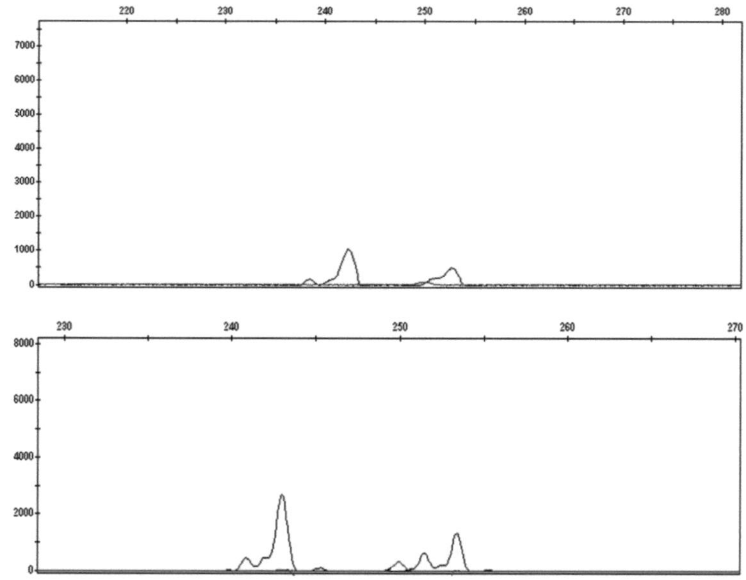

Fig. 4. Electropherograms of Leccino cultivar oil (up) and leaves (down) obtained with SSR DCA03. The reduced height of oil peaks was determined by the lower quality and concentration of template with respect to leaves.

Next, the screening of olive cultivar diversity continued when a total of 30 cultivars was analyzed with 16 SSR (Alba et al., 2009a). A molecular identification key with the minimum number of markers revealed that 2 SSR were able to distinguish all the cultivars. In order to identify the SSR that better respond to this purposes we used two measures of the ability of primers to distinguish between genotypes by calculating Resolving Power (RP) (Prevost & Wilkinson, 1999) and Power of Discrimination (Kloosterman et al., 1993). RP consists in the use of a function strongly correlated to the proportion of genotypes identified by a certain molecular marker, independently of the number of genotypes composing the collection. PD is calculated by considering in the formula the genotype frequency instead of the allele frequency. A correlation between the number of distinguished cultivars (NDC) and the values of PD and RP of each SSR was estimated. The comparison between the two possible indices of diversity revealed a higher suitability of PD, with respect to RP, to predict the

ability of SSR marker in distinguishing among genotypes, suggesting that indices based on genotypic frequency are more suitable than those based on allelic frequency. A Principal Component Analysis (PCA) was performed for this review by using SSR profiles (Fig. 5) and computing data to construct a three-dimensional array of eigenvectors by the DCENTER module of the NTSYS-PC program (Rohlf, 1992).
The first Component accounted for 9.8% of the total allelic variation, while the second for 9.2% and the third for 6.0%, for a total SSR diversity of 25.0%. This low value suggested that more Components should be considered when representing the PCA and indicated that 16 SSR allowed the screening of a minimal part of the total biodiversity contained by the 30 olive cultivars investigated.

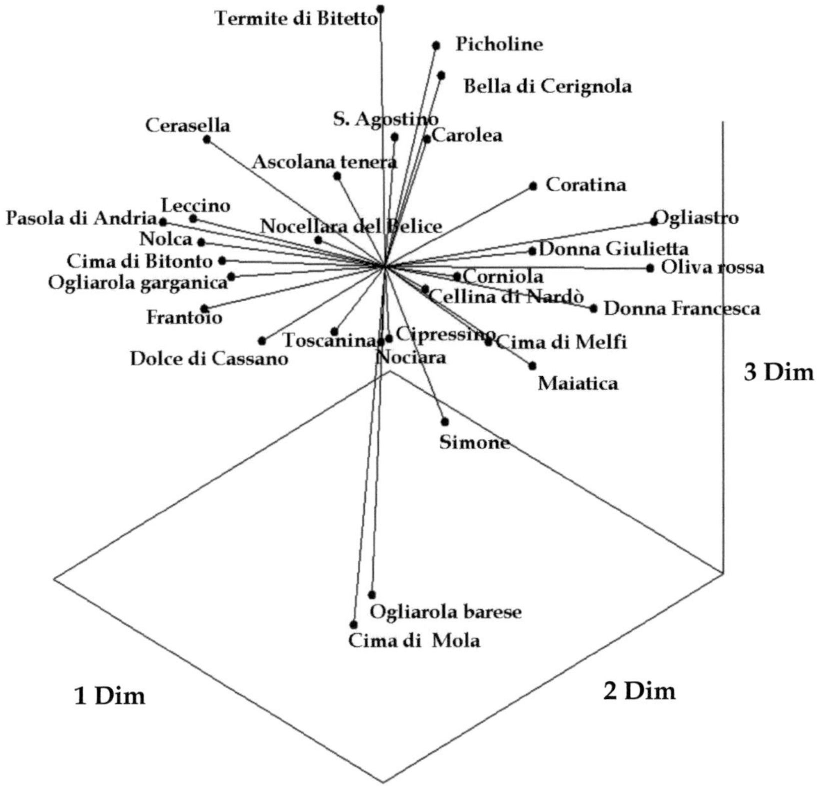

Fig. 5. Principal Component Analysis of 30 Italian olive cultivars analysed with 16 SSRs. Lines represent vectors.

3. Conclusions

Molecular markers have become an indispensable tool in plant breeding for both biodiversity studies and quality selection of products. Marker Assisted Selection (MAS) reduces entrie redundancy in germplasm collections, reveals cases of synonymy and homonymy in several crops, let a early selection of interesting lines in breeding programs Several classes of molecular markers (ISSR, AFLP, SSR) were screened on more than 80 olive cultivars of the Mediterranean basin at the DiBCA in approximately 10 years, passing from

traditional lab protocols laborious, in some cases hazardous and subjected to many environmental variables, and arriving to the use of health-safe reagents and standardized procedures. In this sense, the efficacy and the advantages showed by SSR with respect to ISSR and AFLP is well known. In particular, in olive oil traceability they revealed their ability to amplify highly degraded DNA extracted from oil. SSR generally amplify fragments of 100-300 bp, while ISSR generate amplicons of higher molecular weight (500-2000bp) and it can be argued that a degraded DNA from oil works better with short fragments. Furthermore, the step of enzymatic digestion in the AFLP can represent a severe bottleneck when processing degraded oil DNA. For all these reasons, SSR are now widely used for cultivar identification, while more needs to be done for their use in traceability. As an example, most of the PDO oils are multivarietal, so the traceability of a single cultivar becomes hard to be verified. Thus, approaches based on quantitative Real Time PCR can help to correctly define which cultivars are present in a certain oil and in which percentages. Indeed, if the approach is based on qualitative PCR, independently from the class of molecular marker, only monovarietal oils can be traced.
Genomes of several crops and woody species have been completely sequenced and characterized, while few is known about the olive genome. However, the increasing availability of Expressed Sequence Tags (EST) deduced from mRNA extracted from olive and annotated on the web (www.Oleadb.it) represent an interesting starting point for "candidate gene" approach, with the goal of a cultivar characterization based on gene functionality. It cannot be ignored that functional genomics offer the possibility not only to distinguish cultivars merely on restricted genomic areas often harbouring intronic regions, but it let the knowledge of how gene biodiversity in olive can lead to a certain organoleptic oil profile rather than to another.
Actually, the DiBCA is attempting researches on SSR alleles specific for some Southern Italian olive cultivars, while the approach of the "candidate gene" has been enterprised to isolate and characterize genes involved in the Kennedy pathway that leads to the syntesis or desaturation of long chain fatty acids in olive oil. In the future, the complete genome sequencing of olive will offer new opportunities to better understand the evolution, the existing biodiversity and the physiological aspects of a species that fused its hystory with that of mankind all over the Mediterranean basin in the last 5000 years.

4. Aknowledgments

This work was conducted within the framework of "Progetto interregionale OLVIVA" funded by Programma Interregionale - "Sviluppo rurale - Innovazione e Ricerca" D.M. n. 25279 of 23/12/03" and of "Progetto Strategico PS59" funded by Regione Puglia, Italy .

5. References

Alba, V., Montemurro, C. & Blanco A. (2008) Valorizzazione dell'olio "Terra di Bari" mediante l'applicazione dei marcatori molecolari microsatellite. *Proceedings of the VIII National Congress on Biodiversity – La Biodiversità: Risorsa per Sistemi Multifunzionali*, Lecce, Italy, April 21 - 23, pp. 15 - 17.

Alba, V., Montemurro, C., Sabetta, W., Pasqualone, A. & Blanco, A. (2009a). SSR-based identification key of cultivars of Olea europaea L. diffused in Southern-Italy. *Scientia Horticulturae*, No. 123, pp. 11–16.

Alba, V., Sabetta, W., Blanco, A., Pasqualone, A. & Montemurro, C. (2009b). Microsatellite markers to identify specific alleles in DNA extracted from monovarietal virgin olive oils. *European Food Research and Technology*, N° 229, pp. 375–382.

Angiolillo, A., Mencuccini, M. & Baldoni, L. (1999). Olive genetic diversity assessed using amplified fragment length polymorphisms. *Theoretical and Applied Genetics*, No. 98, pp. 411-421.

Baldoni, L., Cultrera, N.G., Mariotti R., Riccioloni, C., Arcioni S., Vendramin, G.G., Buonamici, A., Porceddu, A., Sarri, V., Ojeda, M.A., Trujillo, I., Rallo, L., Belaj, A., Perri, E., Salimonti, A., Muzzalupo, I., Casagrande, A., Lain, O., Messina, R. & Testolin, R. (2009). A consensus list of microsatellite markers for olive genotyping. *Molecular Breeding*, Vol. 24, No. 3, pp. 213-231.

Belaj, A., Satovic, Z., Cipriani, G., Baldoni, L., Testolin, R., Rallo, L. & Trujillo, I. (2003). Comparative study of the discriminating capacity of RAPD, AFLP and SSR markers and their effectiveness in establishing genetic relationships in olive. *Theoretical and Applied Genetics*, No. 107, pp. 736–744.

Bendini, A., Cerretani, L., Carrasco-Pancorbo, A., Gòmez-Caravaca, A. M., Segura-Carretero, A., Fernàndez-Gutiérrez, A. & Lercker, G. (2007). Phenolic molecules in virgin olive oils: A survey of their sensory properties, health effects, antioxidant activity and analytical methods. An overview of the last decade. *Molecules*, 12, 1679-1719.

Breton, C., Claux, D., Metton, I., Skorski, G. & Berville, A. (2004). Comparative study of methods for DNA preparation from olive oil samples to identify cultivar SSR alleles in commercial oil samples: possible forensic applications. *Journal of Agricultural and Food Chemistry*, No. 52, pp. 531–537.

Carriero, F., Fontanazza, G., Cellini, F. & Giorio, G. (2002). Identification of simple sequence repeats (SSRs) in olive (*Olea europaea* L.). *Theoretical and Applied Genetics*, No. 104, pp. 301–307.

Catalano, P. & Caponio, F. (1996). Machines for olive paste preparation producing quality virgin olive oil. *Fett/Lipid*, No. 98, pp. 408–412.

Cipriani, G., Marrazzo, M.T., Marconi, R., Cimato, A. & Testolin, R. (2002). Microsatellite markers isolated in olive (Olea europaea L.) are suitable for individual fingerprinting and reveal polymorphism within ancient cultivars. *Theoretical and Applied Genetics*, No. 104, pp. 223–228.

Commission Regulation (EU) No 61/2011 of 24 January 2011 amending Regulation (EEC) No 2568/91 on the characteristics of olive oil and olive-residue oil and on the relevant methods of analysis. Official Journal of the European Union L23 of 27.01.2011, pp. 1-14.

Consolandi, C., Palmieri, L., Severgnini, M., Maestri, E., Marmiroli, N., Agrimonti, C., Baldoni, L., Donini. P., De Bellis, G. & Castiglioni, B. (2008). A procedure for olive traceability and authenticvity: DNA extraction, multiplex PCR and LDR-universal array analysis. *European Food Research & Technology*, No. 227, pp. 1429-1438.

Corrado, G., La Mura, M., Ambrosino, O., Pugliano, G., Varricchio, P. & Rao R. (2009). Relationship of Campanian olive cultivars: comparative analysis of molecular and phenotypic data. *Genome*, No. 52, pp. 692-700.

De la Rosa, C., James, M. & Tobutt, K.R. (2002). Isolation and characterization of polymorphic microsatellites in olive (*Olea europaea* L.) and their transferability to other genera in the Oleaceae. *Molecular Ecology Notes*, No. 2, pp. 265–267.

de la Torre, F., Bautista, R., Ca´novas, F. & Claros, G. (2004). Isolation of DNA from olive oil and oil sediments: application in oil fingerprinting. *Food, Agriculture & Environment*, Vol 2, No. 1, pp. 84-89.

Diaz, A., De la Rosa, R.,Martin, A. & Rallo, P. (2006). Development, characterization and inheritance of new microsatellites in olive (*Olea europaea* L.) and evaluation of their usefulness in cultivar identification and genetic relationship studies. *Tree Genetics & Genomes*, No. 2, pp. 165–175.

Doveri, S., O'Sullivan, D.M. & Lee, D. (2006). Non-concordance of genetic profiles between olive oil and fruit: a cautionary note to the use of DNA markers in provenance testing. *Journal of Agricultural and Food Chemistry*, No. 54, pp. 9921-9926.

Ganino, T., Bartolini, G., & Fabbri, A., (2006) The classification of olive germplasm—a review. *Journal of Horticoltural Science and Biotechnology*, Vol. 81, No. 3, pp. 319–334.

Grati-Kamoun, N., Lamy Mahmoud F., Rebaı, A., Gargouri, A., Panaud, O. & Saar, A. (2006). Genetic diversity of Tunisian olive tree (*Olea europaea* L.) cultivars assessed by AFLP markers. Genetic Resources and Crop Evolution, No. 53, pp 265–275.

Harwood, J. L. & Yaqoob, P. (2002). Nutritional and health aspects of olive oil. European Journal of Lipid Science and Technology, 104, 685-697. http://faostat.fao.org/site/567/default.aspx#ancor

Istat, Istituto di Statistica Italiana, (2010). Tav. C27. Available from <http://agri.istat.it/sag_is_pdwout/jsp/consultazioneDati.jsp>

Jaccard, P. (1901). Étude comparative de la distribution florale dans une portion des Alpes et des Jura. *Bulletin de la Société Vaudoise des Sciences Naturelles*, No. 37, pp. 547-579.

Kloosterman, A.D., Budowle, B. & Daselaar, P. (1993). PCR amplification and detection of the human D1S80 VNTR locus. Amplification conditions, population genetics and application forensic analysis. *International Journal of Legal Medicine*, No. 105, pp. 257–264.

Loukas, M. & Krimbas, C.B. (1983). History of olive cultivars based on their genetic distances. *Journal of Horticultural Science*, No. 58, pp. 121–127.

Montemurro, C., Pasqualone, A., Simeone, R., Sabetta W. & Blanco A. (2008a). AFLP molecular markers to identify virgin olive oils from single Italian cultivars. *European Food Research and Technology*, No. 226, pp. 1439–1444.

Montemurro, C., Simeone, R., Blanco A., Saponari, M., Bottalico, G., Savino, V., Martelli, G.P. & Pasqualone, A. (2008b) Sanitary selection and molecular characterization of olive cultivars grown in Apulia. *Acta Horticulturae*, No. 791, pp. 603-609.

Montemurro, C., Simeone, R., Pasqualone, A., Ferrara, E. & Blanco, A. (2005). Genetic relationships and cultivar identification among 112 olive accessions using AFLP and SSR markers*Journal of Horticultural Science & Biotechnology*, Vol. 80, No. 1, pp. 105–110.

Official Journal of the European Communities. Commission Regulation (EC) No 1263/1996 of 1 July 1996 supplementing the Annex to Regulation (EC) No. 1107/96 on the registration of geographical indications and designations of origin under the procedure laid down in Article 17 of Regulation (EEC) No 2081/92. *Official Journal European Community*, 1996, L163, pp. 19-21.

Official Journal of the Italian Republic. Ministero per il Coordinamento delle Politiche Agricole, Alimentari e Forestali. Decreto Ministeriale 4 novembre 1993, n. 573. Regolamento recante norme di attuazione della legge 5 febbraio 1992, n. 169, per la disciplina del riconoscimento delle denominazioni di origine, dell'albo degli oliveti, della denuncia di produzione delle olive e dell'attività delle commissioni di degustazione degli oli a denominazione di origine controllata. *Official Journal Italian Republic* , 1994, 3, pp. 1-14.

Official Journal of the Italian Republic. Ministero per le Politiche Agricole. Decreto 29 settembre 1998. Approvazione del disciplinare di produzione della denominazione di origine controllata dell'olio extravergine di oliva "Collina di Brindisi". *Official Journal Italian Republic*, 1998, 250, pp. 43-45.

Pasqualone, A., Caponio, F. & Blanco, A. (2001). Inter-simple sequence repeat DNA markers for identification of drupes from different *Olea europaea* L. cultivars. *European Food Resesearch Technology*, No 213, pp. 240–243.

Pasqualone, A., Montemurro C., Caponio, F. & Blanco, A. (2004a). Identification of virgin olive oil from different cultivars by analysis of DNA microsatellites. *Journal of Agricultural and Food Chemistry*, No 52, pp. 1068-1071.

Pasqualone, A., Montemurro C., Caponio, F. & Simeone R. (2003). Multiplex amplification of DNA microsatellite markers to fingerprint olive oils from single cultivars. *Polish Journal of Food and Nutrition Science*, No. 12/53, pp. 96-99.

Pasqualone, A., Montemurro C., Caponio, F., Simeone R. & Blanco, A. (2004b). Analisi del DNA in oli di oliva vergini monovarietali mediante marcatori microsatelliti: fattori limitanti dovuti alla tecnologia produttiva. *La Rivista Italiana delle Sostanze Grasse*, No. 81, pp. 221-224.

Pasqualone, A., Montemurro, C., Ashtar, S., Caponio, F., Ferrara, E., Saponari, M., Ibrahem, A. AL, Rahman Kalhout, A. & Blanco, A. (2007b). Set up of a DNA bank of *Olea europaea* L. the objective being cultivar traceability in olive oil. *Proceedungs of the 3rd CIGR International Symposium, Section VI – Food and agricultural products: processing and innovations*, Naples, Italy, September 24-26. Published on *La Rivista Italiana delle Sostanze Grasse*, (2008), No 85, pp. 83-91.

Pasqualone, A., Montemurro, C., Summo, C., Sabetta, W., Caponio, F. & Blanco A. (2007a). Effectiveness of microsatellite DNA markers in checking the identity of Protected Designation of Origin extra virgin olive oil. *Journal of Agricultural and Food Chemistry*, No 55, pp. 3857-3862.

Powell, W., Morgante, M., Andre, C., Hanafey, M., Vogel, J., Tingey, S. & Rafalski, A. (1996). The utility of RFLP, RAPD, AFLP and SSRP (microsatellite) markers for germplasm analysis. *Molecular Breeding*, No. 2, pp. 225-238.

Prevost, A. & Wilkinson, M.J. (1999). A new system of comparing PCR primers applied to ISSR fingerprinting. *Theoretical and Applied Genetics*, No. 98, pp. 107–112.

Rohlf, F.J. (1992). Numerical taxonomy and multivariate analysis system version 1.70. State University of New York, Stony Brook N.Y.

Russell, J.R., Fuller, J.D., Macaulay, M., Hatz, B.G., Jahoor, A., Powell, W. & Waugh, R. (1997). Direct comparison of level of genetic variation among barley accessions detected by RFLPs, AFLPs, SSRs and RAPDs. *Theoretical and Applied Genetics*, No. 95, pp. 714-22.

Sefc, K.M., Lopes, M.S., Mendonca, D., Rodrigues Dos Santos, M., Laimer Da Camara Machado, M. & Da Camara Machado, A. (2000). Identification of microsatellite loci in olive (*Olea europaea*) and their characterization in Italian and Iberian olive trees. *Molecular Ecology*, No. 9, pp. 1171–1173.

Sensi, E., Vignai, R., Scali, M., Masi, E. & Cresti, M. (2003). DNA fingerprinting and genetic relatedness among cultivated varieties of *Olea europaea* L. estimated by AFLP analysis. *Scientia Horticulturae*, No. 97, pp. 379–388.

Sharp, P.J., Kreis, M., Shewry, P.R & Gale, M.D. (1988). Location of β-amylase sequences in wheat and its relatives. *Theoretical and Applied Genetics*, No. 75, pp. 286–290.

Visioli, F., Poli, A. & Galli, C. (2002). Antioxidant and other biological activities of phenols from olives and olive oil. Medicinal Research Reviews, 22, 65-75.

Vos, P., Hogers, R., Bleeker, M., Reijans, M., Van De Lee, T., Hornes, M., Frijters, A., Pot, J., Peleman, J., Kuiper, M. & Zabeau, M. (1995). AFLP: a new technique for DNA fingerprinting. *Nucleic Acids Research*, No. 23, pp. 4407–4414.

4

Systematic Diversity of the Family Poaceae (Gramineae) in Chile

Víctor L. Finot[1], Juan A. Barrera[1], Clodomiro Marticorena[1] and Gloria Rojas[2]
[1]*University of Concepción,*
[2]*National Museum of Natural History,*
Chile

1. Introduction

The role of Systematics in studies of biodiversity is essential to a variety of studies, including species conservation, extinction, biodiversity hotspots, bio-prospecting and ecosystem function (Alroy, 2002; Scotland & Wortley, 2003; Smith & Wolfson, 2004; Wilson, 2000). The analysis of the biodiversity as well as the analysis of the distribution of species richness at different levels (national, regional), the distribution of the endemic species, the detection of areas whose preservation is necessary and many other topics related to the conservation of the biodiversity requires an important collection effort, so that the organized databases constructed by the herbaria become as comprehensive as possible. Herbarium specimens represent a rich source of information for botanists and ecologists, even though data based on herbaria collections have many limitations, since they are geographically and seasonally biased, and taxonomically incomplete (Crawford & Hoagland, 2009; Delisle et al., 2003; Fuentes et al., 2008; Funk & Richardson, 2002; Ponder et al., 2001). Moreover, it has been established that there is a tendency to a decline in the number of specimens of vascular plants collected in the last years (Prather et al., 2004), although taxonomists are aware that there are still many undescribed species (Smith & Wolfson, 2004). In order to know how many species of grasses exist in Chile, as well as their identity and taxonomic distribution, this chapter provides a checklist of the family Poaceae in Chile, taking into account the nomenclatural changes recently proposed. Moreover, we analyze the completeness of the inventory of the family represented in two of the most important national herbaria.

Grasses (Poaceae or Gramineae) are the fifth most diverse family among the flowering plants or Angiosperms and the second most diverse family among the Monocotyledons. Poaceae comprises about 10,000 species in approximately 700 genera (Clayton & Renvoize, 1986; Tzvelev, 1989; Watson & Dallwitz, 1992). Recent evidence suggests that grasses had already diversified during the Cretaceous. The evidence came from phytolith analysis (Prasad et al., 2005), tiny crystals of silica formed in the epidermal cells of leaves or floral bracts of grasses and other plants. The discovery of grass phytoliths in coprolites of titanosaurid sauropods that lived in India 65 to 71 million years ago (Prasad et al., 2005), suggested that grasses and dinosaurs coevolved (Piperno & Sues, 2005). Phylogenetic approach to reveal the evolutionary history of grasses in a biogeographical context suggests that Poaceae originated in the African or South American regions of Gondwana during the late Cretaceous (Bouchenak-Khelladi et al., 2010).

The economic significance of the grass family is undeniable. Grasses are found on all continents, including Antarctica (e.g. *Deschampsia antarctica*) and are ecologically dominant in some ecosystems such as the African savannas (Kellogg, 2000). Grasslands, in which grasses are the most important floristic component, cover about 40% of the earth surface (Peterson et al., 2010). Most people on Earth depend on grasses, such as wheat, corn, oats, rice, sugarcane, and rye, for a large part of their diet. In addition, domestic animals are fed on diets based largely on forage grasses. Moreover, many of the most serious weeds growing on agricultural land are also members of the grass family.

2. Diversity, phylogeny and classification of the grass family

Grasses are unique from a morphological point of view. The grass flower contains a bicarpellary gynoecium surrounded by an androecium composed of three or more stamens in one or two whorls. The perianth is reduced to two or three lodicules situated outside the stamens; the lodicules open the florets during the pollination process. Outside the lodicules, in adaxial position, there is the palea. Subtending the flower there is another bract, the lemma (Kellogg, 2000). The elemental inflorescence is the spikelet, with one to many florets inserted along an axis, the rachilla. Each spikelet has two empty bracts called glumes which protect the immature spikelet. The lemma and the palea enclosing the flower or caryopsis constitute the floret. Lemmas often bear awns or mucros born at the apex or on the back of the body of the lemma. Awns are very common in the family, nearly the half of the genera of Poaceae have awns (GPWG, 2001). The inflorescence (in fact a synflorescence) is a panicle, a raceme or a spike (for a detailed descriptive terminology on grass inflorescence see Allred, 1982). The fruit is a caryopsis or grain; a caryopsis is defined as a one-seeded indehiscent fruit with the seed coat fused with the pericarp (the ovary wall). The embryo is lateral and highly differentiated, with shoot (plumule) and root meristems, leaves and vascular system. The embryo has a scutellum considered to be a modified cotyledon; in many species there is the epiblast, opposite to the scutellum, considered to be a rudiment of a second cotyledon or an outgrowth of the coleorrhiza (Tzvelev, 1983). The epiblast is absent in the subfamilies Arundinoideae and Panicoideae (Clayton & Renvoize, 1986). In the lower part of the embryo the root meristem is covered by the coleorrhiza. In the upper part, the plumule is covered by the coleptile. The embryos of Arundinoideae, Chloridoideae, Centothecoideae and Panicoideae have a mesocotyl (first internode), between the insertion of the scutellum and the coleoptile. The embryos of Arundinoideae, Bambusoideae, Centothecoideae, Chloridoideae, Ehrhartoideae and Panicoideae have a cleft between the coleorrhiza and the scutellum. The plants are annual or perennial, herbaceous or woody. The stems, called culms, are simple or branched and often with rhizomes or stolons. Culms are hollow or more rarely, solid. Roots are fibrous and adventitious (homorriza).

Leaves are distichous; the leaves have the basal portion forming the sheath and the upper portion forming the blade. At the adaxial junction of the sheath and blade there is a membranous ligule, sometimes transformed in a fringe of hairs. Most grasses lack an abaxial ligule; an abaxial ligule is present in Bambuseae, some members of the PACMAD clade and a few Pooideae. The sheath is sometimes auriculate. In bamboos, the leaves have a pseudopetiole, a constriction at the base of the leaf blade

Two main photosynthetic pathways, C3 and C4, are found in Poaceae but C3/C4 intermediates also occur. In C3 photosynthesis CO_2 combines with ribulose 1,5-biphosphate in the Calvin-Benson cycle. The first detectable metabolic product of this process is phosphoglycerate, a compound with three carbon atoms. C3 photosynthesis takes place in the leaf mesophyll. C3 grasses are well adapted to temperate climates. In C4 photosynthesis or Hatch-Slack cycle the first detectable metabolic product is oxalacetate, a compound with four carbon atoms. In C4 grasses, C4 activity is confined to the mesophyll and C3 photosynthesis is displaced to the bundle sheath surrounding the vascular tissue (Kranz syndrome). It is presumed that C4 photosynthesis is an adaptation to low CO_2 levels and high O_2 levels. C4 plants minimize photorespiration sequestering Rubisco in the cells of the bundle sheath making C4 photosynthesis more efficient than C3, especially at high temperatures and arid environments. C4 photosynthesis evolved in four of the 13 subfamilies of Poaceae (Panicoideae, Aristidoideae, Chloridoideae and Micrairoideae). The earliest fossil grass leaves with C4 anatomy is dated 12.5 Ma but Chloridoideae phytoliths have been dated 19 Ma. It has been suggested that C3 photosynthesis is ancestral to the origin of C4 photosynthesis and occurs about 32 Ma during the Oligocene, and that the origin of the C4 pathway is polyphyletic (Vincentini et al., 2008).

The family Poaceae is monophyletic. Characters that unambiguously support the monophyly of the family are the grass-type embryo lateral, peripheral to the endosperm and highly differentiated in the caryopsis, and a *trnT* inversion in the chloroplast genome (GPWG, 2001).

The grass family has been divided in a number of subfamilies ranging from two to 13 (for a review see GPWG, 2001). Traditionally, the family was divided in two major groups: Festucoideae (= Pooideae) and Panicoideae (Hitchcock, 1950). The system of grasses (Tzvelev, 1989) also recognized only two subfamilies: Bambusoideae with 14 tribes, and Pooideae, with 27 tribes. In Tzvelev's system, Panicoideae are embedded in Pooideae. One of the most widely used systems is that of Clayton & Renvoize, which divided the family in six subfamilies: Bambusoideae, Pooideae, Centothecoideae, Arundinoideae, Chloridoideae and Panicoideae (Clayton & Renvoize, 1986). The phenetic system of Watson & Dallwitz recognizes seven subfamilies (the same as Clayton & Renvoize + Stipoideae) (Watson & Dallwitz, 1992). The largest proposed number of subfamilies is 13 (Caro, 1982): Bambusoideae, Streptochaetoideae, Anomochlooideae, Olyroideae, Centostecoideae, Oryzoideae, Ehrhartoideae, Phragmitoideae, Festucoideae (= Pooideae), Eragrostoideae (= Chloridoideae), Aristidoideae, Panicoideae and Micrairoideae.

The evolutionary history of Poaceae has been deciphered using different molecular markers, such as restriction site maps of the chloroplastidial DNA (Soreng & Davis, 1998), sequences of various chloroplast genes such as *ndhF* (Clark et al., 1995; Sánchez-Ken & Clark, 2010), rpoC2 (Barker et al., 1999), *rbcL* (Barker et al., 1995; Sánchez-Ken & Clark, 2010), matK (Hilu et al., 1999), rps4 (Nadot et al., 1994), and sequences of several nuclear genes such as phytochrome B (Mathews et al., 2000), GBSSI (Mason-Gamer et al., 1998), ITS (Hsiao et al., 1999), and 18S rDNA (Hamby and Zimmer, 1988). The Grass Phylogeny Working Group (GPWG, 2001) combined the data from these sources to produce a phylogeny of the family (Kellogg, 2001). They recognized 12 subfamilies: Anomochloideae, Pharoideae, Puelioideae, Bambusoideae*[1], Ehrhartoideae*, Pooideae*, Aristidoideae*, Arundinoideae*, Danthonioideae*, Centothecoideae, Panicoideae*, and Chloridoideae*. Three early diverging lineages (Anomochloideae, Pharoideae and Puelioideae) and two major lineages were recognized: a clade comprising the subfamilies Bambusoideae, Ehrhartoideae and Pooideae, called the BEP clade, and the PACCAD clade, containing the subfamilies Panicoideae,

---

[1] An asterisk indicates the subfamilies present in Chile.

Arundinoideae, Chloridoideae, Centothecoideae, Aristidoideae and Danthonioideae. Later, the resurrection of the subfamily Micrairoideae and the synonymization of Centothecoideae with Panicoideae changed the acronym of the PACCAD clade to PACMAD (Sánchez-Ken et al., 2007; Sánchez-Ken & Clark, 2010). The PACMAD clade comprises ca. 5000 species or about half of the diversity of the family.

Subfamily **Anomochlooideae** is the earliest diverging lineage of Poaceae followed by Pharoideae and Puelioideae (GPWG, 2001). Anomochlooideae and Pharoideae were embedded in Bambusoideae as tribe Anomochloeae and tribe Phareae respectively, but later resurrected as subfamilies (Clark & Judziewicz, 1996). Anomochlooideae includes two genera (*Anomochloa* and *Streptochaeta*) from tropical America characterized by "spikelet equivalent" instead of true grass spikelets. Puelioideae includes two genera (*Gaduella, Puelia*), native to tropical Africa of perennial rhizomatous grasses of shaded forest understory (Clark et al., 2000). Anomochlooideae, Pharoideae and Puelioideae are absent in the Chilean grass flora.

Subfamily **Bambusoideae** includes approximately 115 genera (GPWG, 2001) of herbaceous (Tribe Olyreae, ca. 110 spp.) and woody bamboos (Bambuseae, ca. 1300 spp.). The world most diverse genus of bamboos is *Chusquea*, with 134 described species. Bambusoideae are perennial (rarely annual) usually woody grasses, with pseudopetiolated leaves. Culms are erect or scandent. Anatomically, they are characterized by the presence of fusoid cells, aligned perpendicular to the long axis of the leaf blade. Fusoid cells are large, thin walled and lack chloroplasts and other cell contents; they probably represent internal gas spaces. C3 photosynthesis takes place in the arm cells; these are thin walled cells with well developed invaginations or arm like lobes (Judziewicz et al., 1999; for a detailed leaf anatomical description of Chilean bamboos species see also Matthei, 1997).

Subfamily **Ehrhartoideae** includes three tribes (Oryzeae, Ehrharteae and Phyllorachidae) and approximately 120 species (Barkworth et al., 2007), characterized by spikelets with one fertile floret often with one or two proximal sterile florets, two lodicules, three or six stamens, C3 photosynthesis, with a double bundle sheath around the veins, the outer sheath parenchymatous and the inner with thick walls (Kellogg, 2009).

Subfamily **Pooideae** is the largest of the 13 subfamilies of Poaceae (GPWG, 2001), comprising about 3560 species (Soreng et al., 2007). It includes some of the most economically important species, such as wheat (*Triticum aestivum*), oats (*Avena sativa*), barley (*Hordeum vulgare*) and rye (*Secale cereale*) and many forage and weed species. Plants herbaceous, with hollow culms (rarely solid); leaves distichous with an adaxial membranous ligule, rarely a fringe of hairs; leaf blades narrow; sheaths with or without auricles. Spikelets usually disarticulating above the glumes, laterally compressed. Glumes 2, lemmas 1 to many. Caryopsis with a linear or punctiform hilum. Endosperm solid, soft or liquid, with compounds starch grains (except in Bromeae, Triticeae, and Brachyelytreae). Embryo small, epiblast present, scutellar cleft absent, mesocotyl internode absent, embryonic leaf margins not overlapping. Photosynthetic pathway: C3. Basic chromosome number x = 7. Microhairs absent. Stomata with parallel-sided subsidiary cells. Pooideae includes some 3300 species distributed in temperate climates and in the tropics in the mountains. Most of the Chilean grasses belong to this subfamily.

Subfamily **Aristidoideae** includes three genera (*Aristida* L., *Stipagrostis* Nees and *Sartidia* de Winter) and more than 350 species in tropical, subtropical and temperate zones, most belonging to *Aristida*. Aristidoideae are members of the PACMAD clade, with C4 or C3 photosynthesis (Cerros-Tlatilpa & Columbus, 2009). Plants are annuals or perennial. The abaxial ligule is absent or present; the adaxial ligule membranous or a fringe of hairs. The leaf sheath is non-auriculate. The inflorescence is a panicle. Spikelets with two glumes; lemmas three-awned; lodicules 2, rarely absent. Caryopsis with short or long-linear hilum. Endosperm hard, containing compound starch grains. Embryo small or sometimes large; epiblast absent; scutellar cleft present or absent; mesocotyl internode elongated; embryonic leaf margins not overlapping. Members of Aristidoideae have C4 photosynthesis (except *Sartidia* with non-Kranz anatomy and C3 photosynthesis). Basic chromosome number x = 11, 12. Bicellular microhairs present. Stomata with dome-shaped or triangular subsidiary cells. With only one genus (*Aristida*) and three species, this subfamily is underrepresented among Chilean grasses.

Subfamily **Arundinoideae** includes 33-38 species of temperate and tropical zones (GPWG, 2001). Arundinoideae are perennial (rarely annual), herbaceous or sometimes woody plants, of temperate and tropical zones. Culms usually hollow; leaves distichous usually with adaxial ligule only; ligule membranous or a fringe of hairs; sheath non-auriculate. Inflorescence usually a panicle. Spikelets disarticulating above the glumes; glumes 2; lemma sterile sometimes present; 1 or more female-fertile florets. Lodicules 2. Caryopsis with hilum short or long-linear; endosperm solid, with compound starch grains. Embryo large or small; epiblast absent; scutellar cleft present; mesocotyl internode elongated; embryonic leaf margins meeting or overlapping. Basic chromosome number x = 6, 9, 12. Photosynthetic pathway: C3. Stomata with dome or triangular subsidiary cells. Bicellular microhairs present (sometimes absent).

Subfamily **Danthonioideae** comprises some 250 species (GPWG, 2001). Plants usually perennial, rarely annual, usually herbaceous with rhizomes, stolons or caespitoses and culms most often solid. Leaves with adaxial ligule; ligule membranous or a fringe of hairs. Inflorescence a panicle. Spikelets laterally compressed, with two glumes and one to several fertile florets; rachilla disarticulating above the glumes and between the florets; lodicules two. Caryopsis with short or long-linear hilum; endosperm hard with compound starch grains; embryo large or small, with epiblast, with scutellar cleft, mesocotyl internode elongated and embryonic leaf margins meeting, rarely overlapping. Basic chromosome numbers x = 6, 7, 9. Photosynthetic pathway: C3. Stomata with dome shaped or parallel-sided subsidiary cells. Bicellular microhairs present.

Subfamily **Chloridoideae** comprises approximately 140 genera and more than 1420 species of arid environments (Hilu & Alice, 2001; Peterson et al., 2010). Plants annual or perennial with hollow or solid culms. Leaves distichous, with adaxial membranous ligule. Inflorescences paniculate. Spikelets with two glumes, laterally compressed or sometimes dorsally compressed; rachilla disarticulating above the glumes. Lemmas 1 to many; lodicules 2 (or absent), non-membranous (fleshy). Caryopsis with the pericarp often free or loose; endosperm solid, with simple or compound starch grains; embryo large, with epiblast, with scutellar cleft, mesocotyl internode elongated and embryonic leaf margins meeting, rarely overlapping. Stomata with dome-shaped or triangular subsidiary cells; bicellular microhairs or chloridoid type (inflated, spherical microhairs). Photosynthesis C4 of two main types: NAD-ME (nicotinamide adenine dinucleotide co-factor malic enzyme) and PCK (phosphoenolpyruvate carboxykinase). Basic chromosome numbers x = 9, 10, rarely 7 or 8 (GPWG, 2001). Stomata with dome-shaped or triangular subsidiary cells. Bicellular microhairs present.

Subfamily **Panicoideae** comprises approximately 206 genera and 3300 species, mainly from tropical and warm temperate climates (Giussani et al. 2001; GPWG, 2001). Plants annual or perennial usually with solid culms. Leaves distichous with adaxial membranous ligule, sometimes a fringe of hairs or absent. Inflorescences panicle, racemes or compound inflorescences. Spikelets 2-flowered, with the lower floret staminate or barren, single or paired, dorsally compressed, with two glumes, disarticulating below the glumes (rarely above the glumes). Lodicules 2, fleshy. Caryopsis with short hilum, hard endosperm containing simple (rarely compound) starch grains; embryo large, epiblast absent, scutellar cleft present, mesocotyl internode elongated, embryonic leaf margins overlapping. Stomata with triangular or dome-shaped subsidiary cells; bicellular microhairs of the panicoid type present. Photosynthetic pathway C3, C4 and C3/C4 intermediates (GPWG, 2001). The presence of bifloral spikelets with the lower flower staminate or neuter, simple starch grains and molecular data, both from chloroplast and nuclear DNA, are synapomorphies that support the monophyly of the Panicoideae (Aliscioni et al., 2003).

3. Some geographical features of the Chilean territory

Details of the geographical and evolutionary characteristics of the Chilean flora have been published elsewhere (Arroyo et al., 1993; Grau, 1995, Squeo et al., 2008; Stuessy & Taylor, 1995). Chile stretches from north to south for nearly 4270 km (ca. 39 degrees) along the western coast of South America, between 17°30′S (10 km North of Visviri) and 56°30′S (Diego Ramírez islands) and extends to the Antarctica at the South Pole (Chilean Antarctic Territory). Also belong to Chile the oceanic islands San Félix and San Ambrosio (Desventuradas islands), the archipelago of Juan Fernández and the Polynesian islands Easter Island and Sala and Gómez. Including the Chilean Antarctic Territory, the country's length spans for about 73 degrees. The maximum width occurs in southern Chile (52°21′S), where there is over 400 km between the Pacific Ocean and the Andes, but the overall width of the country usually does not exceed 200 km. Chile borders with Peru on the North, on the East with Argentina and Bolivia, on the West with the Pacific Ocean and on the South with the South Pole. In the north, Chile shares with Bolivia, Peru and Argentina the Altiplano, a plateau which does not descend from 4000 m of altitude. The Andes runs end-to-end the territory, establishing the border with Argentina. The Andes reaches heights of almost 7000 m in the northern zone, while to the south it falls below 2500 m. The Atacama Desert, considered to be one of the driest places in the world, dominates the climate on the north (Pankhurst & Hervé, 2007), with rainfall below 10 mm per year. From the Aconcagua valley to the region of Bío-Bío (approx. 32-38°S), the Mediterranean climate is dominant, with the exception of the high peaks of the Andes with cold weather due to the altitude. The Mediterranean region of Chile is included in the Chilean hotspot of biodiversity (Winter Rainfall Area of Central-Northern Chile Hotspot or Chilean Winter Rainfall and Valdivian Forests). The Chilean hotspot is located between 25 and 40°S (Arroyo et al., 1999), including the regions of Coquimbo and Atacama and extending to the region of Los Lagos in southern Chile. This hotspot contains 3892 vascular plant species, of which 1957 (more than 50%), are endemic to Chile. The Chilean hotspot covers almost 400.000 km^2 including the islands San Félix, San Ambrosio and the archipelago of Juan Fernández (Conservation International, 2011). South of 38°S the mixed deciduous-evergreen temperate forests occur.

The biogeographic sketch of the Chilean vegetation (Cabrera & Willink, 1973) divided the territory into two phytogeographical regions (*Región Neotropical* and *Región Antárctica*) and domains (*Dominio Patagónico* and *Dominio Subantárctico*). A detailed account of the Chilean vegetation can be found in (Gajardo, 1994). To the Patagonian domain belong the Provinces *Provincia Altoandina, Provincia Puneña, Provincia del Desierto, Provincia Chilena Central* and *Provincia Patagónica* (Cabrera & Willink, 1973). *Provincia Altoandina* extends from Venezuela to Tierra del Fuego, in the high Andes. In the Andes of Mendoza (Argentina) and adjacent regions of Chile of *Provincia Altoandina* dominate communities of *Festuca, Poa, Deyeuxia* and *Nasella. Provincia Puneña* extends from 15-27°S, in the high Andes between 3200 and 4000 m of altitude. Communities of *Pappostipa chrysophylla* (coirón amargo) are common in this area. *Provincia del Desierto* is located on the Pacific coast between 5 and 30°S. This province has a warm, dry weather because of the Humboldt Current. The coastal fogs known as "*camanchacas*" allow the growth of vegetation mainly with species of the family Nolanaceae (*Alona, Nolana*). This province (also known as *Subregión del desierto costero*) is very interesting from a floristic point of view due to the high number of endemic species that live there (Gajardo, 1994). South of La Serena (Region of Coquimbo), the coastal fogs provide sufficient moisture to sustain the forest of Fray Jorge, a relict forest of *Aextoxicon punctatum* (olivillo, Aextoxicaceae), which only reappears only in southern Chile (Valdivia). *Provincia Chilena Central* extends between 32°S and 38°S (except in the high mountains), including the regions of Valparaíso, Metropolitana, O'Higgins, Maule and Bío-Bío. This province is dominated by sclerophyllous forests with *Bielschmiedia miersii* (belloto, Lauraceae), *Peumus boldus* (boldo, Monimiaceae), *Cryptocarya alba* (peumo, Lauraceae), *Kageneckia oblonga* (huayu, Rosaceae), *Lithrea caustica* (litre, Anacardiaceae), *Quillaja saponaria* (quillay, Rosaceae), *Colliguaya dombeyana* (colliguay, Euphorbiaceae), etc. In the central valley known as Depresión Intermedia, the espinal of *Acacia caven* (espino, Mimosaceae) is the dominant vegetation community. The espinal has a very high diversity of Poaceae. Species commonly found in the prairie of the espinal are *Agrostis capillaris, Aira caryophyllea, Aristida pallens, Avena barbata, Briza maxima, B. minor, Bromus hordeaceus, B. rigidus, B. stamineus, Chascolytrum subaristatum, Chusquea quila, Cynosurus echinatus, Dactylis glomerata, Danthonia chilensis* var. *aureofulva, Hordeum chilense, H. murinum, Lolium multiflorum, L. perenne, Melica violacea, Nasella gibba, N. neesiana, N. pfisteri, Paspalum dasypleurum, Phalaris amethystina, Poa annua, Piptochaetium montevidense,* and *Vulpia bromoides.* In southern Chile, *Provincia Patagónica* comprises the regions of Aysén and Magallanes. This region is characterized by dry and cold climate with vegetation consisting mainly of grassland steppe (Fig. 1), where communities of *Festuca pallescens* and *F. gracillima* are important in Aysén and Magallanes, respectively (Luebert & Pliscoff, 2006). Other species frequently found are *Festuca argentina, Jarava neaei, Deschampsia antarctica, D. elongata* (Pisano, 1985).

Región Antárctica and *Dominio Subantárctico* include two provinces: *Provincia Subantárctica* and *Provincia de Juan Fernández*. The archipelago of Juan Fernández, situated 670 km west of continental Chile in the Pacific Ocean, comprises three islands of volcanic origin: Masafuera or Alejandro Selkirk (33°37'S, 80°46'W), Masatierra or Robinson Crusoe (33°37'S, 78°50'W), and Santa Clara. Masafuera is located 180 km west of Masatierra whereas the small island Santa Clara is located 1 km SW of Masatierra (Errázuriz et al., 1998; Swenson et al. 1997). The archipelago of Juan Fernández is characterized by a high level of endemism; 31% of the vascular plants are endemic (Baeza et al., 2007; Swenson et al. 1997). The vascular flora of the islands comprises 42 families of flowering plants, including the monotypic and endemic Lactoridaceae. According to Skottsberg, the largest families are Asteraceae, Cyperaceae and Poaceae (Skottsberg, 1956). On the other hand, the high number of alien plants is a serious threat to the native flora of Juan Fernández (Matthei et al., 1993; Swenson et al., 1997).

4. Analysis of the Chilean grass diversity

To analyze the diversity of the family Poaceae in Chile, a database was prepared based on the collections of the herbaria of University of Concepción (CONC), National Museum of Natural History (SGO) and specimens cited in Tropicos database (tropicos.org), and taxonomic literature [Acevedo de Vargas (1959, *Cortaderia* Stapf); Baeza (1996, *Danthonia* DC., *Rytidosperma* Steud.); Cialdella & Arriaga (1998, *Piptochaetium* J. Presl); Cialdella & Giussani (2002, *Piptochaetium*); De Paula (1975, *Anthoxanthum* L. under *Hierochlöe* R.Br.); Finot et al. (2005, *Trisetum* Pers.); Matthei (1965, *Nassella* (Trin.) E. Desv. under *Stipa* L.; 1975, *Briza* L.; 1982, *Festuca*; 1986, *Bromus* L.; 1987[a], *Aristida* L., 1987b; *Panicum* L.); Muñoz-Schick (1985, *Melica* L.; 1990, *Nassella*); Nicora (1998, *Eragrostis* Wolf); Rúgolo de Agrasar (1978, 2006, *Deyeuxia* Kunth; 1982, *Bromidium* Nees et Meyen; 1999, *Corynephorus* P. Beauv.); Rúgolo de Agrasar & Molina (1997, *Agrostis* L.)]; Rúgolo de Agrasar et al. (2009, *Danthonia*); Soreng & Gillespie (2007, *Nicoraepoa*); Soreng & Peterson (2008, *Poa*)]. A total of 13.924 records were included in the database, including the name of the species, subfamily, tribe, subtribe, geographical origin (native, introduced, endemic), collector, latitude, longitude, altitude, locality, date of collection, herbarium, and herbarium number. A checklist of the species based in the database, the Catalogue of the New World Grasses (Judziewicz et al., 2000; Peterson et al., 2001; Soreng et al., 2003; Zuloaga et al., 2003, available online at http://www.Tropicos.org) and the Catálogo de las plantas vasculares del Cono Sur (Zuloaga et al., 2008, available online at http://www2.darwin.edu.ar) was prepared. Recent taxonomic treatments of Chloridoideae (Peterson et al., 2007; Peterson et al., 2010), Danthonioideae (Linder et al., 2010), Panicoideae (Sánchez-Ken & Clark, 2010), *Avenella, Deschampsia, Vahlodea* (Chiapella & Zuloaga, 2010), *Bromus* (Planchuelo, 2010), *Cenchrus* (Chemisquy et al., 2010), *Deyeuxia* (Rúgolo de Agrasar, 2006), *Digitaria* (Vega et al., 2009) and *Trisetum* (Finot, 2010), were followed to update the nomenclature and the classification of the species. Lists of the species of the family Poaceae growing in Chile have been published for different regions (Arroyo et al., 1989, 1990, 1992, 1998; Baeza et al., 1999, 2002, 2007; Finot et al., 2009; Kunkel, 1968; Matthei et al., 1993; Reiche, 1903; Rodríguez et al., 2008a, 2008b; Rundel et al., 1996) or for all the country (Muñoz, 1941; Marticorena & Quezada, 1985; Zuloaga et al., 2008).

Many species of Chilean grasses were collected and described early in the nineteenth century. The collection of grass specimens began at a fairly early stage of the knowledge of the diversity of the Chilean flora, in the early nineteenth century, with the collections made by Eduard Poeppig, Carlo Bertero, Rodulfo Amando and Federico Philippi, and Claudio Gay. In this period, Emile Desvaux published one of the first complete treatments of the Chilean grasses in Gay's "*Historia Física y Política de Chile*" (Desvaux, 1854), including 22 beautiful illustrations of native and endemic Chilean grasses, in Gay's Atlas. In the early twentieth century, new collections were made by Karl Reiche, Félix Jaffuel, Víctor Baeza, Franz Neger, Ernesto Barros, Atanasio Hollermayer, Hugo Gunckel, Erich Werdermann, Karl Junge, and Gilberto Montero, among others. Nonetheless, the number of specimens and, as a result, the number of species collected in the first 10 decades (1828-1917) is quite small (Fig. 2). In the early twentieth century only about 20% of the currently known grass flora had been collected. Since 1918 the collection effort was significantly increased; among 544 and 2063 specimens per decade were collected. These new collections allowed recording more than 430 new species. In order to compare the collection effort in the fifteen political regions in which the country is divided (Table 1), we calculated the collection index (Squeo et al., 1998):

$$CI = \text{Number of species/Number of collections} \quad (1)$$

Lat °S	*Region*	*Number of observed taxa*	*Mean number of estimated taxa**	*Represented percentage*	*Estimated number of unknown taxa*	*NCO*	*CI*
17°30'-19°06'	AP	109.0	150.5	72.4	41.5	429	0,23
18°56'-21°39'	TA	73.0	120.0	60.9	47.0	250	0,34
20°56'-26°05'	AN	125.0	181.0	69.1	56.0	853	0,11
25°17'-29°11'	AT	106.0	173.8	61.0	67.8	385	0,20
29°02'-32°16'	CO	167.0	237.8	70.2	70.8	888	0,16
32°02'-33°57'	VA	186.0	253.6	73.3	67.6	1055	0,16
32°55'-34°19'	ME	198.0	260.6	76.0	62.6	1375	0,13
33°51'-35°01'	OH	89.0	161.7	55.0	72.7	172	0,49
34°41'-36°33'	MA	172.0	261.6	65.8	89.6	648	0,25
36°00'-38°30'	BB	263.0	337.0	78.0	74.0	3118	0,07
37°35'-39°37'	AR	188.0	244.2	77.0	56.2	1254	0,14
39°16'-40°07'	LR	143.0	185.5	77.1	42.5	742	0,18
40°07'-44°04'	LL	131.0	202.3	64.8	71.3	634	0,19
43°38'-49°16'	AY	112.0	179.6	62.4	67.6	373	0,26
48°39'-62°22'	MG	158.0	195.3	80.9	37.3	1829	0,07

Table 1. Poaceae diversity of Chile at national and regional level. NCO = number of collected specimens, CI = collection index, Regions: AP = Arica and Parinacota, TA = Tarapacá, AN = Antofagasta, AT = Atacama, CO = Coquimbo, VA = Valparaíso, ME = Metropolitan, OH = O'Higgins, MA = Maule, BB = Bío-Bío, AR = Araucanía, LR = Los Ríos, LL = Los Lagos, AY = Aysén, MG = Magallanes and Antarctica Chilena. *Mean value of ICE, Chao2, Jacknife1, Jacknife2, Bootstrap and Michaelis-Menten estimators.

The collection index takes values ranging from 1 to near zero. Value 1 indicates poor collection effort while values near to 0 indicate that the region is over-collected (Squeo et al., 1998). Collections are not uniformly distributed along the country; on the contrary, some regions have been collected more or less intensely (e.g. Bío-Bío, Metropolitan, Magallanes, Araucanía, Valparaíso), while others regions (e.g. O'Higgins, Tarapacá and Aysén) have been weakly collected (Table 1, Fig. 5). The species accumulation curves allow the estimation of species richness from a sample and to compare the species richness of different areas. The accumulation curve for the country together with the curves of the five species richness estimators are shown in Fig. 3. The accumulation curve for the all country tends to be asymptotic (Fig. 3) and the mean number of estimated taxa (Table 1) indicated that our database included about 85.5% of the taxa expected to be found in Chile. There are about 70 taxa not yet collected, nevertheless, these indicators show that the overall knowledge of

Fig. 1. Grassland communities in Aysén, Chile (R. Wilckens & F. Silva).

Chilean grasses is good enough. On the other hand, the species accumulation curves at regional level (Fig. 4) show the number of taxa collected as a function of sampling effort (time) for each political region. Figure 4 show that the regions with lower richness are Tarapacá (TA), O'Higgins (OH), Atacama (AT) and Aysén (AY), while those with higher species richness are Bío-Bío (BB), Metropolitan (ME), Araucanía (AR) and Valparaíso (VA). According to Table 1, only 55% of the expected taxa for the region of O´Higgins are represented in the botanical collections.

The species richness of Poaceae along the country is shown in Fig. 5. The regions located in northern Chile, Arica and Parinacota (AP), Tarapacá (TA), and Antofagasta (AN), known collectively as "Norte Grande" (Far North), and the regions of Atacama and Coquimbo, known as "Norte Chico" (Near North), are characterized by its desert climate because of the presence of the Atacama Desert. In the Far North, the diversity of grasses reaches relatively low values when compared with the regions of central and southern Chile.

In Arica and Parinacota, there are 109 taxa (68% non-endemic native, 5.94% endemic, and 15.84% introduced), belonging to six subfamilies (Bambusoideae and Ehrhartoideae are absent in northern Chile), and 42 genera. The taxonomic biodiversity is higher than in the other regions of the Far North. Six endemic species were detected in Arica and Parinacota: *Anatherostipa venusta*, *Bromus gunckelii*, *Cynodon nitidus*, *Festuca panda*, *Nassella pungens*, and *Trisetum johnstonii* subsp. *mattheii*, the latter species is an endemic at regional level.

In Tarapacá there were 73 taxa in six subfamilies and 36 genera; however, the collection effort is lower than in the adjacent regions; four species endemic to Chile were present in this region: *Anatherostipa venusta*, *Bromus gunckelii*, *Cynodon nitidus*, and *Polypogon linearis* Trin.

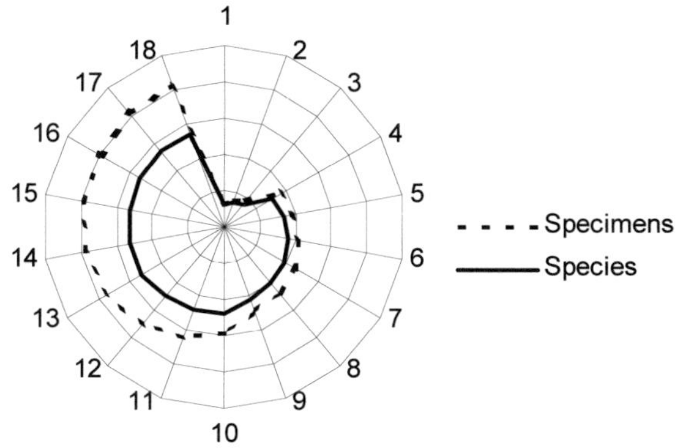

Fig. 2. Accumulated number of specimens and accumulated number of species collected in Chile in the 18 decades between 1828 and 2007.

In Antofagasta there were 125 taxa, and a greater collection effort (Table 1); in comparison with the other two regions of "Norte Grande", in Antofagasta reside a greater number of endemic species: *Anatherostipa venusta, Festuca morenensis, F. tunicata, F. werdermannii, Jarava matthei, J. tortuosa, Nassella pungens, Poa paposana,* and *Polypogon linearis.*

In the near north, we found two regions: Atacama (AT) and Coquimbo (CO), located between the hyper-arid region of Antofagasta and the more fertile region of Central Chile, between the rivers Copiapó and Aconcagua. In Atacama, only 106 taxa were recognized, of which 58 are native or endemic, while other six species were included in the Catalogue of the vascular flora of Atacama (Squeo et al. 2008). Nevertheless, the collection index shows a relatively small collection effort for grasses in Atacama (Table 1). Endemic species in Atacama are *Festuca werdermannii, Jarava tortuosa, Nassella duriuscula, N. pungens,* and *Poa paposana.*

In Coquimbo there is an increase of the taxonomic biodiversity, coupled with an increased collection effort. The number of taxa reaches 167, which corresponds to the highest for Poaceae in northern Chile. According to the Catalogue of the vascular flora of Coquimbo (Marticorena et al., 2001), the flora of this region (native and naturalized) includes 1727 species of which nearly 9% belong to Poaceae. An analysis of the biodiversity of Coquimbo (Squeo et al., 2001) mentions 104 species of Poaceae, the second most diverse family after Asteraceae in Coquimbo. An increase in the number of endemic grasses also takes place (Table 1). There are 26 endemic grasses belonging to 12 genera. The high degree of endemic species of Poaceae coincides with the high degree of endemism present in this region (53.5% of the vascular flora is endemic to Chile) (Squeo et al., 2001). The subfamily Bambusoideae has its septentrional limit of distribution in the region of Coquimbo. In Fray Jorge, there are 11 species of Poaceae, including *Chusquea cumingii* (Arancio et al., 2004), which represents the northernmost record of subfamily Bambusoideae in Chile.

Central Chile includes the regions of Valparaíso, Metropolitana, O'Higgins, Maule and Bío-Bío, approximately between 32° and 36°S. The high degree of anthropic pressure in central Chile is reflected in the greatest number of introduced species (approx. 10-16% of grasses are adventives). Due to the high percentage of endemic species, the high percentage of introduced species is disquieting as most of them are invasive weeds (Matthei, 1995). As has been pointed out, invasive species play a major role in displacing native plants and are the second leading threat to biodiversity following habitat destruction (Holcombe et al., 2010). It has been suggested that approximately 690 species of alien plants have been introduced and became naturalized in Chile, 507 of which reside in the Mediterranean area of central Chile; Poaceae is the most important family, with 151 species (Arroyo et al., 2000).

Easter Island (Isla de Pascua, Rapa Nui) belongs to the Region of Valparaíso in central Chile (27°9′S, 109°27′W) and lies 3700 km off the Chilean coast. The island has a small amount of vegetation. Currently, there are 22 genera and about 200 species of seed plants, most of them introduced (Mann et al., 2003). Introduced grasses (Poaceae) dominate the vegetation while native species number about forty-six. Only eight endemic species remain. All trees are introduced species. The endemic tree *Sophora toromiro* Skottsb. (Fabaceae) disappeared in the 1950's, probably due to animal grazing and anthropogenic impact. A major effort to reintroduce this species on the island has been done by CONAF (Corporación Nacional Forestal) and Jardín Botánico Nacional of Chile. Among grass species *Rytidosperma paschalis* is endemic. Introduced grasses in Easter Island are *Agrostis stolonifera, Austrostipa scabra* (sometimes cited as *Stipa horridula* Pilg.), *Avena fatua, Briza minor, Cenchrus clandestinus, C. echinatus, Chloris gayana, Cynodon dactylon, Dichelachne micrantha, Digitaria ciliaris, D. setigera, D. violascens, Eleusine indica, Eragrostis atrovirens, Gastridium phleoides, Hordeum murinum, Lachnagrostis filiformis, Lolium perenne, Melinis repens, Poa annua, Setaria parviflora, Sorghum halepense, Sporobolus indicus, S. africanus* (cultivated for grazing purpose), and *Vulpia myuros.* Native grasses: *Axonopus compressus* (sometimes cited under the name *A. paschalis*) *Bromus catharticus, Paspalum dasypleurum, P. dilatatum, P. forsterianum,* and *P. scrobiculatum* var. *orbiculare* (Giraldo-Cañas, 2008; Mann et al., 2003; Markgraf, 2003; Matthei, 1995; Skottsberg, 1956; Steadman, 1995).

In the archipelago of Juan Fernández (Region of Valparaíso), Poaceae is represented by approximately 35 genera and 61 species. Most of the species belong to the subfamily Pooideae (44 spp., 72%) and Panicoideae (9 spp., 14%). Only two Bambusoideae are found, *Chusquea fernandeziana,* endemic to Masatierra, and *C. culeou* (Baeza et al., 2007), one cultivated species of the subfamily Arundinoideae (*Arundo donax*), two species of Danthonioideae (*Danthonia chilensis* and *D. malacantha*), and two species of Chloridoideae (*Cynodon dactylon* and *Eleusine tristachya*). Five species are endemic: *Agrostis masafuerana* Pilg. (Masafuera), *Chusquea fernandeziana* (Masatierra), *Megalachne berteroana* and *M. masafuerana* (Masatierra and Masafuera), and *Podophorus bromoides* possibly extinct (Baeza et al, 2002; Baeza et al., 2007). *Polypogon imberbis* is considered endemic to Juan Fernández, but occasionally found in continental Chile (Müller, 1985). *Megalachne* and *Podophorus* are endemic genera.

The regions with greater diversity of Poaceae in Central Chile were Bío-Bío, Metropolitan, and Valparaíso, with 263, 198, and 186 taxa, respectively. According to our data, 57 species of grasses from 21 genera of grasses endemic to Chile are found in the area of the Chilean hotspot. On the other hand, O'Higgins is the most weakly collected region in Central Chile. Our data show that only 55% of the estimated grass flora in this region is represented in the Herbarium collections.

Southern Chile includes the regions of Araucanía, Los Ríos and Los Lagos, from approx. 37°S to 43°S. The Region of Araucanía is the best collected. There are 188 known taxa of Poaceae, of which 57 are introduced and 14 are endemic: *Anthoxanthum altissimum, A. spicatum, Bromidium trisetoides, Chusquea quila, Danthonia araucana, D. chilensis* var. *aureofulva, Deschampsia monandra, Gymnachne koelerioides, Melica violacea, Nassella duriuscula, N. juncea*

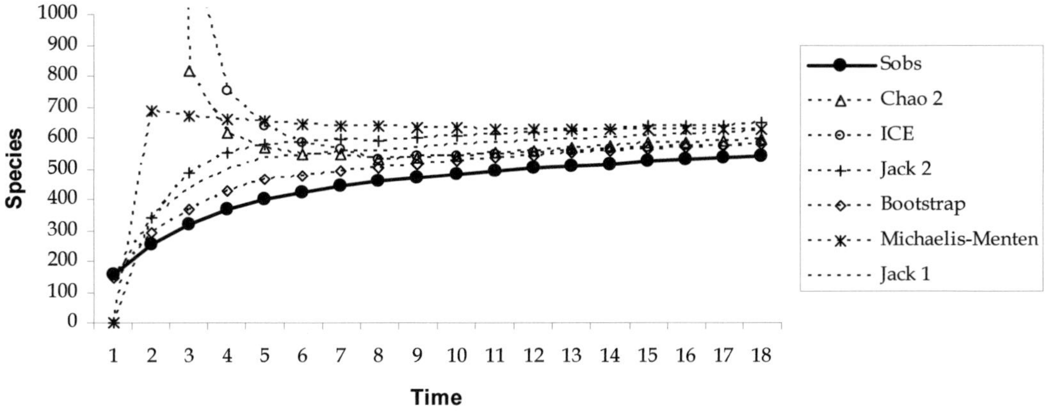

Fig. 3. Species accumulation curve of observed species (Sobs) in Chile and curves of estimated number of species calculated using the indices Chao 2, ICE, Jacknife 1, Jacknife 2, Bootstrap y Michaelis-Menten.

and *N. macrathera, Phalaris amethystina,* and *Poa cumingii*. In the region of Los Ríos 143 taxa were collected, 46 of which are introduced and 13 are endemic to Chile. Endemic species are the following: *Alopecurus lechleri, Anthoxanthum altissimum, Chusquea macrostachya, C. montana, C. quila, C. uliginosa, Danthonia araucana, D. chilensis* var. *aureofulva, Melica violacea, Nassella duriuscula, N. juncea* and *N. macrathera,* and *Poa cumingii*. In Los Lagos, 131 taxa were recorded, 38 introduced species, and eight endemic species (*Agrostis insularis, Anthoxanthum altissimum, Chusquea macrostachya, C. montana, C. quila, C. uliginosa, Poa cumingii,* and *Polypogon linearis*).

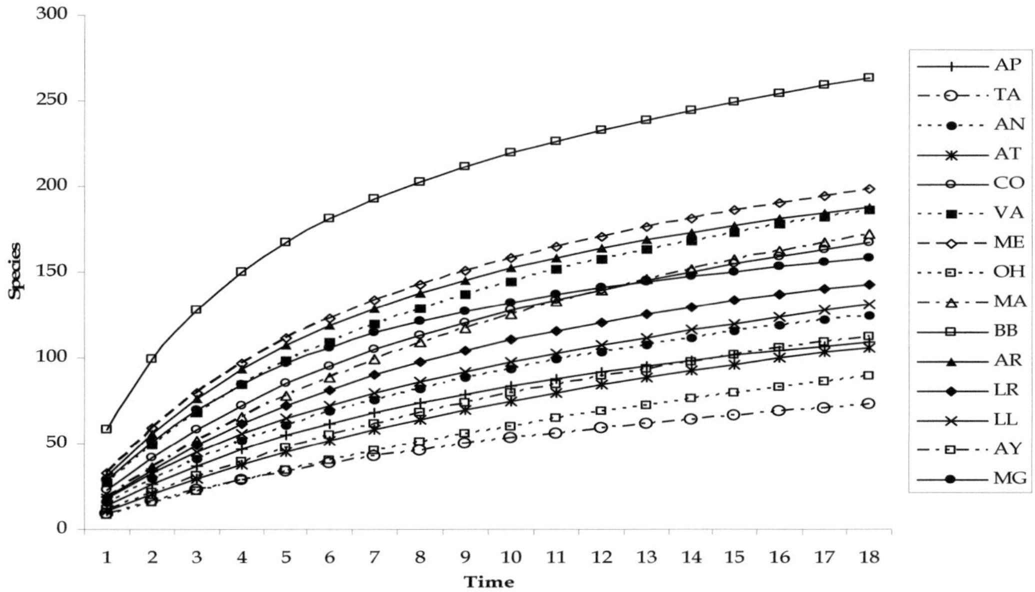

Fig. 4. Species accumulation curves of observed species (Sobs) in the fifteen political regions of Chile. See Table 1 for abbreviations of the regional names.

Austral Chile, also known as Chilean Patagonia, includes the regions of Aysén and Magallanes and extends from about 43°S to the south. While Aysén has been weekly collected, Magallanes is one of the best known from a floristic point of view (Fig. 5). In Aysén 112 grass species have been collected (Table 1), including 77 native species, 19 introduced species and 4 endemic species (*Anthoxanthum altissimum, Chusquea montana, C. quila*, and *C. uliginosa*). In Magallanes 158 species have been collected (Table 1). Endemic species in this region are *Alopecurus heleochloides, Festuca magensiana, Hordeum brachyatherum*. Introduced species number 38 and native species number 117.

5. Taxonomic diversity of Chilean grasses

In Chile, Poaceae number approximately 523 species and 57 infraspecific taxa, distributed in 122 genera (Table 2), representing about 10.1% of the Chilean flora. Poaceae is the second most diverse family of angiosperms in Chile after Asteraceae with about 863 species (Moreira-Muñoz & Muñoz-Schick, 2007). Of the 13 subfamilies of Poaceae, eight are present in Chile. As expected, most of the Chilean grasses (388 spp., 75%) belong to the subfamily Pooideae, followed by a few Panicoideae (59 spp., 10%), and Chloridoideae (43 spp., 7.9%). Species are distributed in 17 tribes and 43 subtribes. Three-hundred and fifty six species (68%) are native, 58 species (11%) are endemic and 109 species and 12 varieties (21%) are introduced. Percentages of native, introduced and endemic species in each political region are shown in Fig. 6. Endemic species belong to 23 genera in four subfamilies: Bambusoideae (7 spp.), Chloridoideae (1 sp.), Danthonioideae (3 spp.), and Pooideae (47 spp.).

Subfamily	Tribes	Subtribes	Genera	Species	Infraspecific taxa	Total taxa	Total %
Aristidoideae	1	1	1	3	0	3	0.52
Arundinoideae	1	1	2	2	0	2	0.35
Bambusoideae	1	1	1	10	1	11	1.89
Chloridoideae	3	9	16	43	3	46	7.93
Danthonioideae	1	1	4	20	4	24	4.13
Ehrhartoideae	2	2	2	2	0	2	0.35
Panicoideae	2	10	19	55	4	59	10.17
Pooideae	6	18	77	388	45	433	74.65
Total	17	43	122	523	57	580	100.00

Table 2. Number of tribes, subtribes, genera, species, and infraspecific taxa in the eight subfamilies of Poaceae present in Chile.

The arrangement of the genera of Poaceae according to the classification of (Soreng et al., 2009) is shown in Table 3. Under each subfamily each species is mentioned. Introduced species are indicated with and asterisk (*), and endemic species are bold faced.

5.1 BEP clade

The three subfamilies of the BEP clade grow in Chile (Bambusoideae, Ehrhartoideae, Pooideae) but only two (Bambusoideae and Pooideae) include native species. The BEP clade comprises the majority of the Chilean grasses (ca. 450 spp, ca. 80%).

5.1.1 Bambusoideae

Bambusoideae are represented in Chile only by the genus *Chusquea* (Bambuseae, Chusqueinae). This diverse genus comprises some 134 described species; notwithstanding, some 70 species remain undescribed (Judziewicz et al. 1999). The genus *Chusquea* is exclusively American, growing from Mexico to Chile and Argentina, from approximately 24°N to 47°S and from see level to approximately 4000 m of altitude. Ten species and one variety grow in Chile, seven species are endemic (Tables 2 and 3). In Chile, the species of *Chusquea* are usually associated to forest margins, from approx. 30°S (Coquimbo) to 46°40'S (Aysén), and from the see level to 2300 m in the Andes. The most widely distributed species in Chile is *C. culeou* (coligüe), living from Choapa (31 ° S) to Aysén (45 °S) between the see level to ca. 2000 m of altitude; this species lives also in Argentina. Common to Chile and Argentina are also *C. andina*, *C. montana* f. *montana* (tihuén) and *C. valdiviensis* (quila del sur). Although *C. quila* (quila) is considered endemic to Chile, some botanists considered it a synonym of *C. valdiviensis* (Nicora, 1978; Parodi, 1945). Even though *C. quila* has been collected between Valparaíso (33°S) and Aysén (44°S) it has its main distribution between Valparaiso and Ñuble (approx. 36°S), while *C. valdiviensis* is distributed mainly from the Araucanía Region (38°S) to the south (Chiloé, 43°S). *Chusquea andina* lives in the central-southern regions of Bío-Bío and Araucanía, above the tree line in the Andes. *Chusquea montana* f. *montana* grows between Ñuble (36°S) and Chiloé (46°S), as well as *C. valdiviensis*. Endemic to Chile are *C. ciliata*, *C. cumingii*, *C. fernandeziana*, *C. macrostachya* (quila), *C. montana* f. *nigricans* (quila enana), *C. quila* and *C. uliginosa*. *Chusquea ciliata* is endemic to the Region of Valparaíso. *Chusquea fernandeziana* is endemic to the Robinson Crusoe or Juan Fernández archipelago, where is found in Masatierra on outcrops between rocks, ravines or forest, usually isolated and sparse (Baeza et al., 2002). When Munro describes *C. ligulata* from Cundinamarca (Colombia), he includes under this name a sterile specimen collected by Bertero in Juan Fernández that possibly corresponds to *C. fernandeziana* (Parodi, 1945). This is, probably, the reason why *C. ligulata* is included sometimes as a synonym of *C. fernandeziana*. However, *C. ligulata* is closer to or conespecific with *C. sneidernii* Asplund of section Longiprophyllae, not to *C. fernandeziana* (Clark, 1990). *Chusquea cumingii* grows between Limarí (Region of Coquimbo, 30 °S) and Ñuble (36 °S), usually below 1500 m of altitude. *Chusquea macrostachya* has been collected between Santiago and Chiloé (33-43°S) but it is found more frequently in the southern regions of the country into the forest under the canopy or in canopy gaps as well as in roadsides. *Chusquea uliginosa* (quila de los ñadis), is found from Valdivia to Aysén (39-44°S), below 1500 m of altitude, in the central valley. The species of *Chusquea* are known in Chile by the vernacular names "coligües" or "quilas". Indigenous people ("mapuches") used the culms of *C. culeou* (coligüe) to build partitions inside their houses ("rucas"), musical instruments ("trutrucas") or fences, and used quilas (*C. quila*, *C. cumingii*), as forage for livestock. Currently, craftsmen use coligües to build furniture. Although coligües are not widely used in the industry, some properties like specific gravity, fiber length and chemical constitution suggest that this plant could be used as raw material for paper and for particle or fiber board production (Poblete et al., 2009).

5.1.2 Ehrhartoideae link

In Chile, Ehrhartoideae are absent from the native grass flora, but both the cultivated (rice) and wild rice (red rice, *Oryza sativa*) are found. Approximately 25,000 ha of rice are grown in a small south-central area of the country, located between O'Higgins (34°S) and Bío-Bío

(36°S), which is the southernmost region of the rice crop in the world. In the same area, wild red rice is found, one of the most problematic weed of rice production in temperate countries (Gealy et al., 2003). *Ehrharta calycina* is native to southern Africa (Barkworth et al., 2007); it has been collected in pastures in Elqui and cultivated in Rinconada de Maipú Experimental Station of University of Chile, Metropolitan Region.

5.1.3 Pooideae benth

In Chile, Pooideae encompasses six tribes and 18 subtribes with nearly 396 species in 76 genera (Table 1). Pooideae are distributed from the north end of Chilean territory (17°30'S) to the Region of Magallanes and Antartica Chilena (62°S), and from see level to 5250 m of altitude. Species of *Anthochloa* (*A. lepidula*), *Catabrosa* (*C. werdermannii*), *Dielsiochloa* (*D. floribunda*), *Deyeuxia* (*D. breviaristata, D. cabrerae* var. *trichopoda*, D. *crispa, D. deserticola*), *Festuca* (*F. orthophylla*), *Nassella* (*N. nardoides*), *Pappostipa* (*P. frigida*), *Poa* (*P. humillima, P. gymnantha*) are found at high altitude (approx. 4000-5250 m), in northern Chile [Region of Arica and Parinacota (17-18°S) and Region of Antofagasta (21-22°S)]. Some genera, like *Anthochloa, Dielsiochloa,* and *Dissanthelium,* are restricted to northern Chile (18-21°S). Some taxa are restricted to central Chile, like the endemic and monotypic *Gymnachne* (*G. koelerioides*) (32°S, Quillota, Region of Valparaíso to 38°S, Cautín, Region of Araucanía), *Helictotrichon bulbosum* (restricted to Ñuble and Concepción, Region of Bío-Bío), *Apera interrupta* (33-38°S, Santiago to Malleco), *Calotheca brizoides* (34-38°S). Genera whose distribution is restricted mainly to southern Chile are *Vahlodea* (Santiago but mainly from Valdivia to Puerto Williams), *Alopecurus* (Choapa but mainly from Santiago to Magallanes), *Anthoxanthum* (mainly from Ñuble to Tierra del Fuego). The largest genera are *Poa* (48 spp.), *Nassella* (ca. 30 spp.), *Agrostis* (28 spp.), *Festuca* (26 spp.), *Bromus* (23 spp.), *Deschampsia* (15 spp), *Jarava* (13 spp.) and *Trisetum* (11 spp.). Several genera are monotypic (*Dielsiochloa, Dichelachne, Leptophyllochloa, Gymnachne, Hainardia, Podagrostis*).

Tribe **Brachypodieae** is represented only by one genus and one species, *Brachypodium distachyon* introduced from southern Europe, this species grows in roadsides in central Chile, from the Region of Valparaíso to the Region of Bío-Bío (32-37°S), as well as in Juan Fernández archipelago. Tribe **Meliceae** comprises the genera *Gyceria* with two species and *Melica* represented by eight species (Table 3). *Glyceria* is represented only by the Eurasiatic *G. fluitans* (L.) R. Br., and a native species *G. multiflora* Steud., inhabiting from Chacabuco (33°S) to Última Esperanza (50°S). *Melica* comprises eight species (including *M. cepacea* sometimes classified in the genus *Bromelica*), seven endemic. Genus *Melica* is distributed mainly in central-southern Chile, between 27°50'S and 40°15'S, and from sea level to 2000 m of altitude (Muñoz-Schick, 1985). Three species are considered vulnerable: *M. longiflora, M. paulsenii,* and *M. poecilantha* (Squeo et al., 2001). Tribe **Stipeae** includes 21 genera and more than 500 species of which 279 live in the New World in temperate regions in both hemispheres, growing mainly in dry open grasslands and steppe communities. Most species of the New World Stipeae (approx. 80%) are South American (Romaschenko et al., 2008). Tribe Stipeae includes nine genera in Chile. *Amelichloa* Arriaga & Barkworth was described recently on the basis of five species segregated from genus *Stipa,* three of which grow in central-southern Chile (Arriaga & Barkworth, 2006). *Anatherostipa* (Hack. ex Kuntze) Peñailillo, was segregated from *Stipa* (Peñailillo, 1996), to include 11 species, four of which grow in Chile, restricted to the Andes of the northern regions (17-25°S) from 3400 to 4600 m of altitude. The South American genus *Jarava* Ruiz & Pav., includes 13 species widely distributed in Chile, mainly from the Andes and Patagonia, between 18°S and 50°S. *Nassella,* being the largest genus of the tribe Stipeae in Chile, encompasses about 30 species, 10 of which are endemic. The genus *Nassella* ranges, approximately, between 17°S and 43°S. Two species of genus *Ortachne, O. breviseta* and *O. rariflora* inhabit the subantartic forests. *Ortachne breviseta* grows between Valdivia and Llanquihue (39-41°S) and *O. rariflora* between Chiloé and Tierra del Fuego (42-55°S), and from see level to 1700 m of altitude. The genus *Pappostipa* includes six species in Chile distributed between Arica (18°S) and Magallanes (53°S). *Pappostipa chrysophylla, P. ibari* and *P. humilis* are restricted to or have its main distribution in southern Chile. On the contrary, *P. atacamensis* and *P. frigida* are distributed primarily in northern Chile. *Pappostipa speciosa* is widely distributed in Chile and it is found also in the northern hemisphere. The genus *Piptatherum* is introduced. *Piptochaetium* includes seven species distributed from Limarí (Coquimbo) to Osorno (Los Lagos). Tribe **Poeae** is by far the largest in species number, with 58 genera and more than 290 species or 72%. The largest genera are *Poa* (48 spp.), *Agrostis* (28 spp.), *Festuca* (27 spp.), *Deyeuxia* (26 spp.), *Deschampsia* (15 spp), *Trisetum* (11 spp.), and *Anthoxanthum* (8 spp.). A large number of genera in this tribe are represented by only one or two species, such as *Bromidium, Calotheca, Chascolytrum, Dielsiochloa, Helictotrichon, Gymnachne, Leptophyllochloa, Megalachne, Podophorus, Relchela, Rhombolytrum,* etc. Twenty-five taxa of Poeae are endemic to Chile: *Agrostis umbellata, A. masafuerana, A. arvensis, A. insularis, Alopecurus heleochloides, A. lechleri, Anthoxanthum altissimum, A. spicatum, Bromidium trisetoides, Deschampsia setacea, D. looseriana, D. monandra, Festuca magensiana, F. morenensis, F. panda, F. tunicata, F. werdermannii, Gymnachne koelerioides, Phalaris amethystina, Poa cumingii, P. paposana, P. pfisteri, Polypogon linearis, Trisetum johnstonii* subsp. *mattheii,* and *T. nancaguense.* Tribe **Triticeae** includes some economically important species, such as wheat, rye, triticale, and barley, grown in the temperate regions of the world. In Chile, Triticeae comprises seven genera: *Agropyron,* represented only by *A. cristatum* an introduce species growing only in Magallanes, *Elymus, Hordeum, Leymus, Secale, Taeniatherum and Triticum***.** Only *Hordeum* and *Elymus* contain native species and only one endemic species is found in this tribe, *Hordeum brachyaterum.* Tribe **Bromeae** includes only the genus *Bromus,* with 24 species. *Bromus gunckelii* is an endemic species of northern Chile. *Bromus mango* was used by indigenous people (mapuches) as a cereal grain. This species is probably extinct (Mösbach, 1999).

5.2 PACMAD clade

Five subfamilies of Chilean Poaceae belong to the PACMAD clade: Aristidoideae, Arundinoideae, Danthonioideae, Panicoideae and Chloridoideae.

5.2.1 Aristidoideae Caro

In Chile, only the genus *Aristida* is present, with three native species. Genus *Aristida* is distributed from Arica and Parinacota (ca. 18°S) to Araucanía (Malleco, 37°50'S). *Aristida adscensionis* is found in northern-central Chile (approx. 18°S-33°S), and *A. pallens* and *A. spegazzinii* in central-southern Chile (approx. 35-37°S). It has been established (Matthei, 1987) that *A. longiseta* previously cited for Chile (Marticorena & Quezada, 1985) is absent in Chile.

5.2.2 Arundinoideae Burmeist

Arundinoideae comprises two species, *Arundo donax* and *Phragmites australis;* both species are tall reeds of wet places. *Phragmites australis* is a weed of rice fields (Matthei, 1995); *A. donax* is sometimes cultivated as ornamental.

5.2.3 Danthonioideae P.H. Linder & N.P. Baker

Subfamily Danthonioideae is represented in Chile by four genera (*Cortaderia, Danthonia, Rytidosperma* and *Schismus*), and 19 species. Some species of *Cortaderia*, known as "cola de zorro" or "Ngerü-quëlen" (Mösbach, 1999) are used as ornamental plants. Species of *Schismus* (*S. arabicus* and *S. barbatus*) are ruderal and agricultural weeds in Central Chile (Matthei, 1995). *Danthonia* comprises four species and two varieties distributed from the region of Coquimbo (32°S) to the region of Los Lagos (Chiloé, 42°50'S). *Rytidosperma* includes five species in Chile, from the region of Coquimbo (Limarí, 30°40'S) to Magallanes (Tierra del Fuego, 52°50'S). *Cortaderia* includes six species distributed from Arica and Parinacota (18°S) to Magallanes (Wollaston island, 55°44'S). *Danthonia araucana, D. chilensis* var. *aureofulva*, and *R. quirihuense* are endemic species.

5.3 Panicoideae Link

Panicoideae consists of two tribes, 10 subtribes, 18 genera and about 50 species in Chile (Table 2). Tribe **Andropogoneae** is represented by the genera *Cymbopogon* (*C. citratus*) in northern Chile, *Schizachyrium* (*S. sanguineum, S. spicatum*) in Central Chile, *Imperata* (*I. condensata*) from Copiapó to Valdivia, *Miscanthus* (*M. sinensis*), cultivated in Central Chile, *Bothriochloa* (*B. laguroides* and *B. saccharoides* in Central Chile, *B. ischaemum* in Easter Island), *Sorghum* (*S. bicolor, S. halepense*), widely distributed in northern and central Chile (*S. halepense* is a serious agricultural weed from Arica and Parinacota to the region of Araucanía and Easter Island) (Matthei, 1995). *Zea mays* is cultivated (corn), but also commonly found in roadsides. Tribe **Paniceae** is represented by some 39 species in 11 genera, most of them have been mentioned as summer weed of agricultural fields in Chile. Genus *Cenchrus* contains seven species, from northern and central Chile, some of them reported as common weeds (*C. incertus, C. clandestinus, C. myosuroides, C. chilensis, C. longisetus*). In addition, all species of genus *Digitaria* in Chile have been reported as weeds in agricultural fields, however, the most important because of its geographical distribution is *D. sanguinalis*, living from Huasco (Atacama) to Valdivia (Los Rios) and Juan Fernández. *Digitaria setigera* and *D. violascens* are introduced weeds in Easter Island (Matthei, 1995). *Eriochloa montevidensis* has been reported only from Valle de Azapa (Arica) as a weed in agricultural and ruderal fields. *Melinis repens* was also reported as a weed species introduced in Easter Island (Matthei, 1995). *Dichanthelium sabulorum* occurs in central-southern Chile (Maule, Bío-Bío). *Echinochloa* includes three species (*E. colona, E. crus-galli*, and *E. crus-pavonis*), all of them important weeds of rice fields. Genus *Panicum* includes four or five species, two of them weeds of ruderal and agricultural places, *P. capillare* and *P. dichotomiflorum*.

Ten species of *Paspalum* have been reported, most of them as summer weeds in corn, sugar beets, rice, and other crops (Matthei, 1995). Genus *Setaria* also includes several important weed species in Chile, growing in vineyards, sugar beet, orchards, etc: *S. parviflora, S. pumila, S. verticillata, S. viridis.*

5.4 Chloridoideae Kunth ex Beilschm

Subfamily Chloridoideae comprises 16 genera and 41 species. Most of the genera are represented in Chile by only 1, 2 or 3 species. The largest genus is *Eragrostis*, comprising 12 species and two varieties (Escobar et al., 2011). Chloridoideae are present in northern, central and Southern Chile but they are rare in Austral Chile. *Bouteloua* comprises only one Arica and Parinacota, in semi-desert slopes of the Andean foothills of Arica, above 2000 m of

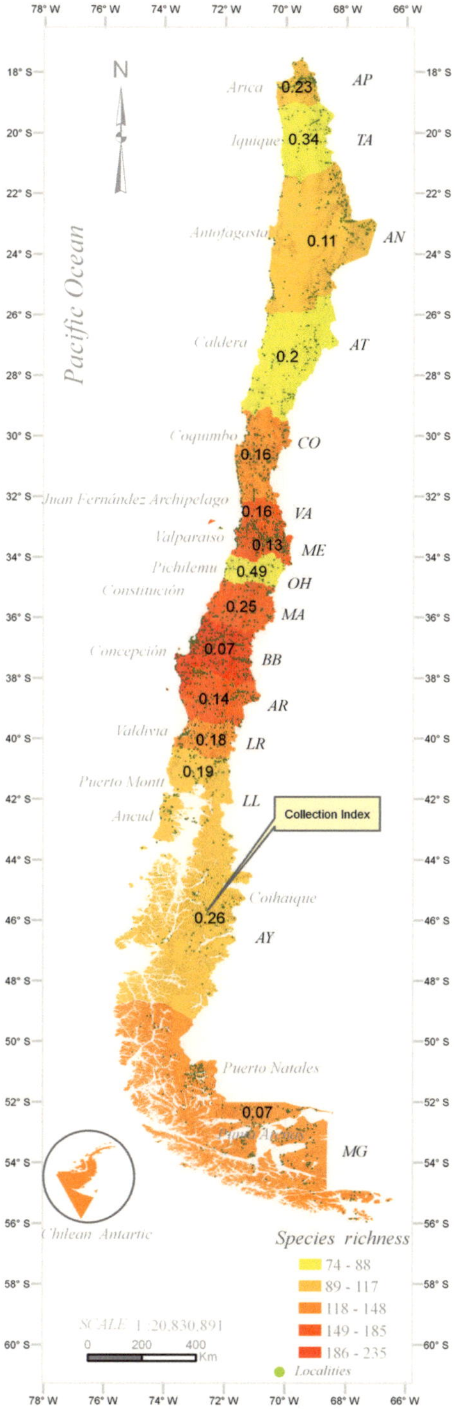

Fig. 5. Collection effort and species richness of Poaceae in the fifteen Chilean political regions. Each point represents at least one specimen.

altitude (Matthei, 1973). Chloris comprises three species; all introduced (*C. gayana, C. radiata* and *C. virgata*). *Chloris gayana* is native from Africa (Senegal); in Chile it is found only in Easter Island (Anderson, 1974; Matthei, 1995). The other two species grow in Northern Chile. *Cynodon nitidus* is endemic to Chile. *Cynodon dactylon* is a common weed in ruderal places, both in continental Chile and Easter Island. *Eleusine indica* is found only in Easter Island. *Eleusine tristachya* is a common weed of gardens as well as orchards and meadows in Central Chile. Genus *Eragrostis* comprises 12 species, eight of them native and four introduced.

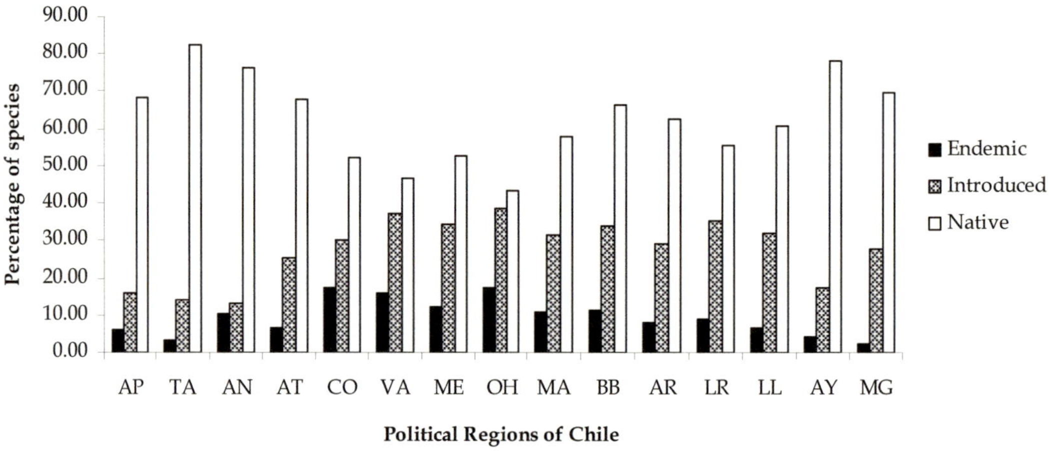

Fig. 6. Number of Endemic, introduced, and native non-endemic species in the fifteen political regions of Chile.

Fig. 7. *Chusquea culeou* in Aysén, Chile (A. Solís).

Eragrostis pycnantha, E. peruviana, E. weberbaueri, E. nigricans, and *E. kuschelii* are restricted to Northern Chile. *Eragrostis peruviana* and *E. kuschelii* grow in Islas Desventuradas (approx. 26°S). *Eragrostis atrovirens* and *E. tenuifolia*, introduced from Eurasia, grow in Easter Island. *Eragrostis polytricha* grows in Central Chile. Spartina densiflora (austral cordgrass), a species probably native from the east coast of South-America (Bortolus, 2006), has been reported as an invasive species in North America, Europe and Africa (Ayres et al., 2004); it grows in

Chile in salt marshes, from Concepción (Bío-Bío) to Chiloé (Los Lagos) as the dominant species of the association Sarcocornio-Spartinetum densiflorae (San Martín et al., 2006). In the same habitats are usually found species of genus Distichlis; D. spicata, can also grow as a weed of orchards and ruderal places (Matthei, 1995).

6. Conclusions

The family Poaceae is represented in Chile by 523 species in 122 genera and eight subfamilies. Fifty eight species from 23 genera are endemic. *Megalachne* and *Gymnachne* are endemic to Chile. Endemic species represent about 11% of the Chilean grass flora. More than 20% are introduced species. Our data, based mostly on the collections of the two most important Chilean herbaria (CONC, SGO), indicate that the present knowledge of the Chilean Poaceae is good enough. The observed species richness reaches over 88% of the estimated species richness. However, the collection index calculated for the different political regions indicates a weak collection effort in some regions (e.g. O'Higgins, Tarapacá and Aysén). New expeditions to these regions are necessary to complete the inventory, because the knowledge of the species richness depends directly on its representation in the herbaria collections.

SUBFAMILY BAMBUSOIDEAE LUERSS.

Tribe Bambuseae Dumort.

Subtribe Chusqueinae Soderstr. & R.P. Ellis

1. ***Chusquea*** Kunth: *C. andina* Phil., ***C. ciliata*** Phil., *C. culeou* E. Desv., ***C. cumingii*** Nees, ***C. fernandeziana*** Phil., ***C. macrostachya*** Phil., *C. montana* Phil. fma. *montana*, **C. *montana*** fma. ***nigricans***, ***C. quila*** Kunth, ***C. uliginosa*** Phil. and *C. valdiviensis* E. Desv.

SUBFAMILY EHRHARTOIDEAE LINK

Tribe Ehrharteae Nevski

2. ***Ehrharta*** Thunb.: **E. calycina* Sm.

Tribe Oryzeae Dumort.

3. ***Oryza*** L.: **O. sativa* L.

SUBFAMILY POOIDEAE BENTH.

Tribe Brachypodieae Harz

4. ***Brachypodium*** P. Beauv.: **B. distachyon* (L.) P. Beauv.

Tribe Meliceae Link ex Endl.

5. ***Glyceria*** R. Br.: **G. fluitans* R. Br., *G. multiflora* Steud.
6. ***Melica*** L.: ***M. argentata*** E. Desv., *M. cepacea* (Phil.) Scribn., ***M. commersonii*** Nees ex Steud., ***M. longiflora*** Steud., ***M. mollis*** Phil., ***M. paulsenii*** Phil., ***M. poecilantha*** E. Desv., ***M. violacea*** Cav.

Tribe Stipeae Dumort.

7. ***Amelichloa*** Arriaga & Barkworth: *A. brachychaeta* (Godr.) Arriaga & Barkworth, *A. brevipes* (E. Desv.) Arriaga & Barkworth, *A. caudata* (Trin.) Arriaga & Barkworth
8. ***Anatherostipa*** (Hack. ex Kunze) Peñailillo: *A. bomanii* (Hauman) Peñailillo, *A. mucronata* (Griseb.) F. Rojas, *A. rigidiseta* (Pilg.) Peñailillo, ***A. venusta*** (Phil.)

Peñailillo

9. ***Autrostipa*** S.W.L. Jacobs & Everett: **A. scabra* (Lindl.) S.W.L. Jacobs & J. Everett

10. ***Jarava*** Ruiz & Pav.: *J. annua* (Mez) Peñailillo, *J. ichu* Ruiz & Pav., *J. ichu* var. *pungens* (Nees & Meyen) Ciald., *J. leptostachya* (Griseb.) F. Rojas, ***J. mattheii*** F. Rojas, *J. neaei* (Nees ex Steud.) Peñailillo, *J. plumosa* (Spreng.) S.W.L. Jacobs & J. Everett, *J. plumosula* (Nees ex Steud.) F. Rojas, *J. pogonathera* (E. Desv.) Peñailillo, *J. psilantha* (Speg.) Peñailillo, *J. pungionata* (Caro & E.A. Sánchez) Matthei, *J. subaristata* (Matthei) Matthei, ***J. tortuosa*** (E. Desv.) Peñailillo

11. ***Nassella*** (Trin.) E. Desv.: *N. arcuata* (R.E. Fr.) Torres, *N. asplundii* Hitchc., *N. chilensis* (Trin.) E. Desv., ***N. coquimbensis*** (Matthei) Peñailillo, *N. depauperata* (Pilg.) Barkworth, ***N. duriuscula*** (Phil.) Brakworth, *N. elata* (Speg.) Torres, *N. filiculmis* (Delile) Barkworth, ***N. gibba*** (Phil.) Muñoz-Schick, *N. gigantea* (Steud.) Muñoz-Schick, ***N. hirtifolia*** (Hitchc.) Barkworth, ***N. juncea*** Phil., ***N. lachnophylla*** (Trin.) Barkworth, *N. laevissima* (Phil.) Barkworth, *N. linearifolia* (E. Fourn.) R.W. Pohl, *N. longiglumis* (Phil.) Brakworth, ***N. macrathera*** (Phil.) Barkworth, *N. manicata* (E. Desv.) Barkworth, *N. nardoides* (Phil.) Barkworth, *N. neesiana* (Trin. & Rupr.) Barkworth, ***N. parodii*** (Matthei) Barkworth, ***N. pfisteri*** (Matthei) Barkworth, *N. phillippii* (Steud.) Barkworth, *N. poeppigiana* (Trin. & Rupr.) Barkworth, *N. pubiflora* (Trin. & Rupr.) E. Desv., ***N. pungens*** E. Desv., *N. rupestris* (Phil.) Torres, *N. tenuis* (Phil.) Barkworth, *N. tenuissima* (Trin.) Barkworth

12. ***Ortachne*** Nees ex Steud.: *O. breviseta* Hitchc., *O. rariflora* (Hook. f.) Hughes

13. ***Pappostipa*** (Speg.) Romasch., P.M. Peterson & Soreng: *P. atacamensis* (Parodi) Romasch., *P. chrysophylla* (E. Desv.) Romasch. var. *chrysophylla*, *P. chrysophylla* var. *cordilleranum* (Parodi) Romasch., *P. frigida* (Phil.) Romasch., *P. humilis* (Cav.) Romasch., *P. ibarii* (Phil.) Romasch., *P. speciosa* (Trin. & Rupr.) Romasch., *P. vaginata* (Phil.) Romasch.

14. ***Piptatherum*** P. Beauv.: **P. miliaceum* (L.) Coss.

15. ***Piptochaetium*** J. Presl: ***P. angolense*** Phil., *P. bicolor* (Vahl) E. Desv., *P. hirtum* Phil., *P. montevidense* (Spreng.) Parodi, *P. panicoides* (Lam.) E. Desv., *P. setosum* (Trin.) Arechav., *P. stipoides* (Trin. & Rupr.) Hack. ex Arechav. var. *stipoides*, *P. stipoides* var. *equinulatum* Parodi

Supertribe Poodae L. Liou

Tribe Poeae R. Br.

Subtribe Agrostidinae Fr.

17. ***xAgropogon*** P. Fourn.: **xA. lutosus* (Poir.) P. Fourn.

18. ***Agrostis*** L.: ***A. arvensis*** Phil., *A. brachyathera* Steud., *A. breviculmis* Hitchc., **A. capillaris* L., **A. castellana* Boiss. & Reut., **A. gigantea* Roth, *A. glabra* (J. Presl) var. *glabra*, *A. glabra* var. *melanthes* (Phil.) Rúgolo et De Paula, *A. idahoensis* Nash, *A. imberbis* Phil., *A. inconspicua* Kunze ex E. Desv., ***A. insularis*** Rúgolo & A.M. Molina, *A. kuntzei* Mez, *A. leptotricha* E. Desv., *A. magellanica* Lam., ***A. masafuerana*** Pilg., *A. mertensii* Trin., *A. meyeni* Trin., **A. nebulosa* Boiss. & Reuter, *A. montevidensis* Spreng., *A. perennans* (Walter) Tuck., *A. philippiana* Rúgolo & De Paula, *A. scabra* Willd., *A. serranoi* Phil., **A. stolonifera* L. var *stolonifera*, **A. stolonifera* var. *palustris*, A. *tolucensis* Kunth, *A. uliginosa* Phil., ***A. umbellata*** Colla, *A. vidalii* Phil., *A. vinealis* Schreb.

19. ***Ammophila*** Host: **A. arenaria* (l.) Link

20. ***Bromidium*** Nees & Meyen: *B. anomalum* (Trin.) Döll, ***B. trisetoides*** (Steud.) Rúgolo
21. ***Calamagrostis*** Adans.: **C. epigeios* (L.) Roth
22. ***Deyeuxia*** Clarion ex P. Beauv.: *D. breviaristata* Wedd., *D. brevifolia* J. Presl. var. *brevifolia, D. brevifolia* var. *expansa* Rúgolo & Villav., *D. cabrerae* (Parodi) Parodi var. *cabrerae, D. cabrerae* var. *aristulata* Rúgolo & Villav., *D. cabrerae* var. *trichopoda* Parodi ex Rúgolo, *D. chrysantha* J. Presl var. *chrysantha, D. chrysantha* var. *phalaroides* (Wedd.) Villav., *D. chrysophylla* Phil., *D. chrysostachya* E. Desv., D. crispa Rúgolo & Villav., *D. curvula* Wedd., *D. deserticola* Phil. var. *breviaristata* Rúgolo & Villav., *D. deserticola* Phil. var. *deserticola, D. diemii* Rúgolo, *D. eminens* J. Presl var. *eminens, D. eminens* var *fulva* (Griseb.) Rúgolo, *D. eminens* var. *discreta* Rúgolo & Villav., *D. erythrostachya* E. Desv. var. *erythrostachya, D. erythrostachya* var. *neuquenensis* Rúgolo, *D. filifolia* Wedd., *D. hackelii* (Lillo) Parodi, *D. heterophylla* Wedd., *D. patagonica* Speg., *D. poaeoides* (Steud.) Rúgolo, *D. rigescens* (J. Presl) Türpe, *D. rigida* Kunth, *D. spicigera* var. *cephalotes* (Wedd.) Rúgolo & Villav., *D. spicigera* J. Presl var. *spicigera, D. suka* (Speg.) Parodi, *D. velutina* Nees & Meyen var. *velutina, D. velutina* var. *nardifolia* (Griseb.) Rúgolo, *D. vicunarum* Wedd., *D. viridiflavescens* (Poir.) Kunth var. *viridiflavescens, D. viridiflavescens* var. *montevidensis* (Nees) Cabrera & Rúgolo, *D. viridis* Phil.
23. ***Dichelachne*** Endl.: *D. micrantha* (Cav.) Domin
24. ***Gastridium*** P. Beauv.: **G. phleoides* (Nees & Meyen) C.E. Hubb.
25. ***Lachnagrostis*** Trin.: *L. filiformis* (G. Forst.) Trin.
26. ***Lagurus*** L.: **L. ovatus* L.
27. ***Podagrostis*** (Griseb.) Scribn. & Merr.: *P. sesquiflora* (E. Desv.) Parodi ex Nicora
28. ***Polypogon*** Desf.: *P. australis* Brongn., *P. chilensis* (Kunth) Pilger, *P. elongatus* Kunth, *P. exasperatus* (Trin.) Renvoize, *P. imberbis* (Phil.) Johow, *P. interruptus* Kunth, ***P. linearis*** Trin., *P. maritimus* Willd., **P. monspeliensis* (L.) Desf., **P. viridis* (Gouan) Breistr.

Subtribe Airinae Fr.

29. ***Aira*** L.: **A. caryophyllea* L., **A. elegantissima* Schur, **A. praecox* L.
30. ***Avenella*** (Bluff & Fingerh.) Drejer: *A. flexuosa* (L.) Drejer
31. ***Corynephorus*** P. Beauv.: **C. divaricatus* (Pourr.) Breistr.
32. ***Deschampsia*** P. Beauv.: *D. airiformis* (Steud.) Benth. & Hook., *D. antarctica* E. Desv., *D. berteroana* (Kunth) Trin., *D. cespitosa* (L.) P. Beauv. var *cespitosa, D. cespitosa* var. *pulchra* (Nees & Meyen) Nicora, *D. cordillerarum* Hauman, *D. danthonioides* (Trin.) Munro, *D. elongata* (Hook.) Munro, *D. kingii* (Hook. F.) E. Desv., *D. laxa* Phil., ***D. looseriana*** Parodi, *D. monandra* Parodi, *D. parvula* (Hook. f.) E. Desv., *D. patula* (Pilg.) Pilg. ex Skottsb., ***D. setacea*** (Huds.) Hack., *D. venustula* Parodi
33. ***Vahlodea*** Fr.: *V. atropurpurea* (Wahlenb.) Fr. ex Hartm.

Subtribe Alopecurinae Dumort.

34: ***Alopecurus*** L.: *A. geniculatus* L. var. *geniculatus, A. geniculatus* var. *patagonicus* Parodi, ***A. heleochloides*** Hack., ***A. lechleri*** Steud., *A. magellanicus* Lam. var. *magellanicus, A. magellanicus* var *bracteatus* (Phil.) Mariano, *A. myosuroides* Huds., **A. pratensis* L.

Subtribe Aveninae J. Presl

35. ***Arrhenatherum*** P. Beauv.: **A. elatius* (L.) P. Beauv. Ex J. Presl & C. Presl subsp. *elatius*, **Arrhenatherum elatius* subsp. *bulbosum* (Willd.) Schübl. & G. Martens

36. ***Avena*** L.: **A. barbata* Pott ex Link, **A. fatua* L., **A. sativa* L., **A. sterilis* L., **A. strigosa* Schreb.
37. ***Helictotrichon*** Besser: *H. bulbosum* (Hitchc.) Parodi
38. ***Koeleria*** Pers.: *K. fueguina* C.E. Calderón ex Nicora, *K. kurtzii* Hack.
39. ***Leptophyllochloa*** C.E. Calderón: *L. micrathera* (E. Desv.) C.E. Calderón
40. ***Relchela*** Steud.: *R. panicoides* Steud.
41. ***Rostraria*** Trin.: **R. cristata* (L.) Tzvel., *R. trachyantha* (Phil.) Tzvel.
42. ***Trisetum*** Pers.: *T. ambiguum* Rúgolo & Nicora, *T. barbinode* Trin., *T. caudulatum* Trin., *T. cernuum* Trin., *T. flavescens* (L.) P. Beauv., *T. johnstonii* (Louis-Marie) Finot subsp. *johnstonii*, ***T. johnstonii*** **subsp.** ***mattheii*** (Finot) Finot, *T. longiglume* Hack. var. *longiglume*, ***T. nancaguense*** Finot, *T. preslei* (Kunth) E. Desv., *T. pyramidatum* Louis-Marie ex Finot, *T. spicatum* (L.) K. Richt. subsp. *spicatum*, *T. spicatum* subsp. *cumingii* (Nees ex Steud.) Finot, *T. spicatum* subsp. *dianthemum* (Louis-Marie) Finot, *T. spicatum* subsp. *phleoides* (d'Urv.) Macloskie

Subtribe Brizinae Tzvelev
43. ***Briza*** L.: **B. maxima* L., **B. minor* L.
44. ***Calotheca*** Desv.: *C. brizoides* (Lam.) Desv.
45. ***Chascolytrum*** Desv.: *C. subaristatum* (Lam.) Desv.
46. ***Gymnachne*** Parodi: ***G. koelerioides*** (Trin.) Parodi
47. ***Rhombolytrum*** Link: *R. rhomboideum* Link

Subtribe Coleanthinae Rchb.
48. ***Catabrosa*** P. Beauv.: *C. aquatica* (L.) P. Beauv., *C. werdermannii* (Pilg.) Nicora & Rúgolo
49. ***Puccinellia*** Parl.: *P. argentinensis* (Hack.) Parodi, *P. biflora* (Steud.) Parodi, *P. frigida* (Phil.) I.M. Johnst., *P. glaucescens* (Phil.) Parodi, *P. magellanica* (Hook. f.) Parodi, *P. preslii* (Hack.) Ponert, *P. pusilla* (Hack.) Parodi, *P. skottsbergii* (Pilg.) Parodi

Subtribe Cynosurinae Fr.
50. ***Cynosurus*** L.: **C. cristatus* L., **C. echinatus* L.

Subtribe Dactylidinae Stapf
51. ***Dactylis*** L.: **D. glomerata* L.
52. ***Lamarckia*** Moench: **L. aurea* (L.) Moench

Subtribe Holcinae Dumort.
53. ***Holcus*** L.: **H. lanatus* L.

Subtribe Loliinae Dumort.
54. ***Dielsiochloa*** Pilg.: *D. floribunda* (Pilg.) Pilg.
55. ***Festuca*** L.: *F. acanthophylla* E. Desv., *F. argentina* (Speg.) Parodi, *F. chrysophylla* Phil., *F. cirrosa* (Speg.) Parodi, *F. contracta* Kirk, *F. deserticola* Phil., *F. filiformis* Pourr., *F. gracillima* Hook. f., *F. hypsophila* Phil., *F. kurtziana* St.-Yves, *F. magellanica* Lam., *F. magensiana* Potztal, *F. monticola* Phil., *F. morenensis* Matthei, *F. nardifolia* Griseb., *F. orthophylla* Pilg., *F. pallescens* (St.-Yves) Parodi, ***F. panda*** Swallen, *F. purpurascens* Banks & Soland. ex Hook., *F. pyrogea* Speg., *F. rigescens* (J. Presl) Kunth, *F. rubra* L., *F. scabriuscula* Phil., *F. tectoria* St.-Yves, *F. thermarum* Phil., *F. tunicata* E. Desv., *F. weberbaueri* Pilg., *F. werdermannii* St.-Yves.
56. ***Lolium*** L.: **L. multiflorum* Lam., **L. perenne* L., *L. rigidum* Gaudin subsp. *rigidum*, **L. rigidum* subsp. *lepturoides* (Boiss.) Sennen & Mauricio, **L. temulentum* L.
57. ***Megalachne*** Steud.: ***M. berteroniana*** Steud., ***M. masafuerana*** (Skottsb. & Pilg.) Matthei

58. ***Podophorus*** Phil.: ***P. bromoides*** Phil.
59. ***Schenodorus*** P. Beauv.: **S. arundinaceus* (Schreb.) Dumort.
60. ***Vulpia*** C.C. Gmel.: *V. antucensis* Trin., **V. bromoides* (L.) Gray, **V. muralis* (Kunth) Nees, **V. myuros* (L.) C.C. Gmel. var *myuros*, **V. myuros* var. *hirsuta* Hack., **V. myuros* var. *megalura* (Nutt.) Auqiuer, *V. octoflora* (Walter) Rydb.

Subtribe Parapholiinae Caro
61. ***Catapodium*** Link: **C. rigidum* (L.) Dony
62. ***Hainardia*** Greuter: **H. cylindrica* (Willd.) Greuter
63. ***Parapholis*** C.E. Hubb.: **P. incurva* (L.) C.E. Hubb., **P. strigosa* (Dumort.) C.E. Hubb.

Subtribe Phalaridinae Fr.
64. ***Anthoxanthum*** L.: ***A. altissimum*** (Steud.) Veldkamp, *A. gunckelii* (Parodi) Veldkamp, *A. juncifolium* (Hack.) Veldkamp, **A. odoratum* L., *A. pusillum* (Hack. ex Dusén) Veldkamp, *A. redolens* (Vahl) P. Royen, ***A. spicatum*** (Parodi) Veldkamp, *A. utriculatum* (Ruiz & Pav.) Y. Schouten & Veldkamp.
65. ***Phalaris*** L.: ***P. amethystina*** Trin., *P. angusta* Nees ex Trin., **P. aquatica* L., **P. arundinacea* L., **P. canariensis* L., **P. caroliniana* Walter, **P. minor* Retz.

Subtribe Phleinae Benth.
66. ***Phleum*** L.: **P. alpinum* L., **P. pratense* L.

Subtribe Poinae Dumort.
67. ***Anthochloa*** Nees & Meyen: *A. lepidula* Nees & Meyen.
68. ***Apera*** Adans.: **A. interrupta* (L.) P. Beauv.
69. ***Dissanthelium*** Trin.: *D. peruvianum* (Nees & Meyen) Pilg.
70. ***Nicoraepoa*** Soreng & L.J.: Gillespie: *N. andina* (Trin.) Soreng & L.J. Gillespie, *N. chonotica* (Phil.) Soreng & L.J. Gillespie, *N. pugionifolia* (Speg.) Soreng & L.J. Gilliespie, *N. robusta* (Steud.) Soreng & L.J. Gilliespie, *Nicoraepoa subenervis* (Hack.) Soreng & L.J. Gilliespie subsp. *subenervis*, *Nicoraepoa subenervis* subsp. *spegazziana* (Nicora) Soreng & L.J. Gilliespie
71. ***Poa*** Barnhart: *P. acinaciphylla* E. Desv., *P. alopecurus* (Gaudich. ex Mirb.) Kunth subsp. *alopecurus*, *P. alopecurus* ssp *fuegiana* (Hook. f.) D.M. Moore & Dogg., *P. alopecurus* ssp *prichardii* (Rendle) Giussani & Soreng, *P. alopecurus* subsp. *fuegiana* (Hook. f.) D.M. Moore & Dogg., *P. androgyna* Hack., **P. annua* L., *P. atropidiformis* Hack. var. *atropidiformis*, *P. aysenensis* Hack., *P. bonariensis* (Lam.) Kunth, *P. bulbosa* L. subsp. *vivipara* (Koeler) Arcang., *P. compressa* L., ***P. cumingii*** Trin., *P. darwiniana* Parodi, *P. denudata* Steud., *P. flabellata* (Lam.) Raspail, *P. gayana* E. Desv., *P. glauca* Vahl subsp. *glauca*, *P. gymnantha* Pilg., *P. hachadoensis* Nicora var. *hachadoensis*, *P. hachadoensis* var. *pilosa* Nicora, *P. holciformis* J. Presl, *P. humillima* Pilg., **P. infirma* Kunth, *P. kurtzii* R.E. Fr., *P. laetevirens* R.E. Fr., *P. lanigera* Nees, *P. lanuginosa* Poir., *P. ligularis* Nees ex Steud. var. *ligularis*, *P. lilloi* Hack., *P. mendocina* Nicora & F.A. Roig, **P. nemoralis* L., *P. obvallata* Steud., *P. palustris* L., ***P. paposana*** Phil., *P. parviceps* Hack., *P. pearsonii* Reeder, *P. perligulata* Pilg., ***P. pfisteri*** Soreng, *P. planifolia* Kuntze, **P. pratensis* L. subsp. *pratensis*, **P. pratensis* subsp. *alpigena* (Lindm.) Hiitonen, *P. scaberula* Hook. f., *P. schoenoides* Phil., *P. secunda* J. Presl subsp. *secunda*, *P. secunda* subsp. *juncifolia* (Scribn.) Soreng, *P. spiciformis* subsp. *prichardii* (Rendle) Giussani & Soreng, *P. spiciformis* var. *ibari* (Phil.) Giussani, *P. spiciformis* var. *spiciformis*, *P. stenantha* Trin. var. *stenantha*, *P. stepparia* Nicora, *P. superata* Hack., *P. trachyantha* Hack., *P. tricolor* Nees ex Steud., **P. trivialis* L. subsp. *trivialis*, *P. yaganica* Speg.

Subtribe Torretochloinae Soreng

72. ***Amphibromus*** Nees: *A. scabrivalvis* (Trin.) Swallen.

Supertribe Triticodae T.D. Macfarl. & L. Watson

Tribe Triticeae Dumort.

Subtribe Hordeinae Dumort.

73. ***Agropyron*** Gaertn.: **A. cristatum* (L.) Gaertn.

74. ***Elymus*** L.: *E. angulatus* J. Presl, *E. magellanicus* (E.desv.) A. Löve, *E. patagonicus* Speg., **E. repens* (L.) Gould, *E. scabriglumis* (Hack.) A. Löve.

75. ***Hordeum*** L.: ***H. brachyatherum*** Phil., *H. chilense* Roem. & Schult., *H. comosum* J. Presl, *H. fuegianum* Bothmer N. Jacobsen & R.B. Jorg., *H. jubatum* L., *H. lechleri* (Steud.) Schenck, **H. marinum* Huds. subsp. *gussoneanum* (Parl.) Thell., **H. marinum* Huds. subsp. *marinum*, **H. murinum* L. subsp. *murinum*, **H. murinum* subsp. *glaucum* (Steud.) Tzvelev, **H. murinum* subsp. *leporinum* (Link) Arcang., *H. muticum* J. Presl, *H. patagonicum* subsp. *magellanicum* (Parodi ex Nicora) Bothmer, Giles & N. Jacobsen, *H. patagonicum* subsp. *santacrucense* (Parodi ex Nicora) Bothmer, Giles & N. Jacobsen, *H. pubiflorum* Hook.f. subsp. *pubiflorum*, *H. pubiflorum* subsp. *halophilum* (Griseb.) Baden & Bothmer, *H. tetraploideum* Covas, **H. vulgare* L.

76. ***Leymus*** Hochst.: **L. arenarius* (L.) Hochst., *L. erianthus* (Phil.) Dubcovs.

77. ***Secale*** L.: **S. cereale* L.

Subtribe Triticinae Fr.

78. ***Taeniatherum*** Nevski: **T. caput-medusae* (L.) Nevski

79. ***Triticum*** L.: **T. aestivum* L., **T. durum* Desf.

Tribe Bromeae Dumort.

80. ***Bromus*** L.: *Bromus araucanus* Phil., *B. berteroanus* Colla, *B. catharticus* Vahl var. *catharticus*, *B. catharticus* var. *elata* (E. Desv.) Planchuelo, *B. coloratus* Steud., *B. erectus* Huds., ***B. gunckelii*** Matthei, **B. hordeaceus* L., *B. japonicus* Thunb., *B. lanatus* Kunth, *B. lanceolatus* Roth, *B. lithobius* Trin., **B. madritensis* L., *B. mango* E. Desv., *B. pellitus* Hack., *B. racemosus* L., **B. rigidus* Roth, *B. scoparius* L. var. *scoparius*, **B. scoparius* var. *villiglumis* Maire & Weiller, *B. secalinus* L., *B. setifolius* J. Presl, *B. setifolius* var. *brevifolius* Nees, *B. setifolius* var. *pictus* (Hook. f.) Skottsb., **B. squarrosus* L., **B. sterilis* L.. **B. tectorum* L., *B. tunicatus* Phil.

SUBFAMILY ARISTIDOIDEAE CARO

Tribu Aristidae C.E. Hubb.

81. ***Aristida*** L.: *A. adscensionis* L., *A. pallens* Cav., *A. spegazzinii* Arechav.

SUBFAMILY ARUNDINOIDEAE BURMEIST.

Tribe Arundinae Dumort.

82. ***Arundo*** L.: **A. donax* L.

83. ***Phragmites*** Adans.: *P. australis* (Cav.) Trin. ex Steud.

SUBFAMILY DANTHONIOIDEAE P.H. LINDER & N.P. BAKER

Tribe Danthonieae Zotov.

84. ***Cortaderia*** Stapf: *C. araucana* Stapf, *C. atacamensis* (Phil.) Pilg., *C. pilosa* (D'Urv.)

Hack. var. *pilosa*, *C. pilosa* var. *minima* (Conert) Nicora, *C. rudiuscula* Stapf, *C. selloana* (Schult. & Schult.) Asch. & Graebn., *C. speciosa* (Nees & Meyen) Stapf

85. ***Danthonia*** DC.: ***D. araucana*** Phil., *D. boliviensis* Renvoize, *D. californica* Bol. var. *americana* (Scribn.) Hitchc., *D. chilensis* E. Desv. var. *chilensis*, ***D. chilensis* var. *aureofulva*** (E. Desv.) C. Baeza, *D. chilensis* var. *glabrifolia* Nicora, *D. decumbens* (L.) DC., *D. malacantha* (Steud.) Pilg.

86. ***Rytidosperma*** Steud.: *R. lechleri* Steud., ***R. paschalis*** (Pilg.) Baeza, *R. pictum* (Nees & Meyen) Nicora var. *pictum*, *R. pictum* var. *bimucronatum* Nicora, ***R. quirihuense*** C. Baeza, *R. violaceum* (E. Desv.) Nicora, *R. virescens* (E. Desv.) Nicora

87. ***Schismus*** P. Beauv.: **S. arabicus* Nees, **S. barbatus* (L.) Thell.

SUBAMILY PANICOIDEAE LINK

Tribe Andropogoneae Dumort.

Subtribe Andropogoninae J. Presl

88. ***Cymbopogon*** Spreng.: **C. citratus* (DC.) Stapf

89. ***Schizachyrium*** Nees: **S. sanguineum* (Retz.) Alston, *S. spicatum* (Spreng.) Herter

Subtribe Saccharinae Griseb.

90. ***Imperata*** Cirillo: *I. condensata* Steud.

91. ***Miscanthus*** Andersson: **M. sinensis* Andersson

Subtribe Sorghinae Clayton & Renvoize

92. ***Bothriochloa*** Kuntze: *B. ischaemum* (L.) Keng, *B. laguroides* (DC.) Herter, *B. saccaroides* (Sw.) Rydb. var. *saccharoides*

93. ***Sorghum*** Moench: **S. bicolor* (L.) Moench, *S. x drumondii* (Nees ex Steud.) Millsp. & Chase, **S. halepense* (L.) Pers.

Subtribe Tripsacinae Dumort.

94. ***Zea*** L.: **Z. mays* L.

Tribe Paniceae R.Br.

Subtribe Cenchrinae Dumort.

95. ***Cenchrus*** L.: **C. americanus* (L.) Morrone, *C. chilensis* (E. Desv.) Morrone, **C. clandestinus* (Hochst. ex Chiov.) Morrone, **C. echinatus* L., **C. incertus* M.A. Curtis, **C. longisetus* M.C. Johnst., **C. myosuroides* Kunth, **C. purpureus* (Schumach.) Morrone

Subtribe Digitariinae Butzin

96. ***Digitaria*** Haller: *D. aequiglumis* (Hack. & Arechav.) Parodi, *D. ciliaris* (Retz.) Koeler, **D. ischaemum* (Schreb.) Schreb. ex Muhl. var. *ischaemum*, **D. sanguinalis* (L.) Scop., *D. setigera* Roth ex Roem. & Schult., *D. violascens* Link

Subtribe Melinidinae Pilg.

97. ***Eriochloa*** Kunth: *E. montevidensis* Griseb.

98. ***Melinis*** P. Beauv.: **M. minutiflora* P. Beauv., **M. repens* (Willd.) Zizka

Subtribe Panicinae Fr.

99. ***Dichanthelium*** (Hitchc. & Chase) Gould.: *D. sabulorum* (Lam.) Gould & Clark var. *sabulorum*, *D. sabulorum* var. *polycladum* (Ekman) Zuloaga

100. ***Echinochloa*** P. Beauv.: **E. colona* (L.) Link, **E. crus-galli* (L.) P. Beauv. var. *crus-galli*, **E. crus-galli* var. *mitis* (Pursh) Peterm., **E. crus-galli* var. *zelayensis* (Kunth) Hitchc., **E. crus-pavonis* (Kunth) Schult.

101. ***Panicum*** L.: **P. capillare* L., *P. dichotomiflorum* Michx., *P. racemosum* (P. Beauv.) Spreng., *P. urvilleanum* Kunth

Subtribe Paspalinae Griseb.

102. ***Axonopus*** P. Beauv.: *A. compressus* (Sw.) P. Beauv.

103. ***Paspalum*** L.: *P. candidum* (Humb. & Bonpl. Ex Flügge) Kunth, *P. dasypleurum* Kunze ex E. Desv., *P. dilatatum* Poir., **P. distichum* L., *P. ekmanianum* Henrard, *P. flavum* J. Presl, ***P. forsterianum*** **Flüggé**, *P. minus* E. Fourn., *P. scrobiculatum* L. var. *orbiculare* (G. Forst.) Hack., *P. urvillei* Steud., *P. vaginatum* Sw.

Subtribe Setariinae Dumort.

104. ***Paspalidium*** Stapf: **P. geminatum* (Forssk.) Stapf

105. ***Setaria*** P. Beauv.: **S. parviflora* (Poir) Kerguelen var. *parviflora, S. parviflora* var. *brachytricha* Pensiero, **S. pumila* (Poir.) Roem. & Schult., **S. verticillata* (L.) P. Beauv., **S. viridis* (L.) P. Beauv.

106. ***Stenotaphrum*** Trin.: **S. secundatum* (Walter) Kuntze

SUBFAMILY CHLORIDOIDEAE KUNTH EX BEILSCHM.

Incertae sedis

107. ***Trichoneura*** Andersson: *T. weberbaueri* Pilg.

Tribe Cynodonteae Dumort.

Subtribe Boutelouinae Stapf

108. ***Bouteloua*** Lag.: *Bouteloua simplex* Lag.

Subtribe Eleusininae Dumort.

109. ***Chloris*** Sw.: **Chloris gayana* Kunth, *C. radiata* (L.) Sw., *C. virgata* Sw.

110. ***Cynodon*** Rich.: *Cynodon affinis* Caro & E.A. Sánchez, **C. dactylon* (L.) Pers., ***C. nitidus*** Caro & E.A. Sánchez

111. ***Eleusine*** Gaertn.: **Eleusine indica* (L.) Gaertn., **E. tristachya* (Lam.) Lam.

112. ***Eustachys*** Desv.: **Eustachys distichophylla* (Lag.) Nees

113. ***Leptochloa*** P. Beauv.: *L. fusca* subsp. *uninervia* (J. Presl) N. Snow

114. ***Microchloa*** R. Br.: *M. indica* (L. f.) P. Beauv., *M. kunthii* Desv.

Subtribe Monanthochloinae Pilg. ex Potztal

115. ***Distichlis*** Raf.: *D. humilis* Phil., *D. scoparia* (Nees ex Kunth) Arechav., *D. spicata* (L.) Greene var. *spicata, D. spicata* var. *mendocina* (Phil.) Hack., *D. spicata* ver. *stricta* (Torr.) Thorne

Subtribe Muhlenbergiinae Pilg.

116. ***Muhlenbergia*** Schreb.: *M. asperifolia* (Nees & Meyen ex Trin.) Parodi, *M. fastigiata* (J. Presl) Henrard, *M. peruviana* (P. Beauv.) Steud.

Subtribe Scleropogoninae Pilg.

117. ***Munroa*** Torr.: *M. andina* Phil., *M. argentina* Griseb., *M. decumbens* Phil.

Subtribe Tripogoninae Stapf

118. ***Tripogon*** Roem. & Schult.: *T. nicorae* Rúgolo & A.S. Vega, *T. spicatus* (Nees) Ekman

Tribe Eragrostidae Stapf

Subtribe Cotteinae Reeder

119. ***Enneapogon*** Desv. ex P. Beauv.: *E. devauxii* P. Beauv.

Subtribe Eragrostidinae J. Presl

120. ***Eragrostis*** Wolf: **E. atrovirens* (Desf.) Trin. ex Steud., *E. attenuata* Hitchc., **E. curvula* (Schrad.) Nees, ***E. kuschelii*** Skottsb., *E. mexicana* (Hornem.) Link subsp. *virescens* (C. Presl) S.D. Koch & Sánchez-Vega, *E. nigricans* (Kunth) Steud. var. *nigricans, E. nigricans* var. *punensis* Nicora, *E. peruviana* (Jacq.) Trin., **E. pilosa* (L.)

P. Beauv., *E. polytricha* Nees, ***E. pycnantha*** (Phil.) Nicora, **E. tenuifolia* (A. Rich.) Hochst. ex Steud., *E. weberbaueri* Pilg.

Tribe Zoysiae Benth.

Subtribe Sporobolinae Benth.

121. ***Spartina*** Schreb.: *S. densiflora* Brongn.

122. ***Sporobolus*** R. Br.: *S. africanus* (Poir) Robyns & Tournay, **S. indicus* (L.) R. Br., *S. rigens* (Trin.) E. Desv., **S. virginicus* (L.) Kunth

Table 3. List of the species of Poaceae registered in Chile, ordered according to (Soreng et al., 2009). * = introduced; bold = endemic.

7. Acknowledgements

The financial support of DIUC 210.122.012-ISP and DIUC 210.212.014.1.0 are gratefully aknowledged. Special thanks to Robert J. Soreng and Paul M. Peterson from Smithsonian Institution, Washington D.C. for sharing important data about the presence of species in Chile. Our thanks to Rosemarie Wilckens, Susan Fisher, Alejandro Solis, and Fernán Silva who kindly provide us some photos of their trip to Aysén held in January 2011.

8. References

Acevedo de Vargas, R. (1959). Las especies de Gramineae del género *Cortaderia* en Chile. *Boletín del Museo Nacional de Historia Natural,* Vol. 27, pp. 205-246, ISSN-0027-3910.

Aliscioni, S.S., Giussani, L.M., Zuloaga, F.O. & Kellogg, E.A. (2003). A molecular phylogeny of *Panicum* (Poaceae: Paniceae): Tests of monophyly and phylogenetic placement within the Panicoideae. *American Journal of Botany,* Vol. 90, No. 5 (May 2005), pp. 796-821, ISSN 1537-2197.

Allred, K.W. (1982). Describing the grass inflorescence. *Journal of Range Management,* Vol. 35, No. 5 (Sep. 1982), pp 672-675, ISSN 0022-409X.

Alroy, J. (2002). How many named species are valid? *Proceedings of the National Academy of Science,* Vol. 99, No. 6 (Mar. 2002), pp. 3706-3711, ISSN 1094-6490.

Anderson, D.E. (1974). Taxonomy of the genus *Chloris* (Gramineae). Brigham Young University Science Bulletin Vol. 19, No. 2 (Mar. 1974), pp. 1-133, ISSN 0068-1024.

Arancio, G., Jara, P., Squeo, F.A. & Marticorena, C. (2004). Riqueza de especies de plantas vasculares en los Altos de Talinay,Parque Nacional Bosque Fray Jorge. In, *Historia Natural del Parque Nacional Bosque Fray Jorge,* pp. 189-204. F.A. Squeo, J.R. Gutiérrez & I.R. Hernández (Eds.). Ediciones Universidad de La Serena, La Serena.

Arriaga, M.O. & Barkworth, M.E. (2006). *Amelichloa*: a new genus in the Stipeae (Poaceae). *Sida Contributions to Botany,* Vol. 22, No. 1, pp. 145-149, ISSN 1934-5259.

Arroyo, M.T.K., Armesto, J.J., Squeo, F. & Villagrán, C. (1993). Global change: Flora and vegetation of Chile. In: *Earth system responses to global change: contrats between North and South America,* H.A. Mooney, E.R. Fuentes & B.I. Kronberg (Eds.), pp. 239-263, Academic Press, San Diego.

Arroyo, M.T.K., Castor, C., Marticorena, C., Muñoz, M. Cavieres, L. & Matthei, O. (1998). La flora del parque nacional Llullaillaco ubicado en la zona de transición de las lluvias de invierno-verano en los Andes del norte de Chile. *Gayana Botánica* Vol. 55, No. 2 (May 1999), ISSN 0016-5301.

Arroyo, M.T.K., Marticorena, C., Matthei, O. & Cavieres, L. (2000). Plant invasions in Chile: present patterns and future predictions. In, *Invasive species in a changing world*, H.A. Mooney & R.J. Hobbs (Eds.), pp. 385-421. Island Press, USA.

Arroyo, M.T.K., Marticorena, C., Miranda, P., Matthei, O. (1989). Contribution to the high elevation flora of the Chilean Patagonia: A checklist of species on mountains on an East-West transect in the Sierra de los Baguales, Latitude 50°S. *Gayana Botánica* Vol. 46, No. 1-2 (Jul. 1989), pp. 121-151, ISSN 0016-5301.

Arroyo, M.T.K., Marticorena, C., Muñoz-Schick, M. (1990). A checklist of the native annual flora of continental Chile. *Gayana Botánica* Vol. 47, No. 3-4 (Apr. 1991), pp. 119-135, ISSN 0016-5301.

Arroyo, M.T.K., Rozzi, R., Simonetti, J.A., Marquet, P. & Salaberry, M. (1999). Central Chile. In *Hotspots: Earth's Biologically Richest and Most Endangered Terrestrial Ecorregions*, R.A. Mittermeier, N. Myers, P. Robles Gil & C. Goettsch Mittermeier (Eds.), pp. 161-171. CEMEX, México.

Arroyo, M.T.K., von Bohlen, Ch., Cavieres, L. & Marticorena, C. (1992). Survey of the alpine flora of Torres del Paine National Park, Chile. *Gayana Botánica* Vol. 49, No. 1-4 (Dec. 1992), pp. 47-70, ISSN 0016-5301.

Ayres, D.R., Smith, D.L., Zaremba, K., Klohr, S. & Strong, D.R. (2004). Spread of exotic cordgrasses and hybrids (*Spartina* sp.) in the tidal marshes of San Francisco Bay, California, U.S.A. *Biological Invasions*, Vol. 6, No. 2 (Jun. 204), pp 221-231, ISSN 1573-1464.

Baeza, C.M. 1996. Los géneros *Danthonia* DC. y *Rytidosperma* Steud. (Poaceae) en América-Una revisión. *Sendtnera* Vol. 3, pp. 11–93, ISSN 0944-0178.

Baeza, C.M., Marticorena, C. & Rodríguez, R. (1999). Catálogo de la flora del monumento natural Contulmo, Chile. *Gayana Botánica* Vol. 56, No. 2 (Dec. 1999), pp. 125-135, ISSN 0016-5301.

Baeza, C.M., Marticorena, C., Stuessy, T., Ruiz, E. & Negritto, M. (2007). Poaceae en el archipiélago de Juan Fernández (Robinson Crusoe). *Gayana Botánica* Vol. 64, No. 2 (Dec. 2007), pp. 125-174, ISSN 0016-5301.

Baeza, C.M., Stuessy, T. & Marticorena, C. (2002). Notes on the Poaceae of Robinson Crusoe (Juan Fernández) islands, Chile. *Brittonia* Vol. 54, No. 3 (Jul. 2002), pp. 154-163, ISSN 0007-196X.

Barker, N.P., Linder, H.P. & Harley, E. (1995). Polyphyly of Arundinoideae (Poaceae): Evidence from *rbcL* sequence data. *Systematic Botany* Vol. 20, No. 4, pp. 423–435, ISSN 0363-6445.

Barker, N.P., Linder, H.P. & Harley, E.H. (1999). Sequences of the grass-specific insert in the chloroplast rpoC2 gene elucidate generic relationships of the Arundinoideae (Poaceae). *Systematic Botany* Vol. 23, No. 3, pp. 327–350, ISSN 0363-6445.

Barkworth, M.E., Capels, K.M., Long, S., Anderton, L.K. & Piep, M.B. 2007. *Flora of North America north of Mexico*, Vol. 24. Poaceae, part 1. Oxford University Press, New York.

Bortolus, A. (2006). The austral cordgrass *Spartina den*siflora Brongn.: its taxonomy, biogeography and natural history. *Journal of Biogeography*, Vol. 33, No. 1 (Jan. 2006), pp. 158-168, ISSN 1365-2699.

Bouchenak-Khelladi, Y., Verboom, G.A., Savolainen, V. & Hodkinson, T.R. (2010). Biogeography of the grasses (Poaceae): a phylogenetic approach to reveal evolutionary history in geographical space and geological time. *Botanical Journal of the Linnean Society*, Vol. 162, No. 4 (Apr. 2010), pp. 543-557, ISSN 1095-8339.

Cabrera, A.L. & A. Willink. (1973). *Biogeografía de América Latina*. The General Secretariat of the Organization of American States, Washington D.C., U.S.A.

Caro, J.A. (1982). Sinopsis taxonómica de las gramíneas Argentinas. *Dominguezia*, Vol. 4, No. 1 (Feb. 1982), pp. 1-51, ISSN 1669-6859.

Cerros-Tlatilpa, R. & Columbus, J.T. (2009). C3 photosynthesis in *Aristida longifolia*: Implication for photosynthetic diversification in Aristidoideae (Poaceae). *American Journal of Botany*, Vol. 96, No. 8 (Jul. 2009), pp. 1379-1387, ISSN 1537-2197.

Chemisquy, A., Giussani, L.M., Scataglini, M.A., Kellogg, E.A. & Morrone, O. (2010). Phylogenetic studies favour the unification of *Pennisetum*, *Cenchrus* and *Odontelytrum* (Poaceae): a combined nuclear, plastid and morphological analysis, and nomenclatural combinations in *Cenchrus*. *Annals of Botany*, Vol. 106, N°.1 (Jul. 2010), pp. 107-130, ISSN 1095-8290.

Chiapella, J. & Zuloaga, F. (2010). A revision of the genera *Deschampsia*, *Avenella*, and *Vahlodea* (Pooideae, Poeae, Airinae) in South America. *Annals of the Missouri Botanical Garden*, Vol. 97, No. 2, (Jul. 2010), pp. 141-162, ISBN 0026-6493.

Cialdella, A.M. & Arriaga, M. (1998). Revisión de las especies sudamericanas del género *Piptochaetium* (Poaceae, Pooideae, Stipeae). *Darwiniana* Vol. 36, No. 1-4, pp. 107-157, ISSN 0011-6793.

Cialdella, A.M. & Giussani, L.M. (2002). Phylogenetic relations of the genus *Piptochaetium* (Poaceae: Pooideae: Stipeae): Evidence from morphological data. *Annals of the Missouri Botanical Garden* Vol. 89, No. 3 (Summer 2002), pp. 305-336, ISSN 0026-6493.

Clayton, W.D & Renvoize, S.A. (1986). *Genera graminum. Grasses of the world*. Kew Bull. Additional Series XIII, Royal Botanic Gardens, Kew.

Clark, L.G. (1990). *Chusquea* sect. *Longiprophyllae* (Poaceae: Bambusoideae): A new Andean section and new species. *Systematic Botany* Vol. 15, No. 4 (Oct.-Dec. 1990), pp. 617-634, ISSN 0363-6445.

Clark, L.G. & Judziewicz, E.J. (1996). The grass subfamilies Anomochlooideae and Pharoideae (Poaceae). *Taxon* Vol. 45, No. 4 (November, 1996), pp. 641-645, ISSN 0040-0262.

Clark, L.G., Kobayashi, M., Mathews, S., Spangler, R.E & Kellogg, E. (2000). The Puelioideae, a new subfamily of Poaceae. *Systematic Botany*, Vol. 25, No 2, pp. 181-187, ISSN 0363-6445.

Clark, L.G., Zhang, W. & Wendel, J.F. (1995) A phylogeny of the grass family (Poaceae) based on *ndhF* sequence data. *Systematic Botany*, Vol. 20, No 4 (October-December 1995), pp. 436–460. ISSN 0363-6445.

Conservation International. (2011). Biodiversity hotspots. 20.01.2011. Available from http://www.biodiversity hotspots.org/xp/Hotspots/hotspotsScience/pages/hotspots_defined.aspx.

Crawford, P.H.C. & Hoagland, B.W. (2009). Can herbarium records be used to map alien species invasion and native species expansion over the past 100 years? Journal of Biogeography, Vol. 36, No. 4 (April 2009), pp. 651-661, ISSN 1365-2699.

Delisle, F., Lavoie, C., Jean, M. & Lachance, D. (2003). Reconstructing the spread of invasive plants: taking into account biases associated with herbarium specimens. *Journal of Biogeography*, Vol. 30, No. 7 (July 2003), pp. 1033-1042, ISSN 1365-2699.

De Paula, M.E. (1975). Las especies del género *Hierochloë* (Gramineae) de Argentina y Chile. *Darwiniana*, Vol. 19, No. 2-4, pp. 422-457, ISSN 1850-1702.

Desvaux, E. (1854). Gramíneas. In, Gay, C. *Historia Física y Política de Chile* Vol. 6, pp. 233-469, lám. 74-83 in Atlas.

Escobar, I., Ruiz, E., Finot, V.L., Negritto, M.A. & Baeza, C.M. (2011). Revisión taxonómica del género *Eragrostis* Wolf en Chile, basado en análisis estadísticos multivariados. *Gayana Botánica* Vol. 68, No. 1, pp. 47-81, ISSN 0016-5301.

Finot, V.L. (2010). Analysis of species boundaries in *Trisetum* Pers. sect. *Trisetaera* Asch. & Graebn. (Poaceae) in America using statistical multivariate methods. *Current Topics in Plant Biology*, Vol. 11 (December 2010), pp. 39-74, ISSN 0972-4575.

Finot, V.L., Marticorena, C., Barrera, J.A., Muñoz-Schick, M. & Negritto, M.A. (2009). Diversidad de la familia Poaceae (Gramineae) en la región del Bío-Bío, Chile, basada en colecciones de herbario. *Gayana Botánica*, Vol. 66, No.2 (December 2009), pp. 134-157. ISSN 0016-5301.

Finot, V.L., Peterson, P.M., Zuloaga, F.O., Soreng, R.J. & Matthei, O. (2005). A revision of *Trisetum* (Poaceae: Pooideae: Aveninae) in South America. *Annals of the Missouri Botanical Garden* 92: 533-568.

Fuentes, N., Ugarte, E., Kühn, I. & Klotz, S. (2008). Alien plants in Chile: inferring invasion periods from herbarium records. *Biological Invasions* 10: 649-657.

Funk, V.A. & Richardson, K.S. (2002). Systematic data in biodiversity studies: Use it or lose it. *Systematic Biology* 51(2): 303-316.

Gajardo, R. (1994). *La vegetación natural de Chile. Clasificación y distribución geográfica*. Editorial Universitaria, Santiago de Chile.

Gealy, D.R., Mitten, D.H. & Rutger, J.N. (2003). Gene flow between red rice (Oryza sativa) and herbicide-resistant rice (O. sativa): implications for weed management. *Weed Technology* Vol. 17, No. 3 (Jule-September 2003), pp. 627-645.

Giraldo-Cañas, D. (2008). Revisión del género *Axonopus* (Poaceae: Paniceae): primer registro del género en Europa y novedades taxonómicas. *Caldasia* Vol. 30, No. 2 (July-December), pp. 301-314, ISSN 0366-5232.

Giussani, L.M:, Cota-Sánchez, J.H., Zuloaga, F.O. & Kellogg, E.A. (2001). A molecular phylogeny of the grass subfamily Panicoideae (Poaceae) shows multiple origins of C4 photosynthesis. *American Journal of Botany*, Vol. 88, No. 11 (Nov. 2001), pp. 1993-2012, ISSN 1537-2197.

Giussani, L.M., Zuloaga, F.O., Quarín, C.L., Cota-Sánchez, J.H., Ubayasena, K. & Morrone, O. (2009). Phylogenetic relationships in the genus Paspalum (Poaceae: Panicoideae: Paniceae): an assessment of the Quadrifaria and Virgata informal groups. *Systematic Botany*, Vol. 34, No. 1 (Jan.-Mar. 2009), pp. 32-43, ISSN 1548-2324.

GPWG (Grass Phylogeny Working Group). (2001). Phylogeny and subfamilial classification of the grasses (Poaceae). *Annals of the Missouri Botanical Garden*, Vol. 88, No. 3 (September 2001), pp. 373-457, ISBN 0026-6493.

Grau, J. (1995). Aspectos geográficos de la flora de Chile. In, *Flora de Chile, Vol. I. Pteridophyta-Gumnospermae*, pp. 63-83, C. Marticorena & R. Rodríguez (Eds.). Ed. Universidad de Concepción, Concepción, Chile.

Hamby, R.K. & Zimmer, E.A. (1988). Ribosomal RNA sequences for inferring phylogeny within the grass family (Poaceae). *Plant Systematics and Evolution* 160: 29–37.

Hilu, K.W. & Alice, L.A. (2001). A phylogeny of Chloridoideae (Poaceae) based on *mat*K sequences. *Systematic Botany* Vol. 26, No. 2, pp. 386-405.

Hilu, K.W., Alice, L.A. & Liang, H. (1999). Phylogeny of Poaceae inferred from matK sequences. *Annals of the Missouri Botanical Garden* 86: 835–851

Hitchcock, A.S. (1950). *Manual of the grasses of the United States.* U.S.D.A. Miscellaneous Publication No. 200, Washington.

Holcombe, T., Stohlgren, T.J. & Jarnevich, C.S. (2010). From points to forecasts: Predicting invasive species habitat suitability in the near term. *Diversity*, Vol. 2, No. 5 (May 2010), pp. 732-767.

Hsiao, C., Jacobs, S.W.L., Chatterton, N.J. & Asay, K.H. (1999). A molecular phylogeny of the grass family (Poaceae) based on the sequences of nuclear ribosomal DNA (ITS). *Australian Systematic Botany* 11: 667–688.

Judziewicz, E.J., Clark, L.G., Londoño, X. & Stern, M.J. (1999). *American bamboos.* Smithsonian Institution, Washington DC.

Judziewicz, E.J., Soreng, R.J., Davidse, G., Peterson, P.M., Filgueiras, T.S. & Zuloaga, F.O. (2000). Catalogue of New World Grasses (Poaceae): I. Subfamilies Anomochloideae, Bambusoideae, Ehrhartoideae and Pharoideae, *Contributions from the United States National Herbarium* Vol. 39, pp. 1-128, ISSN 0097-1618.

Kellogg, E.A. (2000). Grasses: A case study in macroevolution. *Annual Review of Ecology and Systematics*, Vol. 31, (November 2000), pp. 217-238, ISSN 0066-4162.

Kellogg, E.A. (2001). Evolutionary history of the grasses. *Plant Physiology*, Vol. 125, No. 3 (Mar. 2001), pp. 1198-1205, ISSN 1532-2548.

Kellogg, E.A. (2009). The evolutionary history of Ehrhartoideae, Oryzeae, and *Oryza*. *Rice* Vol. 2, (Jan. 2009), pp 1-14, ISSN 1939-8425.

Kunkel, G. (1968). Die Phanerogamen der Insel La Mocha (Chile). *Wildenowia* Vol. 4, No. 3, pp. 329-400, ISSN 0374-8960.

Linder, P., Baeza, M., Barker, N.P., Galley, Ch., Humphreys, A.M., Lloyd, K.M., Orlovich, D.A., Pirie, M.D., Simon, B.K., Walsh, N. & Verboom, G.A. (2010). A generic classification of the Danthonioideae (Poaceae). *Annals of the Missouri Botanical Garden* Vol. 97, No.3 (Oct. 2010), pp. 306-364, ISBN 0026-6493.

Luebert, F. & Pliscoff, P. (2006). *Sinopsis bioclimática y vegetacional de Chile.* Editorial Universitaria, Santiago de Chile.

Mann, D., Chase, J., Edwards, J., Beck, W., Reanier, R., and Mass, M. (2003). Prehistoric destruction of the primeval soils and vegetation of Rapa Nui (Isla de Pascua, Easter Island). In, *Easter Island.* Loret, J. & Tanacredi, J.T. (eds.), pp. 133-154, Kluwer Academic/Plenum Publishers, ISBN 0-306-47494-8, New York, USA.

Markgraf, W. (2003). Anthropogen beeinflusste Landschaftsentwicklung im Osten der Osterinsel. Diplomarbeit zur Diplomprüfung im Fach Geographie Universität zu Kiel.

Marticorena, C. & Quezada, M. (1985). Catálogo de la flora vascular de Chile. *Gayana Botánica* Vol. 42, No. 1-2 (Dec. 1985), pp. 5-157, ISSN 0016-5301.

Marticorena, C., Squeo, F.A., Arancio, G. & Muñoz, M. (2001). Catálogo de la flora de la IV Región. In, *Libro rojo de la flora nativa y de los sitios prioritarios para su conservación:*

Región de Coquimbo (Squeo, F.A., Arancio, G. & Gutiérrez, J.R., Eds.), pp. 105-142. Ed. Universidad de La Serena, La Serena, Chile.

Mason-Gamer, R.J., Weil, C.F. & Kellogg, E.A. (1998). Granule-bound starch synthase: structure, function, and phylogenetic utility. *Molecular Biology and Evolution*, Vol. 15, No. 12 (Dec. 1998), pp. 1658–1673, ISSN 1537-1719.

Mathews, S., Tsai, R.C. & Kellogg, E. (2000). Phylogenetic structure in the grass family (Poaceae): evidence from the nuclear gene phytochrome B. *American Journal of Botany*, Vol. 87, No. 1 (Jan. 2000), pp. 96–107, ISSN 1537-2197.

Matthei, O. (1965). Estudio crítico de las gramíneas del género *Stipa* en Chile. *Gayana Botánica*, Vol. 13 (Jun. 1965), 1-137, ISSN 0016-5301.

Matthei, O. (1973). *Trichoneura* Andersson (Gramineae), nuevo género para la flora de Chile. *Boletín de la Sociedad de Biología de Concepción* Vol. 46, pp. 37-39, ISSN 0037-850X.

Matthei, O. (1975). Der *Briza*-Kimplex in Südamerika: *Briza, Calotheca, Chascolytrum, Poidium* (Gramineae). Eine Revision. *Wildenowia* Vol. 8, pp. 1-168, ISSN 0374-8960.

Matthei, O. (1982). El género *Festuca* (Poaceae) en Chile. *Gayana Botánica* Vol. 37 (Jun. 1982), pp. 1-64, ISSN 0016-5301.

Matthei, O. (1986). El género *Bromus* L. (Poaceae) en Chile. *Gayana Botánica*, Vol. 43, No. 1-4 (Dec. 1986), pp. 47-110, ISSN 0016-5301.

Matthei, O. (1987a). Las especies del género *Aristida* L. (Poaceae) en Chile. *Gayana Botánica* Vol. 44, No. 1-4 (Sep. 1987), pp. 17-23, ISSN 0016-5301.

Matthei, O. (1987b). Las especies del género *Panicum* (Poaceae) en Chile. *Gayana Botánica* Vol. 44, No. 1-4 (Sep. 1988), pp. 25-32, ISSN 0016-5301.

Matthei, O. (1995). *Manual de las malezas que crecen en Chile*. Alfabeta Imp., Santiago de Chile.

Matthei, O. (1997). Las especies del género *Chusquea* Kunth (Poaceae: Bambusoideae), que crecen en la X Región, Chile. *Gayana Botánica*, Vol. 54, No. 2, (April 1998), pp. 199-220, ISSN 0016-5301.

Matthei, O., Marticorena, C. & Stuessy, T.F. (1993). La flora adventicia del Archipiélago de Juan Fernández. *Gayana Botánica* Vol. 50, No. 2 (Dec. 1993), pp. 69-102, ISSN 0016-5301.

Mösbach, E.W. (1999). *Botánica indígena de Chile*. Editorial Andrés Bello, Santiago de Chile.

Moreira-Muñoz, A. & Muñoz-Schick, M. (2007). Classification, diversity, and distribution of Chilean Asteraceae: implications for biogeography and conservation. *Diversity and* Distributions, Vol. 13, No. 6 (Nov. 2007), pp. 818-828, ISSN 1472-4642.

Müller, C. 1985. Zur verbreitung und taxonomie der gattung *Polypogon* (Poaceae) in Südamerika. *Wiss. Z. Karl-Marx-Univ. Leipzig, Math.-Naturwiss. R.*, Vol. 34, No. 4, pp. 437-449.

Muñoz, C. (1941). *Indice bibliográfico de las gramíneas chilenas*. Boletín Técnico No. 2, Ministerio de Agrocultura, Santiago de Chile.

Muñoz-Schick, M. (1985). Revisión de las especies del género *Melica* L. (Gramineae) en Chile. *Boletín del Museo Nacional de Historia Natural*, Vol. 40, pp. 41-89, ISSN-0027-3910.

Muñoz-Schick, M. (1990). Revisión del género *Nassella* (Trin.) E. Desv. en Chile. *Gayana Botánica* Vol. 47, No. 1-2 (Oct. 1990), pp. 9-35, ISSN 0016-5301.

Nadot, S., Bajon, R. & Lejeune, B. (1994). The chloroplast gene rps4 as a tool for the study of Poaceae phylogeny. *Plant Systematic & Evolution*, Vol. 191, No. 1-2 (Mar. 1994), pp. 27–38, ISSN 1615-6110.

Nicora, E.G. (1978). Gramineae. In, Flora Patagónica, M.N. Correa (Ed.), pp. 1-563. Colecciones Científicas INTA, Buenos Aires, Argentina.

Nicora, E.G. (1998). Revisión del género Eragrostis Wolf. (Gramineae: Erahrostideae) para Argentina y países limítrofes. Boissera Vol. 54 (May 1998), pp. 1-109, ISSN 0373-2975.

Pankhurst, R.J. & Hervé, F. (2007). Introduction and overview. In, *The geology of Chile*, T. Moreno & E. Gibbons (Eds), pp. 1-4. The Geological Society of London. Version: 21 January 2011. http://www.amazon.com/Geology-Chile-Teresa-Moreno/dp/186239220X#reader_186239220X

Parodi, L.R. (1945). Sinopsis de las gramíneas chilenas del género *Chusquea*. *Revista Universitaria (Santiago)*, Vol. 30, No. 1, pp. 331-345.

Peñailillo, P. (1996). *Anatherostipa*, un nuevo género de Poaceae (Stipeae). *Gayana Botánica*, Vol. 53, No. 2, pp. 277-284, ISSN 0016-5301.

Peterson, P.M., Columbus, J.T. & Pennington, S.J. (2007). Classification and biogeography of New World Grasses: Chloridoideae. *Aliso* Vol. 23, pp. 580-594, ISSN 0065-6275.

Peterson, P.M., Romaschenko, K. & Johnson, G. (2010). A classification of the Chloridoideae (Poaceae) based on multi-gene phylogenetic trees. *Molecular Phylogenetics and Evolution* Vol 55 (June 2010), pp. 580-598, ISSN 1095-9513.

Peterson, P.M., Soreng, R.J., Davidse, G., Filgueiras, T.S., Zuloaga, F.O. & Judziewicz, E.J., (2001). Catalogue of New World Grasses (Poaceae): II. Subfamily Chloridoideae, *Contributions from the United States National Herbarium* Vol. 41, pp. 1-255, ISSN 0097-1618.

Piperno, D. & Sues, H.D. (2005). Dinosaurs dined on grass. *Science*, Vol. 310, No. 5751 (Nov. 2005), pp. 1126-1128, ISSN 1095-9203.

Pisano, E. (1985). La estepa patagónica como recurso pastoral en Aysén y Magallanes. *Ambiente y Desarrollo*, Vol. 1, No. 2 (Jun. 1985), pp. 45-59, ISSN 0716-1476.

Planchuelo, A.M. (2010). Nuevas citas y distribución de especies de *Bromus* L. (Poaceae: Bromeae) para Bolivia. *Kempffiana*, Vol. 6, N°.1, (Jun. 2010), pp. 3-15, ISSN 1991-4652.

Poblete, H., Cuevas, H. & Días-Vaz, J.E. (2009). Property characterization of *Chusquea culeou*, a bamboo growing in Chile. *Maderas. Ciencia y Tecnología*, Vol. 11, No. 2, pp. 129-138, ISSN 0718-211X.

Ponder, W.F., Carter, G.A., Flemons, P. & Chapman, R.R. (2001). Evaluation of museum collections data for use in biodiversity assessment. *Conservation Biology*, Vol. 15, No. 3 (Jun. 2001), pp. 648-657, ISSN 1523-1739.

Prasad, V., Strömberg, C.A.E., Alimohammadian, H. & Sahni, A. (2005). Dinosaurs coprolites and the early evolution of grasses and grazers. *Science*, Vol. 310, No. 5751 (Nov. 2005), pp. 1177-1180, ISSN 1095-9203.

Prather, L.A., Alvarez-Fuentes, O., Mayfield, M.H. & Ferguson, C.J. (2004). The decline of plant collecting in the United States: A threat to the infrastructure of biodiversity studies. *Systematic Botany*, Vol. 29, No. 1 (Jan. 2004), pp. 15-28, ISSN 1548-2324.

Reiche, K. (1903). La isla de la Mocha. Anales del Museo Nacional de Chile Vol. 16, pp. 1-104.

Rodríguez, R., Grau, J., Baeza, C. & Davies, A. 2008a. Lista comentada de las plantas vasculares de los Nevados de Chillán, Chile. *Gayana Botánica*, Vol. 65, No. 2 (Dec. 2008), ISSN 0016-5301.

Rodríguez, R., Marticorena, A. & Teneb, E. (2008b). Plantas vasculares de los ríos Baker y Pascua, región de Aisén, Chile. Gayana Botánica, Vol. 65, No. 1 (Jun. 2008), pp. 39-70, ISSN 0016-5301.

Rúgolo de Agrasar, Z.E. (1978). Las especies australes del género *Deyeuxia* Clar. (Gramineae) de la Argentina y Chile. *Darwiniana*, Vol. 21, No. 2-4, pp. 417-453, ISSN 1850-1702.

Rúgolo de Agrasar, Z.E. (1982). Revalidación del género *Bromidium* Nees et Meyen emend. Pilger (Gramineae). *Darwiniana*, Vol. 24, No. 1-4 (Jul. 1982), pp. 187-216, ISSN 1850-1702.

Rúgolo de Agrasar, Z.E. (1999). Corynephorus (Gramineae), género adventicio para la flora de Chile. Hickenia, Vol. 2, No. 64 (Jan. 1999), pp. 299-302, ISSN 0325-3732.

Rúgolo de Agrasar, Z.E. (2006). Las especies del género *Deyeuxia* (Poaceae, Pooideae) de la Argentina y notas nomenclaturales. *Darwiniana*, Vol. 44, N°.1, (Jul. 2006), pp. 131-293, ISSN 1850-1702.

Rúgolo de Agrasar, Z.E., García, N. & Mieres, G. (2009). *Danthonia decumbens* (Danthoniae, Poaceae), nueva especie adventicia para Chile continental. *Gayana Botánica*, Vol. 66, No. 1 (jun. 2009), pp. 92-94, ISSN 0016-5301.

Rúgolo de Agrasar, Z.E. & Molina, A.M. (1997). Las especies del género *Agrostis* L. (Gramineae, Agrostideae) de Chile. *Gayana Botánica*, Vol. 54, No. 2, (Apr. 1997), pp. 91-156.

Rundel, P.W., Dillon, M.O. & Palma, B. (1996). Flora and vegetation of Pan de Azúcar National Park in the Atacama desert of northern Chile. *Gayana Botánica*, Vol. 53, No. 2 (Dec. 1996), pp. 295-315.

Sánchez-Ken, J.G. & Clark, L.G. (2010). Phylogeny and a new tribal classification of the Panicoideae s.l. (Poaceae) based on plastid and nuclear sequence data and structural data. *American Journal of Botany* Vol. 97, No. 10 (Sep. 2010), pp. 1732-1748, ISSN 1537-2197.

Sánchez-Ken, J.C., Clark, L.G., Kellogg, E.A. & Kay, E.E. (2007). Reinstatement and emendation of the subfamily Micrairoideae (Poaceae). *Systematic Botany*, Vol. 32, No. 1 (Jan.-Mar. 2007), pp. 71-80. ISSN 1548-2324.

San Martín, C., Subiabre, M. & Ramírez, C. (2006). A floristic and vegetational study of a latitudinal gradient of salt-marshes in South-Central Chile. *Ciencia e Investigación Agraria*, Vol. 33, No. 1 (Jun.-Apr. 2006), pp. 33-40, ISSN 0718-1620.

Scotland, R.W. & A.H. Wortley. (2003). How many species of seed plants are there? *Taxon* Vol. 52, No. 1 (Feb. 2003), pp. 101-104, ISSN 0040-0262.

Skottsberg, C. (1956). *The natural history of Juan Fernandez and Easter Island*. Vol. I. Geography, Geology origin of island life. Almquist & Wiksells Boktrycheri Ab, Uppsala.

Smith, G.F. & M.M. Wolfson. (2004). Mainstreaming biodiversity: the role of taxonomy in bioregional planning activities in South Africa. *Taxon*, Vol. 53, No. 2 (May 2004), pp. 467-468. ISSN 0040-0262.

Steadman, D.W. (1995). Prehistoric extinctions of Pacific Island birds: Biodiversity meets zooarchaeology. *Science*, Vol. 267, No. 5201 (Feb. 1995), pp. 1123-1131, ISSN 0036-8075.

Soreng, R.J. & Davis, J.I. (1998). Phylogenetics and character evolution in the grass family (Poaceae): Simultaneous analysis of morphological and chloroplast DNA restriction site character sets. *The Botanical Review*, Vol. 64, No. 1 (Jan. 1998), pp. 1-85, ISSN 0006-8101.

Soreng, R.J., Davidse, G., Peterson, P.M., Zuloaga, F.O., Judziewicz, E.J., Filgueiras, T.S. & Morrone, O. (2009). Catalogue of New World Grasses (Poaceae). In, *Tropicos*, 27.12.2010, http://www.tropicos.org.

Soreng, R.J., Davis, J.I. & Voionmaa, M.A. (2007). A phylogenetic analysis of Poaceae tribe Poeae sensu lato based on morphological characters and sequence data from three plastid-encoded genes: evidence for reticulation, and a new classification for the tribe. *Kew Bulletin*, Vol. 62, pp. 425-454, ISSN 1874-933X.

Soreng, R.J. & Gillespie, L.J. (2007). Nicoraepoa (Poaceae: Poeae), a new South American genus based on Poa subg. Andinae, and emmendation of Poa sect. Parodiochloa of the Sub-Antarctic islands. *Annals of the Missouri Botanical Garden*, Vol. 94, No. 4 (Dec. 2007), pp. 821-849, ISSN 0026-6493.

Soreng, R.J. & Peterson, P.M. (2008). New records of Poa (Poaceae) and Poa pfisteri: a new species endemic to Chile. *Journal of the Botanical Research Institute of Texas* Vol. 2, No. 2, pp. 847-859, ISSN 1934-5259.

Soreng, R.J., Peterson, P.M., Davidse, G., Judziewicz, E.J., Zuloaga, F.O., Filgueiras, T.S. & Morrone, O. (2003). Catalogue of New World Grasses (Poaceae): IV. Subfamily Pooideae, *Contributions from the United States National Herbarium* Vol. 48, pp. 1-730, ISSN 0097-1618.

Squeo, F.A., Arancio, G., Marticorena, C., Muñoz-Schick, M., & Gutiérrez, J.R. (2001). Diversidad vegetal de la IV región de Coquimbo, Chile. In, *Libro rojo de la flora nativa y de los sitios prioritarios para su conservación: Región de Coquimbo,* F.A. Squeo, G, Arancio. & J.R. Gutiérrez (Eds.), pp. 149-158. Ed. Universidad de La Serena, La Serena, Chile.

Squeo, F.A., Arroyo, M.T.K., Marticorena, A., Arancio, G., Muñoz-Schick, M., Negritto, M., Rojas, G., Rosas, M. Rodríguez, R., Humaña, A.M., Barrera, E. & Marticorena, C. (2008). Catálogo de la flora vascular de la región de Atacama. In, *Libro rojo de la flora nativa y de los sitios prioritarios para su conservación: Región de Atacama,* F.A. Squeo, G. Arancio & J.R. Gutiérrez (Eds.), pp. 97-120. Ed. Universidad de La Serena, Concepción, Chile.

Squeo, F.A., Cavieres, L., Arancio, G., Novoa, J., Matthei, O., Marticorena, C., Rodríguez, R., Arroyo, M.T.K. & Muñoz, M. (1998). Biodiversidad de la flora vascular en la región de Antofagasta, Chile. *Revista Chilena de Historia Natural*, Vol. 71, No. 4 (Dec. 1998), pp. 571-591, ISSN 0717- 6317.

Stuessy, T.F. & Taylor, C. (1995). Evolución de la flora chilena. In, *Flora de Chile, Vol. I. Pteridophyta-Gumnospermae*, pp. 85-118, C. Marticorena & R. Rodríguez (Eds.). Ed. Universidad de Concepción, Concepción, Chile.

Swenson, U., Stuessy, T.F., Baeza, M. & Crawford, D. (1997). New and historical plant introductions and potential pests in the Juan Fernández islands, Chile. ISSN 0016-5301., Vol. 52, No. 3, pp. 233-252. E-ISSN: 1534-6188.

Tzvelev, N.N. (1983). *Grasses of the Soviet Union*, Oxonian Press, New Dehli.

Tzvelev, N.N. (1989). The system of grasses (Poaceae) and their evolution. *The Botanical Review*, Vol. 55, No. 3 (Jul.-Sep. 1989), pp. 141-203, ISSN 0006-8101.

Vega, A.S., Rua, G.H., Fabbri, L.T. & Rúgolo, Z.E. (2009). A morphology-based cladistic analysis of *Digitaria* (Poaceae, Panicoideae, Paniceae). *Systematic Botany*, Vol. 34, No. 2 (Apr.-Jun. 2009), pp. 312-323, ISSN 1548-2324.

Vincentini, A., Barber, J.C., Aliscioni, S.S., Gissani, L.M. & Kellogg, E. (2008). The age of grasses and clusters of origins of C4 photosynthesis. *Global Change Biology*, Vol. 14, No. 6 (Jun. 2008), pp. 2963-2977, ISSN 1365-2486.

Watson, L., & Dallwitz, M.J. (1992 onwards). *The grass genera of the world: descriptions, illustrations, identification, and information retrieval; including synonyms, morphology, anatomy, physiology, phytochemistry, cytology, classification, pathogens, world and local distribution, and references*. Version: 23rd Apr. 2010. http://delta-intkey.com'.

Wilson, E.O. (2000). On the future of conservation biology. *Conservation Biology*, Vol. 14, No. 1 (Feb. 2000), pp. 1-3, ISSN 1523-1739.

Zuloaga, F.O., Morrone, O. & Belgrano, M. (2008). *Catálogo de las plantas vasculares del Cono Sur. Vol. 1: Pteridophyta, Gymnospermae, Monocotyledoneae*. Monographs in Systematic Botany 107, Missouri Botanical Garden Press, St. Louis, Missouri, U.S.A.

Zuloaga, F.O., Morrone, O., Davidse, G., Filgueiras, T.S., Peterson, P.M., Soreng, R.J. & Judziewicz, E.J., (2003). Catalogue of New World Grasses (Poaceae): III. Subfamilies Panicoideae, Aristidoideae, Arundinoideae, and Danthonioideae, *Contributions from the United States National Herbarium* Vol. 46, pp. 1-662, ISSN 0097-1618.

5

Arboreal Diversity of the Atlantic Forest of Southern Brazil: From the Beach Ridges to the Paraná River

Maurício Bergamini Scheer[1] and Christopher Thomas Blum[2]
[1]*Research and Development Assistance, Sanepar*
[2]*Sociedade Chauá*
Brazil

1. Introduction

The Atlantic Forest (a hotspot for conservation) used to be the second largest tropical moist forest of South America (Oliveira-Filho & Fontes, 2000). It originally covered 20% (around 1.5 million km^2) of the Brazilian territory and occupied highly heterogeneous environmental conditions (Ribeiro et al., 2009). Nowadays, however, the Atlantic Forest covers less than 1% of the country or 7.5% of remnants (Myers et al., 2000). When intermediate secondary forests and small fragments (< 100 ha) are included, this contribution increases, ranging from 11.4 to 16% (Ribeiro et al. 2009). The Atlantic Forest still comprises 20000 plant species, 8000 being endemic (Myers, 2000).

For instance, in the 1950s, 40% of the Atlantic Forest still covered almost all the state of Paraná in Southern Brazil (Fundação SOS Mata Atlântica et al., 1998). During the following three decades, habitats that took thousands of years to evolve continued to be destroyed. What remains today covers an area 9.5 times smaller (10.58%) than the original area (Fundação SOS Mata Atlântica & INPE, 2010). This coincided with a period when the government aimed at an "economic development". However, this erroneous decision caused an irreparable loss of ecosystem functions (e.g. amount and quality of water, soil, carbon stocks, biodiversity, etc). Once more, immediate "development" led to the loss of a valuable ecosystem that had potential to bring true and sustainable development. The respect for the natural dynamics of ecosystems and species evolution is rarely presented in environmental discussions. Therefore it is timely to study biodiversity and to promote its management and restoration.

This chapter aims at discussing tree diversity of different formations of the Atlantic Forest in Southern Brazil, using data from phytosociological studies carried out in the state of Paraná. This study attempts to assess the following questions:

What ecosystems are understudied in terms of phytosociological surveys? Are the tree species richness and diversity of the different formations of these Atlantic Forests sufficiently known? What are the tree species richness and the diversity when considering only individuals found in phytosociological surveys? What species are the most abundant in each ecosystem evaluated? What are the differences and similarities among these formations?

2. Data analysis

Besides presenting a brief literature review, this study compiles abundance data and updated floristic information from the most representative tree phytosociological studies of the Atlantic Forest formations in the state of Paraná, Southern Brazil. The data were obtained from scientific papers, doctoral thesis, master dissertations and from our own surveys performed for at least 10 years in the region (Table 1). The floristic information from 39 studies encompassing 58 forest sites was included in the present study.

There was a difficulty in finding studies with same inclusion criteria (same diameter at breast height - DBH) for different formations. The minimum DBH value available in the original dataset ranged from 3.1 to 10.0 cm. This range was therefore considered in the present study. Additionally, the different sample sizes were not standardized among surveys.

The altitudes of the sites range from 5 to 1750 m a.s.l. The most distant sites (separated by 590 km) are located in the following geographic coordinates: 25°23′ S; 48°13′ W, near the Atlantic Ocean and 22°43′ S; 53°18′ W, near the Upper Paraná River (Figure 1).

Sites in early and middle sucessional stages, as well as undetermined *taxa* and exotic species were not included. *Taxa* identified only to the family or genus level were grouped according to taxonomic hierarchy. For example: the Myrtaceae group included many undetermined *taxa* of this family, and the *Lonchocarpus* group comprised undetermined species of this genus. For the richness estimation, each taxonomic group of undetermined *taxa* was considered as a unique species. The data underwent a detailed review to check all accepted species names and synonymy according to the "Species List of the Brazilian Flora" (Forzza et al., 2010). The compilation of the surveys on forest structure comprised 29 hectares of sampled area and 36627 measured individuals. The diversity indexes were calculated according to Magurran (1988). Canonical correspondence analyses (CCA) processed by the program CANOCO 4.5 (Ter Braak & Smilauer, 2002) were used to assess the relationship between abundance of the tree species of 58 sites comprising nine Atlantic Forest formations, and geo-climatic variables. The matrix with abundances per forest site includes 631 species. The geo-climatic matrix includes the following variables: distance from the ocean, annual temperature, altitude and annual rainfall. Data not presented in the original studies were obtained from climatic maps of IAPAR (Caviglione et al., 2000). Major approximations of mean annual temperatures (decrease of 0.54 °C for every 100 m of increased altitude) were used following recommendations in Roderjan & Grodski (1999). The Brazilian official vegetation classification (Veloso et al., 1991; IBGE, 1992) was used to group the sites into each Atlantic Forest formation (see below).

3. Atlantic Forest Biome, environmental and vegetational features

The Atlantic Forest in the state of Paraná has three distinct types of forest ecosystems: the Dense Rainforest (Atlantic "Ombrophilous" Dense Forest), the Araucaria Rainforest (Mixed "Ombrophilous" Forest) and the Semideciduous Seasonal Forest (IBGE, 1992). Each one of these forests also comprises distinct formations and associated or ingrown ecosystems, resulting from geomorphological and climatological features (Figure 1). The five main categories of formations (IBGE, 1992) were included, namely: Lowland (Coastal Plain Forest), Alluvial (Floodplain Forest), Submontane and Montane (both can be also considered Lower Montane), and Upper Montane. The Dense Rainforest presents all of these categories, whilst the Araucaria Rainforest and the Semideciduous Seasonal Forest Comprise mainly the Alluvial and Montane and the Alluvial and Submontane formations, respectively.

N	Forest formation	Municipality	Location (site)	Alt. (m a.s.l.)	Reference
1	Lowland DRF	Guaraqueçaba	Ilha do Superagui	12	Jaster (1995)
2	Lowland DRF	Guaratuba	Guaratuba	10	Galvão et al. (2002)
3	Lowland DRF	Matinhos	Matinhos	10	Galvão et al. (2002)
4	Lowland DRF	Paranáguá	Ilha do Mel	5	Menezes Silva (1990)
5	Alluvial DRF	Guaraqueçaba	Itaqui Reserve/site1	20	Zacarias (2008)
6	Alluvial DRF	Guaraqueçaba	Itaqui Reserve/site2	20	Zacarias (2008)
7	Submontane DRF	Antonina	Cachoeira Reserve	350	Liebsch et al. (2007)
8	Submontane DRF	Guaraqueçaba	Morro do Quitumbe	200	Athayde (1997)
9	Submontane DRF	Guaraqueçaba	Morro do Superagui	20-170	Jaster (1995)
10	Submontane DRF	Guaratuba	Rio Cubatãozinho	400	Guapyassu (1994)
11	Submontane DRF	Morretes	Morretes	485	Silva (1994)
12	Submontane DRF	Morretes	Serra da Prata/site1	400-600	Blum (2006)
13	Montane DRF	Guaratuba	Morro dos Perdidos	800	Blum et al. (2001)
14	Montane DRF	Morretes	Serra da Prata/site2	800-1100	Blum (2006)
15	Montane DRF	Piraquara	Mananciais da Serra	1030	Reginato & Goldenberg (2007)
16	Montane DRF	Quatro Barras	Morro Anhangava/site2	1150	Roderjan (1994)
17	Montane DRF	São José dos Pinhais	Guaricana/site1	500-700	Schorn (1992)
18	Montane DRF	São José dos Pinhais	Guaricana/site2	500-700	Schorn (1992)
19	Upper Montane DRF	Morretes	Pico Marumbi	1385	Rocha (1999)
20	Upper Montane DRF	Quatro Barras	Morro Anhangava/site3	1300	Portes et al. (2001)
21	Upper Montane DRF	Quatro Barras	Morro Anhangava/site1	1350	Roderjan (1994)
22	Upper Montane DRF	Morretes	Serra da Prata/site3	1400	Scheer et al. (in press a)
23	Upper Montane DRF	Antonina	Serra do Ibitiraquire	1700	Scheer et al. (in press a)
24	Upper Montane DRF	Morretes	Serra da Igreja	1300	Scheer et al. (in press a)
25	Upper Montane DRF	Guaraqueçaba	Serra Gigante	950	Scheer et al. (in press a)
26	Upper Montane DRF	Morretes	Morro Mãe Catira	1300	Koehler (2001)
27	Upper Montane DRF	Tijucas do Sul	Morro Araçatuba	1400	Koehler (2001)
28	Upper Montane DRF	São José dos Pinhais	Serra do Salto	1300	Koehler (2001)
29	Upper Montane DRF	Morretes	Morro Vigia	1300	Koehler (2001)
30	Alluvial ARF	Araucária	Distr.General Lúcio/site1	897	Pasdiora (2003)
31	Alluvial ARF	Araucária	Distr.General Lúcio/site2	897	Pasdiora (2003)
32	Alluvial ARF	Araucária	Rio Barigui	875	Barddal et al. (2004)
33	Alluvial ARF	Pinhais	Pinhais/site2	850	Seger et al. (2005)
34	Alluvial ARF	São José dos Pinhais	Rio Miringuava/site1	880	Jaster et al. (2002)
35	Montane ARF	Curitiba	Barigui Park	900	Kozera et al. (2006)
36	Montane ARF	Curitiba	Capão do Tigre	900	Rondon Neto et al. (2002)
37	Montane ARF	General Carneiro	Fazenda Pizzato	990	Watzlawick et al. (2005)
38	Montane ARF	Guarapuava	Fazenda 3 Capões/site1	950	Cordeiro (2010)
39	Montane ARF	Guarapuava	Fazenda 3 Capões/site2	950	Cordeiro (2010)
40	Montane ARF	Guarapuava	Fazenda 3 Capões/site3	950	Cordeiro (2010)
41	Montane ARF	Guarapuava	Fazenda 3 Capões/site4	950	Cordeiro (2010)
42	Montane ARF	Guarapuava	Araucárias Park	1070	Cordeiro & Rodrigues (2007)
43	Montane ARF	Jaguariaíva	Paredão da Santa	1195	Blum (ongoing study)
44	Montane ARF	Pinhais	Pinhais/site1	850	Seger et al. (2005)
45	Montane ARF	São João do Triunfo	São João do Triunfo	780	Durigan (1999)
46	Montane ARF	São José dos Pinhais	Rio Miringuava/site2	910	Jaster et al. (2002)
47	Montane ARF	Tijucas do Sul	Tijucas do Sul	850	Geraldi et al. (2005)
48	Montane ARF	Tijucas do Sul	Fazenda Bührer	850	Geraldi et al. (2005)
49	Alluvial SSF	Diamante do Norte	Caiuá Ecol.Station/site1	250	Borghi et al. (2004)
50	Alluvial SSF	Diamante do Norte	Caiuá Ecol.Station/site2	250	Costa Filho et al. (2006)
51	Alluvial SSF	Porto Rico	Alto Paraná	240	Campos et al. (2000)
52	Alluvial SSF	Londrina	Ribeirão dos Apertados	500	Bianchini et al. (2003)
53	Submontane SSF	Astorga	Ribeirão Aurora	550	Veiga et al. (2003)
54	Submontane SSF	Diamante do Norte	Caiuá Ecol.Station/site3	280	Del Quiqui et al. (2007)
55	Submontane SSF	Londrina	Mata dos Godoy Park	500	Soares-Silva et al. (1998)
56	Submontane SSF	Sapopema	Fazenda Bom Sucesso	780	Silva et al. (1995)
57	Submontane SSF	Tomazina	Rio das Cinzas	500	Blum et al. (2003)
58	Submontane SSF	Umuarama	Estrada Boiadeira	450	Blum & Petean (2008)

Table 1. List of the analyzed fores sites in Atlantic Forest formations in the state of Paraná, Brazil. (N - site number used in this study; DRF - Dense Rainforest; ARF - Araucaria Rainforest; SSF - Semidecidual Seasonal Forest).

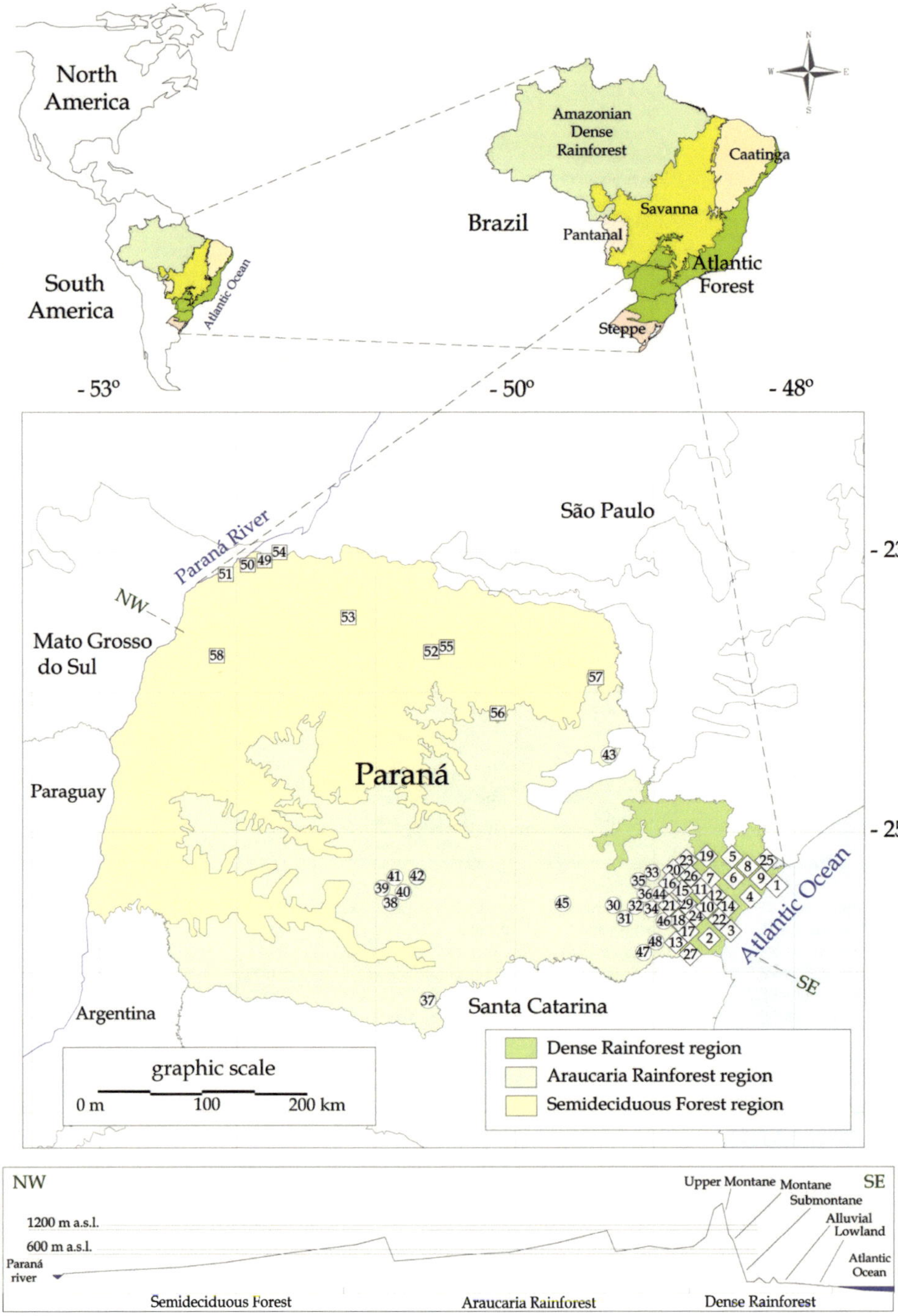

Adapted from: MMA (2011) and Roderjan et al. (2002).

Fig. 1. Location of the 58 sites in the Atlantic Forest types in the state of Paraná, Southern Brazil. Numbers are related to sites presented in Table 1.

3.1 Dense Rainforest

The Dense Rainforest, or Atlantic Forest *sensu stricto* (Oliveira-Filho & Fontes, 2000), is characterized by the dominance of large trees (25-30 m height) associated with many other biological forms, mainly epiphytes and woody lianas, that gives it its tropical appearance (Klein, 1979). Evergreen trees dominate the dense canopy (Veloso et al., 1991; IBGE, 1992).

Different plant communities of this forest type are found from the beach ridges near the Atlantic Ocean to the upper montane ridges of the Serra do Mar (Sea Mountain Range) and its western slopes towards the ecotone with Araucaria Rainforest (Figure 1). The altitudinal gradient ranges from 1 to 1887 m a.s.l., resulting in this region in a decrease of 0.54 °C for every 100 m increase in altitude (Roderjan & Grodski, 1999). While the coastal plain formations are in soils derived from Cenozoic (mainly Quaternary) marine sediments, alkali granites, embedded in high-grade metamorphic terrains, form the mineral soil horizons of the upper portions of the landscape. According to Mineropar (2001), intrusive igneous rocks from Serra do Mar were originated nearly 550 million years ago in the Upper Proterozoic to Cambrian. Faults from the Brasiliano (or Pan African) Cycle (events of the end of the late Proterozoic) and the Ponta Grossa Arch, cut the landscape in the NE-SW and NW-SE axes, respectively. Therefore, the soils of the Dense Rainforest have a high variety in parent material, genesis, depth, horizons, layers, structure, texture and organic matter.

The climate of the lower portions of the Dense Rainforest gradient, up to the transition between the Submontane and Montane formations, 600 - 800 m a.s.l. (Blum, 2010), can be classified as Cfa, or meso-thermic, according to the Köppen classification system. Daily mean annual temperature is 21 °C, the mean temperature in the coldest month is about 16 °C and, in the warmest month is higher than 22 °C, with hot and wet summers (December-March), and a no-pronounced dry season. Maack (2002) considered the region as having a transitional tropical climate (Af). The climate of upper portions, above 700 m a.s.l., is classified as Cfb (humid subtropical, meso-thermic, with cool summers, frequent frosts and no pronounced dry season). The mean temperature of the coldest month (July) in the region is less than 16 °C (reaching 12.5 °C) and of the warmest month (February) is less than 22 °C (20.5 °C). Measurements in the coastal region exceed 2000 mm of rain per year, and on the slopes of the mountains these values reach 3500 mm of rain per year (Caviglione et al., 2000; Maack, 2002).

As a result of the high environmental variety the Dense Rainforest is the most heterogeneous, complex and rich forest ecosystem of Southern Brazil (Leite & Klein, 1990).

A compiled list of the tree flora of the Dense Rainforest of Southern Brazil, using data of the botanical collection of the Barbosa Rodrigues Herbaria, revealed 708 species. More than 50% of those species occur exclusively in this type of Atlantic Forest (Leite & Klein, 1990).

Dense Rainforest communities in the advanced stages of succession cover an area of 3937.5 km^2 or 51,5% of the original distribution area as interpreted from satellite images from 1999 (Pires et al., 2005).

Forward we present the five cattegories of the Atlantic Rainforest.

3.1.1 Lowland formation

In South Brazil, the Lowland formation (Coastal Plain Forest) is restricted to Quaternary coastal plains growing on tsites near to sea level to about 20 m a.s.l.. Poorly developed soils and the high susceptibility to flooding during the rainiest periods are remarkable factors that led to its typical floristic and structural composition (Leite & Klein, 1990; Veloso et al., 1991; IBGE, 1992; Roderjan et al., 2002; Pires et al., 2005).

This formation presents a continuous canopy (about 20-25 m height) and two physiognomies can be distinguished. *Calophyllum brasiliense* trees dominate the canopy of areas with soils subject to waterlogging (Histosols, Spodosols and Entisols). This species is generally associated with *Tabebuia cassinoides, Tapirira guianensis, Ficus luschnatiana, Ilex pseudobuxus, Clusia criuva* and *Pouteria beaurepairei*. In better-drained lands, non-hydromorphic Entisols (Quartzipsamments/Arenosols) and Spodosols support higher diversity. There are common in the canopy *Tapirira guianensis, Ocotea pulchella, Ficus organensis, Manilkara subcericea, Pera glabrata, Alchornea triplinervia, Andira anthelmia, Ilex theezans, Ternstroemia brasiliensis*, besides many Myrtaceae such as *Psidium cattleianum* and *Myrcia multiflora* (Leite & Klein, 1990; Silva, 1990; Jaster, 1995; Jaster, 2002; Roderjan et al., 2002; Pires et al., 2005).

3.1.2 Alluvial formation

The Alluvial Dense Rainforest grows in Fluvisols and Gleysols in alluvial plains influenced by mountain range sediments carried by rivers (Roderjan et al., 2002).

The canopy is usually 20-25 m tall and some expressive species of this formation are *Pseudobombax grandiflorum, Alchornea triplinervia, Ficus organensis, Andira anthelmia* and *Syagrus romanzoffiana. Inga sessilis, Coussapoa microcarpa, Psidium cattleianum, Ocotea pulchella, Myrcia insularis* and *Marlierea tomentosa* are also important in these communities (Roderjan et al., 2002; Pires et al., 2005; Zacarias, 2008).

3.1.3 Submontane formation

This ecosystem comprises the lower portions of slopes of the mountain ranges and the Ribeira River Valley. According to IBGE (1992), this formation occurs between 30 and 400 m a.s.l.. However, Roderjan et al. (2002) adapted the upper limit of the Submontane formation to 600 m a.s.l., considering the regional scale. Results of a survey performed by Blum & Roderjan (2007) agree well with this limit. The Submontane Atlantic Rainforest generally occurs on Argisols, Oxisols and Cambisols, mainly in colluvial fans (Roderjan et al., 2002).

The dense canopy varies between 25 and 30 m in height and is characterized by high tree diversity and richness. *Virola bicuhyba, Sloanea guianensis, Aspidosperma pyricollum, Cedrela fissilis, Cariniana estrellensis, Pseudopiptadenia warmingii* and *Schyzolobium parahyba* are frequent in the canopy. *Bathysa australis, Pausandra morisiana, Euterpe edulis, Geonoma gamiova* and *Psychotria nuda* are common in the dominated strata (Leite & Klein, 1990; Maack, 2002; Roderjan et al., 2002; Pires et al., 2005; Blum, 2006).

This formation presents the highest floristic diversity of Southern Brazil due to the combination of factors like soils with good physical support and nutritional capacity, higher temperatures and well distributed rainfalls (Leite & Klein, 1990; Roderjan et al., 2002). These features also promote the development of dense and large-sized arboreal communities, associated with terrestrial and epiphytic strata, extremely rich and abundant (Blum, 2010).

3.1.4 Montane formation

The forest communities distributed over the intermediate slopes of the mountain ranges at elevations above the Submontane limits are classified as Montane formations. According to Roderjan et al. (2002) and Blum (2006), in the state of Paraná these communities are situated between 600 and 1200 m a.s.l..

It is noteworthy that the upper limit is also variable depending on specific soil and climate and, in many cases, the Upper Montane formation can already occur below 1200 m a.s.l. (Pires et al., 2005). Floristic differences are observable in relation to the lower level, but structurally and physiognomic, the Montane and Submontane formations are similar (Roderjan et al., 2002; Pires et al., 2005). Cambisols (with no textural gradient) and Entisols are very common in the Montane belt (Schorn, 1992; Roderjan, 1994, Blum, 2006).

The main environmental factors that affect the differentiation between the Montane and Submontane formations (Lower Montane Forests) are the climate, the topography and the soils. In the Montane Atlantic Forest it can occur occasional frosts, which are extremely limiting for many typical species of Submontane formation, that are subject to milder climate. It should be noted that the geomorphological differences result in distinct pedologies. The Montane terrains are steep and dissected while the Submontane sites are usually understated (Roderjan et al., 2002; Pires et al., 2005; Blum, 2006; 2010).

The canopy of the Montane Dense Rainforest is regular, varying about 20-25 m height. Several species of Lauraceae dominate in the upper strata, especially *Ocotea catharinensis, Ocotea odorifera, Ocotea bicolor* and *Cryptocarya aschersoniana. Aspidosperma pyricollum, Pouteria torta, Cabralea canjerana, Sloanea lasiocoma, Guapira opositta, Ilex paraguariensis* and *Guatteria australis* are also relevant. The lower strata are characterized by Myrtaceae, Rubiaceae and Monimiaceae families. Ferns (tree ferns) such as *Cyathea phalerata* and another species of Cyatheaceae are common in the understory (IBGE, 1992; Roderjan, 1994; Blum et al., 2001; Roderjan et al., 2002; Pires et al., 2005; Blum, 2006).

3.1.5 Upper Montane formation

In the state of Paraná, faults belonging to the Brasiliano (or Pan African) Cycle and the Ponta Grossa Arch currently confine the Upper Montane Rainforests (or Cloud Forests), allowing such vegetation to reach areas close to the main tops of the Sea Mountain Range (Scheer et al., in press b). This formation generally occurs from 1200 m a.s.l (Roderjan et al., 2002), even though it can be found at 900 m a.s.l., in small isolated mountains due to geomorphological conditions and the "*Massenerhebung* effect" (Grubb, 1971). In larger mountains, such as the Paraná Peak, the typical Upper Montane Rainforest ranges from 1400 to 1850 m a.s.l.., interspersed with high altitude grasslands. The changes in vegetation from forests to grasslands are abrupt and include ecotonal areas with "dwarf forests" or shrubby physiognomy with species of both formations across a gradient of a few meters (2-5 m).

Although typical Upper Montane Forests are composed by simplified tree associations, 346 vascular plant species have been detected in four mountain ranges (Scheer & Mocochinski, 2009). Small-sized trees ranging from 3 to 7 m tall, are subject to more restrict environmental conditions, such as low temperatures, strong winds and constant and heavy cloudiness, intense light radiation and shallow soils with low fertility and substantial histic horizons (Histosols and Leptosols). *Ilex microdonta, Siphoneugena reitzii, Myrceugenia seriatoramosa, Citronella paniculata, Weinmannia humilis, Ocotea porosa, Podocarpus sellowii* and *Drimys brasiliensis* are typical species in such areas (Leite & Klein, 1990; Roderjan, 1994, Koehler et al., 2002; Roderjan et al., 2002; Pires et al., 2005; Scheer, 2010; Scheer et al., in press a).

3.2 Araucaria Rainforest

Also called "Mixed Ombrophilous Forest" (IBGE, 1992), this forest physiognomy is characterized by merging elements from two distinct flora origins: the Tropical Afro-Brazilian and the Temperate Austro-Brazilian (Veloso et al., 1991). Classified as a Subtropical Forest, this ecosystem occurs mainly in the First and Second Plateaus of Paraná (mainly in the Center and the South of the state) at altitudes generally varying between 800 and 1000 m a.s.l. (Figure 1). In this region, temperatures are relatively low and frosts are common. In many locations, these formations share the landscape with natural grasslands.

According to Köppen System, the climate of the Araucaria Rainforest region is Cfb, with annual average temperatures between 16 and 18 °C and average annual rainfall generally between 1400 and 1600 mm (Caviglione et al., 2000).

This forest formation shows structural variations related to environmental diversification, varying from dense formations with trees of ca. 25 to 35 m tall, to stunted formations consisting of variable density of trees and shrubs, associated with terrestrial ferns and bamboos (Leite & Klein, 1990). According to Maack (2002), *Araucaria angustifolia*, commonly known as the "Brazilian pine" or "Paraná pine" is the dominant tree of this region, distinguishing this landscape.

A compiled list of the Araucaria Rainforest tree flora of Southern Brazil, using data of the botanical collection of the Barbosa Rodrigues Herbaria, revealed 352 species. Almost 50% of those species occur exclusively in this type of Atlantic Forest (Leite & Klein, 1990).

Castella & Britez (2004) analyzed satellite images from 1998 and concluded that Araucaria Rainforest communities at intermediate or advanced succession stages still covered 13420.6 km^2 of the state of Paraná, representing about 16.2% of the original cover (Castella & Britez, 2004). However, primary remnants are much less representative.

Two formations are presented in this section: Alluvial and Montane Araucaria Rainforests.

3.2.1 Alluvial formation

The Alluvial Araucaria Rainforest is associated mainly with the Montane formation and is easily distinguished by its typical physiognomy. This is a riparian forest that always occupies alluvial lands adjacent to watercourses (IBGE, 1992; Roderjan et al., 2002).

The physiognomy is structurally characterized by a high density of medium and small individuals, with the canopy ranging between 10 and 20 m in height. Communities can present different degrees of development. In the fairly homogeneous associations, subject to soils with considerable hydromorphy, such as some Fluvisols and Gleysols, *Sebastiania commersoniana* is the most relevant species. In more developed associations *Vitex megapotamica, Schinus terebinthifolius, Allophylus edulis, Luehea divaricata, Symplocos uniflora, Blepharocalyx salicifolius, Myrrhinium atropurpureum, Myrciaria tenella* and *Daphnopsis racemosa* are present. Even *Araucaria angustifolia* can be observed where lower hydromorphy allows its growth (Leite & Klein, 1990; Roderjan et al., 2002; Barddal, 2002; Pires et al., 2005).

3.2.2 Montane formation

According to IBGE (1992), the altitudinal range of the Montane Araucaria Rainforest occurrence is between 400 and 1000 m a.s.l.. However, some authors, such as Leite & Klein (1990) and Roderjan et al. (2002), rightly argue that typical communities occur at altitudes higher than 800 m a.s.l. Below this altitudinal belt there is the beginning of the transition between the Montane Araucaria Rainforest and the Dense Atlantic Rainforest (on the east) or the Semideciduous Seasonal Forest (on the west).

The typical physiognomy is marked by the dominance of *Araucaria angustifolia*, standing out over a continuous canopy that reaches on average 25-30 m height. The continuous strata is characterized by *Ocotea porosa, Nectandra lanceolata, Matayba elaeagnoides, Casearia decandra, Podocarpus lambertii, Cinnamodendron dinisii, Sloanea lasiocoma, Campomanesia xanthocarpa, Cedrela fissilis, Nectandra grandiflora, Jacaranda puberula, Drimys brasiliensis, Ilex paraguariensis* and *Lithraea brasiliensis*. Species of Myrtaceae and Monimiaceae prevail in the dominated strata. Ferns, especially *Dicksonia sellowiana,* are also common (Leite & Klein, 1990; Durigan, 1999; Roderjan et al., 2002; Rondon Neto et al., 2002; Cordeiro & Rodrigues, 2007).

3.3 Semideciduous Seasonal Forest

The Semideciduous Seasonal Forest region in the the state of Paraná occurs mainly in the Third Plateau (almost the entire North and West of the state) in altitudes generally between 200 and 600 m a.s.l. (Roderjan et al., 2002).

According to the Köppen system, the climate can be classified as Cfa, or meso-thermic, wet. Daily mean temperatures in the coldest month are under 18 °C and the mean temperature of the warmest month is over 22 °C (Maack, 2002). However, in this ecosystem the year can be divided into two distinct seasons: one tropical with intense summer rainfalls and short dry periods, and another subtropical with low winter temperatures and scarce precipitation. During this unfavorable cold and dry period, between 20 and 50% of the canopy trees are deciduous (Veloso et al., 1991; IBGE, 1992). In very specific locations this forest has as ingrown ecosystem the Savanna (Cerrado), which reaches its austral limit in this specified region, covering otherwise a major part of Brazil's Mid-West (Figure 1).

The Semideciduous Seasonal Forest shows succinct variations ranging from the evergreen to deciduous trees, which reach heights close to 30-40 m without forming a continuous superior canopy. This feature allows a great deal of sunlight to reach the forest ground, turning possible the development of a vigorous lower stratum (Silva & Soares Silva, 2000). There are also shrubs, lianas and epiphytes, although in lower abundance and richness compared to rainforests (Leite et al., 1986; Lamprecht, 1990; Leite & Klein, 1990; Roderjan et al., 2002).

A compiled list of the tree flora of the Semideciduous Seasonal Forest of Southern Brazil, using data of the botanical collection of the Barbosa Rodrigues Herbaria, presents at least 213 tree species (Leite & Klein, 1990). Silva & Soares Silva (2000), found 206 arboreal species in Godoy State Park.

Satellite images taken in 1998, indicate that 4174.7 km^2 of Semideciduous Seasonal Forest communities at intermediate or advanced stages of succession covered only 5.4% of the original area of distribution in that year (SEMA, 2002). Some forests of the northern region were reduced to less than 1% in Maringá and 0.8% in Assaí municipalities (IPARDES, 1986).

Forward are presented the characterizations for the Alluvial and Submontane formations of this forest type.

3.3.1 Alluvial formation

This formation is distributed over riverine floodplains and some islands in the Paraná river, and also around some of its tributaries. The Alluvial formation occurs predominantly in soils with considerable hydromorphy, such as some Fluvisols, Entisols (Quartzipsamments) and Gleysols (Roderjan et al., 2002).

The Alluvial Semideciduous Seasonal Forest is characterized by a canopy about 15 to 20 m height and lower floristic diversity. Among the main species are *Cecropia pachystachya, Triplaris americana, Calophyllum brasiliense, Gallesia integrifolia* and *Chrysophyllum gonocarpum. Sebastiania commersoniana, Anadenanthera colubrina, Acrocomia aculeata* and *Inga uruguensis* are also common (Leite et al. 1986; IBGE, 1992, Roderjan et al., 2002; SEMA, 2002; Costa Filho et al., 2006).

3.3.2 Submontane formation

In the state of Paraná, this formation can be found under 600 m a.s.l.. The Submontane Semideciduous Seasonal Forest occurs in soils of different lithologies (sandstone and igneous extrusive rocks) that result in Oxisols, Ultisols, Regosols (Psamments), Inceptisols, Leptosols and Arenosols (Roderjan et al., 2002). Depending on the lithology, different textures and fertility levels can be found in these soils.

The emergent irregular canopy can reach around 35-40 m height. In the upper stratum are common *Aspidosperma polyneuron, Handroanthus heptaphyllus, Gallesia integrifolia, Balfourodendron riedelianum, Peltophorum dubium, Astronium graveolens, Diatenopteryx sorbifolia, Parapitadenia rigida, Cariniana estrellensis, Cedrela fissilis, Albizia hasslerii, Lonchocarpus guilleminianus, Machaerium stipitatum, Holocalyx balansae, Rauvolfia sellowii,* and *Nectandra megapotamica,* among others. The dominated strata are characterized by *Guarea macrophylla, Actinostemon concolor, Metrodorea nigra, Sorocea bomplandii* and *Pilocarpus pennatifolius* (Hueck, 1972; Leite et al., 1986; Leite & Klein, 1990; Maack, 2002; Roderjan et al., 2002; SEMA, 2002).

4. The studied Atlantic Forest formations

4.1 Analysis of the sampled sites in the Atlantic Forests

Among the 58 sites, comprising 29 ha of plots and 36627 sampled individuals, selected from 39 studies, the Dense Rainforest has the highest number of sampled sites (29 out of a total of 58) and the highest number of tree individuals found (Table 2). This great amount of data available is due to the high number of Upper Montane Dense Rainforest sampled sites, which present relatively many areas with primary and well preserved vegetation. In spite of the relatively smaller sampled area, small-sized trees present in high abundance make up for a high number of individuals found.

On the other hand, the Alluvial formation of the Dense Rainforest is the least sampled one (Table 2). The lack of phytosociological studies conducted in this formation may be related to its actual small cover area (representing only 0.89% of the remnants of Dense Rain Forest) (Pires et al., 2005), and also to its level of degradation, which make the search for typical and representative remnants difficult. According to this data, the studied area is almost six times smaller than the area of the neighbor Lowland formation. Therefore, this understudied vegetation needs more studies.

The low number of studied sites in the Semideciduous Seasonal Forest in the state of Paraná is also remarkable, especially when the great covering area (Figure 1) and the considerable latitudinal extension (22° 30′ - 26° 30′ S) of this ecosystem are considered. The lack of studies is related to the advanced stage of degradation of this forest type. In a few decades this ecosystem was reduced to scarce and fragmented remnants, generally in bad conservation conditions. We did not find phytosociological surveys in the Southwestern region of the state of Paraná, where the Semideciduous Forest is also found (Figure 1). Moreover, forest structure studies are not available even in the most representative remnant of the Submontane Semideciduous Forest of Southern Brazil, located in the Iguaçu National Park, a protected area of approximately 1852.6 km^2. This fact points out to the urgency of knowing better this important ecosystem and its resources.

Besides the similar sampled areas shown by the three Atlantic Forest types (Table 2), and highest number of studied sites for the Dense Rainforest, the sum of the 10 sites of the Semideciduous Forest comprises the largest sampled area (11.8 ha). The two most extensively sampled formations are the Montane Araucaria Rainforest and the Submontane Semideciduous Forest. However, these sampling areas are very small compared to the sampling area of other surveys in different ecosystems. As an example, a single study in the Amazonian Dense Forest, could easily cover an area of 20 ha (Pitman et al., 2002; Laurance et al., 2010).

4.2 Alfa diversity

In the 58 selected phytosociological surveys, 700 species, 256 genera and 83 families were sampled (Table 2). Of the total number of species, 10 *taxa* represent groups with undetermined species at the family level and 58 *taxa* represent groups with undetermined species at the genus level. It is important to mention that the Atlantic Forest is habitat for many other tree species that were not found due to the criterion of inclusion, or due do the area needed to sample, that neither checklists can cover completely.

Biome	Atlantic Forest types	Atlantic Forest formations	Number of sites	Sampled area (ha)	Number of measured individuals	Number of species	Number of genera	Number of families	Shannon-Wiener Diversity Index	Simpson Diversity Index 1-D	Evenness
ATLANTIC FOREST	Dense Rainforest		29	9.01	14165	469	174	72	5.11	0.99	-
		Alluvial	2	0.32	766	78	59	32	3.38	0.94	0.77
		Lowland	4	1.86	2808	148	87	49	3.93	0.96	0.79
		Submontane	6	2.38	3251	265	132	57	4.61	0.98	0.83
		Montane	6	3.48	3140	210	101	52	4.64	0.98	0.87
		Upper Montane	11	0.97	4200	88	45	30	3.20	0.93	0.71
	Araucaria Rainforest		19	8.30	9196	220	101	51	4.29	0.97	-
		Aluvial	5	0.81	1792	79	54	33	2.48	0.74	0.57
		Montane	14	7.49	7404	211	99	50	4.38	0.98	0.82
	Semideciduous Forest		10	11.80	13265	282	154	60	4.64	0.99	-
		Alluvial	4	4.34	5354	157	107	43	4.26	0.98	0.84
		Submontane	6	7.46	7911	252	140	56	4.46	0.98	0.81
	ATLANTIC FOREST		**58**	**29.11**	**36627**	**700**	**256**	**83**	**5.53**	**0.99**	**-**

Table 2. Richness and other diversity parameters of the analyzed Atlantic Forest formations in the state of Paraná, Southern Brazil.

Among the Atlantic Forest types analyzed in this study, the Dense Rainforest is the richest in tree species (469). These species are distributed in 174 genera and 72 families. This tropical forest presents the highest tree diversity according to the Shannon-Wiener index, being its Montane and Submontane formations the most diverse (Table 2; Figure 3). Even though the analysis shows the highest diversity value in the Montane formation, according to many studies, the Submontane formation shows a tendency of being the most diverse (Guapyassú, 1994; Roderjan, 1994; Jaster, 1995; Athayde, 1997; Blum, 2006).

Tabarelli & Mantovani (1999) compiled phytosociological studies on the Dense Rainforest in Southeastern Brazil, which comprised 432 species measured in 2.3 ha and considered 2640 trees measured through the quarter-plot method. According to these authors the richness of these forests are low when compared to other Neotropical Forests of South America. However, more studies considering similar conditions (*e.g.* larger plots - 1 ha - with homogeneous sites) are needed to corroborate these results.

However, some diversity indexes such as Eveness presented for the Atlantic Forest types can be influenced by the disproportionate sampling among their formations.

According to Table 2, the Araucaria Rainforest presents less than half of the Dense Rainforest tree species richness (220), in part due to its lower environmental heterogeneity and altitudinal range. In addition, the colder climate in the Araucaria Rainforest region probably restricts the occurrence of a substantial number of species.

The Semideciduous Seasonal Forest shows intermediate values of species richness (282) and diversity when compared with the two types of Rainforests. (Table 2; Figure 3).

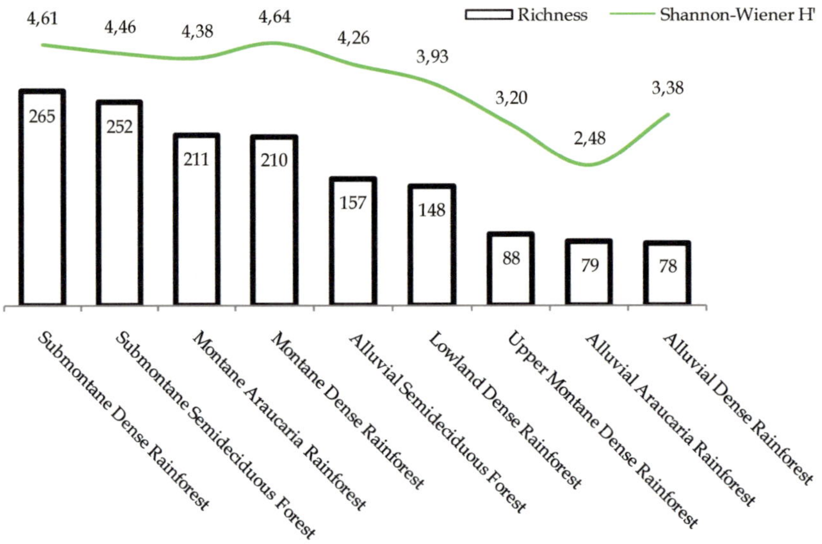

Fig. 3. Tree richness and Shannon-Wiener diversity indexes of the Atlantic Forest formations in the state of Paraná, Southern Brazil.

The higher richness when compared with the Araucaria Rainforest is due to some distinct environmental conditions, such as lower altitude, which result in higher temperatures. On the other hand, the restriction imposed by the seasonal climate, especially by the unfavorable dry season, prevents the Seasonal Forest to reach values of species richness similar to those observed in the Dense Rainforest (i.e. 66% higher). A lower difference (31%) in this parameter between these two types of forest was observed by Oliveira Filho & Fontes (2000) in floristic surveys in Southeastern Brazil.

The Submontane formations present the highest richness of tree species, followed by the Montane formations (Figure 3). The Lowland, Upper Montane and Alluvial formations present lower richness. From the data available, the Alluvial Dense Rainforest has the lowest number of species. However, this is certainly related to the lack of inventories in this type of forest. Due its proximity and environmental similarity with the coastal Lowland formation, it is probable that the real values of tree species richness of both forests are similar.

Therefore, among the Atlantic Forest formations, the Alluvial Araucaria Rainforest can be considered the least rich in tree species. This is due to two important environmental factors: the milder climate, with frequent frosts; and the hydromorphic feature of its soils, with high susceptibility to flooding during the rainiest periods.

According to Table 3, Myrtaceae, Fabaceae and Lauraceae are the families with the highest tree species richness of the Atlantic Forest, making up to 40% of the measured Rainforest species and 35% of the Semideciduous Seasonal Forest species.

Myrtaceae presents the highest number of tree species in the two Rainforests, reaching 112 species in the Dense Rainforest, while Fabaceae is the richest family of the Semideciduous

Forest. Some tree families, such as Fabaceae, Moraceae, Sapotaceae and Meliaceae, show a tendency of preferring warmer environments. On the other hand, Aquifoliaceae, Asteraceae and Myrsinaceae present comparatively high floristic expression in the Rainforest environments. In addition, Melastomataceae presents high species richness only in the Dense Rainforest. According to Gentry (1995) and Oliveira-Filho et al. (2000), with increasing altitude there is a decrease in Fabaceae richness and an increase in Aquifoliaceae and Asteraceae richness.

At the genus level, *Eugenia* is the richest in the tropical forests (mainly in the Dense Rainforest), even though the Araucaria Rainforest (subtropical) presents many species of this taxon. Oliveira-Filho et al. (2000) observed the same tendency in an extensive survey on floristic differentiation patterns among Atlantic Forests in Southeastern Brazil. The richness of *Eugenia* is relatively higher in slope rain forests, and its relative position is reversed with *Myrcia* in lowland (including alluvial) and plateau (Araucaria Rainforest) formations. *Marlierea* has a considerable number of species in Lowland and Submontane formations of the Dense Rainforest, whilst *Myrceugenia* has many species in Upper Montane and Montane formations of both rainforests (Table 4).

Ocotea is among the three genera with the highest number of tree species in the three Atlantic Forest types in the state of Paraná. *Ilex* and *Myrceugenia* have a considerable number of species in the two rainforests, whilst *Trichilia* and *Cordia* have many species in the Semideciduous Forest (Tables 3 and 4).

Dense Rainforest (S=469; N=29)				**Araucaria Rainforest (S=220; N=19)**				**Semideciduous Forest (S=282; N=10)**			
Family	**S**	**Genus**	**S**	**Family**	**S**	**Genus**	**S**	**Family**	**S**	**Genus**	**S**
Myrtaceae	112	*Eugenia*	39	Myrtaceae	49	*Myrcia*	14	Fabaceae	47	*Eugenia*	13
Lauraceae	39	*Myrcia*	25	Lauraceae	23	*Ocotea*	11	Myrtaceae	32	*Nectandra*	8
Fabaceae	35	*Ocotea*	18	Fabaceae	18	*Eugenia*	10	Lauraceae	19	*Ocotea*	8
Rubiaceae	24	*Miconia*	12	Asteraceae	10	*Myrceugenia*	8	Meliaceae	13	*Trichilia*	8
Melastomataceae	21	*Ilex*	11	Rubiaceae	10	*Ilex*	6	Rubiaceae	11	*Cordia*	7
Moraceae	12	*Myrceugenia*	11	Salicaceae	8	*Myrsine*	6	Solanaceae	11	*Solanum*	7
Sapotaceae	12	*Calyptranthes*	10	Solanaceae	8	*Solanum*	6	Salicaceae	10	*Inga*	6
Aquifoliaceae	11	*Inga*	9	Aquifoliaceae	6	*Symplocos*	6	Euphorbiaceae	9	*Myrcia*	6
Euphorbiaceae	11	*Nectandra*	9	Myrsinaceae	6	*Casearia*	5	Rutaceae	9	*Casearia*	5
Asteraceae	9	*Marlierea*	8	Sapindaceae	6	*Maytenus*	5	Boraginaceae	7	*Ficus*	5
Myrsinaceae	9	*Myrsine*	8	Symplocaceae	6	*Allophylus*	4	Moraceae	7	*Lonchocarpus*	5
Annonaceae	8	*Symplocos*	8	Annonaceae	5	*Lonchocarpus*	4	Sapotaceae	7	*Machaerium*	5
Meliaceae	8	*Coccoloba*	7	Celastraceae	5	*Machaerium*	4	Annonaceae	6	*Pouteria*	5
Salicaceae	8	*Psychotria*	7	Euphorbiaceae	4	*Nectandra*	4	Malvaceae	6	*Sloanea*	4
Sapindaceae	8	*Ficus*	6	Rutaceae	4	*Piptocarpha*	4	Sapindaceae	6	*Zanthoxylum*	4

Table 3. The 15 best represented families and genera measured in the three Atlantic Forest types of the state of Paraná, Southern Brazil. S = number of species, N = number of sites.

The most abundant species found in phytosociological studies in the Atlantic Forest formations in the state of Paraná are given in Table 5. Except for the Alluvial, the sampling for all the other formations can be considered very representative of what can be found in these forests. This compilation obtained data for at least 2800 individuals and four sites for each formation, whereas phytosociological studies for the Semideciduous Seasonal forest in the Southwest of the state of Paraná cannot be found.

Dense Rainforest (S=469; N=29)										Araucaria Rainforest (S=220; N=19)				Semideciduous Forest (S=282; N=10)			
Alluvial* (N=2)	**S 78**	**Lowland (N=4)**	**S 148**	**Submontane (N=6)**	**S 265**	**Montane (N=6)**	**S 210**	**Upper Montane (N=11)**	**S 88**	**Aluvial (N=5)**	**S 79**	**Montane (N=14)**	**S 211**	**Alluvial (N=4)**	**S 157**	**Submontane (N=6)**	**S 252**
Myrtaceae	18	Myrtaceae	35	Myrtaceae	53	Myrtaceae	47	Myrtaceae	26	Myrtaceae	22	Myrtaceae	45	Fabaceae	28	Fabaceae	45
Fabaceae	7	Fabaceae	13	Fabaceae	23	Lauraceae	28	Lauraceae	8	Fabaceae	8	Lauraceae	23	Myrtaceae	16	Myrtaceae	31
Melastomataceae	6	Lauraceae	9	Lauraceae	19	Fabaceae	14	Aquifoliaceae	7	Aquifoliaceae	4	Fabaceae	14	Meliaceae	11	Lauraceae	19
Rubiaceae	6	Myrsinaceae	6	Rubiaceae	15	Rubiaceae	13	Asteraceae	5	Salicaceae	4	Asteraceae	10	Lauraceae	8	Meliaceae	13
Aquifoliaceae	3	Rubiaceae	6	Melastomataceae	13	Aquifoliaceae	7	Symplocaceae	4	Euphorbiaceae	3	Rubiaceae	10	Euphorbiaceae	6	Salicaceae	10
Bignoniaceae	3	Euphorbiaceae	5	Sapotaceae	11	Melastomataceae	7	Cunoniaceae	3	Lauraceae	3	Salicaceae	8	Sapindaceae	6	Rubiaceae	9
Clusiaceae	3	Melastomataceae	5	Moraceae	10	Proteaceae	5	Melastomataceae	3	Myrsinaceae	3	Solanaceae	8	Sapotaceae	6	Rutaceae	9
Annonaceae	2	Moraceae	5	Euphorbiaceae	8	Sapindaceae	5	Rubiaceae	3	Rubiaceae	3	Aquifoliaceae	6	Rubiaceae	5	Euphorbiaceae	8
Arecaceae	2	Aquifoliaceae	4	Meliaceae	8	Asteraceae	4	Celastraceae	2	Sapindaceae	3	Myrsinaceae	6	Rutaceae	5	Solanaceae	8
Lauraceae	2	Arecaceae	4	Annonaceae	6	Cunoniaceae	4	Clethraceae	2	Anacardiaceae	2	Sapindaceae	6	Annonaceae	4	Boraginaceae	7
Meliaceae	2	Bignoniaceae	4	Cyatheaceae	5	Euphorbiaceae	4	Ericaceae	2	Symplocaceae	2	Symplocaceae	6	Apocynaceae	4	Moraceae	7
Sapindaceae	2	Sapotaceae	4	Monimiaceae	5	Monimiaceae	4	Myrsinaceae	2	Annonaceae	1	Annonaceae	5	Boraginaceae	4	Annonaceae	5
Sapotaceae	2	Annonaceae	3	Salicaceae	5	Myrsinaceae	4	Polygonaceae	2	Araucariaceae	1	Celastraceae	5	Salicaceae	4	Malvaceae	5
Urticaceae	2	Clusiaceae	3	Apocynaceae	4	Salicaceae	4	Proteaceae	2	Arecaceae	1	Euphorbiaceae	4	Anacardiaceae	3	Sapindaceae	5
Anacardiaceae	1	Cyatheaceae	3	Arecaceae	4	Solanaceae	4	Winteraceae	2	Bignoniaceae	1	Rutaceae	4	Malvaceae	3	Apocynaceae	4
Myrcia	9	*Myrcia*	11	*Eugenia*	20	*Eugenia*	16	*Eugenia*	7	*Myrcia*	8	*Myrcia*	14	*Eugenia*	8	*Eugenia*	12
Miconia	4	*Eugenia*	8	*Myrcia*	10	*Ocotea*	15	*Myrcia*	7	*Ilex*	4	*Ocotea*	11	*Trichilia*	6	*Nectandra*	8
Eugenia	3	*Calyptranthes*	5	*Miconia*	9	*Myrcia*	9	*Ilex*	6	*Eugenia*	3	*Eugenia*	10	*Cordia*	4	*Ocotea*	8
Ilex	3	*Marlierea*	5	*Ocotea*	9	*Ilex*	7	*Myrceugenia*	6	*Myrceugenia*	3	*Myrceugenia*	7	*Nectandra*	4	*Trichilia*	8
Andira	2	*Myrsine*	5	*Marlierea*	6	*Myrceugenia*	6	*Ocotea*	5	*Myrsine*	3	*Ilex*	6	*Pouteria*	4	*Cordia*	7
Inga	2	*Ficus*	4	*Calyptranthes*	5	*Nectandra*	5	*Symplocos*	4	*Casearia*	2	*Myrsine*	6	*Campomanesia*	3	*Inga*	6
Psychotria	2	*Ilex*	4	*Cyathea*	5	*Calyptranthes*	4	*Persea*	3	*Erythrina*	2	*Solanum*	6	*Casearia*	3	*Myrcia*	6
Tibouchina	2	*Inga*	4	*Inga*	5	*Casearia*	4	*Agarista*	2	*Inga*	2	*Symplocos*	6	*Guarea*	3	*Casearia*	5
Abarema	1	*Nectandra*	4	*Mollinedia*	5	*Inga*	4	*Clethra*	2	*Machaerium*	2	*Casearia*	5	*Inga*	3	*Ficus*	5
Alchornea	1	*Alchornea*	3	*Psychotria*	5	*Mollinedia*	4	*Coccoloba*	2	*Ocotea*	2	*Maytenus*	5	*Lonchocarpus*	3	*Lonchocarpus*	5
Amaioua	1	*Miconia*	3	*Trichilia*	5	*Myrsine*	4	*Daphnopsis*	2	*Sebastiania*	2	*Allophylus*	4	*Machaerium*	3	*Machaerium*	5
Aniba	1	*Ocotea*	3	*Brosimum*	4	*Roupala*	4	*Drimys*	2	*Symplocos*	2	*Lonchocarpus*	4	*Ocotea*	3	*Solanum*	5
Blepharocalyx	1	*Pouteria*	3	*Campomanesia*	4	*Rudgea*	4	*Maytenus*	2	*Allophylus*	1	*Machaerium*	4	*Sloanea*	3	*Zanthoxylum*	4
Brosimum	1	*Andira*	2	*Nectandra*	4	*Alchornea*	3	*Miconia*	2	*Araucaria*	1	*Nectandra*	4	*Allophylus*	2	*Albizia*	3
Cabralea	1	*Byrsonima*	2	*Chrysophyllum*	4	*Cinnamomum*	3	*Myrsine*	2	*Banara*	1	*Zanthoxylum*	4	*Aspidosperma*	2	*Campomanesia*	3

Table 4. The 15 best represented families and genera measured in the nine Atlantic Forest formations of the state of Paraná, Southern Brazil. S = number of species, N = number of sites. *Data from Zacarias (2008).

Dense Rainforest (N=29; I=14288)									
Alluvial* (N=2)	**I (766)**	**Lowland (N=4)**	**I (2808)**	**Submontane (N=6)**	**I (3251)**	**Montane (N=6)**	**I (3140)**	**Upper Montane (N=11)**	**I (4200)**
Canopy species									
Tabebuia cassinoides	112	*Tabebuia cassinoides*	287	*Sloanea guianensis*	121	*Nectandra reticulata*	126	*Ilex microdonta*	700
Psidium cattleianum	75	*Clusia criuva*	183	*Calyptranthes grandifolia*	117	*Alchornea triplinervia*	85	*Siphoneugena reitzii*	637
Pera glabrata	55	*Ocotea pulchella*	183	*Guapira opposita*	100	*Ilex paraguariensis*	79	*Drimys angustifolia*	280
Syagrus romanzoffiana	44	*Calophyllum brasiliense*	178	*Hieronyma alchorneoides*	100	*Ocotea odorifera*	76	*Myrcia pulchra*	278
Tibouchina trichopoda	34	*Ternstroemia brasiliensis*	100	*Eugenia cerasiflora*	52	*Cryptocarya aschersoniana*	67	*Ocotea porosa*	206
Andira anthelmia	16	*Ilex pseudobuxus*	88	*Marlierea tomentosa*	50	*Syagrus romanzoffiana*	63	*Myrceugenia seriatoramosa*	180
Handroanthus umbellatus	14	*Pera glabrata*	81	*Virola bicuhyba*	43	*Cabralea canjerana*	56	*Handroanthus catarinensis*	177
Ocotea pulchella	12	*Psidium cattleianum*	81	*Marlierea silvatica*	42	*Myrsine umbellata*	55	*Ilex chamaedryfolia*	151
Inga edulis	11	*Tapirira guianensis*	59	*Trichilia lepidota*	39	*Ocotea porosa*	53	*Citronella paniculata*	139
Maytenus robusta	11	*Syagrus romanzoffiana*	57	*Cupania oblongifolia*	33	*Weinmannia paulliinifollia*	51	*Blepharocalyx salicifolius*	129
Ilex pseudobuxus	10	*Byrsonima ligustrifolia*	49	*Protium kleinii*	28	*Ocotea catharinensis*	47	*Weinmannia humilis*	111
Amaioua guianensis	9	*Guapira opposita*	47	*Cecropia pachystachya*	27	*Guatteria australis*	46	*Ouratea vaccinioides*	96
Jacaranda puberula	9	*Myrcia pubipetala*	46	*Quiina glazovii*	26	*Ocotea bicolor*	45	*Eugenia neomyrtifolia*	87
Calophyllum brasiliense	8	*Myrcia racemosa*	45	*Calyptranthes lucida*	26	*Guapira opposita*	43	*Pimenta pseudocaryophyllus*	70
Myrcia glabra	8	*Pouteria beaurepairei*	38	*Alchornea triplinervia*	25	*Cupania vernalis*	37	*Myrcia dichrophylla*	63
Understory species								*Myrcia guianensis*	61
Psychotria nuda	97	*Cyathea atrovirens*	182	*Psychotria nuda*	176	*Cyathea phalerata*	187	*Ternstroemia brasiliensis*	61
Eugenia blastantha	24	*Erythroxylum amplifolium*	64	*Rudgea jasminoides*	161	*Cordiera concolor*	93	*Myrcia hartwegiana*	60
Euterpe edulis	23	*Myrcia multiflora*	60	*Garcinia gardneriana*	115	*Alsophila setosa*	87	*Podocarpus sellowii*	50
Guarea macrophylla	17	*Marlierea tomentosa*	53	*Euterpe edulis*	114	*Dicksonia sellowiana*	79	*Laplacea fructicosa*	46
Myrcia insularis	12	*Hedyosmum brasiliense*	33	*Bathysa australis*	113	*Ocotea teleiandra*	54	*Ocotea vaccinioides*	45

Table 5. Species with the highest number of individuals measured for each formation of Atlantic Forest in the state of Paraná, Southern Brazil. I = number of sampled individuals, N = number of sites. *Data from Zacarias (2008).

Araucaria Rainforest (N=19; I=9196)				**Semideciduous Seasonal Forest (N=10; I=13264)**			
Alluvial* (N=5)	**I (1792)**	**Montane (N=14)**	**I (7404)**	**Alluvial (N=4)**	**I (5354)**	**Submontane (N=6)**	**I (7911)**
Canopy species							
Sebastiania commersoniana	903	*Araucaria angustifolia*	514	*Cecropia pachystachya*	393	*Astronium graveolens*	263
Allophylus edulis	62	*Campomanesia xanthocarpa*	364	*Triplaris americana*	245	*Parapiptadenia rigida*	232
Schinus terebinthifolius	56	*Podocarpus lambertii*	293	*Gallesia integrifolia*	174	*Campomanesia xanthocarpa*	226
Myrrhinium atropurpureum	52	*Matayba elaeagnoides*	291	*Peltophorum dubium*	152	*Machaerium stipitatum*	224
Blepharocalyx salicifolius	45	*Allophylus edulis*	200	*Sloanea guianensis*	152	*Luehea divaricata*	224
Vitex megapotamica	42	*Casearia decandra*	200	*Calophyllum brasiliense*	147	*Casearia gossypiosperma*	205
Myrsine umbellata	37	*Cupania vernalis*	171	*Alchornea triplinervia*	144	*Aspidosperma polyneuron*	187
Campomanesia xanthocarpa	27	*Nectandra grandiflora*	157	*Croton floribundus*	135	*Balfourodendron riedelianum*	169
Symplocos uniflora	24	*Cinnamodendron dinisii*	150	*Ocotea diospyrifolia*	127	*Nectandra megapotamica*	167
Casearia decandra	20	*Styrax leprosus*	138	*Astronium graveolens*	118	*Apuleia leiocarpa*	101
Erythrina crista-galli	18	*Ilex paraguariensis*	136	*Hymenaea courbaril*	114	*Senegalia polyphylla*	96
Matayba elaeagnoides	15	*Ocotea porosa*	127	*Guarea guidonia*	106	*Holocalyx balansae*	94
Myrcia splendens	15	*Nectandra megapotamica*	105	*Parapiptadenia rigida*	99	*Nectandra reticulata*	93
Prunus myrtifolia	13	*Lonchocarpus campestris*	86	*Nectandra reticulata*	95	*Lonchocarpus campestris*	93
Erythroxylum deciduum	12	*Luehea divaricata*	85	*Lonchocarpus cultratus*	94	*Peltophorum dubium*	92
Understory species							
Myrciaria tenella	141	*Eugenia uniflora*	345	*Chrysophyllum gonocarpum*	152	*Metrodorea nigra*	471
Myrceugenia glaucescens	36	*Sebastiania commersoniana*	271	*Zygia cauliflora*	148	*Chrysophyllum gonocarpum*	393
Myrcia laruotteana	30	*Casearia sylvestris*	191	*Actinostemon concolor*	118	*Plinia rivularis*	270
Sebastiania brasiliensis	28	*Dicksonia sellowiana*	69	*Guarea kunthiana*	91	*Zygia cauliflora*	244
Daphnopsis racemosa	20	*Brunfelsia pilosa*	60	*Metrodorea nigra*	70	*Eugenia uniflora*	154

Table 5. (Continued)

5. Beta diversity of Atlantic Forest formations

According to the Venn diagrams (Figure 4), the two Rainforest types share at least 99 tree species, most of them very common in the Montane formations of both forest types, like *Cinnamodendrum dinisii, Ocotea porosa, Drimys brasiliensis* and *Ocotea odorifera*. The surveys carried out in the Dense Rainforest shared 94 species with the Semideciduous Forest, which in turn shared 88 species with the Araucaria Rainforest. At least 50 arboreal species (8% of the species measured) occur in the three Atlantic Forest types, such as *Campomanesia xanthocarpa, Casearia sylvestris, Alchornea triplinervia, Nectandra megapotamica, Sloanea guianensis, Cupania vernalis, Casearia decandra, Syagrus romanzoffiana, Blepharocalyx salicifolius, Myrsine umbellata, Ocotea pulchella* and *Ilex paraguariensis*, among the most abundant.

A total of 272 species were found exclusively in the Dense Rainforest, e.g. *Tibouchina trichopoda, Andira anthelmia, Handroanthus umbellatus, Tabebuia cassinoides, Clusia criuva, Pera glabrata, Ternstroemia brasiliensis, Virola bicuhyba, Marlierea tomentosa, Cupania oblongifolia, Protium kleinii, Quiina glazovii, Aspidosperma pyriccolum, Myrcia freyreissiana, Siphoneugena reitzii, Drimys angustifolia* and *Handroanthus catarinensis*.

At least 51 tree species were found only in Araucaria Rainforest surveys, such as *Podocarpus lambertii, Nectandra grandiflora, Lithraea brasiliensis, Guettarda uruguensis, Curitiba prismatica, Symplocos celastrinea, Myrrhinium atropurpureum, Myrcianthes pungens, Ocotea nutans, Erythroxylum deciduum, Cinnamomum amoenum and Zanthoxylum kleinii.*

Among the 127 species found only in the Semideciduous Seasonal Forest (Figure 4), the most abundant are *Chrysophyllum gonocarpum, Metrodorea nigra, Astronium graveolens, Parapiptadenia rigida, Plinia rivularis, Casearia gossypiosperma, Triplaris americana, Balfourodendron riedelianum, Peltophorum dubium, Holocalyx balansae, Aspidosperma polyneuron and Gallesia integrifolia.*

The dendrogram using Sorensen's similarity coefficients for tree species (Figure 5) shows two major distinct groups. The first one comprises the five Dense Rainforest formations, and the second comprises the Araucaria Rainforest and the Semideciduous Seasonal Forest, reinforcing the data given in Venn diagram (Figure 4). These two types of Atlantic Forest

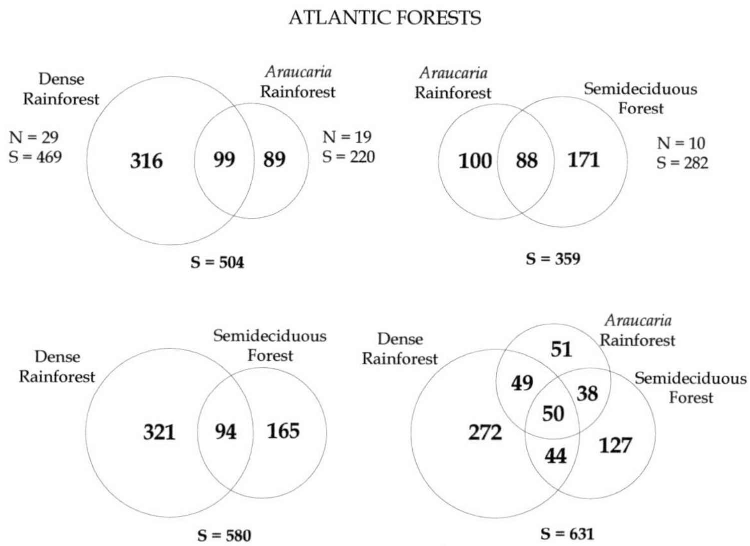

Fig. 4. Venn diagrams presenting the number of measured tree species shared in the 58 sites of the three types of Atlantic Forest in the state of Paraná, Southern Brazil.

share 24.5% of the total species. The two Rainforests share 19.6% of the species and the Semideciduous Seasonal Forest share 16.2% of the species with the Dense Rainforest. These values differ from those found by Oliveira-Filho & Fontes (2000), in which Dense Rainforest and Semideciduous Forest of Southeastern Brazil shared a high proportion of tree species in their checklists: 50% and 66% respectively.

The comparatively high floristic similarity within the Southeastern region than in the Southern region, can be related to the absence of Araucaria Rainforests between these two tropical forests further north.

In the group of the Dense Rainforest, "slope forests" were separated from the "coastal and alluvial plain" forests (Figure 5). The second main group divided the Araucaria Rainforest and the Semideciduous Seasonal Forest.

Pair wise comparisons of the Sorensen's similarity coefficients show values ranging from 0.09 between the Alluvial Semideciduous Forest and the Upper Montane Dense Rainforests to 0.61 between the two formations of the Semideciduous Forest.

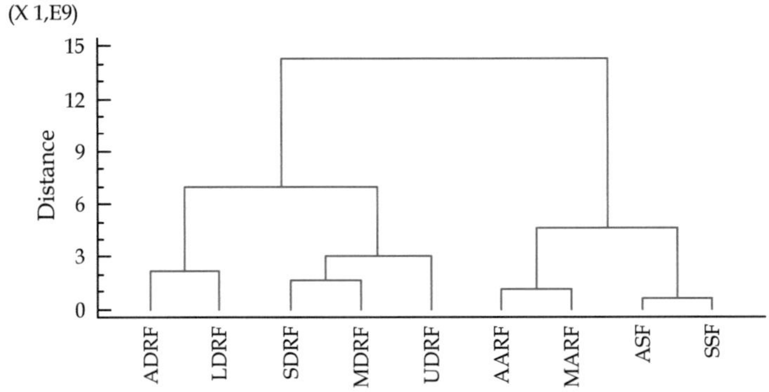

Fig. 5. Dendrogram showing the similarity between tree species measured in nine main formations of the Atlantic Forest of the state of Paraná, Southern Brazil. The cluster analysis was carried out using Sorensen's similarity coefficients, squared euclidean distances and the Ward's method of agglomeration (DRF - Dense Rainforest; ARF - Araucaria Rainforest; SF - Seasonal Forest; A - Alluvial; L - Lowland; S - Submontane; M - Montane; U - Upper Montane).

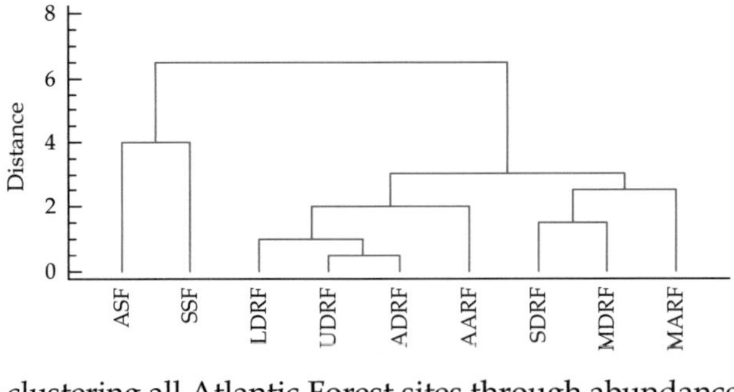

Fig. 6. Dendrogram clustering all Atlantic Forest sites through abundance similarities for tree species with squared euclidean distances and the Ward's method of agglomeration (DRF - Dense Rainforest; ARF - Araucaria Rainforest; SF - Seasonal Forest; A - Alluvial; L - Lowland; S - Submontane; M - Montane; U - Upper Montane).

The dendrogram based on abundance data (Figure 6) shows some different results as compared with the dendrogram based on the Sorensen's similarity. In the first group are the two Semideciduous Forests, while in the second are all studied Rainforests. This group is divided in two subgroups: one presenting the highest tree species richness and diversities (Shannon-Wiener index) among the rainforests analyzed and other with the lowest richness and diversities.

The first axis resulting from the CCA (eigenvalue = λ = 0.637) showed a gradient associated with altitude on one hand and annual temperature an rainfall on the other, separating Montane Rainforests (on the left) and Lowland and Alluvial Rainforests (on the right) (Figure 7). Oliveira-Filho & Fontes (2000) and Scudeller et al. (2001) found similar patterns for these variables in studies in Southeastern Brazil.

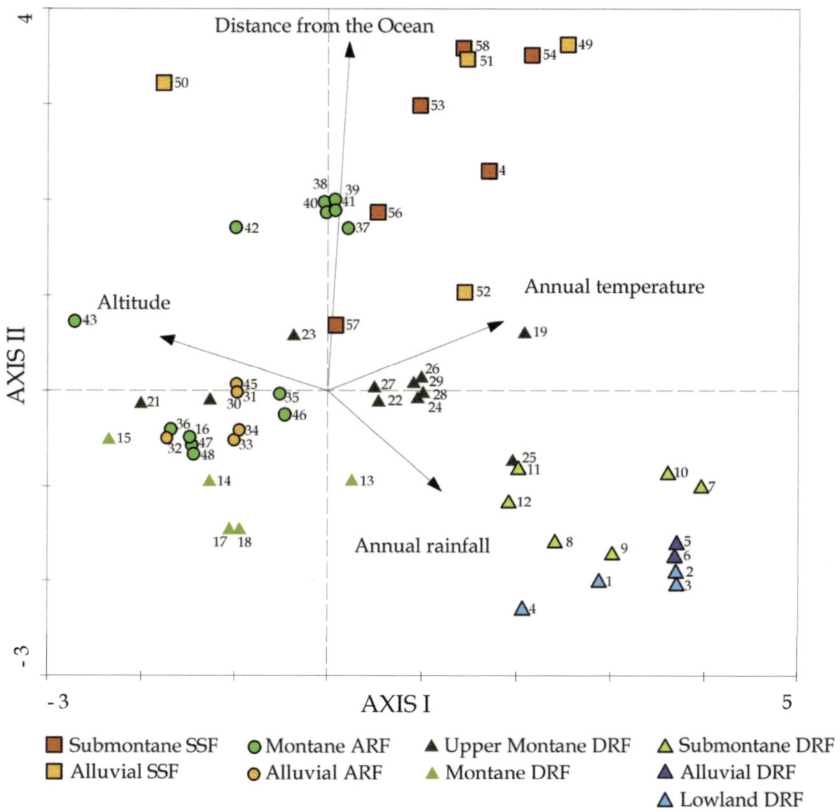

	Axis 1	Axis 2	Axis 3	Axis 4
Eigenvalues (species)	0.637	0.425	0.311	0.212
Cumulative percentage variance:				
• of species data	4.4	7.3	9.4	10.8
• of species-environment relation	40.2	67.0	86.7	100.0
Species-environment correlations	0.899	0.860	0.826	0.212
Significance of species-environment correlation (Monte Carlo test)	0.001	0.001	0.001	0.001

Fig. 7. Biplot of a Canonical Correspondence (CCA) applied to species found in 58 sites in Atlantic Forests in the state of Paraná, Southern Brazil. Numbers are related to sites presented in Table 1.

The second axis ($\lambda = 0.425$) shows the Semideciduous Seasonal Forest tree species and their abundance change with increasing the distance from the ocean, and decreasing rainfall (and probably increasing rainfall seasonality). The opposite occurred with Submontane, Alluvial and Lowland Dense Rainforests.

The relatively high eigenvalues found (> 0.3), indicate considerable abundance and species turnover along gradients mainly in axis 1 and 2. As also found by Scudeller et al. (2001), the low variance explained by the first axis indicates that other variables not investigated or methodological restrictions in this study probably influenced the abundance distribution.

6. Final considerations

The literature review and the data compilation resulted in an important database to understand how the Atlantic Forest is being studied in terms of forest structure and tree diversity. The most abundant tree species of this compilation probably corroborate with those found in studies on the Atlantic Forest further South in Brazil (including the state of Santa Catarina), and in Southeastern Brazil (*e.g.* São Paulo state formations). However, studies analyzing these relationships are needed.

Although many formations are relatively well studied in this respect, some formations such as the alluvial forests, mainly in the Dense Rainforest domain, are understudied. The same situation can be observed for the entire Semideciduous Seasonal Forest. Even more interesting is the lack of phytosociological data about this forest type in the Southwestern region of the state of Paraná. Therefore, future research is needed to build up a database for other and more specific studies. Knowing the species that share (or should share) the environment with ourselves is crucial to those who want to observe the habitat, interpret it and to promote conservation, preservation and true sustainable development.

7. Acknowledgements

We would like to thank Kelly Gutseit and Cesar B. Daniel for their valuable comments and suggestions. We are grateful to many colleagues for their substantial help with discussions, support in the office or in the field in some of the our surveys (Alan Y. Mocochinski, Daros A. T. da Silva, Joachim Graf Neto, Juarez Michelotti, Marcelo Brotto, Marília Borgo, Pablo M. Hoffmann, Rafael D. Zenni, Renata C. de Sousa, Ruddy T. Proença, Kelly G. Martins, Franklin Galvão, Carlos V. Roderjan, Otávio A. Bressan and Charles Carneiro).

8. References

Athayde, S.F.de. (1997). Composição florística e estrutura fitossociológica em quatro estágios sucessionais de uma Floresta Ombrófila Densa Submontana como subsídio ao manejo ambiental - Guaraqueçaba - PR. M.Sc Dissertation. Pós-graduação em Botânica, Universidade Federal do Paraná. Curitiba, Paraná, Brasil

Barddal, M.L. (2002). Aspectos Florísticos e Fitossociológicos do Componente Arbóreo-Arbustivo de uma Floresta Ombrófila Mista Aluvial - Araucária, PR. M.Sc Dissertation. Pós-graduação em Engenharia Florestal, Universidade Federal do Paraná. Curitiba, Paraná, Brasil

Barddal, M.L., Roderjan, C.V., Galvão, F. & Curcio, G.R. (2004). Caracterização florística e fitossociológica de um trecho sazonalmente inundável de floresta aluvial, em Araucária, PR. *Ciência Florestal,* Vol. 14, No. 2, ISSN 0103-9954

Bianchini, E., Silveira, R.P., Dias, M.C. & Pimenta, J.A. (2003). Diversidade e estrutura de espécies arbóreas em área alagável do município de Londrina, sul do Brasil. *Acta Botanica Brasilica,* Vol. 17, No. 3, pp. 405-419, ISSN 0102-3306

Blum, C.T., Santos, E.P., Hoffmann, P.M. & Socher, L.G. (2001). Análise Florística e Estrutural de um Trecho de Floresta Ombrófila Densa Montana no Morro dos Perdidos, Serra de Araçatuba, PR. *Proceedings of VI Encontro Regional de Botânicos do Paraná e Santa Catarina,* Curitiba, november 2001

Blum, C.T., Silva, D.A.T., Hase, L.M. & Miranda, D.L.C. (2003). Caracterização Florística e Ecológica de Remanescentes Florestais no Rio das Cinzas, Norte Pioneiro/PR. *Proceedings of Seminário Nacional Degradação e Recuperação Ambiental - Perspectiva Social,* Foz do Iguaçú, november 2003

Blum, C.T. (2006). A Floresta Ombrófila Densa na Serra da Prata, Parque Nacional Saint-Hilaire/Lange, PR - caracterização florística, fitossociológica e ambiental de um gradiente altitudinal. M.Sc Dissertation. Pós-graduação em Engenharia Florestal, Universidade Federal do Paraná. Curitiba, Paraná, Brasil

Blum, C.T. & Roderjan, C.V. (2007). Espécies indicadoras em um gradiente da Floresta Ombrófila Densa na Serra da Prata, Paraná, Brasil. *Revista Brasileira de Biociências,* Vol. 5, No. 2, pp. 873-875, ISSN 1679-2343

Blum, C.T. & Petean, M.P. (2008). Flora. In: *Estudo de Impacto Ambiental da BR-487, trecho Cruzeiro do Oeste - Porto Camargo, PR.* ENGEMIN / DNIT, Curitiba.

Blum, C.T. (2010). Os componentes epifítico vascular e herbáceo terrícola da Floresta Ombrófila Densa ao longo de um gradiente altitudinal na Serra da Prata, Paraná. Ph.D Thesis. Pós-graduação em Engenharia Florestal, Universidade Federal do Paraná. Curitiba, Paraná, Brasil

Borghi, W.A., Martins, S.S., Del Quiqui, E.M. & Nanni, M.R. (2004). Caracterização e avaliação da mata ciliar à montante da hidrelétrica de Rosana, na Estação Ecológica do Caiuá, Diamante do Norte, PR. *Cadernos da Biodiversidade,* Vol. 4, No. 2, pp. 9-18, ISSN 1415-9112

Campos, J.B., Romagnolo, M.B. & Souza, M.C. (2000). Structure, composition and spatial distribution and dynamics of tree species in a remnant of Semideciduous Seasonal Alluvial Forest of the Upper Paraná River floodplain. *Brazilian Archives of Biology and Technology,* Vol. 43, No. 2, pp. 185-194, ISSN 1516-8913

Castella, P.R., Britez, R.M. (2004). *A Floresta com Araucária no Paraná: conservação e diagnóstico dos remanescentes florestais,* FUPEF/PROBIO/MMA, Brasília

Caviglione, J.H., Kiihl, L.R.B., Caramori, P.H. & Oliveira, D. (2000). *Cartas climáticas do Paraná,* IAPAR - Instituto Agronômico do Paraná, Retrieved from: www.iapar.br/modules/conteudo/conteudo.php?conteudo=677

Cordeiro, J. & Rodrigues, W.A. (2007). Caracterização Fitossociológica de um remanescente de Floresta Ombrófila Mista em Guarapuava, PR. *Revista Árvore,* Vol. 31, No. 3, pp. 545-554, ISSN 0100-6762

Cordeiro, J. (2010). Compartimentação pedológico-ambiental e sua influência sobre a florística e estrutura de um remanescente de Floresta Ombrófila Mista na região de

Guarapuava, PR. Ph.D Thesis. Pós-graduação em Engenharia Florestal, Universidade Federal do Paraná. Curitiba, Paraná, Brasil

Costa Filho, L.V., Nanni, M.R. & Campos, J.B. (2006). Floristic and phytosociological description of a riparian forest and the relationship with the edaphic environment in Caiuá Ecological Station - Paraná - Brazil. *Brazilian Archives of Biology and Technology*, Vol. 9, No. 5, pp. 785-798, ISSN 1516-8913

Del Quiqui, E.M., Martins, S.S., Silva, I.C., Borghi, W.A., Silva, O. H. da & Pacheco, R.B. (2007). Estudo fitossosiológico de um trecho da Floresta Estacional Semidecidual em Diamante do Norte, Estado do Paraná, Brasil. *Acta Scientiarum*, Vol. 29, No. 2, pp. 283-290, ISSN 1679-9275

Durigan, M.E. (1999). Florística, dinâmica e análise protéica de uma Floresta Ombrófila Mista em São João do Triunfo - PR. M.Sc Dissertation. Pós-graduação em Engenharia Florestal, Universidade Federal do Paraná. Curitiba, Paraná, Brasil

Forzza, R.C. *et al.* (2010). Introdução, In: *Lista de Espécies da Flora do Brasil*, Acessed april 22, 2011, Available from: <http://floradobrasil.jbrj.gov.br/2010/>

Fundação SOS Mata Atlântica, INPE & ISA. (1998). *Atlas da evolução dos remanescentes florestais da Mata Atlântica e ecossistemas associados período 1990-1995*, Fundação SOS Mata Atlântica, São Paulo.

Fundação SOS Mata Atlântica & INPE (2010). *Atlas da evolução dos remanescentes florestais da Mata Atlântica e ecossistemas associados período 2008-2010*, Fundação SOS Mata Atlântica, Retrieved from:
http://mapas.sosma.org.br/site_media/download/atlas-relatorio2008-2010parcial.pdf

Galvão, F., Roderjan, C.V., Ziller, S.R. & Kuniyoshi, Y.S. (2002). Composição florística e fitossociológica de caxetais do litoral do Estado do Paraná - Brasil. *Floresta*, Vol. 32, No. 1, pp. 19-42, ISSN 0015-3826

Gentry, A.H. (1995). Patterns of diversity an floristic composition in Neotropical montane forests. In: *Biodiversity and conservation of Neotropical montane forests. Neotropical Montane Forest Biodiversity and Conservation Symposium 1*, Churchill, S.P., Balslev, H., Forero, E. & Luteyn. J.L., pp. 103-126, New York Botanical Garden, ISBN 0893274003, New York

Geraldi, S.E., Koehler, A.B. & Kauano, E.E. (2005). Levantamento Fitossociológico de dois framentos da Floresta Ombrófila Mista em Tijucas do Sul, PR. *Revista Acadêmica*, Vol. 5, No. 2, pp. 27-36, ISSN 0013-989X

Grubb, P.J. (1971). Interpretation of the "Massenerhebung" effect on tropical mountains. *Nature*, Vol. 229, pp. 44-45, ISSN 0028-0836

Guapyassú, M.S. (1994). Caracterização fitossociológica de três fases sucessionais de uma Floresta Ombrófila Densa Submontana Morretes - Paraná. M.Sc Dissertation. Pós-graduação em Engenharia Florestal, Universidade Federal do Paraná. Curitiba, Paraná, Brasil

Hueck, K. (1972). *As florestas da América do Sul*, Polígono, São Paulo.

IBGE (1992). *Manual Técnico da Vegetação Brasileira - Manuais Técnicos de Geociências nº1*, Fundação Instituto Brasileiro de Geografia e Estatística - DERNA, Rio de Janeiro

IPARDES (1986). *Algumas Características físicas e cobertura arbórea do Estado do Paraná*, CODESUL, Curitiba

Jaster, C.B. (1995). Análise estrutural de algumas comunidades florestais no litoral do estado do Paraná, na área de domínio da Floresta Ombrófila Densa - Floresta Atlântica. M.Sc Dissertation. Forstwissenschaftlicher Fachbereich, Institut für Waldbau - Abteilung II - Tropen und Subtropen, Georg-August-Universität. Göttingen, Deutschland

Jaster, C.B. (2002). A estrutura como indicadora do nível de desenvolvimento sucessional de comunidades arbóreas da restinga - uma proposta metodológica. Ph.D Thesis. Pós-graduação em Engenharia Florestal, Universidade Federal do Paraná. Curitiba, Paraná, Brasil

Jaster, C.B., Iantas, R., Blum, C.T. (2002). Flora. In: *Estudo de Impacto Ambiental da Barragem do Rio Miringuava, São José dos Pinhais - PR,* SANEPAR - Companhia de Saneamento do Paraná, Curitiba

Klein, R.M. (1979). Ecologia da Flora e Vegetação do Vale do Itajaí. *Sellowia - Anais Botânicos do Herbário Barbosa Rodrigues,* No. 32

Koehler, A., Galvão & F., Longhi, S.J. (2002). Floresta Ombrófila Densa Altomontana: aspectos florísticos de diferentes trechos na Serra do Mar, PR. *Ciência Florestal,* Vol. 12, No. 2, pp. 27-39, ISSN 0103-9954

Kozera, C., Dittrich, V.A.O. & Silva, S.M. (2006). Fitossociologia do componente arbóreo de um fragmento de Floresta Ombrófila Mista Montana. *Floresta,* Vol. 36, No. 2, pp. 225-237, ISSN 0015-3826

Lamprecht, H. (1990). *Silvicultura nos Trópicos,* Deutsche Gesellschaft für Technische Zusammenarbeit - GTZ, ISBN 3-88085-425-4, Eschborn

Laurance, S.G.W., Laurance, W.F., Andrade, A., Fearnside, P.M., Harms, K.E., Vicentini, A. & Luizao, R.C.C. (2010). Influence of soils and topography on Amazonian tree diversity: a landscape-scale study. *Journal of Vegetation Science* Vol. 21, pp. 96–106, ISSN 1100-9233

Leite, P.F., Klein, R.M., Pastore, U. & Coura Neto, A.B. (1986). *A vegetação da área de influência do reservatório da usina hidrelétrica de Ilha Grande (PR/ MS): levantamento na escala 1:250.000,* ELETROSUL - IBGE, Brasília

Leite, P. & Klein, R.M. (1990). Vegetação. In: *Geografia do Brasil: região Sul,* Instituto Brasileiro de Geografia e Estatística, pp. 113-150, Rio de Janeiro

Liebsch, D., Goldenberg, R. & Marques, M.C.M. (2007). Florística e estrutura de comunidades vegetais em uma cronoseqüência de Floresta Atlântica no Estado do Paraná, Brasil. *Acta Botanica Brasilica,* Vol. 21, No. 4, pp. 983-992, ISSN 0102-3306

Maack, R. (2002). *Geografia Física do Estado do Paraná,* Imprensa Oficial, Curitiba

Magurran, A.N. (1988). *Ecological Diversity and its Measurement,* Princeton University Press, ISBN 0-691-08485-8, New Jersey

Melo, M.S., Giannini, P.C.F. & Pessenda, L.C.R. (2000). Gênese e Evolução da Lagoa Dourada, Ponta Grossa, PR. *Revista do Instituto Geológico,* Vol. 21, No. 1-2, pp. 17-31, ISSN 0100-929X

Menezes-Silva, S. (1990). Composição florística e fitossociologia de um trecho de floresta de restinga na Ilha do Mel, município de Paranáguá, PR. M.Sc Dissertation. Pós-graduação em Biologia Vegetal, Universidade Estadual de Campinas. Campinas, São Paulo, Brasil

MINEROPAR (2001). *Atlas geológico do Estado do Paraná*. Acessed april 22, 2011, Available from: http://www.mineropar.pr.gov.br/arquivos/File/MapasPDF/atlasgeo.pdf

MMA (2011). Mapas de Cobertura Vegetal dos Biomas Brasileiros. Acessed may 06, 2011, Available from http://mapas.mma.gov.br/mapas/aplic/probio/datadownload.htm

Myers, N., Mittermeier, R.A., Mittermeier, C.G., Fonseca, G.A. & Kent, J. (2000). Biodiversity hotspots for conservation priorities. *Nature*, Vol. 403, pp. 853-858, ISSN 0028-0836

Oliveira-Filho, A.T. & Fontes, M.A.L. (2000). Patterns of Floristic Differentiation among Atlantic Forests in Southeastern Brazil and the Influence of Climate, *Biotropica*, Vol. 32, No. 4b, pp. 793-810, ISSN 1744-7429

Pasdiora, A.L. (2003). Florística e Fitossociologia de um trecho de Floresta Ripária em dois compartimentos ambientais do rio Iguaçu, Paraná, Brasil. M.Sc Dissertation. Pós-graduação em Engenharia Florestal, Universidade Federal do Paraná. Curitiba, Paraná, Brasil

Pires, P.T.L., Zilli, A.L. & Blum, C.T. (2005). *Atlas da Floresta Atlântica no Paraná – área de abrangência do Programa Proteção da Floresta Atlântica*, SEMA/Programa Proteção da Floresta Atlântica - Pró-Atlântica, Curitiba

Pitman, N.C.A, Terborgh, J.W., Silman, M.R., Núñez, P., Neill, D.A., Cerón, C.E., Palacios, W.A. & Aulestia, M. (2002). A comparison of tree species diversity in two upper Amazonian forests. *Ecology*, Vol. 83, No. 11, pp. 3210–3224, ISSN 0012-9658

Portes, M.C.G.de O., Galvão, F. & Koehler, A. (2001). Caracterização florística e estrutural de uma Floresta Ombrófila Densa Altomontana do Morro do Anhangava, Quatro Barras - PR. *Floresta*, Vol. 31, No, 1/2, pp. 9-18, ISSN 0015-3826

Reginato, M. & Goldenberg, R. (2007). Análise florística, estrutural e fitogeográfica da vegetação em região de transição entre as Florestas Ombrófilas Mista e Densa Montana, Piraquara, Paraná, Brasil. *Hoehnea*, Vol. 34, No. 3, pp. 349-364, ISSN 0073-2877

Ribeiro, M.C., Metzger, J.P., Martensen, A.C., Ponzoni, F.J. & Hirota, M.M. (2009). The Brazilian Atlantic Forest: How much is left, and how is the remaining forest distributed? Implications for conservation. *Biological Conservation* Vol. 142, No. 6, pp. 1141-1153, ISSN 0006-3207

Roderjan, C.V. (1994). A Floresta Ombrófila Densa Altomontana do Morro do Anhangava, Quatro Barras, PR - Aspectos climáticos, pedológicos e fitossociológicos. Ph.D Thesis. Pós-graduação em Engenharia Florestal, Universidade Federal do Paraná. Curitiba, Paraná, Brasil

Roderjan, C.V. & Grodski, L. (1999). Acompanhamento meteorológico em um ambiente de Floresta Ombrófila Densa Altomontana no morro Anhangava, Município de Quatro Barras - PR, no ano de 1993. *Cadernos da Biodiversidade*, Vol. 2, No. 1, pp. 27-34, ISSN 1415-9112

Roderjan, C.V., Galvão, F., Kunizoshi, Y.S. & Hatschbach, G.G. (2002). As Unidades Fitogeográficas do Estado do Paraná. *Ciência & Meio Ambiente – Fitogeografia do Sul da América*, No. 24, pp. 75-92

Rondon Neto, R.M. (2002). Caracterização florística e estrutural de um fragmento de Floresta Ombrófila Mista, em Curitiba, PR - Brasil. *Floresta*, Vol. 32, No. 1, pp. 3-16, ISSN 0015-3826

Scheer, M.B. (2010). Ambientes altomontanos no Paraná: florística vascular, estrutura arbórea, relações pedológicas e datações por ^{14}C. Ph.D Thesis. Pós-graduação em Engenharia Florestal, Universidade Federal do Paraná. Curitiba, Paraná, Brasil

Scheer, M.B. & Mocochinski, A.Y. (2009). Floristic composition of four tropical upper montane rain forests in Southern Brazil. *Biota Neotropica,* Vol. 9, No. 2, pp. 51-70, ISSN 1676-0603 (*in portuguese*)

Scheer, M.B., Mocochinski, A.Y. & Roderjan, C.V. (in press a). Tree component structure of tropical upper montane rain forests in Southern Brazil. *Acta Botanica Brasilica,* ISSN 0102-3306 (*in portuguese*)

Scheer, M.B., Curcio, G.R. & Roderjan, C.V. (in press b). Environmental functionalities of upper montane soils in Serra da Igreja, Southern Brazil. *Revista Brasileira de Ciência do Solo,* ISSN 1806-9657 (*in portuguese*)

Schorn, L.A. (1992). Levantamento florístico e análise estrutural em três unidades edáficas em uma Floresta Ombrófila Densa Montana no Estado do Paraná. M.Sc Dissertation. Pós-graduação em Engenharia Florestal, Universidade Federal do Paraná. Curitiba, Paraná, Brasil

Scudeller, V.V., Martins, F.R. & Shepherd, G.J. (2001). Distribution and abundance of arboreal species in the atlantic ombrophilous dense forest in Southeastern Brazil, *Plant Ecology,* Vol. 152, pp. 185-199, ISSN 1385-0237

Seger, C.D., Dlugosz, F.L., Kurasz, G, Martinez, D.T., Ronconi, E., Melo, L.A.N. de, Bittencourt, S.M. de, Brand, M.A., Carniatto, I., Galvão, F. & Roderjan, C.V. (2005). Levantamento florístico e análise fitossociológica de um remanescente de Floresta Ombrófila Mista localizado no Município de Pinhais, Paraná - Brasil. *Floresta,* Vol. 35, No. 2, p. 291-302, ISSN 0015-3826

SEMA (2002). *Mapeamento dos Remanescentes de Floresta Estacional Semidecidual,* SEMA/FUPEF, Curitiba.

Silva, F.das C. (1994). Composição florística e estrutura fitossociológica da Floresta Tropical Ombrófila da encosta Atlântica no Município de Morretes, Estado do Paraná. *Acta Biológica Paranaense,* Vol. 23 No. 1-4, pp. 1-54, ISSN: 0301-2123

Silva, F.das C., Fonseca, E.P., Soares-Silva, L.H., Muller, C. & Bianchini, E. (1995). Composição florística e fitossociologia do componente arbóreo das florestas ciliares da Bacia do Rio Tibagi. 3. Fazenda Bom Sucesso, Município de Sapopema, PR. *Acta Botanica Brasilica,* Vol. 9, No. 2, pp. 289-302, ISSN 0102-3306

Silva, F.das C. & Soares-Silva, L.H. (2000). Arboreal flora of the Godoy Forest State Park, Londrina, Pr. Brazil. *Edinburgh Journal of Botany,* Vol. 57, pp. 107-120, ISSN 0960-4286

Soares-Silva, L.H., Kita, K.K. & Silva, F.C. (1998). Fitossociologia de um trecho de floresta de galeria no Parque Estadual Mata dos Godoy, Londrina, PR, Brasil. *Boletim do Herbário Ezechias Paulo Heringer* Vol. 3, pp. 46-62

Tabarelli, M. & Mantovani, W. (1999). A riqueza de espécies arbóreas na floresta atlântica de encosta no Estado de São Paulo (Brasil). *Revista Brasileira de Botânica,* Vol. 22, No. 2, pp. 217-223, ISSN 0100-8404

Ter Braak, C.J.F. & Smilauer, P. (2002). *CANOCO Reference manual and CanoDraw for Windows user's guide: Software for Canonical Community Ordination (version 4.5),* Microcomputer Power, Ithaca

Veiga, M.P. da , Martins, S.S., Silva, I.C., Tormena, C.A. & Silva, O.H. da. (2003). Avaliação dos aspectos florísticos de uma mata ciliar no Norte do Estado do Paraná. *Acta Scientiarum*, Maringá, Vol. 25, No. 2, pp. 519-525, ISSN 1679-9275

Veloso, H.P., Rangel Filho, A.L. & Lima, J.C. (1991). *Classificação da Vegetação Brasileira adaptada a um Sistema Universal*, Instituto Brasileiro de Geografia e Estatística/Departamento de Recursos Naturais e Estudos Ambientais, Rio de Janeiro

Watzlawick, L.F., Sanquetta, C.R., Valério, A.F. & Silvestre, R. (2005). Caracterização da composição florística e estrutura de uma Floresta Ombrófila Mista, no Município de General Carneiro (PR). *Ambiência*, Vol. 1, No. 2, p. 229-237, ISSN 1808-0251

Zacarias, R.R. (2008). O componente arbóreo de dois trechos de Floresta Ombrófila Densa Aluvial em solos hidromórficos, Guaraqueçaba, Paraná. M.Sc Dissertation. Pós-graduação em Ciências Biológicas, Universidade Federal do Paraná. Curitiba, Paraná, Brasil

6

Structure and Floristic Composition in a Successional Gradient in a Cloud Forest in Chiapas, Southern Mexico

Miguel Ángel Pérez-Farrera, César Tejeda-Cruz,
Rubén Martínez-Camilo, Nayely Martínez-Meléndez, Sergio López,
Eduardo Espinoza-Medinilla and Tamara Rioja-Paradela
Universidad de Ciencias y Artes de Chiapas
Mexico

1. Introduction

Southern Mexico is well known for its high biodiversity (CONABIO, 2008). This biodiversity is a result of several factors like its geographic position, geographic diversity, and physiographic richness (Ferrusquilla-Villafranca, 1998). In particular, Chiapas, Mexico's southernmost state holds seven physiographic zones, including valleys, mountain chains, plateaus, and coastal plains (Müllerried, 1957). Most of this biological richness is to be found in the eastern moist forest, northern mountains, central plateau, and Sierra Madre (Breedlove, 1981). The Sierra Madre mountain chain harbors some of the very last patches of Cloud Forest, which is one of the most endangered ecosystems both in Mexico and at a global scale (Challenger, 1998; Toledo-Aceves et al., 2010). Fortunately, three existing biosphere reserves namely El Triunfo, La Sepultura and Volcán Tacaná, aim to protect and maintain this highly threatened ecosystem.

As elsewhere, natural areas compete for land with human activities such as agriculture and cattle ranching, recently, climate change has added up to the list of threats. Only at El Triunfo reserve between 1983 and 1993 were lost 8,946 ha, including 5,084 ha of Cloud Forest (March & Flamenco, 1996). As a region, the Sierra Madre de Chiapas, was between 1998 and 2005, the region that suffered the greatest impact related to climate change in the form of massive landslides. For example, more than 15,000 ha of Cloud Forest were affected in Chiapas by Hurricane Isis (Richter, 2000), while this phenomenon has also occurred in other parts of the Americas (Restrepo & Alvarez, 2006).

The loss of forest cover in the upper parts of the mountain chain generates a reduction of water retention and filtering capability which results in soil loss and consequently in river sedimentation. This also has occasioned an increment of water flow volume which augments flood risk. One way to help to reduce flood risk is through the ecological restoration of forest systems in the upper basin, and subsequent recovery of the ecological services associated to forest. Hence, information on the structure of natural plant communities, including structure and floristic composition, is central to establish sound ecological restoration strategies and policies. Our research objective, was thus, to evaluate and analyze the natural successional process in a cloud forest along a successional gradient

(20-25 years old, 30-35 years old, and mature forest), and to determine the floristic composition, vegetation structure, and species replacement along this gradient.

2. Methods

Our study took place at El Triunfo Biosphere Reserve (ETBR) in Chiapas, Mexico. El Triunfo lays on the Sierra Madre de Chiapas Mountain range (Figure 1), which runs parallel to the Pacific coast. ETBR has an extension of 119,177 ha and its altitudinal range goes from 450 to 3000 meters above sea level (Arreola-Muñoz et al., 2004). Along this range, several climates occur; from hot humid in the low parts to temperate humid in the high lands (García, 2004). Several vegetation types are present, but upper parts are dominated by Cloud Forest (Rzedowski, 1978).

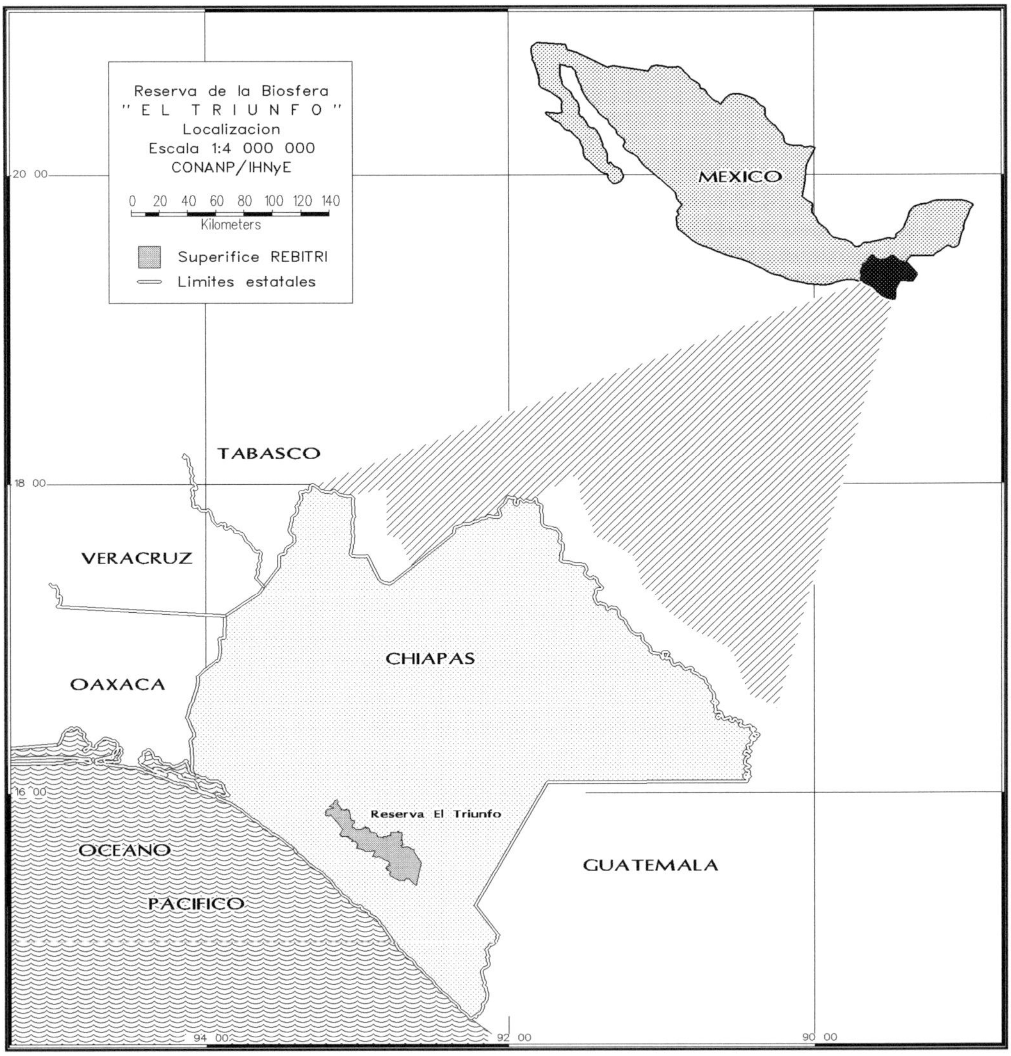

Fig. 1. Map showing El Triunfo Biosphere Reserve boundaries in Chiapas, southern Mexico.

We selected three study sites at ETBR, all localized in the core zone I, based on accessibility and disturbance history. We established sampling plots in three different successional stages according to the time elapsed since disturbance: 1.- Early growth, 20-25 years old (n = 10); 2.- old growth 30-35 years old (n = 7); and 3.- Mature forest (n = 10).

We modified the method proposed by Ramírez-Marcial et al. (2001) to describe the physiognomy and structure of plant communities. We used 0.1 ha circular plots, with a radius of 17.8 m. Inside each circular plot we set smaller circular sub-plots (Figure 2). To avoid border effect, plots were placed at least 50 meters from the border. We randomly selected a point to be used as the center of the plot, and then we placed four straight lines 17.8 m long to each cardinal direction with marks at 5.6, 12.6 y 17.8 m. Then, beginning from the plot center, we traced a circle at the 5.6 m mark (circle "A"), another at the 12.6 m mark (circle "B"), and finally one at the 17.8 mark (circle "C"). Then we measured different vegetation features in each circle. Circle A: all trees with DBH ≥5 cm and ≤ 10 cm; circle B: all trees with a DBH >10 cm and ≤ 30 cm; circle C: all trees with a DBH > 30 cm.

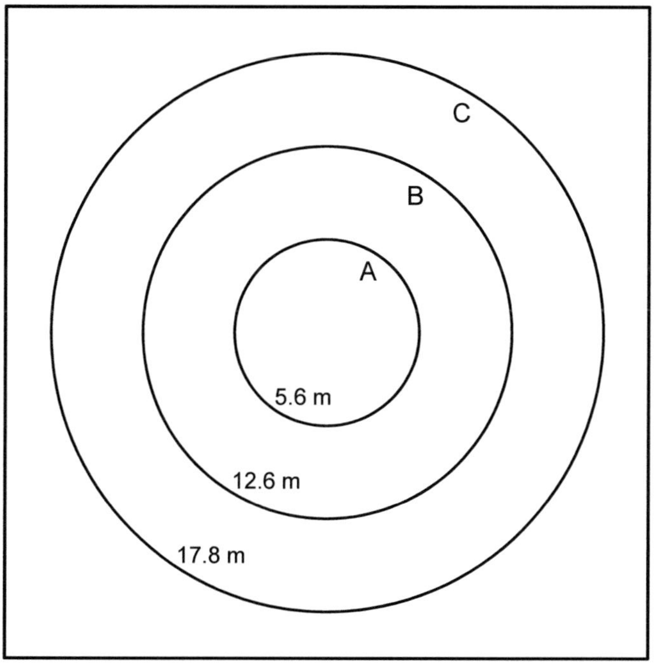

Fig. 2. Plot sampling layout (modified from Ramírez-Marcial et al., 2001)

2.1 Data analysis

Dominance (D): Due to the difficulty of measuring horizontal crown projection to estimate dominance (Lamprecht, 1990), we used basal area, expressed in m², for each species to estimate Absolute Dominance (AD):

$$\mathbf{AD} = (\pi/4) \times dbh^2$$

Where:

π = 3.1416, dbh = diameter at breast height

Relative dominance (DR): Is the percentage of the contribution of each species to the total basal area:

$$\boldsymbol{DR} = \frac{Absolut\ Dominance}{Total\ Absolut\ Dominance} \times 100 \quad (1)$$

Absolut Density (De): Is a parameter that allow us to know the density of each species or family

$$\boldsymbol{De} = \frac{N}{a} \quad (2)$$

Where:
N = Number of individuals of each species or family
a = a given area
Relative Density (Der): Is the percentage of the contribution of each species or family to the total number of individuals per hectare:

$$\boldsymbol{Der} = \frac{Density\ of\ a\ given\ species}{Sum\ of\ the\ density\ of\ all\ species} \times 100 \quad (3)$$

Frecuency (FA): Is the distribution regularity of each species in a given area. It is measured as the percentage of the number of sub-plots where a given species occurs in relation to the total of sampled sub-plots

$$\boldsymbol{FA} = \frac{number\ of\ sub-plots\ where\ the\ species\ occurs}{total\ number\ of\ sub-plots} \quad (4)$$

Relative frequency (FR): Is the percentage of the absolute frequency of a given species in relation to the sum of the frequencies of all present species

$$\boldsymbol{FR} = \frac{absolut\ frequency\ of\ a\ given\ species}{total\ absolut\ frequency} \times 100 \quad (5)$$

Importance Value Index (IVI): Is the arithmetic sum of the relative relative abundance (AR), relative frequency (FR), and relative dominance (DR). Being the sum of these three parameters for an ecosystem equal to 300, we divide the result by three. This value could be used for species or families:

$$\textbf{IVI (F)} = (AR + FR + DR) / 3 \quad (6)$$

Floristic diversity: For each parcel we measured the Shannon-Wiener index, Fisher's α, and Jaccard's equitability. Floristic similarity was estimated using binary coefficients Krebs (1999) based in the presence/absence of families, expressed as:

$$CC_i = \frac{c}{s_1 + s_2 - c} \quad (7)$$

Where:
CC_i= Jaccard's coefficient
s_1 y s_2= NUmber of species in community 1 and 2
c = number of species for both communities
This index is designed to equal 1 when similarity is total and 0 when no species are shared (Magurran, 1988). We then constructed a dendrogram to show using the UPGMA method. Floristic groups from the dendrogram were then compared using the Hutcheson t test (P *(P<0.05)* (Magurran, 1988), as described by the equation:

$$t = \frac{H'_1 - H'_2}{(Var\ H'_1 + Var\ H'_2)^{1/2}} \qquad (8)$$

And with the degrees of freedom defined by the equation:

$$t = \frac{(Var\ H'_1 + Var\ H'_2)^2}{[(Var\ H'_1)^2/N_1] + [(Var\ H'_2)^2/N_2]} \qquad (9)$$

Where:
Hi = Shannon-Wiener index for area i.
Var Hi = Shannon-Wiener index variance for area i.
Ni = Total number of individuals in area i.

3. Results

We found a total of 1416 individuals belonging to 105 species, 74 genera and 48 families. The old growth was the successional stage that presented the highest number of species, genera and families (Table 1). The families with the greatest number of species in all successional stages were Lauraceae, Asteraceae and Rubiaceae (Table 2). The Genera better represented were *Quercus, Clethra, Eugenia, Saurauia,* and *Cyathea.*

Successional stage	**Species**	**Genera**	**Families**
Old growth (30-35 years)	56	45	31
Early growth (20-25 years)	43	39	30
Mature forest	54	40	29
Total	105	74	48

Table 1. Number of species, genera and families in a Cloud Forest successional gradient at El Triunfo, Chiapas, Mexico

Families	Mature Forest		Old growth		Early growth		G	S
	G	S	G	S	G	S		
Lauraceae	4	7	4	5	2	2	5	10
Fagaceae	1	6	1	3	0	0	1	7
Asteraceae	0	0	5	5	3	3	6	6
Myrtaceae	1	6	1	3	0	0	1	6
Rubiaceae	5	5	2	2	2	2	5	5
Clethraceae	1	2	1	2	1	2	1	5
Fabaceae	2	2	0	0	2	2	4	4
Myrsinaceae	2	3	2	3	1	1	2	4
Araliaceae	2	2	2	3	2	2	2	4
Melastomataceae	1	1	1	2	1	2	2	3
Rest of families	21	20	26	28	25	27	45	51

Table 2. Wood plant families with a DBH >5 cm families in a Cloud Forest successional gradient at El Triunfo, Chiapas, Mexico. Species is denoted by (S) and Genera by (G).

3.1 Mature forest

We found 54 species in the mature forest, which also presented the greatest basal area (46.88 ±280.86 cm). Five species contributed with 175% of the IVI: *Matudaea trinervia, Symplococarpon flavifolium, Glossostipula concinna, Amphitecna montana,* and *Licaria excelsa* in order of importance (Table 3). Total tree density was 7.65 ± 13.65 ind/ha.

Mature forest presented a well-developed tree stratum, defined by trees with a DBH > 400 cm, dominated by *Matudaea trinervia, Symplococarpon flavifolium, Quercus laurina, Quercus benthamii, Glossostipula concinna, Licaria excelsa, Quercus peduncularis* and *Amphitecna montana.* While the understory is defined by a great variety of wooded plants with small DBH (> 10 cm) like *Geonoma undulata, Piper subsessilifolium* and *Psychotria galeottiana.*

SPECIE	Relative Frequency	Relative Density	Relative Basal Area	IVI
Mature Forest				
Matudaea trinervia	6.43	18.4	81.50	106.33
Symplococarpon flavifolium	7.14	14.53	6.82	28.49
Glossostipula concinna	7.14	6.54	1.71	15.39
Old Growth				
Quercus benthamii	4.42	11.09	27.50	43.01
Asteraceae sp2.	0.88	6.33	25.40	32.62
Matudaea trinervia	4.42	5.66	21.45	31.53
Early Growth				
Saurauia madrensis	6.25	13.55	41.24	61.04
Crossopetalum parviflorum	5.56	12.12	11.70	29.38
Hedyosmum mexicanum	6.25	10.87	10.93	28.05

Table 3. The three most important species for each stage in a Cloud Forest successional gradient at El Triunfo, Chiapas, Mexico.

3.2 Old growth

We found 56 species in the 30-35 years old successional stage, accounting for more than 50% of the total number of species. This stage presented the lowest total, basal area, while tree density was 10.85 ± 18.22 ind/ha. The more important tree species were *Quercus benthamii, Matudaea trinervia, Ardisia compressa, Ocotea acuminatissima,* and *Clethra nicaraguensis,* which accounted for an IVI of 113% (Table 3). Arboreal elements with a DBH between 100 and 400 cm were: *Fuchsia paniculata, Crossopetalum parviflorum, Glossostipula concinna, Cyathea myosuroides, Styrax glabrescens, Desmopsis lanceolata, Rhamnus capraeifolia, Nectandra sinuata, Symplococarpon flavifolium, Saurauia kegeliana, Quercus conspersa, Clethra mexicana, Trophis cuspidata,* and *Nectandra globosa.* While species with a DBH < 10 cm were dominated by *Turpinia paniculata, Eugenia* aff. *uliginosa, Meliosma matudae, Malvaviscus arboreus,* and *Cestrum elegantissimum.* Total tree density was 11.28 ± 10.72 ind/ha.

3.3 Early growth

A total of 43 species and above 50% of the total genera were found in the 20-25 years old successional stage. This stage presented the lowest basal area and a tree density of 16.96±50.75 ind/ha. The species with the highest importance values were *Saurauia madrensis, Crossopetalum parviflorum, Hedyosmum mexicanum, Heliocarpus donnellsmithii,* and *Cestrum elegantissimum,* which account for an added IVI of 162% (Table 3). Trees with a DBH > 400 cm were represented by *Saurauia madrensis, Crossopetalum parviflorum, Hedyosmum mexicanum, Heliocarpus donnellsmithii, Cestrum elegantissimum, Rhamnus capraeifolia,*
Saurauia aff. *oreophila,* and *Liquidambar styraciflua;* arboreal elements with a DBH between 100 and 400 were *Saurauia kegeliana, Fuchsia paniculata, Verbesina apleura, Lepidaploa polypleura, Clethra lanata, Brunellia mexicana, Arachnothryx buddleioides, Wigandia urens* and *Comarostaphylis arbutoides;* finally, species with a DBH < 100 cm included *Pinus oocarpa, Ocotea acuminatissima, Citharexylum mocinoi, Clethra hondurensis, Clusia flava, Piper pseudo-lindenii, Trichillia havanensis, Matudaea trinervia, Prunus annularis, Myriocarpa longipes, Licaria excelsa, Cyathea sp, Conostegia volcanalis, Ardisia compressa, Pterocarpus* aff. *rohrii, Glossostipula concinna, Ostrya virginiana,* and *Dendropanax arboreus* (Table 3). Total tree density was 13.05 ± 19.61 ind/ha.

3.4 Diversity patterns

The greates richnness and diversity were found in the intermediate successional stage (30-35 years growth). This stage presented species that are characteristic both of the mature forest and the early growth (Table 4). The nature of this intermediate state is supported by statistical differences in diversity between the early growth and mature forest (Δ = - 0.21 p = 0.01), while there were no differences between early growth and old growth (Δ = -0.56 p > 0.05), nor between old growth and mature forest (Δ = 0.35 p > 0.05).

Successional stage	Mature forest	Old growth	Early growth
# Plots	10	7	10
Area (ha)	1	0.70	1
Density	413	442	561
Species	54	56	43
H'	3.127	3.48	2.92
α	16.59	16.98	10.86
D	13.48	23.79	13.71

Table 4. Number of plots, total sampled area (ha), plant density and estimated values for species richness(S), Shannon-Wiener diversity index (H'), Simpson's diversity index (D) and Fisher's alpha (α) for each successional stage.

In all three successional stages there was a high abundance of plants in early development stages and a low abundance of bigger sizes, so, according to Peter (1996), the class of structure found in the system is Type I (Figure 3). In all three successional stages, short size individuals are predominant, while in the early successional stage, no individual reaches a DBH of 200 cm.
Similarity analysis results showed that mature forest has a higher similarity with the old growth (26%). Figure 4 shows dominant species of each successional stage and shared species between habitats. Eight species were found in all three stages, while mature forest presented the highest number of exclusive species.

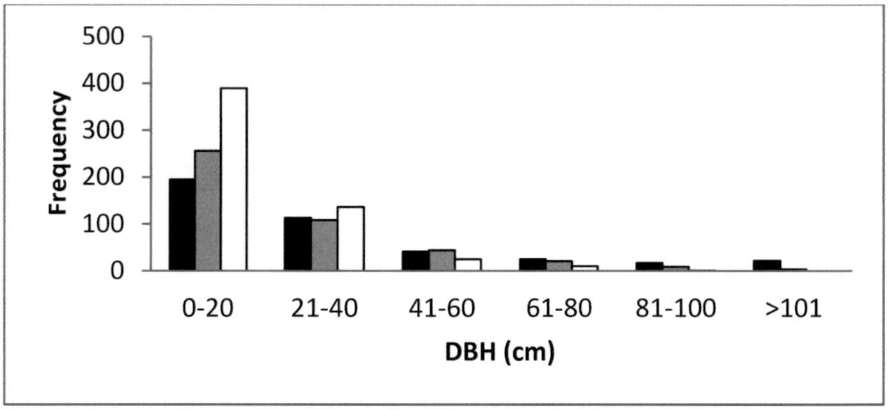

Fig. 3. Diametric class structure in three successional stages in El Triunfo, Chiapas, Mexico. Black columns denote Mature Forest, gray columns Old Growth and blank columns Early Growth.

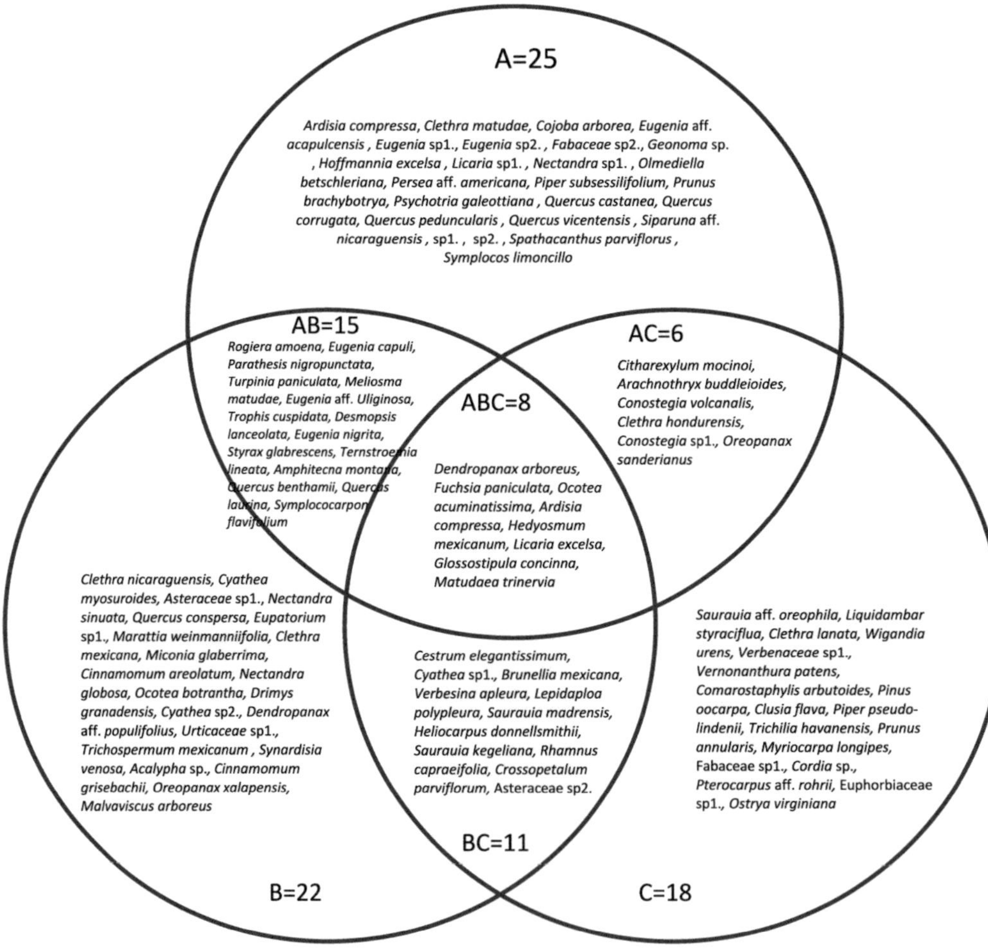

Fig. 4. Unique and shared species for each successional stage and their transitional phases between them. A=Mature forest, B=Old Growth, C=Early growth.

In contrast with Kappelle (1996) who found the highest richness and diversity in early successional stages, our results indicate that intermediate stage (old growth) has the highest richness and diversity. Nowadays, the old idea that the highest diversity is to be found in mature undisturbed systems has been challenged. The intermediate disturbance hypothesis, originally presented by Connell (1978) states that at intermediate levels of disturbance, diversity is maximized because both competitive and opportunistic species can coexist. The highest diversity at intermediate levels of disturbance that we found may support this notion. It is probable that our intermediate successional stage has spatial and temporal heterogeneity within the system, favoring the coexistence of numerous species. Denslow (1980) suggests that alpha diversity in successional stages in tropical forests is due to the high number of plantules of disturbance adapted species. According to Bazzaz (1975) the high species diversity in secondary forests may be explained by the high degree of horizontal and vertical micro-environmental complexity in early successional stages.

Williams-Linera (1991) found for mature Cloud Forest IVI values very similar to what we report in this study for *Matudaea trinervia, Quercus peduncularis,* and *Hedyosmum mexicanum* in a study done in the same area. Old growth had a 20% similarity with early growth, which suggests a high rate of species substitution. Availability, dispersion and germination of these species may play a key role in forest recovery and restauration in early phases after a disturbance (Guariguata & Kattan, 2002; Ten Hoopen & Kappelle, 2006; Wilms & Kappelle, 2006).

We may consider our system to be healthy because all class structures in all habitats had a DBH between 0-20 cm in more than 40% of individuals. This, according to Peter (1996) suggests that these structures are more stable and are considered healthy.

Finally, we compared our results with several other studies on successional gradients elsewhere in the neotropics (table 5). Some Costa Rica mature forest has twice the number of species than El Triunfo. However, secondary forest in Costa Rica showed similar numbers to

Forest stage	Country	Age (years)	Plot size (ha)	Average altitude	Species richness	Shannon Index	Reference
Tropical subalpine forest							
Primary	Peru	Mature	0.10	3400	9	-	Young, 1992
Tropical upper montane forest							
Primary	Ecuador	Mature	1	3280	32	-	Valencia & Jørgensen, 1993
Primary	Colombia	Mature	0.80	2850	33	-	Bazuin et al 1993
Primary	Costa Rica	Mature	0.1	2975	20	3.31	Kappelle et al 1996
Secundary	Costa Rica	15	0.10	2975	21	3.39	Kappelle et al 1996
Secundary	Costa Rica	30	0.10	2975	20	3.14	Kappelle et al 1996
Tropical montane cloud forest							
Primary	Mexico	Mature	1	1920	54	3.13	This study
Secundary	Mexico	30-35	0.7	1860	56	3.48	This study
Secundary	Mexico	20-25	1	1820	43	2.92	This study
Tropical lower montane forest							
Primary	Costa Rica	Mature	0.18	1050	105	4.06	Kuzee et al, 1994
Secundary	Costa Rica	11	0.19	1030	54	2.84	Kuzee et al, 1994
Secundary	Costa Rica	35	0.10	950	69	3.51	Kuzee et al, 1994

Table 5. Comparison of species richness and diversity between some primary and secondary forests in the Neotropics.

what we found in El Triunfo. These results suggest that secondary forest at El Triunfo could be more diverse than primary forest.

The patterns on species diversity and species replacement along a successional gradient we obtained from this study would be of great help to design sound strategies for Cloud Forest restoration. This is very important since little is known on Cloud Forest dynamics and because this habitat is considered one of the most endangered all over the world.

5. Acknowledgement

We are thankful to El Triunfo Biosphere Reserve personnel for granting permits to work in the reserve. This research is part of the Cloud Forest Restoration Project funded by CONACyT-FOMIX-CHIAPAS: CHIS-2007-07-77710.

6. References

Arreola-Muñoz A.V., G. Cuevas-García, R. Becerril-Macal, L. Noble-Camargo and M. Altamirano. (2004). El medio físico y geográfico de la Reserva de la Biosfera El Triunfo, Chiapas. In: La Reserva de la Biosfera El Triunfo, tras una década de conservación. Pérez-Farrera M.A., N. Martínez-Meléndez, A. Hernández-Yáñez and A.V. Arreola-Muñóz (Eds). pp. 29-52, Universidad de Ciencias y Artes de Chiapas. México. ISBN:968-5149-34-8, Tuxtla Gutiérrez

Bazzaz F.A. (1975). Plant species diversity in old-field successional ecosystems in southern Illinois. *Ecology* Vol. 56, No. 2, (Early Spring 1975), pp. 485-488, ISSN 0012-9658

Bazuin T., E. Gerritzen, and K. Stelma. (1993). Structure analysis of an Andean oak forest in the south-west Colombia. Thesis. Larenstein International Scchool for Higher Agricultural Education Agricultural Education Velp. 64 pp

Breedlove D.E. (1981). *Flora of Chiapas. Part I. Introduction to the Flora of Chiapas.* California Academy of Sciences, San Francisco California. ISBN 094-022-8009

Cayuela L., D. J. Golicher, J. M. Rey Benayas, M. González-Espinosa and N. Ramírez-Marcial. (2006). Fragmentation, disturbance and tree diversity conservation in tropical montane forests. *Journal of Applied Ecology,* Vol. 43, No. 6. (December 2006), pp. 1172-1181, ISSN 1365-2664

Challenger, A. 1998. Utilización y conservación de los ecosistemas terrestres de México: pasado, presente y futuro. Comisión Nacional para el Uso y Conocimiento de la Biodiversidad, Instituto de Biología de la UNAM y Agrupación Sierra Madre S.C. ISBN 970-900-0020, México City.

CONABIO. (2008). *Capital Natural de Mexico. Volumen I. Conocimiento actual de la biodiversidad.* Comisión Nacional para el Conocimiento y Uso de la Biodiversidad. ISBN 978-607-7607-03-8, Mexico City.

Connell, J.H. (1978). Diversity in tropical rain forests and coral reefs. *Science,* Vol. 199, No. 4335 (March 1978), pp. 1302-1310, ISSN 1095-9203

Denslow, J.S. (1980). Gap partitioning among tropical rainforest trees. *Biotropica,* Vol. 12, No. 2, (June 1980), pp. 47–55. ISSN 1744-7429

Ferrusquilla-Villafranca, I. (1998). Geología de México. In: *Diversidad Biologica de Mexico: Origenes y Distribucioón.* T. P. Ramamoorthy, R. Bye, A. Lot, and J. Fa, (eds). Instituto de Biologia, UNAM, Mexico City

García, E. A., (2004), *Modificaciones al Sistema de Clasificación climática de Köeppen,* Universidad Autónoma de México, Instituto de Geografía, Mexico City.

Guariguata, M. and G.H. Kattan. (2002). Ecología y conservación de bosques neotropicales (eds). Editorial Tecnológica, ISBN 9968-801-11-9, San José, Costa Rica

Jardel P.E.J. (1991). Perturbaciones naturales y antropogénicas y su influencia en la dinámica sucesional de los bosques de Las Joyas, Sierra de Manantlán, Jalisco. *Tiempos de Ciencia,* Vol. 22, pp. 9-26

Jardel P.E.J., A.L. Santiago-Pérez and C. Cortés-Montaño. (1998). Dinámica de rodales en bosque de pino y mesófilo de montaña en la sierra de Manantlán, México. VII Congreso Latinoamericano de Botánica XIV congreso Mexicano de Botánica. Diversidad y conservación de los recursos naturales en Latinoamérica. Libro de resúmenes.

Kappelle M., T. Geuze, M. E. Leal and A.M. Cleef. (1996). Successional age and forest structure in a Costa Rican upper montane *Quercus* forest. *Journal of Tropical Ecology.* Vol. 12, No. 5, (September 1996), pp. 681-698, ISSN 0266-4674

Krebs, C.J. (2009). Ecology: The experimental analysis of distribution and abundance. 6th ed. Benjamin Cummings, ISBN: 0321507436, San Francisco, USA.

Kuzze M., S. Wijdeven and T. De Haan. (1994). Secondary forest and succession: analysis of structure and species composition of abandoned pastures in the Monteverde cloud forest reserve, Costa Rica. International Agricultural College Larenstein, Velps and Agricultural University Wageningen, Wageningen

Lamprecht, H. (1990). Silviculture in the tropics: tropical forest ecosystems and their tree species; possibilities and methods for their long-term utilization. Gesellschaft für Technische Zusammenarbeit, Eschborn. Germany.

March I. & A. Flamenco. (1996). Evaluación rápida de la deforestación en las áreas naturales protegidas de Chiapas (1970-1993). Informe técnico para TNC y USAID. El Colegio de la Frontera Sur. San Cristobal de las Casas, México.

Magurran, A.E. (1988). *Ecological Diversity and Its Measurement.* Princeton University Press. Princeton, ISBN13: 978-0-691-08491-6

Müllerried, F.K. (1957). *Geología de Chiapas.* Gobierno del Estado de Chiapas. Tuxtla Gutiérrez.

Ramírez-Marcial N., M. González-Espinosa and G. Williams-Linera. (2001). Anthropogenic disturbance and tree diversity in Montane Rain Forests in Chiapas, Mexico. *Forest Ecology and Management,* Vol. 154, No. 1-2, (November 2001), pp. 311-326, ISSN 0378-1127

Restrepo, C. and N. Álvarez. (2006). Landslides and their impact on land-cover in the mountains of Mexico and Central America. *Biotropica,* Vol. 38, No. 4, (July 2006), pp. 446–457. ISSN 1744-7429

Richardson D.M. and J. Bond. (1991). Determinants of plant distribution: evidence from pine invasions. *The American Naturalist,* Vol. 137, No. 5, (May 1991), pp. 639-668, ISSN 00030147

Richardson D.M. (1998). Forestry trees as invasive aliens. *Conservation Biology,* Vol. 12, No. 1, (February 1998), pp. 18-26, ISSN 1523-1739

Richter M. (2000). The ecological crisis in Chiapas: A case study from Central America. Mountain Research and Development, Vol. 20, No. 4, pp. 332-339, ISSN: 0276-4741

Rzedowski J. (1978). *Vegetación de México.* Limusa México City, ISBN 968-1800-028

Rzedowski J. (1996). Análisis preliminar de la flora vascular de los bosques mesófilos de montaña de México. *Acta Botanica Méxicana,* Vol. 35, pp. 59-85, ISSN 0187-7151

Saldaña-Acosta A. and E.J. Jardel P. (1991). Regeneración natural del estrato arbóreo en bosques subtropicales de montaña en la Sierra de Manantlán, México. Estudios preliminaries. *Biotam,* Vol. 3, pp. 36-50, ISSN 0187-847

Sugden A.M., E.V.J.Tanner and V. Kapos. (1985). Regeneration following clearing in a Jamaican montane forest: result of a ten-year study. *Journal of Tropical Ecology,* Vol. 1, No. 4, (November 1985), pp. 329-351, ISSN 0266-4674

Ten Hoopen M. and M. Kappelle. (2006). Soil Seed Bank Changes Along a Forest Interior-Edge-Pasture Gradient in a Costa Rican Montane Oak Forest. In: *Ecology and Conservation of Neotropical Montane Oak Forests.* M. Kappelle (ed), pp. 299-308, Springer-Verlag, ISBN 978-3-642-06695-5, Berlin, Germany.

Toledo-Aceves, T., J.A. Meave, M. González-Espinosa and N. Ramírez-Marcial. (2010). Tropical montane cloud forests: Current threats and opportunities for their conservation and sustainable management in Mexico. *Journal of Environmental Management,* Vol. 92, No. 3, (March 2011), pp. 974-98, ISSN 0301-4797

Valencia R., and P.M. Jörgense. (1992). Composition and structure of a humid montane forest on the Pasachoa Volcano, Ecuador. *Nordic Journal of Botany* 12:239-247

Williams-Linera, G. (1991). Nota sobre la estructura del estrato arbóreo del bosque mesófilo de montaña en los alrededores del campamento "El Triunfo", Chiapas. *Acta Botánica Mexicana,* Vol. 13, pp. 1-7, ISSN 0187-7151

Wilms, J.J.A.M. and M. Kappelle. (2006). Frugivorous birds and seed dispersal in disturbed and old growth montane oak forests in Costa Rica. In: *Ecology and Conservation of Neotropical Montane Oak Forests.* M. Kappelle (ed), pp. 309–324, Springer-Verlag, ISBN 978-3-642-06695-5, Berlin, Germany.

Young K. R. (1993). Woody and scandent plant on the edges of an Andrean timberline. *Bulletin of the Torrey Botanical Club* 120:1-18

7

Spatial Patterns of Phytodiversity - Assessing Vegetation Using (Dis) Similarity Measures

S. Babar[1,2], A. Giriraj[3,4], C. S. Reddy[2], G. Jurasinski[5],
A. Jentsch[4] and S. Sudhakar[2]

[1]University of Pune, Pune
[2]National Remote Sensing Centre, Hyderabad
[3]International Water Management Institute (IWMI), Colombo
[4]University of Bayreuth, Bayreuth
[5]University of Rostock, Rostock
[1,2]India
[3]Sri Lanka
[4,5]Germany

1. Introduction

Patterns in vegetation can be expressed through the variation in species composition between plots, which has been termed 'beta-diversity' by Whittaker (1960, 1972). The general importance of beta-diversity has been emphasized in recent years (e.g. Gering et al., 2003; Olden et al., 2006; Sax and Gaines 2003; Srivastava 2002). However, even though Legendre et al., (2005) postulate that "beta diversity is a key concept for understanding the functioning of ecosystems, for the conservation of biodiversity, and for ecosystem management", implementation is still scarce compared to measures of species richness, alpha-diversity and its derivates (but see e.g. Condit et al., 2002; Kluth and Bruelheide, 2004; Pitkänen 2000).

However, the term 'beta-diversity' is not very clear and has multiple meanings (e.g. Jurasinski and Retzer, 2009; Lande, 1996; Qian et al., 2005). Most often 'beta-diversity' is defined as the compositional similarity between vegetation samples (sites, habitats, plots) expressed through (dis)similarity or distance measures. Generally, compositional similarity decreases with distance between plots. This phenomenon is called 'distance decay' of similarity (Tobler, 1970) and can be seen as a characteristic of all geographic systems. Nekola and White (1999) investigated species compositional similarity between fir and spruce stands across North America (Qian et al., 1998, 2005) and concluded that distance decay of similarity might be useful as a descriptor of diversity distribution as well as for the study of factors influencing the spatial structures of communities. The few comparative studies investigating distance decay for more than one group of organisms show that it heavily depends on organism groups as well as on the region under study (Ferrier et al., 1999; Oliver et al., 2004).

Although it is far from being well covered, the spatial change in species composition has received some more attention in recent years (Jurasinski and Retzer 2009). The phenomenon was utilized to evaluate the validity of the dispersal theory (Hubbell 2001) against niche concepts in the Neotropic: Chust et al., (2006), Condit et al., (2002), Duivenvoorden et al., (2002) and Ruokolainen et al., (2002), investigated distance decay in terra firme forests in Panama, Ecuador, and Peru, whereas Duivenvoorden et al., (1995), Ruokolainen et al., (1997), and Tuomisto et al., (2003) concentrated on rain forests in the Amazon basin. However, there are no studies in this regard in the Paleotropic. This hinders the overall generalization of the findings. The data set of the Eastern Ghats provides a unique opportunity as we can calculate distance decay rates in paleotropic ecosystems. Unfortunately there were only few environmental variables recorded. Furthermore these are categorical and assign relatively coarse classes to the plots.

Our aim is to develop a spatially explicit, widely applicable method to assess phytodiversity encompassing species richness, spatial and temporal heterogeneity, and functional diversity and to relate it to environmental conditions (including site conditions, disturbance regime and biological richness value). There is an urgent need for standardized and comparable data in order to detect changes of biodiversity. Such methods are required to be representative as well as pragmatic due to the simple fact that there is insufficient time to obtain complete data sets relating to temporal trends. If biodiversity is lost rapidly at the landscape level, frequent re-investigations have to be done in order to detect and analyze such changes. Thus, our objective was to provide a method that allows for the tracking of changes in biodiversity at the landscape scale. Here, we investigate the distance decay of compositional similarity whilst accounting for the change of the relationship between compositional similarity and its drivers with geographical distance between plots.

Steinitz et al., (2006) highlight that the studies are limited to which extent patterns of distance decay depend on the position along environmental gradients. Accordingly Jones et al., (2006) emphasize that explanatory power of variables might increase with the length of the gradient covered. Based on this, we hypothesize for the present study (Eastern Ghats part of Andhra Pradesh, India) as:

1. Similarity in plant species composition decreases continuously with distance. Due to the small scale of the study (largest geographical distance covered is 8 km) the rate of distance decay will be relatively low compared to large-scale studies.
2. The correlation between compositional similarity of vegetation and the dissimilarity of predictor variables (vegetation type, disturbance - biological richness value, abiotic environmental conditions) is changing with geographical distance between multiple plots. We expect the correlation to increase with spatial scale/distance between sampling units because the sampled environmental gradient is likely to increase as well.
3. Disturbance is the main driver of vegetation patterns in the regarded transitional ecosystem. We especially consider disturbances because deciduous ecosystems of the Eastern Ghats have a long-lasting disturbance history and stress tolerators are clearly favoured compared to competitors (Rawat, 1997).

1.1 Previous studies

1.1.1 Measuring and analyzing similarity

Compositional similarity or differentiation diversity between sampling plots is an important basis for most numerical analyses in vegetation ecology. It is at the heart of ordination methods and has general importance regarding the testing of ecological theory (Legendre et al., 2005). Moreover, it represents the basis for most of the analyses in the present study and shall therefore be discussed in some detail in this chapter. Data on species composition is generally of multivariate character. Thus, hypothesis testing regarding the relation between species composition and its drivers can hardly be achieved with normal statistics. This led to a specific set of methods for vegetation ecologists (Jongman et al., 1987; Legendre & Legendre, 1998; Leyer and Wesche 2007; Sokal and Rohlf, 1981). The majority of this method is based on the calculation of biological resemblance and ecological distance in data space. Resemblance can be calculated with a wide range of coefficients and indices, measuring similarity, dissimilarity, proximity, distance, association or correlation (Orlóci 1978; Tamas et al., 2001).

1.1.2 Measuring similarity

Compositional similarity measures differentiation diversity (Jurasinski and Retzer, 2009) and can be calculated with resemblance measures. These are available in two primary groups (Legendre and Legendre 1998): (1) Quantitative or distance indices are used to calculate the proximity of two samples in data space from quantitative measurements or abundance data. (2) Similarity/Dissimilarity measures can handle binary data as it is typically found in presence/absence data sets. There are a large number of different indices and coefficients available and comparative reviews can be found in Cheetham and Hazel (1969), Hubalek (1982), Janson and Vegelius (1981), Koleff et al., (2003), Shi (1993) and Wolda (1981). All binary similarity/dissimilarity measures are based on the same set of variables. Whether a coefficient measures similarity or dissimilarity depends on the implementation of the formula. However, most of them can easily be transformed from a similarity to a dissimilarity measure by calculating 1-S (with S being similarity, Legendre and Legendre, 1998; Shi, 1993). From some similarity coefficients a dissimilarity measure following the Euclidean metric can be obtained by calculating sqrt $1 - S$ instead (for details see Legendre and Legendre, 1998).

Indices incorporating the species not present in both of the compared samples (d, see Fig. 1) are controversially discussed (e.g. Simple Matching by Sokal & Michener (1958) or the coefficient of Russel and Rao (1940)). Legendre and Legendre (1998) call these symmetrical (see Janson and Vegelius, 1981 for another definition of symmetry) and discuss the "double-zero problem" at the beginning of their chapter on similarity indices, because it "is so fundamental with ecological data" (Legendre and Legendre, 1998, p. 253): Species are supposed to have unimodal distributions along environmental gradients. If a species is absent from two compared sampling units (which is expressed by zeros in the species matrix), it is not discernable on which end of the gradient the both sites are with respect to a certain environmental parameter (Field, 1969). Both might be above the optimal niche value for that species, or both below, or on opposite tails of the gradient. Thus, the incorporation of unshared species (d) might lead to wrong conclusions when the relation between environment and species composition is under study (Legendre and Legendre, 1998).

Additionally, Shi (1993) states that the status of d in similarity coefficients for paleoecological studies is not clear and cannot be assessed directly because of its great dependence on the less common taxa absent from both sites: "In palaeontology, absence of a taxon, particularly a rare one, may have been derived from differential preservation and/or sampling errors rather than some ecological factors". Finally, Field (1969) shall be cited as he found a very well fitting metaphor: "No marine ecologist would say that the intertidal and abyssal faunas were similar because both lacked the species found on the continental shelf".

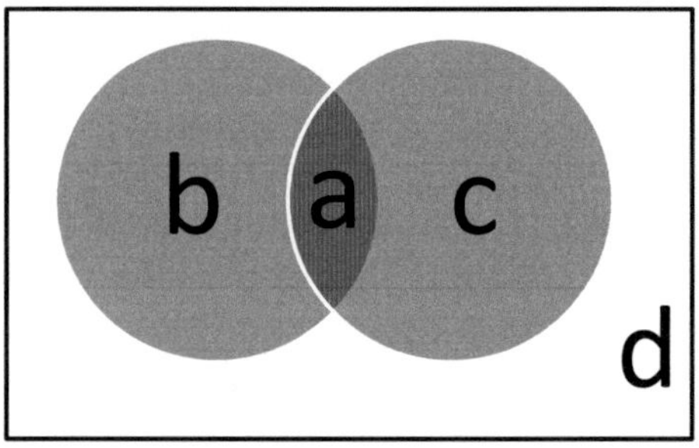

Fig. 1. Illustration of the matching components providing the basis for binary similarity measures a) the number of species shared by two compared units, b) the number of species unique to one of the compared plots, c) the number of species unique to the other one of the compared plots, d) the number of species not found in the two compared plots but in the whole dataset (unshared species).

(Dis) similarity indices, which take unshared species into account, mingle different ideas of differentiation diversity (additive partitioning, multiplicative partitioning, turnover, see Jurasinski, 2007). Furthermore, they tend to be less specific as the values show less variance because d is far bigger than a, b, or c for most of the datasets recorded in the field. "Including double-zeros in the comparison between sites would result in high values of similarity for the many pairs of sites holding only a few species; this would not reflect the situation adequately" (Legendre and Legendre, 1998). Furthermore is the total inventory diversity (gamma) as a background for the calculation of d often difficult to define. When temporal changes are addressed, the question arises, whether the species pool of one time step or the whole species pool as recorded over several time steps should be regarded.

1.1.3 Required features for a (Dis) similarity measures

From the previous studies mentioned above it is obvious that a measure of diversity should possess the following three properties:

- It compares the similarity of a focal plot to several other plots, e.g. its surrounding neighbors taking species identity into account.
- It yields a single value as result, which can be directly attributed to the investigated focal plot.
- Its values should range between 0 and 1 for the sake of standardization and ease of interpretation.

From the multi-plot similarity measures found in the literature and introduced above, none meets all these properties. Thus, we propose a new multi-plot similarity measure, which is discussed in the method section. We call it simply the coefficient of multi-plot similarity. The performance of this new measure regarding the detection of typical pattern is tested for the random and continuous datasets carried out in the Eastern Ghats of Andhra Pradesh (India).

2. Materials and methods

2.1 Study site

957 random plots spread across Eastern Ghats of Andhra Pradesh were used in this study (Fig. 2). To typify the Eastern Ghats of Andhra Pradesh at landscape level it was classified in

two zones viz., northern Eastern Ghats (Zone-1) and southern Eastern Ghats (Zone-2). Following the gradient in the geology, soil, bioclimatic (temperature and precipitation), altitude and degree of disturbance, the zones were characterized and analysed. For continuous enumeration a total of six transects were laid and data collected was used for analysis. These transects were laid both in zone-1 and zone-2 and distributed evenly in deciduous ecosystems.

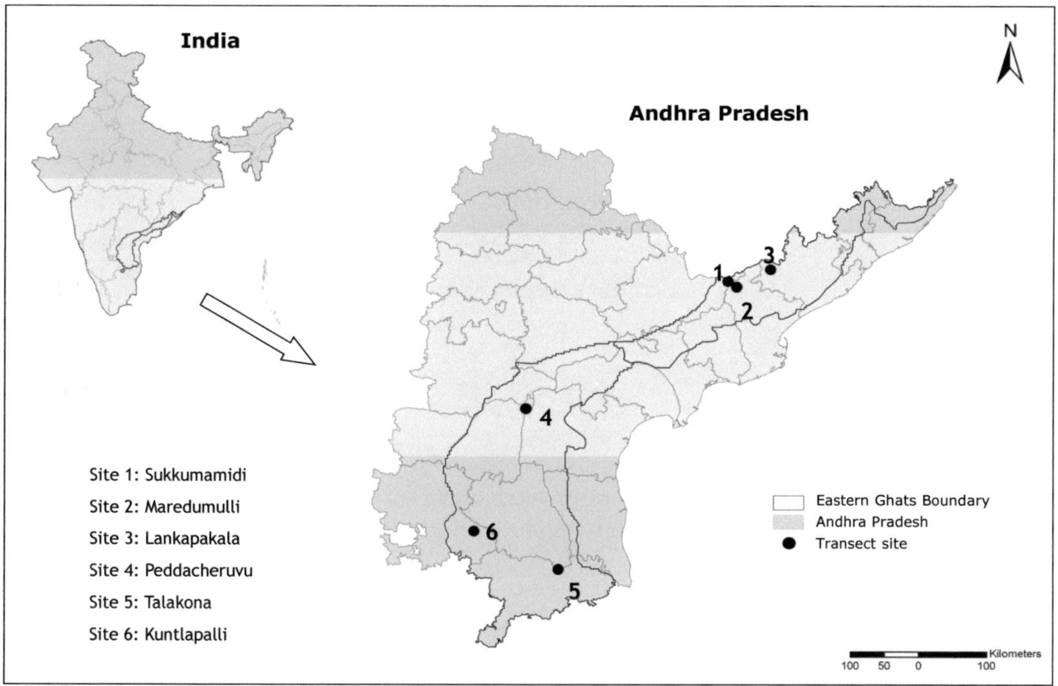

Fig. 2. Location map of the six transects and zones of the northern and southern Eastern Ghats for random plots in the state of Andhra Pradesh, India

The areas studied are three 0.5-ha plots, which are located in the Sileru-Maredumilli ranges of the Northern Eastern Ghats of Andhra Pradesh (zone-1) and Nallamalais-Seshachalam-Nigidi hill ranges of Southern Eastern Ghats of Andhra Pradesh (zone-2) (Fig. 2). These forests are classified as South Indian Moist Deciduous and Orissa Semi evergreen forests (Champion & Seth, 1968). Three 0.5ha plots area was established at three different sites: Site 1 is located about 2 km from Sukkumamidi, a tribal hamlet in Khammam district, which receives mean annual rainfall about 1200-1400mm and elevation ranging from 400-600m. Site 2 is located about 6 km from Maredumilli tribal village in East Godavari district, receives mean annual rainfall about 1400-1600mm with an elevation of 600-800m. Site 3 is located about 2 km from Lankapakala tribal hamlet in Visakhapatnam district receives mean annual rainfall 1600-1800mm and an elevation of 800-1100m is shown in Table 1. All the three study sites were undisturbed. There are no records on the intensity and the extent of disturbance.

The next set of sample plots were laid in Southern Eastern of Andhra Pradesh (zone-2). These forests are classified as Tropical Dry Deciduous by Champion & Seth (1968). Three 0.5ha plots area were established at three different sites: Site 4 is located about 3km from

Peddacheruvu, a chenchu tribal hamlet in Nallamalais of Kurnool district which receives mean annual rainfall about 900-1000mm. Site 5 is located about 4km from Talakona, a Yanadi tribal hamlet in Seshachalam hills of Chittoor district which receives mean annual rainfall 800-900mm. Site 6 is located about 2km from Kuntlapalli village, Anantapur district receives mean annual rainfall about 600-700mm. The rocks are of Kurnool-Cuddapah formations (quartzite and slate formation predominate) and altitude ranges from 400-600m. Thus, these sites show variability in rainfall pattern even though phytogeographical range is contiguous.

Sites	Location	Forest type	Elevation (m)	Rainfall (mm)
1	Sukkumamidi	Moist Deciduous	500	1200-1400
2	Maredumulli	Moist Deciduous	700	1400-1600
3	Lankapakala	Moist Deciduous - Semievergreen	900	1600-1800
4	Peddacheruvu	Dry Deciduous	600	900-1000
5	Talakona	Dry Deciduous	700	800-900
6	Kuntlapalli	Dry Deciduous	800	400-600

Table 1. Study area detail for the continuous plots on its forest type, elevation and rainfall pattern in Northern and Southern Eastern Ghats of Andhra Pradesh, India

2.1.1 Geographical extent

The Eastern Ghats are a discontinuous range of mountains along eastern coast of Peninsular India extending over 1750 km with the average width of about 100 km and extends from $10^{0}05'$ to $22^{0}30'$ N Latitude and $76^{0}23'$ to $86^{0}50'$ E longitude (Fig. 2). The Eastern Ghats are 'tors' of geological antiquity and are geologically older than Himalayas and Western Ghats. Eastern Ghats cover four states (Orissa, Andhra Pradesh, Tamil Nadu and small portion of Karnataka) and present study was undertaken in the Eastern Ghats part of Andhra Pradesh lying approximately between 12 - 19 N latitude and 76 - 84 E longitudes. It is bounded by Eastern coast on the East, Deccan plateau on the West, South and North covers the Eastern Ghats part of Tamil Nadu and Orissa state respectively.

Study area includes Eastern Ghats districts of Andhra Pradesh viz., parts of Srikakulam, Vizianagaram, Visakhapatnam, East Godavari, West Godavari, Khammam, Krishna, Mehbubnagar, Nalgonda, Guntur, Kurnool, Anantapur, Prakasam, Kadapa, Nellore and Chittoor (Fig. 2). The total geographical area covered under Eastern Ghats of Andhra Pradesh is 98,662 km^2 having 23,894 km^2 of forest area. Major rivers like Godavari, Krishna and Pennar cut the range into discontinuous blocks of hills along the East coast in Andhra Pradesh. Major forest range includes the Upper Sileru range in north, lower Velikonda Range lies to the east, and the higher Palikonda-Lankamalla-Nallamalla ranges in the west.

Eastern Ghats also harbours wide range of wild crops (millet, rice) and economic and medicinal plants. Endemic plants of this region are basically Palaeo-endemics and are localized. They have very narrow distribution range and several studies indicate that they are under gradual process of extinction (Reddy et al., 2006). Nearly, 54 tribal communities inhabit Eastern Ghats region (MoEF, 1997). The Eastern Ghats also harbours one of the richest Bauxite deposits in the world.

2.1.2 Vegetation distribution

Andhra Pradesh ranks first amongst the states and Union Territories in terms of area under tree cover (SFR, 2001). The total forest area of the state is 44,637 km^2, which occupies 16% of the total geographical area of 2,75,068 km^2 (SFR, 2001). The forests in the region are broadly classified into Semi Evergreen, Moist Deciduous, Dry Deciduous, Thorn and Scrub forests and are comparable with the existing (Champion & Seth, 1968) classification.

2.1.3 Geology, Soil and bioclimate

Eastern Ghats is not formed of one particular geological formation but consist of rocks varying in origin and structure according to the location. Geologically, zone-1 is mainly of Charnockites and Khondalites (Krishnan, 1960) having red and black soil, while, zone-2 is made up of Quartzite and Slate formations with red, mixed red, black and lateritic soil.
Climate of zone-1 is warm and humid with an annual precipitation of 1200-1700 mm compared to zone-2 which is hot and dry with lesser precipitation of 600-1000 mm. Topographically, zone-1 has higher altitudinal range (100-1672m) compared to zone-2 (100-1000m).

2.1.4 Protected area

Eastern Ghats of Andhra Pradesh has 1 National Park (i.e., Sri Venkateshwara National Park) and 6 wildlife sanctuaries (viz., Papikonda Wild Life Sanctuary (WLS), Kolleru WLS, Nagarjunsagar Srisailam Tiger Reserve (NSTR) WLS, Gundla Brahmeshwarm WLS, Sri Lankamalleshwara WLS and Sri Penunsila Narsimha WLS) of these only Kolleru WLS is the largest fresh water lake in the country and also treated as one of the Ramsar convention wetland sites in the world.

2.2 Methodology

2.2.1 Phytosociology data analysis

For the analysis of data the methodology is same as discussed in chapter-2. Six transects were laid (10 x 500m) and the individuals with $\geq$ 10cm were enumerated. For each of these plots the GPS location was collected using handheld Garmin E-trex GPS and other biotic and abiotic environmental parameters (slope, aspect and altitude) were gathered. The fieldwork was conducted in January 2006 to March 2007.

2.2.2 General considerations for statistical analysis

For evaluating distance decay relationships one basically plots similarity between sites against their geographical distance (Nekola & White, 1999). So the first to calculate are similarities between sample locations. There exists a multitude of coefficients for the calculation of compositional resemblance of species samples.
Sørensen similarity (Sørensen, 1948) is used to calculate compositional similarity based on plot inventories of all tree species throughout the presented study. Sørensen similarity does satisfy the criteria of linearity, homogeneity (if all values are multiplied by the same factor the value is not changing), symmetry (independence from calculation direction, after (Janson and Vegelius, 1981) and scaling between 0 and 1 (Koleff et al., 2003). It is well

established and extensively used especially in vegetation ecology (e.g. Condit et al., 2002; Kluth & Bruelheide, 2004). This guarantees comparability with other studies. Sørensen is favored over Jaccard because the latter is more important in zoological studies (Koleff et al., 2003). Sørensen differs insofar that it does weight the shared species double which is seen as advantageous by Legendre & Legendre (1998) since shared species have more explanatory power regarding the underlying processes of the found patterns (Watt, 1947). Geographic distances between plots were obtained through the calculation of Euclidean distances between the x- and y-coordinates with the function dist () of the R package base (R Development Core Team, 2005).

2.2.3 Slope and aspect

Slope aspect and slope inclination may have a significant effect on species richness (Badano et al., 2005) and species composition especially in semi-arid vegetation (Sternberg & Shoshany, 2001). To obtain a distance measure integrating aspect and inclination, we use the model of a unit sphere and calculate great-circle distances between virtual locations. This allows for the generation of continuous rather than class variables as e.g. found in Kjällgren and Kullman (1998). For each plot a virtual location on the sphere is defined using the values for aspect as longitude and 90°-inclination as latitude. Therefore the virtual points are located in the pole region as long as inclination is low which leads to small (virtual) distances between them. The idea behind is that solar radiation; wind or other factors highly depending on aspect and inclination (Wilkinson & Humphreys, 2006) are not considerably different on plots with varying aspects as long as inclination is low. The longitude values on the unit sphere are derived from the directional reference made in the field. The equator of the sphere is thought as the compass circle. The Prime Meridian of the virtual sphere is the great circle through North and South of the compass. As in geographic terms longitude counts positive in Eastern and negative in Western direction. With Phi = latitude = 90°-inclination and Lambda = longitude = aspect the great circle distance between A and B can be calculated with formula 1. As we use a unit sphere the maximum distance between two inclination/aspect pairs is perimeter/2 of the sphere, which is by definition Pi. To scale the possible distances between 0 and 1 the results of formula [1] are divided by Pi. Thus, a great-circle distance of 1 is rather scarce in the real world; however, two vertical rock walls with opposite aspect would share it.

$$\zeta = \arccos(\sin(\phi_A)\cdot\sin(\phi_B) + \cos(\phi_A)\cdot\cos(\phi_B)\cdot\cos(\lambda_B - \lambda_A)) \quad (1)$$

2.2.4 Statistical analysis

All statistical analyses were performed using functions of the packages base, stats, vegan and simba of the R statistics system (R Development Core Team, 2005). For better reading it is referred to the functions in the form 'function [package]' when the package is not mentioned before or 'function ()' when it is clear which package is meant in the following.

2.2.5 Compositional similarity

A common way to evaluate the structuring of compositional similarity within a data set is to use NMDS (Non-Metric multidimensional Scaling) plots. Non-metric multidimensional scaling (Kruskal, 1964) differs from other known ordination methods in that it does not build on a specific distance measure. Whereas PCA (principal components analysis), RDA (redundancy analysis) rely on Euclidean distance, and DCA (detrended correspondence analysis), CA (correspondence analysis), CCA (canonical correspondence analysis) rely on ChiSquare distance, NMDS (as well as PCoA - principal coordinates analysis) is open to any kind of distance measure. Therefore it allows the implementation of measures, which have been proven adequate for ecological data as Bray-Curtis distance (Faith et al., 1987; Minchin, 1989). A good dissimilarity measure for communities has a good rank order relation to distance along environmental gradients. Because NMDS uses only rank information and maps these ranks non-linearly onto ordination space, it can handle non-linear species responses of any shape and effectively and robustly find the underlying gradients (Oksanen, 2005). It is an iterative ordination method that attempts to minimize a stress function, which measures the difference between the original floristic distances among sampling units and their new distances in the ordination space. Since NMDS is a non-metric method, it optimizes the rank order of distances rather than their actual values (Legendre & Legendre, 1998).

NMDS has been shown to be a very robust method regarding its reliability even when certain assumptions (like Gaussian species responses or sampling pattern) are violated (Minchin, 1989). NMDS was able to deal with any kind of response model, which was not the case for its strongest competitor, the DCA (ibid.). On the other hand the scaling of the axis scores does not allow drawing any conclusions regarding the position on the axis and ecological implications thereof. However, the positioning of sampled sites in a NMDS plot allows interpretation regarding their neighbors. These are similar in their species composition. When sites occur clumped in the NMDS this might be attributable to specific geographic positions or environmental conditions.

DCA bases on CA, which can be seen as a weighted principal coordinates analysis (PCoA), computed on ChiSquare distances (Faith et al., 1987). It therefore depends on the relationship between ChiSquare distance and ecological distance. This is the most important difference between the two methods: DCA is based on an underlying model of species distributions, the unimodal model, while NMDS is not. However, not all species exhibit the same response curve (e.g. Gaussian responses) (Minchin, 1989). Thus, it is preferable to use NMDS especially with no specific hypothesis regarding species environment interaction in mind. Furthermore Legendre & Legendre (1998) state that detrending should be avoided, except for the specific purpose of estimating the lengths of gradients. De'ath (1999) formulates that there are two classes of ordination methods - 'species composition representation' (e.g. NMDS) and 'gradient analysis' (e.g. the various flavors of CA). This means that NMDS rather is a mapping method, which allows for projecting the multivariate data-space onto a two-dimensional map whereas PCA, CA and its relatives base on projection and rotation (Oksanen, 2004).

In the present case the alteration of species composition through time was in focus and not the gradient representation in species composition. Therefore NMDS is the method of choice as it furthermore enabled the use of Bray-Curtis distance which is the quantitative one-complement of the Sørensen index. Thus, the results are easily comparable and interpretation may not be hindered due to the implementation of different metrics.

2.2.6 Distance decay

Distance decay or spatial auto-correlation of quantitative univariate variables is usually calculated using semi-variograms (Legendre & Legendre, 1998). For multivariate data Mantel correlograms can be applied (Legendre & Legendre 1998; Sokal & Rohlf 1981). A simple possibility for vegetation data is to regress similarity of units regarding species composition against their geographical separation (Nekola and White, 1999; Steinitz et al., 2006). To test for the influence of different vegetation types on patterns of compositional similarity data was divided into subsets. As the plots can be assigned to 2 geographically distinct regions, the distance decay of different vegetation types and subsets based on other categorical variables (fragmentation, slope, disturbance, etc.) is compared between the two regions.

The similarity values of the subsets are compared with an ANOVA-like function (mrpp[vegan]), (Oksanen et al., 2007)) and tested for significant differences using a permutation procedure (diffmean[simba]). Normal tests and ANOVA might fail here because the similarities are not independent from each other.

The Multiple Response Permutation Procedure (mrpp) allows testing whether there is significant difference between two or more groups of sampling units. The method is insofar similar to anova, in that it compares dissimilarities within and among groups. If two groups of sampling units are really different (e.g. in their species composition), then average within-group compositional dissimilarities ought to be less than the average of the dissimilarities between two random collections of sampling units drawn from the entire population. The mrpp statistic delta gives the overall weighted mean of within-group means of the pairwise dissimilarities among sampling units. The mrpp[vegan] algorithm first computes all pairwise distances in the entire dataset, then calculates delta. Then the sampling units and their associated pairwise distances are permuted, and delta is recalculated based on the permuted data. The last steps are repeated N times. N defaults to 1000 which provides a possible significance-level of $p<0.001$ as significance is tested against the distribution of the permuted deltas.

After testing for significant differences between subsets, the differences in mean similarity are tested with a permutation procedure (diffmean[simba]). The difference in mean similarity between two sets is calculated. The two sets are joints together and two random sets the same size as the original sets are selected and their difference in mean is calculated. Then the sampling units and their associated pairwise distances are permuted, and the difference in mean is recalculated based on the permuted data. The last steps are repeated N times. N defaults to 1000 which provides a possible significance-level of $p<0.001$ as significance is tested against the distribution of the permuted values.

To answer the question if distance decay is significantly different between the various evaluated subsets of the data, the slopes of the distance decay relationship have been calculated for the three subsets and compared. A permutation procedure following Nekola & White (1999) has been implemented as an R function (diffslope[simba]) to test for significance. For each subset compositional similarity between plots is regressed against their geographical separation. Before calculating the difference in slope between two subsets the values of compositional similarity are rescaled to a common mean. So testing the difference in slope of the distance decay relationship is independent of differences in the mean (Nekola & White, 1999; Steinitz et al., 2006). Linear regression is carried out on both of the subsets, and the difference in slope is calculated and stored. Then the variable pairs (geographical separation, compositional similarity) are randomly reassigned to the two data-subsets. Regression is calculated for each of the random subsets and the difference in slope is obtained again. The last steps are repeated 1000 times. Finally the difference between the observed slopes is compared to the differences based on random reassignment. Number of times when randomization are being produced differences in slope which is higher than the original data are summed up and divided by the number of permutations to get a p-value.

3. Results

3.1 Phytosociology analysis

3.1.1 Tree species richness and diversity indices

Results on the phytosociological parameters for the random plots are presented in Table 2. A total of 278 (zone-1) and 679 plots (zone-2) covering major vegetation types were enumerated to assess species richness, composition and diversity patterns. A sum of 25,621 individuals belonging to 963 species, 512 genera and 133 families were observed in the entire study area. It was observed that species richness was higher in zone-2 with 818 species belonging to 124 families as compared to 372 species of 95 families in zone-1. Herb species richness in zone-2 (350) was higher than in zone-1 having 101 species (Table 2). A total of 57 grass species majority of which belong to Poaceae followed by Cyperaceae were observed in this study.

Site	Vegetation type	No. of Plots	Trees	Shrubs	Herbs	Climbers	Grass	Species Richness	No. of tree Individuals	Families	Species Diversity (H')	Stand Density (ha^{-1})	Basal Area (m^2/ha)
Northern Eastern ghats (Zone-1)	Semievergreen	18	68	18	31	12	3	148	299	58	5.3	415.3	40.95
	Moist Deciduous	77	76	31	66	20	7	227	794	69	5.2	257.8	15.79
	Dry Deciduous	108	84	30	70	27	6	240	2100	75	5.0	486.1	19.04
	Dry Evergreen	17	52	17	37	12		135	181	57	5.2	266.2	8.65
	Thorn Forest	10	17	14	27	12	1	100	50	60	3.7	156.3	2.89
	Degraded Forest	48	64	19	44	18	4	165	863	59	4.2	449.5	12.94
	Total	278	137	51	101	44	10	372	4287	95			17.39
Southern Eastern Ghats (Zone-2)	Moist Deciduous	102	167	71	184	61	22	520	1511	104	6.6	370.3	13.28
	Dry Deciduous	325	191	87	281	73	37	680	5024	117	6.6	386.5	14.67
	Hardwickia Mixed	85	145	73	166	51	21	466	952	103	6.1	290.2	11.13
	Red Sanders Mixed	65	118	62	157	52	15	415	937	93	5.9	360.4	12.64
	Dry Evergreen	47	125	46	128	39	15	365	629	88	6.2	334.6	11.10
	Thorn Forest	36	112	55	93	48	11	326	427	86	5.7	296.5	8.61
	Degraded Forest	19	106	36	66	21	11	247	235	75	6.2	309.2	10.45
	Total	679	205	119	339	90	51	818	9715	124			13.15

Table 2. Species Richness in different vegetation type and habits in Northern and Southern Eastern Ghats of Andhra Pradesh, India

Parameters	Site					
	1	2	3	4	5	6
No. of Species	43	65	72	53	61	55
No. of Individuals	308	525	495	407	307	415
No. of Families	24	33	39	27	30	35
No. of Genera	37	59	61	46	54	48
Species Diversity	2.9	3.0	3.5	3.3	3.6	3.4
Basal area (m^2/ha)	14.71	17.61	24.42	4.35	10.11	6.82
No. of Endemics	2	0	0	1	0	138
No. of RET	0	0	4	9	2	2
Stand Density (ha^{-1})	616	1050	990	814	614	830

Table 3. Consolidated results for the six transect sites in Northern and Southern Eastern Ghats of Andhra Pradesh, India

For the six transects in both the zones results are presented in Table 3. A total of 6 transect of 0.5ha covering major vegetation types were enumerated to assess species richness, species composition and species diversity patterns. A total of 2,457 individuals belonging to 197 species, 139 genera and 57 families were observed in the entire study area. It was observed that species richness was higher in site 3 with 72 species belonging to 39 families as compared to other sites (Table 3). Least species diversity was found in site 1 and highest in site 5.

3.1.2 Species density, dominance, rarity

The first ten dominant species explains the dominance and structure of the forests. In zone-1, semi-evergreen forests are predominated by *Mangifera indica* (40.15), *Garuga pinnata* (15.98), *Xylia xylocarpa* (15.39) and *Stereospermum suaveolens* (15.38) (Table 4a) and this forest type doesn't exist in zone-2. In the case of Moist deciduous forests of the zone-1, it was represented by *Terminalia alata* (39.99), *Xylia xylocarpa* (22.22), *Pterospermum xylocarpum* (19.26) and *Bursera serrata* (9.15). Contrastingly zone-2 was characterized by *Limonia acidissima* (10.47), *Cassia fistula* (10.15), *Terminalia alata* (9.42) and *Madhuca indica* (7.28). Deciduous forest are the dominated forests among all the forest type, which is composed of *Xylia xylocarpa* (45.89), *Anogeissus latifolia* (24.48), *Terminalia alata* (12.42) and *Stereospermum suaveolens* (11.71) in zone-1, while in zone-2 *Chloroxylon sweitenia* (14.83), *Anogeissus latifolia* (9.50), *Cochlospermum religiosum* (9.27) and *Sterculia urens* (8.15) dominates. The difference in species composition is mainly due to the variations in altitudinal and precipitation gradients and also certain degree of disturbance prevailing in this forest type.

Zone	Semi-Evergreen Forest	IVI	Moist Deciduous Forest	IVI	Dry Deciduous Forest	IVI	Degraded Forest	IVI
Zone-1	Mangifera indica	40.15	Terminalia alata	39.99	Xylia xylocarpa	45.89	Cleistanthus collinus	53.14
	Garuga pinnata	15.98	Xylia xylocarpa	22.22	Anogeissus latifolia	24.48	Anogeissus latifolia	40.89
	Buchanania lanzan	15.86	Pterospermum xylocarpum	19.26	Terminalia alata	12.42	Xylia xylocarpa	36.66
	Xylia xylocarpa	15.39	Bursera serrata	9.15	Stereospermum suaveolens	11.71	Lannea coromandelica	17.00
	Stereospermum suaveolens	15.38	Cassia fistula	8.53	Lannea coromandelica	10.97	Terminalia alata	14.51
	Terminalia alata	15.18	Bombax ceiba	7.74	Grewia tilaefolia	9.96	Grewia tilaefolia	14.50
	Macaranga peltata	15.15	Terminalia bellirica	7.38	Tectona grandis	9.78	Garuga pinnata	11.98
	Memecylon edule	10.20	Anogeissus latifolia	7.36	Garuga pinnata	8.78	Stereospermum suaveolens	7.55
	Syzygium cumini	9.66	Syzygium cumini	6.81	Pterospermum xylocarpum	7.29	Dalbergia paniculata	7.29
	Pterospermum xylocarpum	8.10	Lannea coromandelica	6.67	Lagerstroemia parviflora	6.75	Diospyros melanoxylon	6.44

Zone			Moist Deciduous Forest	IVI	Dry Deciduous Forest	IVI	Degraded Forest	IVI
Zone-2			Limonia acidissima	10.47	Chloroxylon swietenia	14.83	Cleistanthus collinus	23.55
			Cassia fistula	10.15	Anogeissus latifolia	9.50	Madhuca indica	7.52
			Terminalia alata	9.42	Cochlospermum religiosum	9.27	Mangifera indica	6.99
			Madhuca indica	7.28	Sterculia urens	8.15	Vitex peduncularis	6.94
			Sterculia urens	7.19	Tectona grandis	8.10	Butea monosperma	6.57
			Polyalthia cerasoides	6.98	Pterocarpus marsupium	7.55	Melia dubia	6.08
			Mangifera indica	6.71	Cassia fistula	7.42	Anogeissus latifolia	5.82
			Chloroxylon swietenia	6.60	Terminalia alata	7.09	Wrightia arborea	5.69
			Cochlospermum religiosum	6.40	Madhuca indica	6.80	Holoptelea integrifolia	5.49
			Mallotus philippensis	6.23	Boswellia serrata	5.96	Cassia fistula	5.48

Zone	Red Sanders Mixed Forest	IVI	Hardwickia Mixed Forest	IVI	Southern Thorn Forest	IVI
Zone-2	Pterocarpus santalinus	36.72	Hardwickia binata	41.50	Acacia chundra	30.67
	Albizia odoratissima	30.13	Anogeissus latifolia	17.24	Chloroxylon swietenia	21.32
	Anogeissus latifolia	12.87	Pterocarpus marsupium	8.15	Lannea coromandelica	10.75
	Hardwickia binata	9.08	Cochlospermum religiosum	8.08	Diospyros chloroxylon	9.26
	Terminalia alata	7.80	Terminalia alata	6.98	Acacia leucophloea	8.08
	Polyalthia cerasoides	6.22	Sterculia urens	6.76	Albizia amara	7.70
	Cassia fistula	5.99	Soymida febrifuga	6.19	Gyrocarpus americanus	7.04
	Cochlospermum religiosum	5.95	Bauhinia racemosa	5.10	Hardwickia binata	6.65
	Sterculia urens	5.90	Limonia acidissima	4.64	Bridelia montana	6.36
	Chloroxylon swietenia	5.67	Boswellia serrata	4.50	Atalantia monophylla	6.03

Table 4a. Important value index for the dominant tree species based on random for different vegetation types analyzed in Northern and Southern Eastern Ghats of Andhra Pradesh, India

Site 1		Site 2		Site 3	
Species	**IVI**	**Species**	**IVI**	**Species**	**IVI**
Xylia xylocarpa	72.46	Xylia xylocarpa	67.23	Schleichera oleosa	39.98
Dillenia pentagyna	20.97	Bursera serrata	26.43	Pterocarpus marsupium	26.05
Lagerstroemia parviflora	20.82	Terminalia alata	20.09	Bauhinia vahlii	22.52
Dendrocalamus strictus	18.04	Ougenia oojeinensis	18.04	Grewia tilaefolia	21.41
Anogeissus latifolia	17.98	Grewia tilaefolia	13.43	Mangifera indica	17.18
Buchanania lanzan	15.68	Pterocarpus marsupium	11.64	Mallotus philippensis	14.54
Cleistanthus collinus	12.81	Syzygium cumini	7.96	Gmelina arborea	13.53
Terminalia alata	12.39	Diospyros sylvatica	7.89	Cassia fistula	12.69
Terminalia bellirica	10.35	Semecarpus anacardium	7.04	Garuga pinnata	11.39
Pterocarpus marsupium	10.22	Bombax ceiba	6.47	Terminalia alata	9.43

Site 4		Site 5		Site 6	
Species	**IVI**	**Species**	**IVI**	**Species**	**IVI**
Chloroxylon swietenia	41.84	Pterocarpus marsupium	27.96	Terminalia pallida	35.19
Albizia amara	28.47	Lannea coromandelica	24.97	Pterocarpus santalinus	30.96
Diospyros chloroxylon	19.95	Anogeissus latifolia	20.86	Anogeissus latifolia	22.00
Ixora arborea	17.09	Dalbergia paniculata	20.39	Terminalia chebula	14.79
Premna tomentosa	16.65	Mitragyna parvifolia	13.72	Gardenia gummifera	14.45
Anogeissus latifolia	16.40	Grewia tilaefolia	12.01	Dolichandrone atrovirens	12.14
Acacia chundra	11.26	Lagerstroemia parviflora	10.98	Buchanania lanzan	10.15
Hildegardia populifolia	10.72	Bambusa arundinacea	10.81	Phyllanthus emblica	9.70
Commiphora caudata	9.59	Terminalia alata	10.68	Chloroxylon swietenia	9.04
Santalam album	9.52	Madhuca indica	8.46	Madhuca indica	8.88

Table 4b. Important value index for the dominant tree species based continuous plots in Northern and Southern Eastern Ghats of Andhra Pradesh, India

In the Southern Eastern Ghats (Zone-2) of Andhra Pradesh, we observed locale-specific formation, which is, characterized as Red-sander (*Pterocarpus santalinus*) mixed forest and Hardwickia (*Hardwickia binata*) mixed forest. Red-sander mixed forests are primarily deciduous system having *Pterocarpus santalinus* (36.72) as dominant species and secondary composition of *Anogeissus latifolia* (30.13), *Hardwickia binata* (9.08) and *Terminalia coriacea* (7.80). In Hardwickia mixed forest, *Hardwickia binata* (41.50) forms a community with *Anogeissus latifolia* (17.24), *Pterocarpus marsupium* (8.15) and *Cochlospermum religiosum* (8.08). These forests in both the zones are mainly dominated by *Cleistanthus collinus* species and also existences of some primary and secondary succession species.

For the transect of the 6 sites the dominance (IVI) varied a lot with very less species common in all the sites. Site 1 and 2 was predominated by *Xylia xylocarpa* (72.46 and 67.23 respectively), followed by *Dillenia pentagyna* (20.97), *Lagerstroemia parviflora* (20.82) and *Anogeissus latifolia* (17.9) in site 1 (Table 4b). In the case of site 3 the structure was predominantly of *Schleichera oleosa* (39.98), *Pterocarpus marsupium* (26.05), *Grewia tilaefolia* (21.41) and *Mangifera indica* (17.18). Contrastingly zone-2 was characterized by dry deciduous forest with Chloroxylon sweitenia (41.84), *Pterocarpus marsupium* (27.96) and *Terminalia pallida* (35.19) dominating site 4, 5 and 6 respectively.

3.1.3 Tree family dominance

For the random plots, a total of 95 families having 8,323 individuals were recorded from zone-1 and 17,383 individuals belonging to 124 families in zone-2. Taxonomically well-represented families include Fabaceae (101), Euphorbiaceae (58), Rubiaceae (51) and Acanthaceae (48). Dominant families with respect to number of individuals are Euphorbiaceae (1786), Mimosaceae (1669), Combretaceae (1606), Fabaceae (1577), Rubiaceae (1368) and Caesalpinaceae (1282). Families with rare occurrences represented by single and double species were 66 for the whole study area.

For the continuous plot of the six sites, a total of 57 families belonging to 2,457 individuals were recorded from the present study. Taxonomically well-represented families include Rubiaceae (18), Euphorbiaceae (16), Fabaceae (11) and Caesalpiniaceae (9). Dominant families with respect to number of individuals are Mimosaceae (337), Combretaceae (274), Fabaceae (191), Rubiaceae (154) and Caesalpinaceae (138). Families with rare occurrences represented by single and double species were 36 for both the study sites. Top ten families explain the species characteristics and found to be 66% (1620 individuals out of 2,457 individuals) dominant for the study site.

3.1.4 Species accumulation curve

Species accumulation curve for the random plots showed different flattening levels in vegetation types of both the zones. The species accumulation as a function of number of individuals is given in Fig. 3a. Deciduous forest in both the zones showed a clear tendency towards flattening of the curve; however, zone-2 shows further increment due to its high species richness. Contrastingly, moist deciduous in both the zones doesn't show a typical flattening compared to the deciduous system. It seems that moist deciduous forest need to

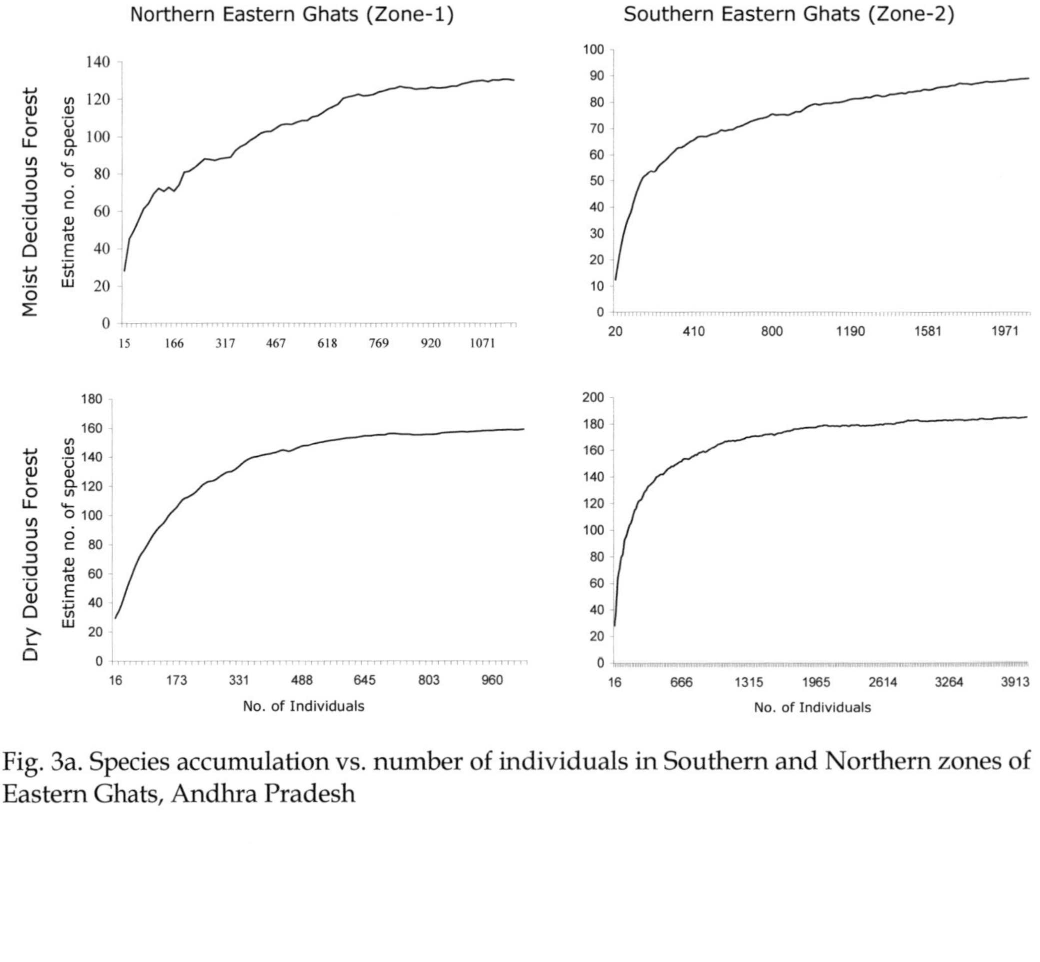

Fig. 3a. Species accumulation vs. number of individuals in Southern and Northern zones of Eastern Ghats, Andhra Pradesh

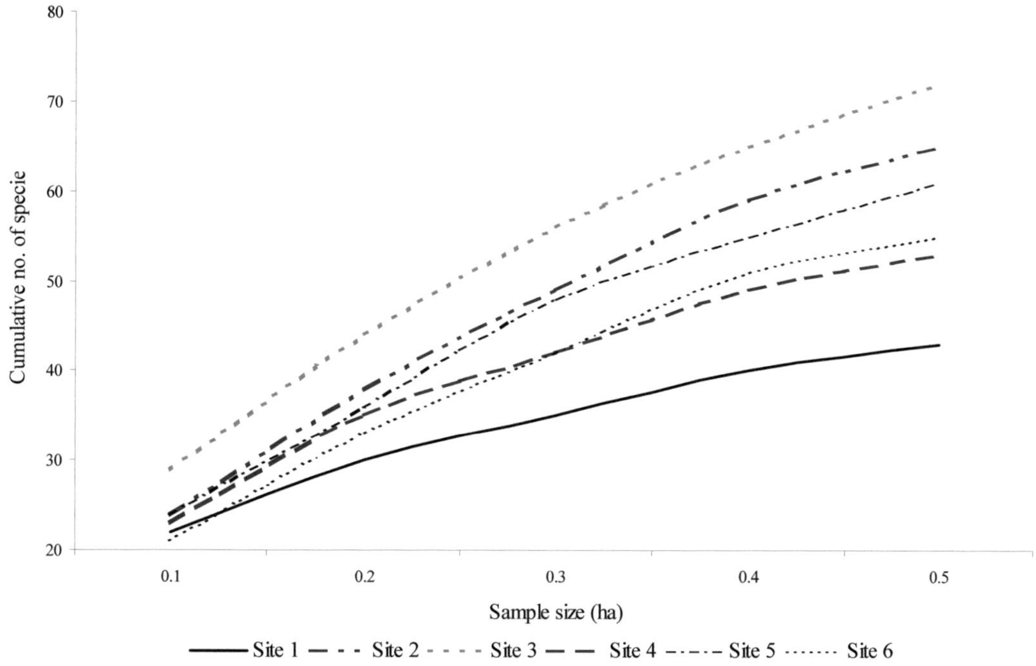

Fig. 3b. Species accumulation curve for the six transects in Northern and Southern Eastern Ghats of Andhra Pradesh

be characterized and analysed based further on its altitudinal and climatic conditions to derive a species-area relation. The species-accumulation curves for the continuous plots of the six sites are given in Fig 3b. Site 1 and 4 were initially steep, and later we observed a tendency towards flattening and similar such pattern was observed for the Site 5 & 6. Site 2 & 3 didn't reach an asymptote due to high species richness and as well landscape heterogeneity.

3.1.5 Stand density and basal area

Stand density and basal area for the random plots in both the zones of deciduous forest is higher compared to other forest classes (Table 2). In zone-1 of deciduous forest stand density was much higher (486 stems ha^{-1} and 19.04 m^2ha^{-1} basal area), when compared to zone-2 having 386 stems ha^{-1} and 14.67 m^2ha^{-1} basal area. Moist deciduous forest of the zone-1 & 2 are in the range of 257-370 stems ha^{-1}. Similarly such patterns were also observed in dry evergreen forests (266-334 stems ha^{-1}). Degraded forest in zone-1 was relatively higher (449 stems ha^{-1}) due to the selective felling for its timber extraction when compared to zone-2. Thorn forest in both the zones is having least stand density class compared to other types.
Stand density and basal area (BA) for the continuous plots in *site* 1, 2 and 3 is higher compared to *site* 4, 5 and 6 (Table 2). *Site* 2 has high stand density of 1050 stems ha^{-1} (BA 17.61 m^2ha^{-1}) as compared to *site* 3 having high basal area of 24.42 m^2ha^{-1} (990 stems ha^{-1}). Least basal area is seen in *site* 4 (4.35 m^2ha^{-1}) and least stand density in *site* 5 (614 stems ha^{-1}). This is due to the dry deciduous forest in zone-2 having less density and less biomass. The girth class distribution pattern for all the sites except *site* 3 and 5 is basically "L" shaped for

the total number of individuals (i.e. G ≥10cm) (Fig. 4). We observed girth class of 10-30cm was having high number of individuals (39%). These species itself doesn't support larger biomass and density as a result there is very less representation in girth class above 200cm.

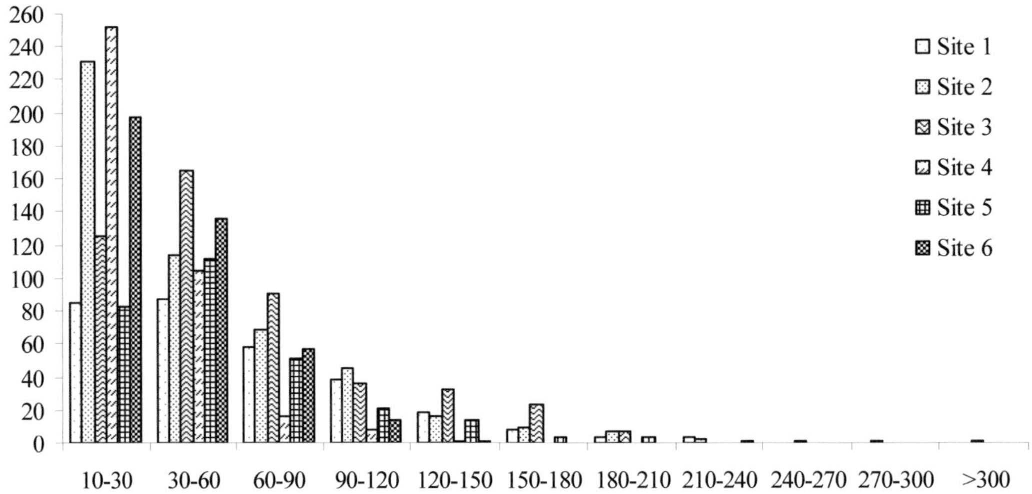

Fig. 4. Girth Class distribution for the six transects in Northern and Southern Eastern Ghats of Andhra Pradesh

3.1.6 Statistical analysis on Dis(Similarity measures)

In the following significance levels are represented with asterisks: * = $p \leq 0.05$, ** = $p \leq 0.01$, *** = $p \leq 0.001$, not significant values simply sport no asterisks.

3.1.6.1 Species richness

Apparently species richness is only very slightly influenced by fragmentation and disturbance (Fig. 5). There is even less impact of the fragmentation and disturbance categories a plot falls in onto species richness when the zones are considered separately (Fig. 6). In the light of this finding it is interesting whether beta-diversity is in contrast to species richness - influenced by fragmentation and disturbance and whether this effect might be also only visible when all data are considered together.

3.1.6.2 Compositional similarity

Richness differs only slightly between the zones (Fig. 7) and although it is not apparent in the Fig. 7, the difference in mean is significant according to a regular t-Test applied to the data. However, a NMDS plot of the data suggests an apparent and direct relationship between zone membership and species composition (Fig. 8). The plots of zone-1 are clearly separated (along the NMDS axes 2) from the plots of zone-2. Fig. 8 also hints at the reasons for the faster decay of similarity in zone-1 (as seen in Fig. 9). The majority of the plots of zone-1 is closely clumped together which means that they are all relatively similar in their species composition. However, there is also some spread in direction of the NMDS-axes 2. In geographic terms a higher likelihood of two plots further apart is being more different than close plots results. The three other tested grouping variables (which have been found to

explain most of the explainable variation in species richness in multiple regressions with backward selection) do not show a clear pattern in the NMDS plots (Fig. 8).
The mrpp() function of package vegan (Oksanen *et al.*, 2007) for the R Statistics System has been implemented to evaluate whether the plots of the two zones can be attributed to different vegetation types. The two zones are clearly distinct in their vegetation composition: The observed delta is significantly lower (0.89) than the expected delta (0.93, as determined by permutation) although the difference is not very large. Furthermore A = 0.036. A is a chance-corrected estimate of the distances explained by group identity. It can be compared to a coefficient of determination of a linear model (R2). Thus, it shows that the grouping into the two zones based on species composition is not very clear. With other words they are less distinct in species composition than the NMDS suggests (Fig. 8, upper left panel). Therefore another, more robust test has been employed as well. The function anosim [vegan] provides basically the same test but acts on ranks instead of the original data. It reports an R of -0.28***: The similarity among plots of one zone is significantly higher than the similarity between plots of different zones.
Because the zones showed considerable grouping in the NMDS, further NMDS plots were drawn for each zone separately, to evaluate whether the species composition could than be more clearly attributed to the membership to categories of fragmentation, disturbance and richness. Fig. 10 shows that this is not the case. When the zones are considered separately no clear groupings according to the categories of the mentioned variables occur.

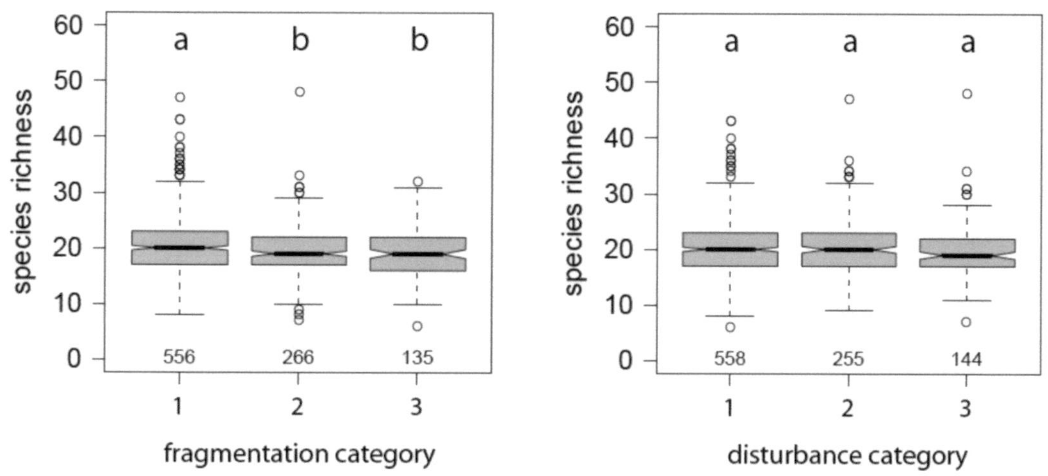

Fig. 5. Richness (species number inside plots) is significantly lower in highly fragmented plots. With disturbance there is no influence. The categorization is very coarse. Note, that there the numbers of plots involved in the categories differ considerably. Overall significance was tested with simple anova (for fragmentation: F = 0.13**, for disturbance: F = 0.83ns). Inference regarding the difference between the classes was obtained with pairwise t-Tests (α = 0.05). Bonferroni correction was applied.

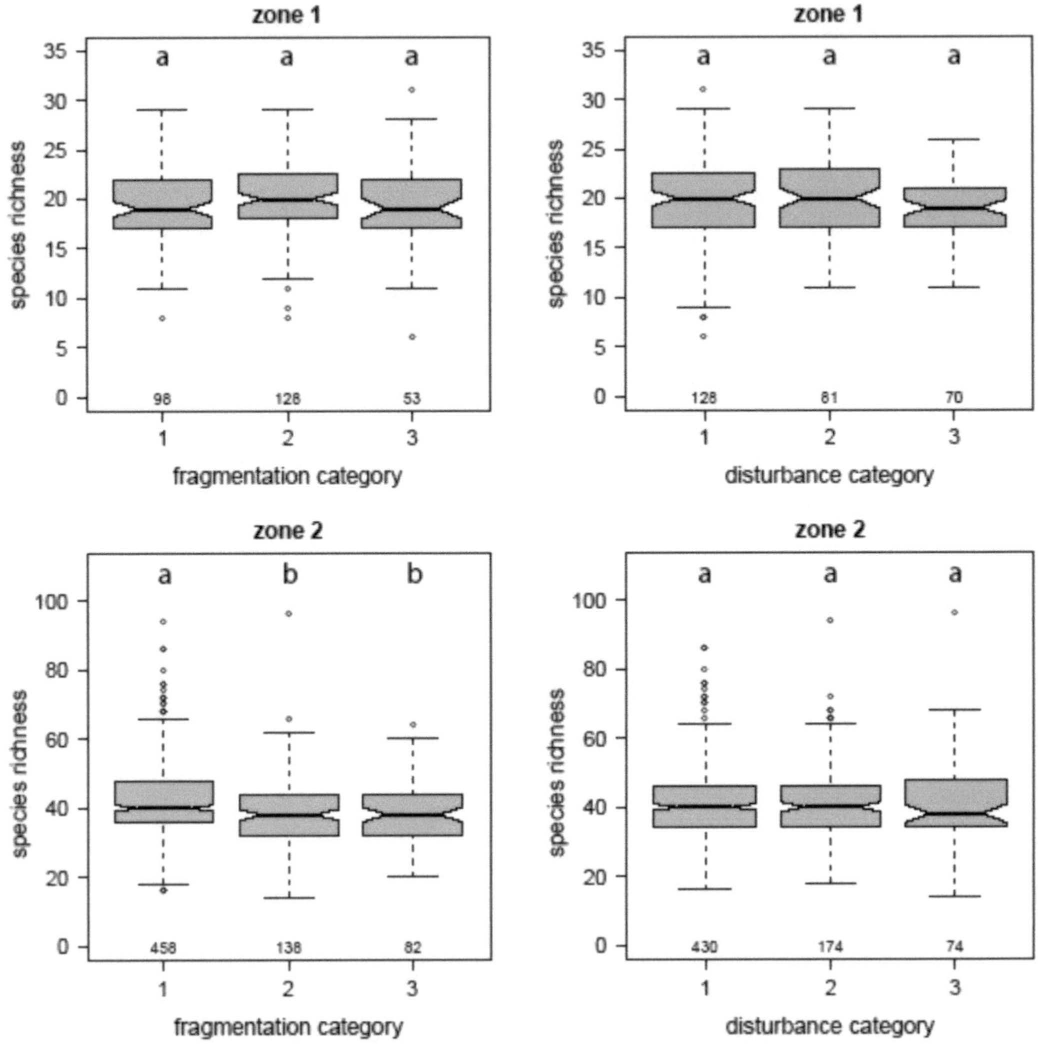

Fig. 6. The relation between richness (species number inside plots) and fragmentation/disturbance is much less clear when the zones are considered separately: Only fragmentation classes in zone 2 artly show a significance impact on species richness. Note, that here as well the numbers of plots involved in the categories differ considerably. Overall significance was tested with simple anova. Zone 1: fragmentation: F = 1.25ns, disturbance: F = 0.80ns; Zone 2: fragmentation: F = 7.13***, disturbance: F = 0.027ns. Inference regarding the difference between the classes was obtained with pairwise t-Tests (a = 0.01) and Bonferroni correction was applied.

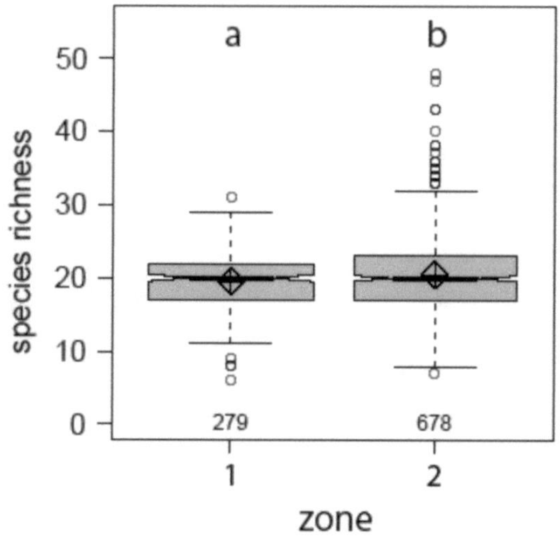

Fig. 7. Difference in richness between zones 1 and 2 (Inference obtained with t-Test)

3.1.6.3 Combined index of species richness, fragmentation, and disturbance

The 3d-plot of the factor variables fragmentation (x-axis), disturbance (y-axis) and the continuous variable richness (z-axis) shows that there is no clear relation between these three. It was also experimented with just putting the categories together (leads to 21 different possibilities). However, these categories explain nothing, because there is no explainable way in which they influence for instance the distance decay relations (see Fig. 11).

3.1.6.4 Distance decay of similarity

There is only relatively slow distance decay of similarity when all data is analyzed together for each zone. However, beta-diversity structure is different for the two zones (Fig. 9). In Zone-1 similarity decreases much faster (-0.00022/km) compared to zone-2 (-0.000088/km). This is more than one order of magnitude and is also reflected in the intercept. The linear regression line of the distance decay relationship intersects the y-axis at a similarity value (Sørensen) of 0.23 for zone-1, whereas the intercept is only 0.11 for zone-2. Not only the distance decay after linear regression but also the spline regression lines show considerable differences between zone-1 and 2. In zone-1 the rate of decay changes heavily and after a rapid decrease from 0 to 100 km distance, the similarity declines much slower. This holds also true when the subsets are further subsetted and e.g. different fragmentation classes are considered (Fig. 12).

Within the zones the slope of the distance decay relationship differs only slightly but significantly between different fragmentation categories (Fig. 12). Only the slopes of the linear regression lines describing the distance decay in the fragmentation classes 1 and 2 of zone-1 are not significantly different (Table 5). The smoothed regression lines cannot be tested in this way but from the illustrations (Fig. 12) it is apparent that there are important differences between the different fragmentation classes.

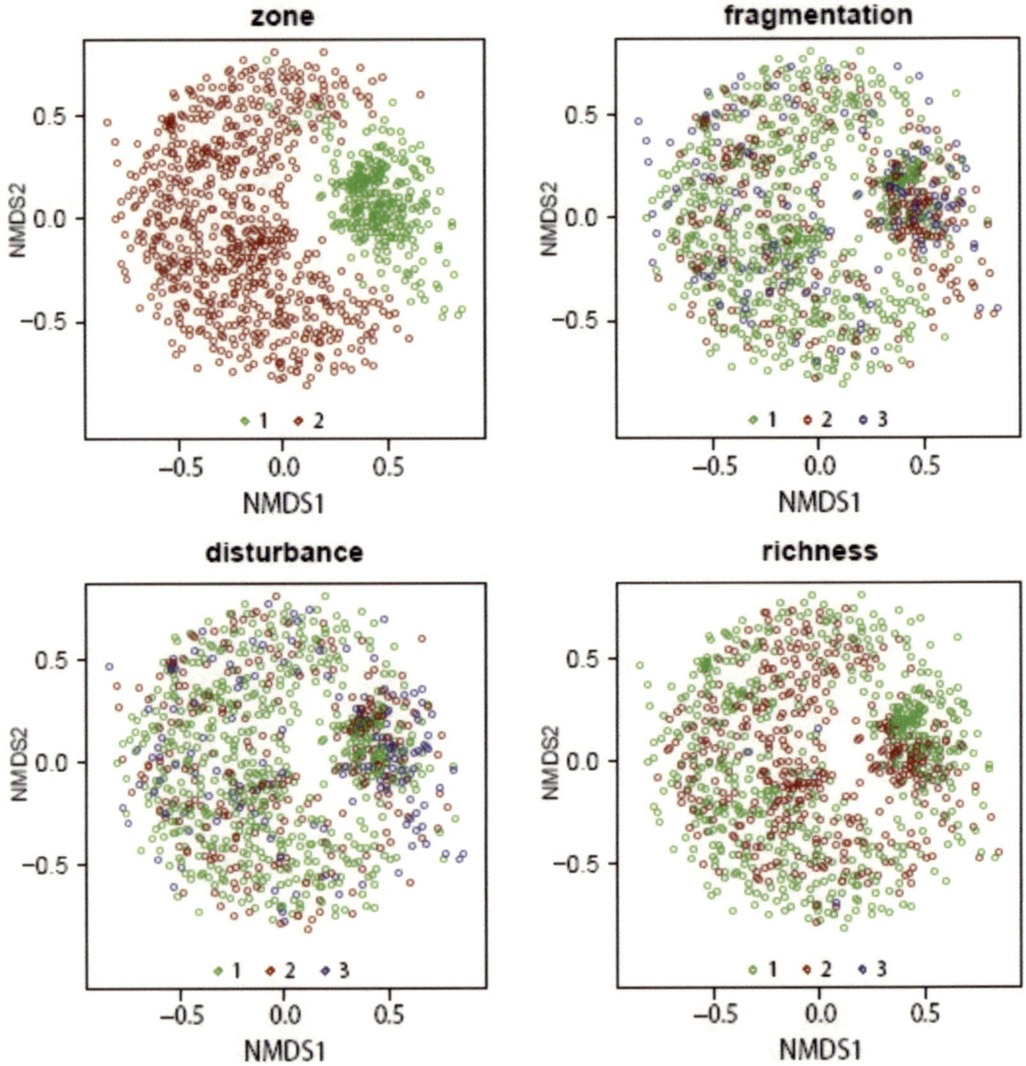

Fig. 8. NMDS based on all data, with different coloring regarding the membership of plots to subsets of data. The subsetting factor names each plot. Apparently the plots of the two zones show a clear difference regarding their vegetation composition compared to the plots of zone-1. Such a clear pattern rarely can be achieved. When the coloring is done for the categories of fragmentation, disturbance and richness, it is apparent that these do not impact species composition. NMDS statistic: stress = 36.10.

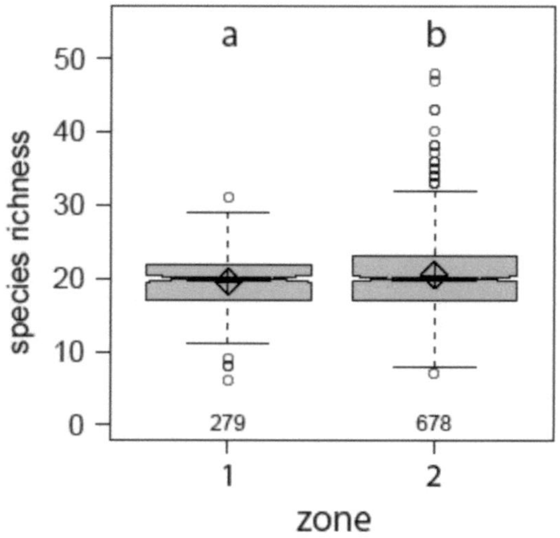

Fig. 9. Difference in richness between zones 1 and 2 (Inference obtained with t-Test)

Fragmentation Class				
Zone	nbx	nby	dsl	ρ
1	1	2	0.0000063	0.405
1	1	3	0.00012	0.001
1	2	3	0.00012	0.001
2	1	2	-0.000019	0.001
2	1	3	0.000027	0.001
2	2	3	0.000046	0.001

Table 5. Differences in slope of the distance decay relationship between subsets of the data. Each of the subsets comprises all plots that fall into the respective fragmentation class. The slopes differ significantly between fragmentation classes with the exception of the comparison between fragmentation classes 1and 2 in zone 1. Abbreviations: nbx - one of the compared subsets, nby - the other of the compared subsets, dsl - difference in slope, calculated by slnbx - slnby, ρ - p value.

The picture does not change much when disturbance classes are considered instead (Table 6). In this case in zone-1 the slope of the linear regression line that describes the distance decay relationship in disturbance class 3 (high) very clearly is significantly steeper compared to the slopes for the classes 1 and 2. The latter two do not differ significantly from each other and the slope is even (very slightly) steeper for disturbance class 2. In zone-2 the slope of class 2 is much steeper than in the disturbance classes 1 and 3 of this zone. This is even apparent in Fig. 13 and as well reflected in the significance tests represented in Table 6.

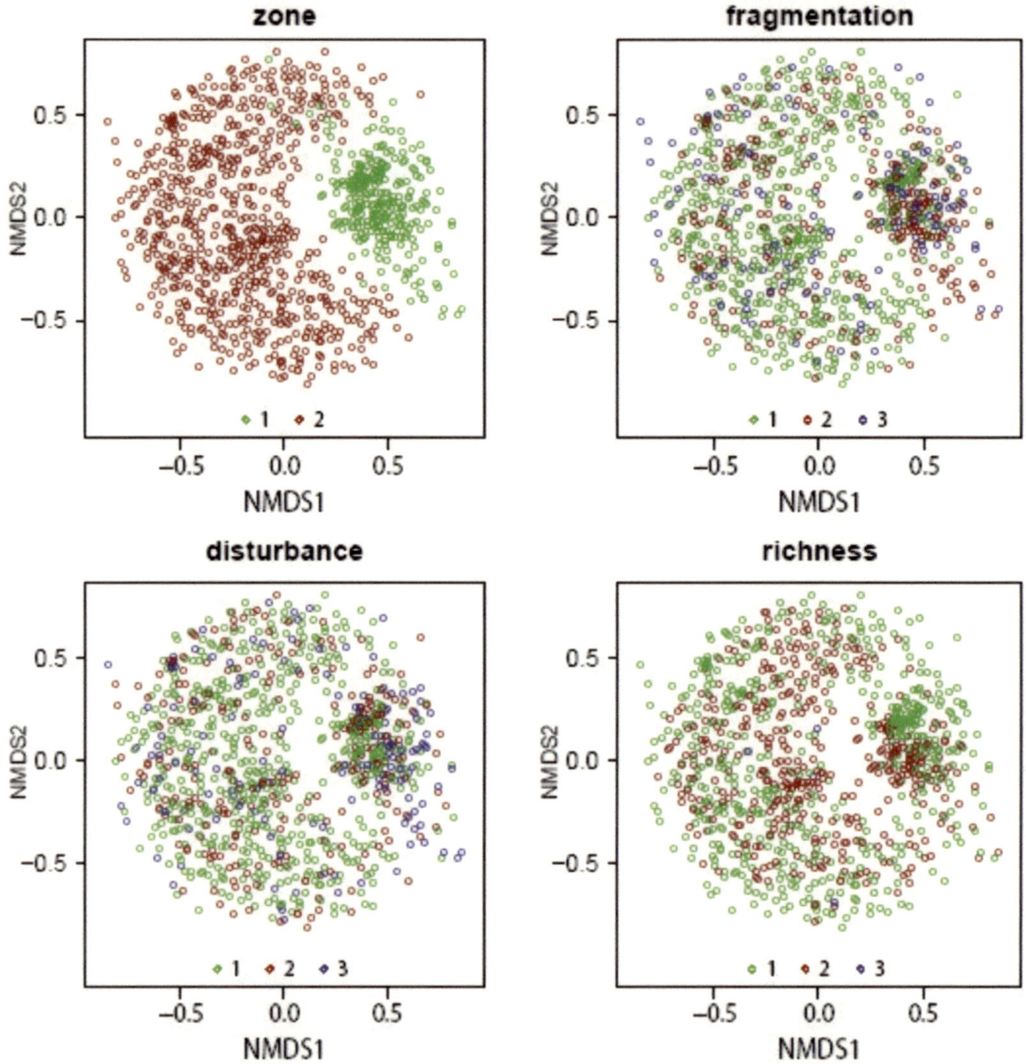

Fig. 10. NMDS based on all data, with different coloring regarding the membership of plots to subsets of data. The subsetting factor names each plot. Apparently the plots of the two zones show a clear difference regarding their vegetation composition compared to the plots of zone-1. Such a clear pattern rarely can be achieved. When the coloring is done for the categories of fragmentation, disturbance and richness, it is apparent that these do not impact species composition. NMDS statistic: stress = 36.10.

Disturbance Class				
Zone	nbx	nby	dsl	ρ
1	1	2	-0.000018	0.236
1	1	3	0.000079	0.004
1	2	3	0.000096	0.007
2	1	2	0.000032	0.001
2	1	3	0.0000025	0.328
2	2	3	-0.000029	0.026

Table 6. Differences in slope of the distance decay relationship between subsets of the data. Each of the subsets comprises all plots that fall into the respective disturbance class. The slopes differ significantly between disturbance classes with the exception of the comparison between disturbance classes 1 and 2 in zone 1 and the comparison between classes 1 and 3 in zone 2. Abbreviations: nbx - one of the compared subsets, nby - the other of the compared subsets, dsl - difference in slope, calculated by slnbx – slnby, ρ - p-value.

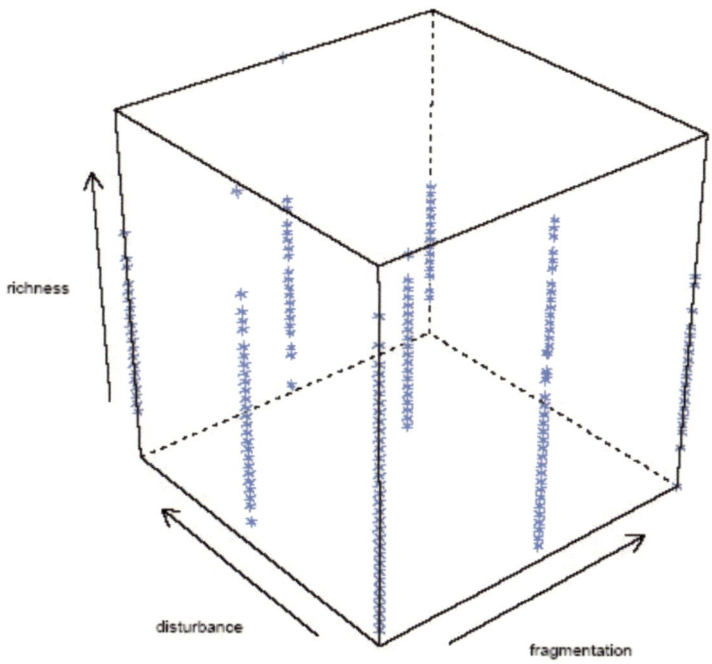

Fig. 11. 3d-plot of disturbance (x-axis), fragmentation (y-axis), and richness (z-axis). The latter is a continuous variable whereas the other two are factor variables. There is no joint relation between these three

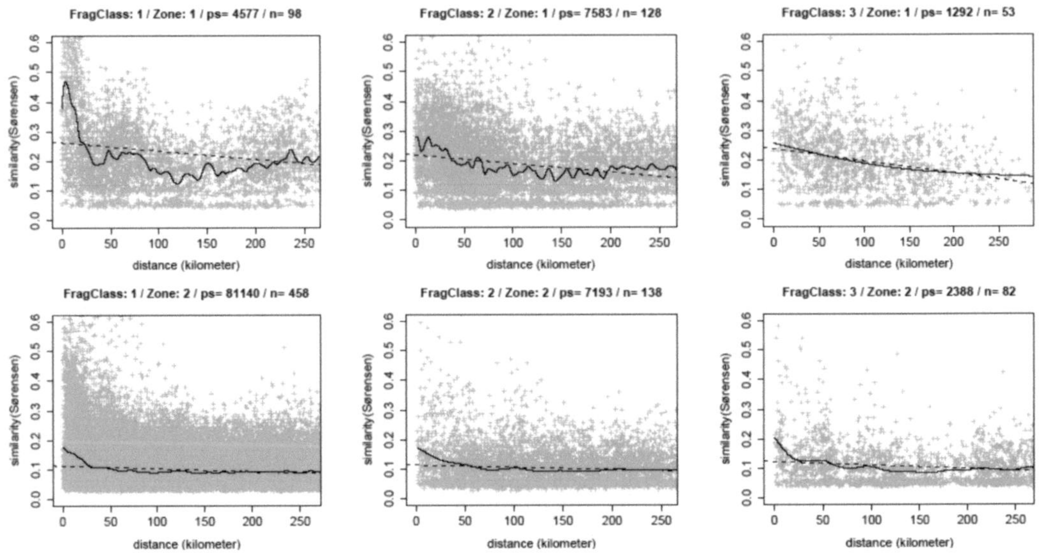

Fig. 12. Slopes of the distance decay relationship for subsets of the data. Each of the subsets comprises all comparisons between the plots of one zone and one fragmentation class therein. Besides the fact that similarity decreases much faster with distance in zone-1; there are apparent differences between fragmentation classes. The spline regression was obtained with a lowess smoothing algorithm as offered by the function lowess() with f=0.2 of the R package stats.

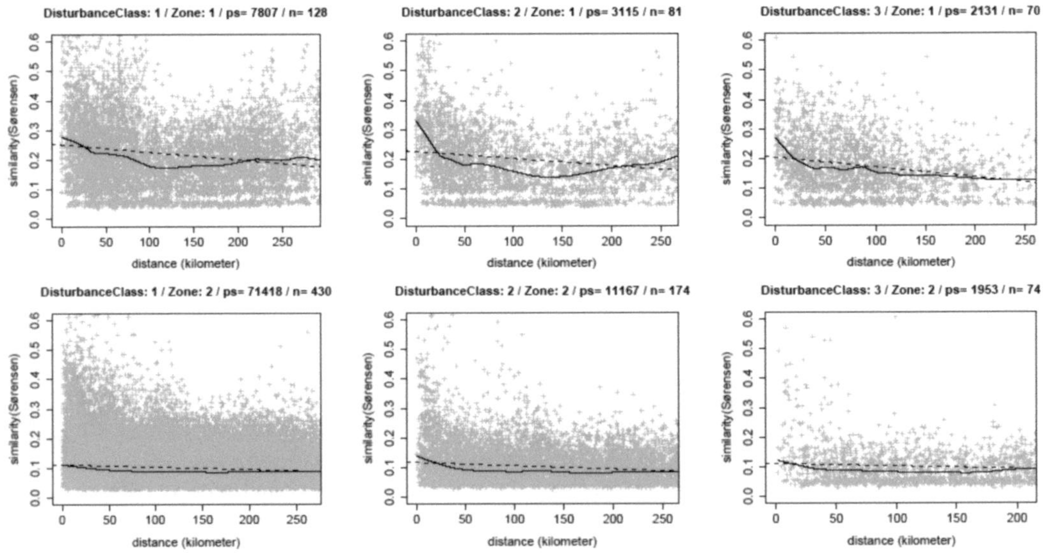

Fig. 13. Slopes of the distance decay relationship for subsets of the data. Each of the subsets comprises all comparisons between the plots of one zone and one disturbance class therein. Besides the fact, that similarity decreases much faster with distance in zone-1; there are apparent differences between disturbance classes. The spline regression was obtained with a lowess smoothing algorithm as offered by the function lowess() with f=0.2 of the R package stats.

3.1.6.5 The influence of slope and aspect on similarity and distance decay

Slope and aspect show no influence on species composition. The similarity in species composition has been regressed against the dissimilarity regarding slope and aspect. No correlation (Mantel) has been found neither for zone-1(r = -0.0068) nor for zone-2 (r = 0.01). This holds even when we controlled for the influence of altitude on this relation (partial Mantel test). In the latter case Mantel correlation values amount to r = -0.019 (zone-1) and r = 0.01341 (zone-2) respectively.

3.1.6.6 Continuous plots for the six sites

The vegetation composition of the 6 sites show considerable differences. This is clearly apparent in the DCA drawn from the data in Fig. 14 (a & b). Because there are very few species shared between plots (see Fig. 15 for more detail regarding this issue) and because there are much more species than plots (when each of the 6 sites is considered as a single sample) the metaMDS algorithm very fast finds a stable solution. After 2 runs a very low stress value of 3.21 results (see also Figure 14a).

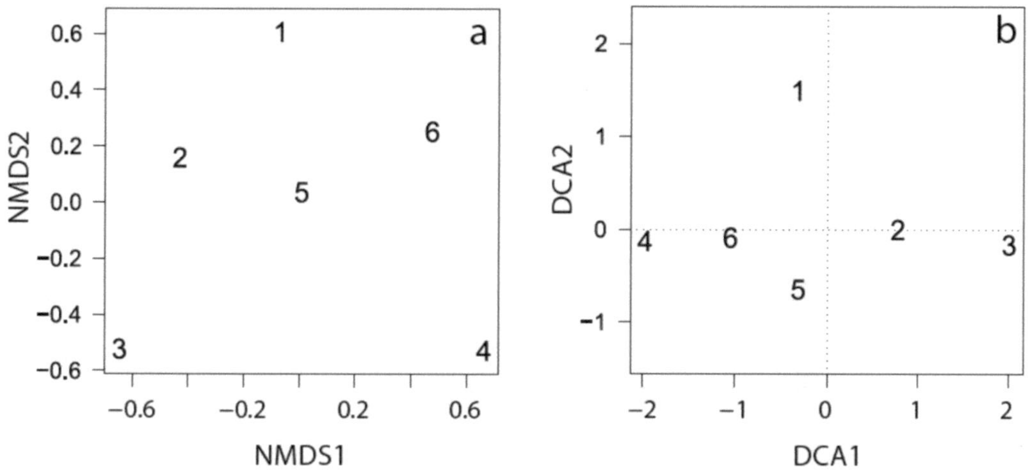

Fig. 14. a and b: Ordination plots of the 6-sites data. The six transects vary considerable in their species composition. Both methods come to comparable results. Sites 3 and 4 are highly dissimilar. Sites 2, 5, and 6 build something like a cluster (but is relatively vague), whereas Site 1 is far from 3 and 4 and closer to the other three but at the same time relatively distinct from all other Sites.

When the quadrats (that make up the sites) are considered separately, it is not possible to calculate a NMDS with the metaMDS algorithm because a majority of the compared quadrats have no species in common. Nevertheless, this means that overall beta-diversity in trees is very high. A solution can be achieved with the slightly less robust isoMDS[MASS] function. It accepts any distance matrix achieved in advance (rather than calculating a distance matrix during the process) and does not feature several random starts with different starting configurations but only a simple iteration algorithm. The result is displayed in Fig. 16. The sites are relatively clearly separated, which indicates that the similarity among quadrats of a site is much higher than similarity across sites. This can be tested with the mrpp() function of package vegan for R (see above). Applied to the data the function returns the following statistics: A = 0.11***, delta_obs = 0.83, delta_exp = 0.94. This indicates that the groups are distinct in their species composition

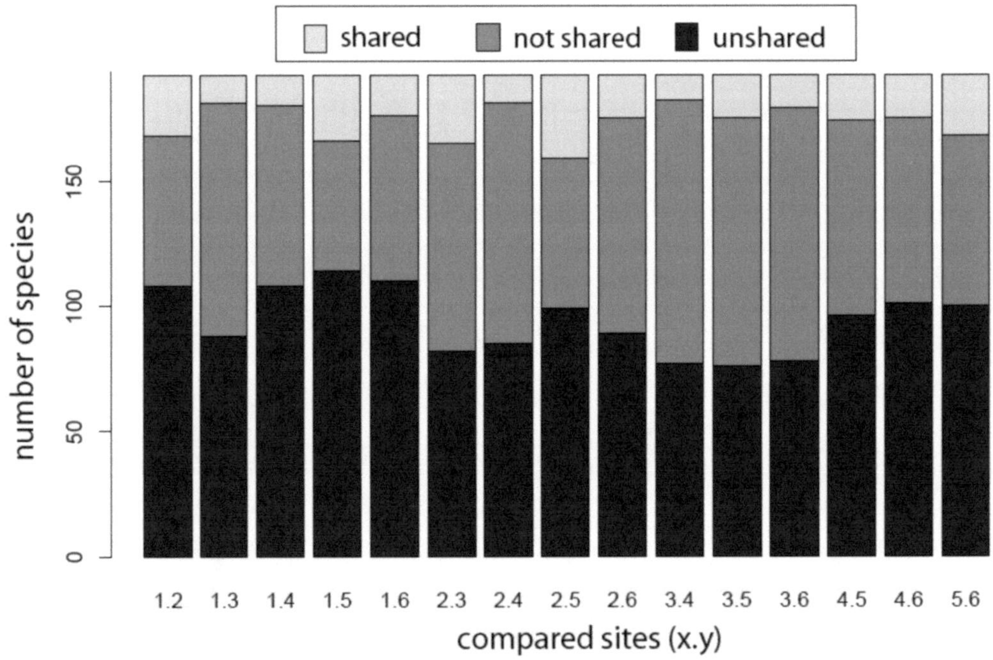

Fig. 15. Pairwise comparison of shared, not shared, and unshared species for all possible pairs of sites from the 6-sites data. Explanation: shared species occur on both of the compared sites, not shared species occur on only one of the compared sites, unshared species do not occur on both the compared sites. Note, that the fraction of shared species is always relatively small

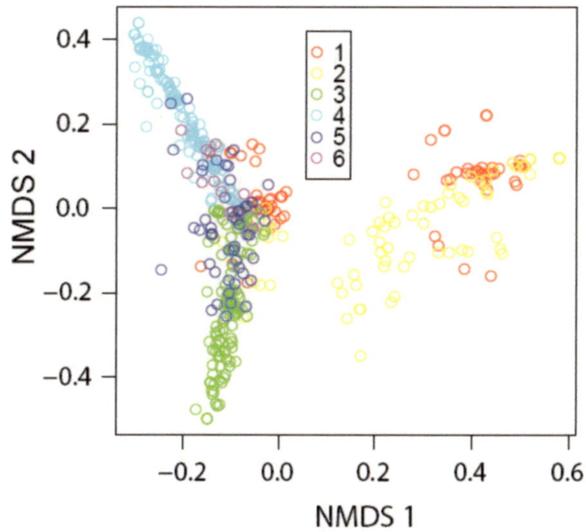

Fig. 16. NMDS plot of the 6-sites data. Colors represent the different sites as indicated in the legend. Sites are relatively clearly separated. Especially sites 3 and 4 show almost no species overlap. The sites of zone 2 (4, 5, and 6) clump much closer together (compared to the sites of zone 1 (1, 2, and 3). Furthermore interesting is the very clear separation of some quadrats in sites 1 and 2 (on the right hand side of the figure) because it occurs within the sites. The stress value 49.16 is rather high which indicates a not so good final solution.

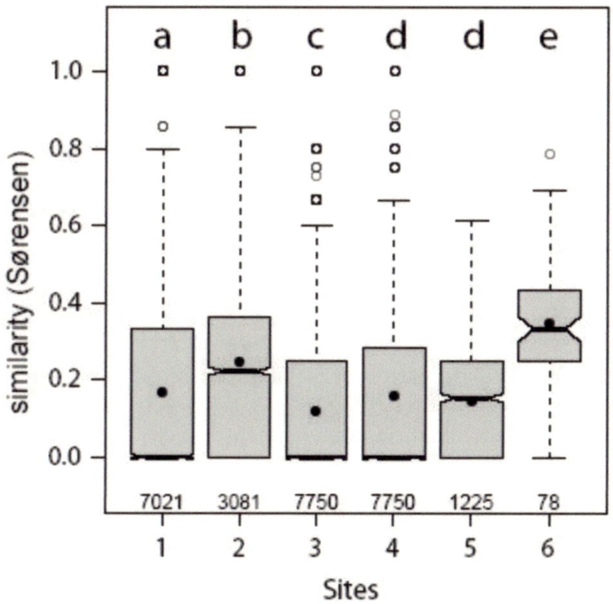

Fig. 17. Comparison of similarity between the 6 sites. The letters above the boxes connects sites not significantly different in mean similarity. Medians are often very low (sites 1, 3, 4) because a lot of quadrats inside do not share any species at all (so similarity is zero)

Also beta-diversity (expressed by compositional similarity) differs significantly between sites (Fig. 17). Only the sites 4 and 5 do not significantly differ in their similarity structure. However they have also been quite similar in species composition (as seen in the NMDS plot in Fig. 16) with site 5 being clearly positioned between site 3 and 4. All of the above can partly be grasped also from the species matrix, wherein - even without ordering - it is obvious that only few species occur on more than one or on even more sites.

A moderate percentage of quadrats (47%) are clustered into the "right" cluster when site membership is (very simplistically) compared to cluster membership. This was obtained by computing a simple Wards clustering with hclust[stats] followed by the application of cutree[stats] which simply cuts the resulting dendogram tree to obtain group (or cluster-) memberships for all tested objects (the quadrats in our case) when the number of groups has been specified.

4. Discussion

4.1 Phytodiversity assessment

4.1.1 Random plots

An analysis of various quantitative inventories of woody species ≥10 cm dbh across the tropics reveals a wide variation in the figures, ranging from 20 species ha^{-1} in flooded Varzea forest of Rio Xingu, Brazil (Campbell *et al.*, 1992) to 300 species ha^{-1} in terra firma, Yanamono, Peru (Gentry, 1988). Within peninsular India, in various quantitative inventories of a comparable area for the same tree girth threshold, in tropical forest sites of the southern Western Ghats, species richness ranged from 30 species ha^{-1} in Nelliampathy (Chandrashekara & Ramakrishnan, 1994), to 57 species ha^{-1} in Mylodai, Courtallum reserve forest (Parthasarathy & Karthikeyan, 1997a), to 64 to 85 species ha^{-1} in Kalakad (Parthasarathy *et al.*, 1992) and 90 species on a 3ha scale in Kalakad-Mundanthurai Tiger reserve (Giriraj, 2006). Compared to various moist tropical forest sites in other parts of the world our present study showed comparable species richness and diversity.

The absolute stand density of both the zones across different vegetation communities are ranging from 156 to 486 trees ha^{-1}, with a whole study area average of 365 trees ha^{-1}. Deciduous forests of the zone-1 had higher stand density of 486 trees ha^{-1} compare to the zone-2 having 386 trees ha^{-1}, the reasons being favorable bioclimatic and edaphic conditions. These stand density are relatively lesser compared to the other sites in the Shervarayan and Kalrayan hills of Eastern Ghats, with the range of 640 to 986 trees ha^{-1} (Kadavul & Parthasarathy, 1999a) and 367 to 667 trees ha^{-1} (Kadavul & Parthasarathy, 1999b) respectively. Whereas the tree densities in various tropical evergreen forests of Western Ghats of peninsular India were: 574 to 915 stems ha^{-1} in medium elevation forest of Kalakad (Parthasarathy, 1999); 852 to 965 stems ha^{-1} in high elevation forest of Kalakad (Parthasarathy, 2001); 583 stems ha^{-1} in Kalakad–Mundanthurai area (Ganesh *et al.*, 1996); 482 stems ha^{-1} in Mylodai, Courtallam reserve forest (Parthasarathy & Karthikeyan, 1997a); a range of 270 to 673 trees ha^{-1} in the 30 ha of Varagalaiar, Anamalais (Ayyappan & Parthasarathy, 1999), all these in southern Western Ghats and 635 stems ha^{-1} in Uppangala forest of central Western Ghats (Pascal & Pelissier, 1996). Density of trees (>30 cm gbh) in tropical forests ranges between 245 and 859 (Ashton, 1964; Campbell *et al.*, 1992; Richards, 1996) with intermediate values of 448 to 617 stems ha^{-1} in Costa Rica (Heaney & Proctor, 1990), 436 stems ha^{-1} in Reserva Forestal de San Ramon of Costa Rica (Wattenberg & Breckle,

1995), 420 to 777 stems ha^{-1} in Brazil (Campbell *et al.*, 1992) and 639 to 713 stems ha^{-1} in Central Amazonia (Ferreira & Prance, 1998).

The basal area for both the zones is ranging from 2.89 and 40.95 m^2ha^{-1} for ≥30 cm girth threshold. Thorn forest of the zone-1 had least basal area and high basal area was observed in the semi-evergreen forest of northern Eastern Ghats (40.95 m^2ha^{-1}). The mean basal area value of the present study is also lesser than the values for the comparable girth threshold of ≥30cm gbh of several other tropical forests: 28.1 and 30.8 m^2ha^{-1} respectively of dry evergreen forest sites of Kuzhanthaikuppam and Thirumanikkuzhi (Parthasarathy & Karthikeyan, 1997b) Puthupet (Parthasarathy & Sethi, 1997) on the Coromandel coast of south India; 24.2 m^2ha^{-1} of Malaysia (Poore, 1968), 27.6 to 32.0 m^2ha^{-1} and 25.5 to 27.0 m^2ha^{-1} of Brazilian Amazon (Campbell *et al.*, 1986, 1992); 27.8 and 41.67 m^2ha^{-1} of Costa Rica (Lieberman & Lieberman, 1987; Watternberg & Breckle, 1995); 32.8 to 40.2 m^2ha^{-1} of Central Amazonian upland forest (Ferreira & Prance, 1998); 42.6 m^2ha^{-1} of Courtallam reserve forest in the Indian Western Ghats (Parthasarathy & Karthikeyan, 1997a); 39.7 m^2ha^{-1} of Uppangala forests, central Western Ghats, India (Pascal & Pelissier, 1996); and 25.91 to 47.75 m^2ha^{-1} in the 30 ha of Varagalaiar, Anamalais, southern Western Ghats (Ayyappan & Parthasarathy, 1999). But a value of present study is lesser than: 53.3 to 94.6 m^2ha^{-1} and 55.3 to 78.3 m^2ha^{-1} of Kalakad, southern Western Ghats, India (Parthasarathy 1999; Parthasarathy, 2001) and the values of a couple of other tropical forests: 47 (for alluvium) to 49.5 m^2ha^{-1} (for slope forest) of New Caledonia (Jaffre & Veillon, 1990), and 62 m^2ha^{-1} of Monteverde, Costa Rica (Nadkarni *et al.*, 1995).

In both the zones, family-wise five predominant families explain the dominance of the forests which includes Combretaceae, Mimosaceae, Euphorbiaceae, Caesalpiniaceae and Rubiaceae. It contributes 39% of the family dominance which characterize the tree community pattern and in close range with other tropical forests regions (Gordon *et al.*, 2004; Linares-Palomino & Ponce-Alvarez, 2005) while the other Indian Eastern Ghats site, where the family Oleaceae (26.6%) dominated (Kadavul & Parthasarathy, 1999a) and in dry evergreen forests in south India, where the Melastomataceae and Rubiaceae with 56% dominated (Parthasarathy & Karthikeyan, 1997b).

The trend of decreasing diversity and density with increasing girth class is in conformity with the studies of Chittibabu & Parthasarathy (2000); Jeffre & Veillon (1990); Kadavul & Parthasarathy (1999a, b); Newbery *et al.*, (1992) and Paijmans (1970). Both the zones had typical reverse J-shaped structure for girth frequency (Fig. 4o). Northern region of the Andhra Pradesh explains mature stands in all the girth-class with good regeneration were in close conformity with other tropical forests around the world (Chittibabu & Parthasarathy, 2000; Kadavul & Parthasarathy, 1999a, b; Manokaran & Kochummen, 1987; Nadkarni *et al.*, 1995; Sukumar *et al.*, 1992).

4.1.2 Continuous plots for the six sites

A total of 197 species, 139 genera and 57 families (Table 3) were stated from six transects covering thee-ha of the tropical forests in Northern and Southern Andhra Pradesh, Eastern Ghats. Species richness (43-72 species ha^{-1}) and species diversity 2.9-3.6 H') are comparable with the other sites in the Eastern Ghats. The mean value of 60 species ha^{-1} recorded in the present study is higher than that of 43 species ha^{-1} in Shervarayan hills (Kadavul & Parthasarathy, 1999a), 57 species ha^{-1} in Mylodai forest of Courtallum (Parthasarathy & Karthikeyan, 1997b). In Mudumalai tropical forest, Western Ghats, the 12 most common species made up to 90.6% while 7 species were represented by only one individual (Sukumar *et al.*, 1992).

Species number per ha found in the present study is smaller in comparison with Malaysian lowland rain forests having 164 and 176 species (Malaysia, Wyatt-Smith, 1966), 150 species (Indonesia, Whitmore, 1990), 223 and 214 species ha^{-1} (Malaysia, Proctor *et al.*, 1983). The wide range of species number 43-72 found in the present study plots can be attributed to the change in elevation, and bioclimatic variations. As compared to the tropics, neo-tropics show a much more complicated situation. In 1 ha plots of tropical rain forests, 91 species (Guiana, Davis & Richards, 1933), 87 species (Brazil, Black *et al.*, 1950) and 83 species (Venezuela, Jordan *et al.*, 1989) with DBH >10cm were reported. These values are lower than in the forest investigated in Xishuangbanna, SW China with 119 species (Cao & Zhang, 1997) and in present study (153 species). Such species diversity pattern may diminish as a function of altitude (Lieberman *et al.*, 1996).

The mean stand density of 409 stems ha^{-1} and range of 307 to 525 stems ha^{-1} in the tropical forests of northern Andhra Pradesh is well within the range of 276 - 905 stems ha^{-1} reported for trees ≥10cm gbh in the tropics (Ghate *et al.*, 1998; Sundarapandian & Swamy, 1997; Sukumar *et al.*, 1997 & Murali *et al.*, 1996). This range of stand density in the present study is comparable with the other Eastern Ghats sites (Shervarayan hills - Kadavul & Parthasarathy, 1999a; Kalrayan hills - Kadavul & Parthasarathy, 1999b; Coromandel coast - Parthasarathy & Sethi, 1997). Low density was observed in other tropical sites across the world, which includes Costa Rica - 448 to 617 ha^{-1} (Heaney & Proctor, 1990); Brazil - 420 to 777 ha^{-1} (Campbell *et al.*, 1992); Malaysia - 250-500 ha^{-1} (Primack & Hall, 1992).

The species-accumulation curve (Fig. 5) for the six different sites varied because of the changes in topography and rainfall. *Site* 1 and 4 were initially steep, and later we observed a tendency towards flattening and similar such pattern was observed for the *Site* 5 & 6. *Site* 2 & 3 didn't reach an asymptote due to high species richness and as well landscape heterogeneity. Similar patterns were noticed in different areas of Eastern and Western Ghats (Kadavul & Parthasarathy, 1999 a, b; Parthasarathy, 1999; Parthasarathy, 2001).

The most obvious variation in tree species and the proportion of dominant species in the six sites can directly be attributed to altitudinal and rainfall distribution. Particularly species richness increase at moderate elevation and beyond the altitude range, there is tendency towards decline (Giriraj *et al.*, 2003); similar pattern was observed in site1. Families with rare occurrences represented by single and double species were 36 for both the study sites.

Current study identified 57 families and the most predominant species rich families are Rubiaceae (18), Euphorbiaceae (16), Fabaceae (11) and Caesalpiniaceae (9) and similar such predominance were recorded from Shervarayan hills (Kadavul & Parathasarthy, 1999a). Steege *et al.*, (2000) and Martin & Aber (1997) reported Leguminosae as the most abundant family in neo-tropical forests. Top ten families explain the species characteristics and found to be 66% (1620 individuals out of 2,457 individuals) dominant for the study site.

Girth class frequency showed L-shaped population structure (Fig. 4) of trees except for *site* 3 and 5. This pattern is in conformity with many other forest stands in Eastern & Western Ghats such as Shervarayan hills (Kadavul & Parathasarthy, 1999a); Kalrayan hills (Kadavul & Parthasarathy, 1999b); Kakachi (Ganesh *et al.*, 1996); Uppangala (Pascal & Pelissier 1996); Mylodai-Courtallum RF (Parthasarathy & Karthikeyan, 1997b). *Site* 3 & 5 didn't have a clear population structure might due to anthropogenic pressure in the form of shifting cultivation for their livelihood. In general the Northern Eastern Ghats (EG) of Andhra Pradesh (AP) exhibit large-scale deforestation as observed in Chapter-3 and southern EG of AP do have pressure on the forests and these region having low altitudinal and precipitation formation resulted to low-level of population structure.

4.2 Spatial patterns of phytodiversity using Dis (Similarity measure)

Generally it is assumed that fragmentation and disturbance have a considerable influence on species richness (Connell, 1978; Huston, 1979). However, in the data set from the Eastern Ghats, this is only partly the case. Disturbance does not seem to drive species richness in the investigated area (Fig. 5 and 6). On the other hand this result might hint at a problem with the disturbance classification.

4.2.1 Compositional similarity

The separation of the plots in zone 1 from the plots of zone-2 is relatively clear (along the NMDS axes 2 in Fig. 8). Such obvious grouping is rarely found in ecological data sets. This means that the two zones are relatively distinct in their vegetation composition. However, astonishingly there is no further grouping within the zones regarding to the categorical parameters fragmentation, disturbance, and richness (also Fig. 8). When the zones are considered separately (Fig. 10) it becomes even more obvious that these parameters (at least in their representation of the actual research) do not drive the differentiation in species composition. Thus, not only richness but also species composition is not driven by disturbance or fragmentation.

Often richness drives compositional similarity of plots because plots with largely different species number very naturally tend to have only very few species in common. However, even that is not the case in the present data (Fig. 10). This holds also when the classification is much finer than displayed in Fig. 10. One reason for that might lay in the overall high beta-diversity in the region: The intercept of the distance decay relationship is comparably low (see e.g. Condit et al., 2002 for comparison data from the Neotropics) which indicates a low similarity (and therewith high beta-diversity) even at short distances between plots.

Species richness, fragmentation and disturbance all have only very minor influence on species composition. Furthermore they are not linearly related to one another. Therefore a joint index cannot be build. If something like a surrogating indicator is the aim, the environmental parameters recorded have to be much more numerous. Furthermore, they should preferably be on continuous scales.

4.2.2 Distance decay

Similar findings are to be stated regarding the rate of the decrease of similarity with distance. Compared to data from the Neotropics (e.g. Condit *et al.*, 2002, 0.0019-0.00055/km) the distance decay rate in zone-1 (0.00022/km) fits nicely in. Astonishingly the rate is much lower in zone-2 (0.000088/km) and as already discussed in the previous paragraph, the intercept is also very low. This means that it doesn't matter how far two plots are from each other. There is always the chance that two plots can be very different regarding their tree species composition.

But even at this low decay rate a phenomenon occurs which seems to be ubiquitous to all distance decay data: There is comparably faster decrease on short distances. This has also been reported by (Jurasinski, 2007) who attributed it to the predominance of dispersal over niche assembly in the short range around vegetation samples. In the investigated data it can be found on all evaluated subsets of the data. This supports the idea of a ubiquitous pattern

in species assembly that bases on the predominance of neutral (Hubbell, 2005) versus niche assembly (Leibold, 1995) on different spatial scales.
The results of the distance decay evaluation indicate that the change of compositional similarity between plots with geographic distance follows different paces depending on fragmentation/disturbance. However the most impressive difference exists between the two zones. So the question arises, what is responsible for the differences in species similarity and distance decay relationship between the two zones? We have already learned above, that this cannot be directly attributed to differences in richness. And in the NMDS the grouping is not as clear with richness as the grouping variable compared to the zones (Fig. 8).

4.2.3 Slope and aspect

The variability of slope and aspect has no influence on species composition. This is in contradiction to findings from semi-arid vegetation (Badano *et al.*, 2005; Jurasinski, 2007; Sternberg & Shoshany, 2001). The difference in radiation which in e.g. Mediterranean ecosystems influences a lot of other factors (heat, moisture, evapotranspiration, etc.) is not an issue in tropical systems. Therefore slope and aspect cannot be used as explaining variables for the species composition of the plots. Thus similarity-distance function might predict the slope of a power-law species area curve (Condit *et al.*, 2002). Based on this characteristic the study concludes that it is an appropriate measure of beta-diversity. Already MacArthur (1965) proposed to use species-area curves as an analytical tool to diversity taking the intercept of the curve as a measure of 'alpha-diversity' and the slope parameter as a measure of 'beta-diversity' (see also Caswell & Cohen, 1993; Ricotta *et al.*, 2002). The present study stated that none of the recorded variables provides a good estimator for species richness or species composition. To evaluate the underlying factors many more and preferably continuous environmental variables should be recorded.

4.2.4 Continuous plots for the six sites

The 6 sites that have been investigated in detail regarding tree species composition can - with the help of ordination techniques be grouped. However, this grouping is not very meaningful because most of the sites cannot be grouped regarding their species composition. Only the *sites* 2, 5, and 6 have some more species in common which would allow the specification of a common vegetation type shared by these three sides.
It may be a problem of small sample size that the quadrats of *site* 5 are intermediate in their species composition between quadrats of *site* 3 and *site* 4 (Fig. 16). Otherwise it is astonishing that some of the quadrats in *site* 5 have a lot in common (species wise) with quadrats in *site* 3, which is in the other zone. This leads to the suggestion that the sample is too small for a classification of vegetation types: From the species matrix it is obvious that only few species occur on more than one or on even more quadrats.
The comparison between the wards clustering and the forced grouping into site membership reveals that there are quite some matches. However, this is relatively simplistic because it is not tested whether a quadrat is ordered together with quadrats of its site. This would be a better test, but this is not easily achieved because it is hard to define rules, which can be applied to such an evaluation. If 30 of the quadrats in on site are clustered into one cluster and 30 into other cluster - which is only a very simple case - the problems already start how to evaluate the assignment to several clusters. The simplistic measure already

shows that the differentiation in species composition is largely driven by the position in sites. However, on smaller scale vegetation types (or better groupings based on tree species data) might be identifiable.

5. Conclusion

The methodology developed for the comparison of multiple plots has been applied to a data set of vegetation in the Eastern Ghats of Andhra Pradesh to assess vegetation dis(similarity) and also evaluate transitional ecosystem. Whittaker's (1960) concept for assessing diversity has triggered a lot of development in ecology. However, especially the term 'beta-diversity' has begun to take on relatively different meanings and thus is a rather confusing concept. The terminological ambiguity is an obstacle to the development in all fields requiring more than inventory data ('alpha'or 'gamma-diversity'sensu Whittaker). Compositional (dis) similarity between samples ('differentiation diversity') and the variation of inventory diversity across scales ('proportional diversity') are important fields for future research, which should not be neglected due to unclear concepts.

Research and data recording in vegetation ecology should be spatially and temporally explicit. Even when no spatial analysis is intended, this might provide for the incorporation of data in later meta-analyses. Different sample designs needs (random, stratified and hexagonal grid) to be validated for different ecosystems prior applying this method, as it is efficient for long-term monitoring purpose. Furthermore it allows tracking temporal changes in spatial patterns through periodically repeated sampling. The changes in spatial patterns can be assessed statistically (Jurasinski & Beierkuhnlein, 2006) and suggested hexagonal grids are efficient method for the investigation and monitoring of spatio-temporal patterns on various scales.

6. References

Ashton P.S. 1964. Ecological Studies in the Mixed Dipterocarp Forests of Brunei State. Oxford Forestry Memoirs.25. Oxford Univ., Oxford.

Ayyappan N. and Parthasarathy N. 1999. Biodiversity inventory of trees in a large-scale permanent plot of tropical evergreen forest at Varagalaiar, Anamalais, Western Ghats, India. Biodiversity and Conservation 8: 1533-1554.

Badano EI, Cavieres LA, Molina-Montenegro MA, Quiroz CL. 2005. Slope aspect influences plant association patterns in the Mediterranean matorral of central Chile. Journal of Arid Environments 62:93-108.

Black G.A., Dobzhansky T., Pavan C. 1950. Some attempts to estimate species diversity and population density of trees in Amazonian forests. Bot. Gaz. 111: 413-25.

Campbell D.G., Daly D.C., Prance G.T., Maciel U.N. 1992. A comparison of the phytosociology and dynamics of three floodplain (varzea. Forest of known ages, Rio Jurna, Western Brazilian Amazon. Botan. Jr. of the Linn. Soc. 108: 213 - 237.

Campbell D.G., Douglas C.D., Prance G.T. and Maciel U.N. 1986. Quantitative ecological inventory of terra firme and varzea tropical forest on the Rio Xingu, Brazilian Amazon. Brittonia 38: 369- 393.

Cao M. and Zhang J. 1997. Tree species diversity of tropical forest vegetation in Xishuangbanna, SW China. Biodiversity and Conservation 6: 995 - 1006.

Caswell, H. & Cohen, J. E. 1993. Local and regional regulation of species-area relations: a patch occupancy model. - In: Ricklefs, R. E. & Schluter, D. (eds.), Species diversity in ecological communities, pp. 99-107, University of Chicago Press, Chicago

Champion H.G. and Seth S.K. 1968. A Revised Survey of the Forest Types of India. Govt. of India Press, Delhi

Chandrashekara U. M. and Ramakrishnan P. S. 1994. Vegetation and gap dynamics of a tropical wet evergreen forest in the Western Ghats of Kerala, India. Journal of Tropical Ecology 10: 337-354.

Cheetham, A. H. & Hazel, J. E. 1969. Binary (presence-absence) similarity coefficients. Journal of Paleontology 43: 1130-1136.

Chittibabu C.V. and Parthasarathy N. 2000. Attenuated tree species diversity in human-impacted tropical evergreen forest sites in Kolli hills, Eastern Ghats, India. Biodiversity and Conservation 9: 1493–1519.

Chust G, Chave J, Condit R, Aguilar S, Lao S, Pérez R. 2006. Determinants and spatial modeling of tree beta-diversity in a tropical forest landscape in Panama. Journal of Vegetation Science 17: 83-92.

Condit R, Pitman N, Leigh EG, Jr., Chave J, Terborgh J, Foster RB, Nunez P, Aguilar S, Valencia R, Villa G, Muller-Landau HC, Losos E, Hubbell SP. 2002. Beta-Diversity in Tropical Forest Trees. Science 295: 666-669.

Connell JH. 1978. Diversity in Tropical Rain Forests and Coral Reefs. Science 199: 1302-1310.

Davis T.A.W. and Richards P.W. 1933. Vegetation of Moraballi Creek, British Guiana: an ecological study of a limited area of tropical rain forest. Journal of Ecology 21: 350-384.

De'ath G. 1999. Extended dissimilarity: a method of robust estimation of ecological distances from high beta diversity data. Plant Ecology 144: 191-199.

Duivenvoorden J. E. 1995. Tree species composition and rain forest-environment relationships in the middle Caquet√° area, Colombia, NW Amazonia. Plant Ecology 120: 91-113.

Duivenvoorden J.F, Svenning J.C, Wright S.J. 2002. Beta-diversity in tropical forests - Response. Science 297: 636-637.

Faith DP, Minchin PR, Belbin L. 1987. Compositional dissimilarity as a robust measure of ecological distance. Plant Ecology 69: 57-68.

Ferreira L.V. and Prance G.T. 1998. Species richness and floristic composition in four hectares in the Jau National Park in upland forests in central Amazonia. Biodiversity and Conservation 7: 1349–1364.

Ferrier S, Gray MR, Cassis GA, Wilke L. 1999. Spatial turnover in species composition of ground-dwelling arthropods, vertebrates and vascular plants in north-east New South Wales: implications for selection of forest reserves. In: Ponder W, Lunney D (Hrsg.) The other 99%: the conservation and biodiversity of invertebrates. Royal Zoological Societs of New South Wales, Sydney, New South Wales, Australia. S. 68-76.

Field, J. G. 1969. The use of the information statistic in the numerical classification of heterogenous systems. Journal of Ecology 57: 565-569.

Ganesh T., Ganesan R., Soubadradevy M., Davidar P., Bawa K.S. 1996. Assessment of plant biodiversity at a mid-elevation evergreen forest of Kalakad-Mudanthurai Tiger reserve, Western Ghats, India. Current Science 71: 379-92.

Gentry A.H. 1988. Changes in plant community diversity and floristic composition on environmental and geographical gradients. Annals of the Missouri Botanical Garden 75: 1-34.

Gering JC, Crist TO, Veech JA (2003) Additive Partitioning of Species Diversity across Multiple Spatial Scales: Implications for Regional Conservation of Biodiversity. Conservation Biology 17: 488-499.

Ghate U., Joshi N.V., Gadgil M. 1998. On the patterns of tree diversity in the Western Ghats of India. Current Science 75: 594 - 603.

Giriraj A. 2006. Spatial characterization and conservation prioritization in tropical evergreen forests of Western Ghats, Tamil nadu using Geoinformatics. Ph.D. Thesis, St.Joseph College, Trichi, India.

Giriraj A., Murthy M.S.R., Britto S., Dutt C.B.S. 2003. Phytodiversity in Intact and Fragmented Evergreen Habitats, Tamil Nadu, India - A conjunctive analysis using RS and ground data. 2003. Proceedings of the 30th Int. Symp. Rem. Sens. Env. 10 - 14th November, Hawaii

Gordon J.E., Hawthorne W.D., Reyes-García A., Sandoval G. and Barrance A.J. 2004. Assessing landscapes: a case study of tree and shrub diversity in the seasonally dry tropical forests of Oaxaca, Mexico and southern Honduras. Biological Conservation 117: 429-442.

Heaney A., Proctor J. 1990. Preliminary studies on forest structure and floristics on volcan Barva, Costa Rica. Journal of Tropical Ecology 6: 307-320.

Hubalek, Z. 1982. Coefficients of association and similarity, based on binary (presence-absence) data: An evaluation. Biological Reviews of the Cambridge Philosophical Society 57: 669-689.

Hubbel SP. 2005. Neutral theory in community ecology and the hypothesis of functional equivalence. Functional Ecology 19: 166-172.

Hubbell SP. 2001. The Unified Neutral Theory of Biodiversity and Biogeography. Princeton University Press, Princeton, 448 S.

Huston M. 1979. A General Hypothesis of Species Diversity. The American Naturalist 113: 81-101.

Jaffré T. and Veillon J.-M. 1990. Etude floristique et structurale de deux forêts denses humides sur roches ultrabasique en Nouvelle Calédonie. Bull. Mus. natn. Hist. nat., Paris, 4e sér. 12, Section B, Adansonia, 3–4, 243-273.

Janson S. and Vegelius J. 1981. Measures of ecological association. Oecologia 49: 371-376.

Jones, M. M., Tuomisto, H., Clark, D. B. & Olivas, P. 2006. Effects of mesoscale environmental heterogeneity and dispresal limitation on floristic variation in rain forest ferns. Journal of Ecology 94: 181-195.

Jongman, R. H. G., ter Braak, C. J. F. & van Tongeren, O. F. R. 1987. Data Analysis in Community and Landscape Ecology. Pudoc, Wageningen.

Jordan T.E., Whigham D.F., Correll D.L. 1989. The role of litter in nutrient cycling in a brackish tidal marsh. Ecology 70: 1906-1915.

Jurasinski G 2007: Spatio-Temporal Patterns of Biodiversity and their Drivers - Method Development and Application, Biogeography, University of Bayreuth, 250.

Jurasinski, G. & Beierkuhnlein, C. 2006. Spatial patterns of biodiversity - assessing vegetation using hexagonal grids. Proceedings of the Royal Irish Academy - Biology and Environment 106B: 401-411.

Jurasinski, G. and Retzer, V. (2009) Inventory, differentiation and proportional diversity - a consistent terminology for quantifying biodiversity. Oecologia, 159(1): 15-26.

Kaduvul K., Parthasarathy N. 1999a. Plant biodiversity and conservation of tropical semi-evergreen forest in the Shervarayan hills of Eastern Ghats, India. Biodiversity and Conservation 8: 421-439.

Kaduvul K., Parthasarathy N. 1999b. Structure and composition of woody species in tropical semi-evergreen forest of Kalrayan hills, Eastern Ghats, India. Tropical Ecology 40: 247-260.

Kjällgren L, Kullman L. 1998. Spatial Patterns and Structure of the Mountain Birch Tree-Limit in the Southern Swedish Scandes - A Regional Perspective. Geografiska Annaler, Series A: Physical Geography 80: 1-16.

Kluth C, Bruelheide H. 2004. Using standardized sampling designs from population ecology to assess biodiversity patterns of therophyte vegetation across scales. Journal of Biogeography 31: 363-377.

Koleff P, Gaston KJ, Lennon JJ. 2003. Measuring beta diversity for presence-absence data. Journal of Animal Ecology 72: 367-382.

Kruskal JB. 1964. Nonmetric multidimensional scaling: a numerical method. Psychometrika 29: 115-129.

Lande R. 1996. Statistics and partitioning of species diversity and similarity along multiple communities. Oikos 76: 25-39.

Legendre P and Legendre L. 1998. Numerical Ecology. Elsevier, Amsterdam, 853

Legendre P, Borcard D, Peres-Neto PR. 2005. Analyzing beta diversity: partitioning the spatial variation of community composition data. Ecological Monographs 75: 435-450.

Leibold MA. 1995. The Niche Concept Revisited: Mechanistic Models and Community Context. Ecology 76: 1371-1382.

Leyer, I. and Wesche, K. 2007. Multivariate Methoden in der Ökologie. Springer, Heidelberg.

Lieberman D. and Lieberman M. 1987. Forest Tree Growth and Dynamics at La Selva, Costa Rica (1969-1982). Journal of Tropical Ecology 3: 347-358.

Lieberman D., Liberman M., Peralta R., Hartshorn G.S. 1996. Tropical forest structure and composition on a large-scale altitudinal gradient in Costa Rica. Journal of Ecology 84: 137±52.

Linares-Palomino R. and Ponce-Alvarez S.I. 2005. Tree community patterns in seasonally dry tropical forests in the Cerros de Amotape Cordillera, Tumbes, Peru. Forest Ecology and Management 209: 261-272.

MacArthur R. H. 1965. Patterns of species diversity. Biological reviews of the Cambridge Philosophical Society 40: 510-533.

Martin M.E. and Aber J.D. 1997. High spectral resolution remote sensing of forest canopy lignin, nitrogen, and ecosystem processes. Ecol. Appl. 7: 431–443.

Minchin PR. 1989. Montane vegetation of the Mt. Field massif, Tasmania: a test of some hypotheses about properties of community patterns. Plant Ecology 83: 97-110.

Ministry of Environment and Forests, 1997. Annual Report. [Online] http://envfor.nic.in/report/ (accessed on 10th May 2001).

Murali K S, Uma S, Shaanker U R, Ganeshaiah K N, Bawa K S. 1996. Extraction of forest products in the forests of Biligirirangan Hills, India. 2: impact of NTFP extraction on regeneration, population structure, and species composition. Economic Botany 50: 252-269

Nadkarni N.M., Matelson T.J. and Haber W.A. 1995. Structural characteristics and floristic composition of a neotropical Cloud Forest, Monteverde, Costa Rica. Journal of Tropical Ecology 11: 481-495.

Nekola JC, White PS. 1999. The distance decay of similarity in biogeography and ecology. Journal of Biogeography 26: 867-878.

Newbery D.Mc.C., Campbell E.J.F., Lee Y.F., Ridsdale C.E. and Still M.J. 1992. Primary lowland dipterocarp forest at Danum valley, Sabah, Malaysia: structure, relative abundance and family composition. Philosophical Transactions of the Royal Society of London, Series B 335: 341-356

Oksanen J, Kindt R, Legendre P, O'Hara RB. 2007. vegan: Community Ecology Package version 1.8-5. http://cran.r-project.org/.

Oksanen J. 2004. Multivariate Analysis in Ecology - Lecture Notes, http://cc.oulu.fi/~jarioksa/opetus/metodi/notes.pdf.

Oksanen J. 2005. Multivariate Analysis of Ecological Communities in R: vegan tutorial.

Olden JD, Poff NL, McKinney ML. 2006. Forecasting faunal and floral homogenization associated with human population geography in North America. Biological Conservation 127: 261-271.

Oliver I, Holmes A, Dangerfield JM, Gillings M, Pik AJ, Britton DR, Holley M, Montgomery ME, Raison M, Logan V, Pressey RL, Beattie AJ. 2004. Land systems as surrogates for biodiversity in conservation planning. Ecological Applications 14: 485-503.

Orlóci, L. 1978. Ordination by resemblance matrices. - In: Whittaker, R. H. (ed.) Ordination of Plant Communities, pp. 239-276., SPB, Den Haag

Paijmans K. 1970. An analysis of four tropical rain forest sites in New Guinea. Journal of Ecology 58: 77-101.

Parthasarathy N. 1999. Tree diversity and distribution in undisturbed and Human-impacted sites of tropical wet evergreen forest in southern Western Ghats, India. Biodiversity and Conservation 8: 1365–1381.

Parthasarathy N. 2001. Changes in forest composition and structure in three sites of tropical evergreen forest around Sengaltheri, Western Ghats. Current Science 80: 389–393.

Parthasarathy N. and Karthikeyan R. 1997b. Biodiversity and population density of woody species in a tropical evergreen forest in Courtallum reserve forest, Western Ghats, India. Tropical Ecology 38: 297-306.

Parthasarathy N., Karthikeyan R. 1997a. Biodiversity and population density of woody species in a tropical evergreen forest in Courtallum reserve forest, Western Ghats, India. Tropical Ecology 38: 297-306.

Parthasarathy N., Kinhal V. and Kumar L.P. 1992. Plant species diversity and human impacts in the tropical wet evergreen forests of Southern Western Ghats. In: Indo-French workshop on tropical forest ecosystems: natural functioning and anthropogenich impact. 26-27 Nov. French Institute Pondicherry

Parthasarathy N., Sethi P. 1997. Tree and liana species diversity and population structure in a tropical dry evergreen forest in south India. Tropical Ecology 38: 19-30.

Pascal J.P., Pelissier R. 1996. Structure and floristic composition of a tropical evergreen forest in south-west India. Journal of Tropical Ecology 12: 191-214.

Pearson 1926. On the coefficient of racial likeness. Biometrika 18: 105-117.

Pitkänen, S. 2000. Classification of vegetational diversity in managed boreal forests in eastern Finland. Plant Ecology 146: Nov 28.

Primack R.B. Hall P. 1992. Biodiversity and forest change in Malaysian Borneo. Bioscience 42: 829-837.

Proctor J., Anderson J.M., Chai P., Vallack H.W. 1983. Ecological studies in four contrasting lowland rain forest in Gunung Mulu National Park, Sarawak. I -Forest environment structure and floristics. Journal of Ecology 71: 237-360.

Qian H, Klinka K, Kayahara GJ. 1998. Longitudinal patterns of plant diversity in the North American boreal forest. Plant Ecology 138: 161-178.

Qian H, Ricklefs RE, White PS. 2005. Beta diversity of angiosperms in temperate floras of eastern Asia and eastern North America. Ecology Letters 8: 15-22.

R Development Core Team 2005: R: A language and environment for statistical computing. R Foundation for Statistical Computing, Vienna, Austria. ISBN 3-900051-07-0, URL http://www.R-project.org, Vienna, Austria.

Rawat G.S. 1997. Conservation status of forests and wildlife in the Eastern Ghats, India. Environmental Conservation 24: 307-315

Reddy C.S., Brahmam M. and Raju V.S. 2006. Conservation Prioritization of Endemic plants of Eastern Ghats, India. Journal of Econ. Taxon. Bot. 30: 755-772.

Richards P.W. 1996. The tropical rain forest: An ecological study. 2nd edition. Cambridge University Press, Cambridge, England

Ricotta C., Carranza M. L. and Avena G. 2002. Computing beta-diversity from species-area curves. Basic and Applied Ecology 3: 15-18.

Ruokolainen K, Linna A, Tuomisto H. 1997. Use of Melastomataceae and Pteridophytes for revealing phytogeographical patterns in Amazonian rain forests. Journal of Tropical Ecology 13: 243-256.

Ruokolainen K, Tuomisto H, Chave J, Muller-Landau HC, Condit R, Pitman N, Terborgh J, Hubbell SP, Leigh EG, Jr., Duivenvoorden JF, Svenning J-C, Wright SJ. 2002. Beta-Diversity in Tropical Forests. Science 297: 1439

Russell, P. F. and Rao, T. R. 1940. On habitat and association of species of anopheline larvae in southeastern Madras. Journal of Malaria Inst. India 3: 153-178.

Sax DF and Gaines SD. 2003. Species diversity: from global decreases to local increases. Trends in Ecology & Evolution 18: 561-566.

Shi G. R. 1993. Multivariate data analysis in palaeoecology and palaeobiogeography--a review. Palaeogeography, Palaeoclimatology, Palaeoecology 105: 199-234.

Sokal R. R. and Michener, C. D. 1958. A statistical method for evaluating systematic relationships. University of Kansas Science Bulletin 38: 1409-1438.

Sokal RR and Rohlf FJ. 1981. Biometry - The principles and practice of statistics in biological research. 2nd edn. Freeman, San Francisco.

Sørensen T. 1948. A method of establishing groups of equal amplitude in plant sociology based on similarity of species content. Biologiske Skrifter 5: 1-34.

Srivastava D.S. 2002. The role of conservation in expanding biodiversity research. Oikos 98: 351-360.

State of Forest Report, 2001. Government of India, Forest Survey of India, Kaulagarh Road, Dehra Dun.

Steege H.T., Sabatier S., Castellanos H., van Andel T., Duivenvoorden J., de Oliveira A.A., Ek R.C., Lilwah R., Maas P.J.M., Mori S.A. 2000. A regional perspective: analysis of Amazonian floristic composition and diversity that includes the Guiana Shield. In: Steege H.T. (ed.), Plant diversity in Guyana. With recommendation for a protected areas strategy. Tropenbos Series 18. Tropenbos Foundation, Wageningen, The Netherlands. Pp.19-34

Steinitz O, Heller J, Tsoar A, Rotem D, Kadmon R. 2006. Environment, dispersal and patterns of species similarity. Journal of Biogeography 33: 1044-1054.

Sternberg M and Shoshany M. 2001. Influence of slope aspect on Mediterranean woody formations: Comparison of a semiarid and an arid site in Israel. Ecological Research 16: 335-345.

Sukumar R., Dattaraja H.S., Suresh H.S., Radhakrishnan J.V., Vasudeva R., Nirmala S., Joshi N.V. 1992. Long-term monitoring of vegetation in a tropical deciduous forest in Mudumalai, Southern India. Current Science 62: 608-616.

Sukumar R., Suresh H.S., Dattaraja H.S., Joshi N.V. 1997. Forest biodiversity research, monitoring and modeling: conceptual background and old world case studies, (eds Dallmeier F., Comiskey J. A.), Parthenon Publishing

Sundarapandian S.M., Swamy P.S. 1997. Plant biodiversity at low-elevation evergreen and moist deciduous forests at Kodayar (Western Ghats, India). Int. J. Ecol. Environ. Sci. 23: 363–379.

Tamas J., Podani J. and Csontos P. 2001. An extension of presence/absence coefficients to abundance data: a new look at absence. Journal of Vegetation Science 12: 401-410.

Tobler WR. 1970. A computer movie simulating urban growth in the Detroit region. Economic Geography 46: 234-240.

Tuomisto H, Ruokolainen K, Aguilar M. and Sarmiento A. 2003. Floristic patterns along a 43-km long transect in an Amazonian rain forest. Journal of Ecology 91: 743-756.

Watt AS. 1947. Pattern and process in the plant community. Journal of Ecology 35: 1-22.

Wattenberg I. and Breckle S.W. 1995. Tree species diversity of a pre-montane rain forest in the Cordillera de TilaraÂ n, Costa Rica. Ecotropica 1: 21-30.

Whitmore T.C. 1990. An introduction to tropical rain forests. Oxford University press

Whittaker RH. 1960. Vegetation of the Siskiyou Mountains, Oregon and California. Ecological Monographs 30: 279-338.

Whittaker RH. 1972. Evolution and measurement of species diversity. Taxon 12: 213-251.

Wilkinson MT, Humphreys GS. 2006. Slope aspect, slope length and slope inclination controls of shallow soils vegetated by sclerophyllous heath—links to long-term landscape evolution. Geomorphology 76: 347-362.

Wolda, H. 1981. Similarity indices, sample size and diversity. Oecologia 50: 296-302.

Wyatt-Smith, J., 1966: Ecological studies on Malayan forests. I. Composition of and dynamic studies in lowland evergreen rain-forest in two 5-acre plots in Bukit Lagong and Sungei Menyala Forest Reserves and in two half-acre plots in Sungei Menyala Forest Reserve, 1947-59. Malayan Forestry Department Research Pamphlet No. 52.

8

Marine Macrophytic Algae of the Western Sector of North Pacific (Russia)

Olga N. Selivanova
Kamchatka Branch of Pacific Institute of Geography, Far Eastern Division of the Russian Academy of Sciences, Russia

1. Introduction

Marine algal flora of the western coasts of Bering Sea is studied non-uniformly, i.e. to a variable degree in different areas. Perhaps, algae of the Commander Islands are studied more thoroughly as compared to the other areas of the Russian Pacific. The data on marine algae of the Islands were presented in many published works by the Russian authors. The comprehensive survey of the literature on the Islands' marine flora was given in our papers (Selivanova & Zhigadlova, 1997 a, b, c). Thereafter we continued our floristic and taxonomic studies on the Commander Islands and published many new papers: Selivanova, 2001 a, b; 2008 a, b, c; 2009; Selivanova & Zhigadlova, 2000; 2003; 2010; Zhigadlova, 2009.

In contrast to Commander Islands algae of the continental part of the Bering Sea are studied rather poorly. Though floristic investigations began there above 200 years ago information on algal flora and structure of benthic communities of this area is still scanty. Remoteness and inaccessibility, severe climate and ice conditions, and a short navigation season make this area very inconvenient for natural studies, so they were episodic and uncoordinated. Practically no seasonal field observations have been conducted there, no marine biological stations have ever existed, and scientific expeditions have been infrequent and sporadic. Therefore information on benthic algae of the area is limited. Special publications on this subject are rare (e.g.Vinogradova, 1973a; 1978; Perestenko, 1988; Zhigadlova & Selivanova, 2004), although data on the marine algae of this area may be found in some general taxonomic, floristic and hydrobiological studies (e.g. Kongisser, 1933; Petrov, 1972, 1973; Vinogradova, 1973b, 1974, 1979; Vinogradova et al., 1978; Vinogradova & Perestenko, 1978; Kussakin & Ivanova, 1978; Perestenko, 1994).

The most detailed and contemporary information on marine algae of the Russian part of Bering Sea is given in our recent publication (Selivanova & Zhigadlova, 2010). This work however was published in Russian in a book with a small number of copies so it is scarcely available to phycologists outside Russia. Partially data on algae of western part of Bering Sea (from Ozernoi Gulf to Dezhnev Cape) was presented earlier in my work (Selivanova, 2002), its electronic version is still available at the address: http://ucjeps.berkeley.edu/constancea/83

Data from this publication are cited in the species list of Bering Sea algae presented in a current paper. However some part of this information has already become out-of date due to considerable and fast changes in the algal systematics caused by the application of molecular-genetic studies in phycology in the latest decade. So our main purpose was to make the inventory of the flora of the western Bering Sea on the basis of our personal collections in conformity with new world data in algal taxonomy and nomenclature.

We studied benthic algae of the western part of Beirng Sea that belongs to the administrative region - Kamchatskii Krai, including Koryak Autonomous District (from Ozernoi Gulf to Dezhnev Bay) (Fig. 1). Water areas northwards of it located on the territory of Chukchi Autonomous District (Anadyrskii Gulf, north of Dezhnev Bay to Bering Strait) were unstudied by us. However data on the algae of this area may be found in the papers of Vinogradova (1973a, b), Tolstikova (1974), Kussakin & Ivanova (1978); Perestenko (1988; 1994) but the inventory of marine algae of this region is uncomplete, and additional floristic and taxonomic studies are necessary.

It should be noted that in spite of better knowledge on algae of the Commander Islands as compared to other areas of western coasts of Bering Sea inventory of their flora is still not finished either because of the difficulties of collection of algae, especially subtidal ones, in this remote hard-to-reach and little-inhabited area. Besides that the general taxonomic base is insufficient and there are still unsolved nomenclatural problems.

That is why the species list of marine benthic algae of the Russian sector of Bering Sea presented in the current work in tables (Table 1-3) should be still treated as preliminary. It is not only due to the permanently renovated information on algal systematics but also due to more careful examination and re-identification of our material and finding of new species especially having sub-microscopic size.

Moreover collections of algae on the studied area were carried out in a short period of time (mostly in August-September) that also makes our list insufficient and obviously its completion needs additional field expeditions in other seasons. Necessity to continue studies on marine algae of the western coasts of Bering Sea is also caused by the threat of extermination of algal species as a result of uncontrolled harvest of marine bioresources. The fact is that rich and diverse vegetation on the shelf of this area attracts increasing attention of the commercial sea fishery organizations in the recent decade of years.

2. Materials and methods

The list of algae presented in the paper is based on phycological material collected by the author during expedition of the Laboratory of Hydrobiology of KBPIG in the Bering Sea in 1988. The material was collected from August through October on the littoral fringe during low tides, with the help of a long hook called "kanza" from the depths of 1 to 3 m, and with usage of SCUBA technique from the depths of 1 to 30 m and with a dredge from deeper waters (up to 120 m). The material from Commander Islands was collected by the author during expeditions of the same laboratory annually in 1986-1992 and incidentally by individual collectors in 1997, 2007-2010. Algae were collected from April through September on the littoral fringe during low tides, and with use of SCUBA from the depths of 1-30 m. Algae cast ashore were also picked up. Material was sectioned freehand with a razor blade, placed in a drop of fresh water on the slides and examined using the light microscope. The sections were studied uncolored or stained with iodine Lugol solution. The processing of collections was conducted at Kamchatka Branch of Pacific Institute of Geography (Petropavlovsk-Kamchatskii, Russia). Material is stored in the herbarium of the above mentioned institute.

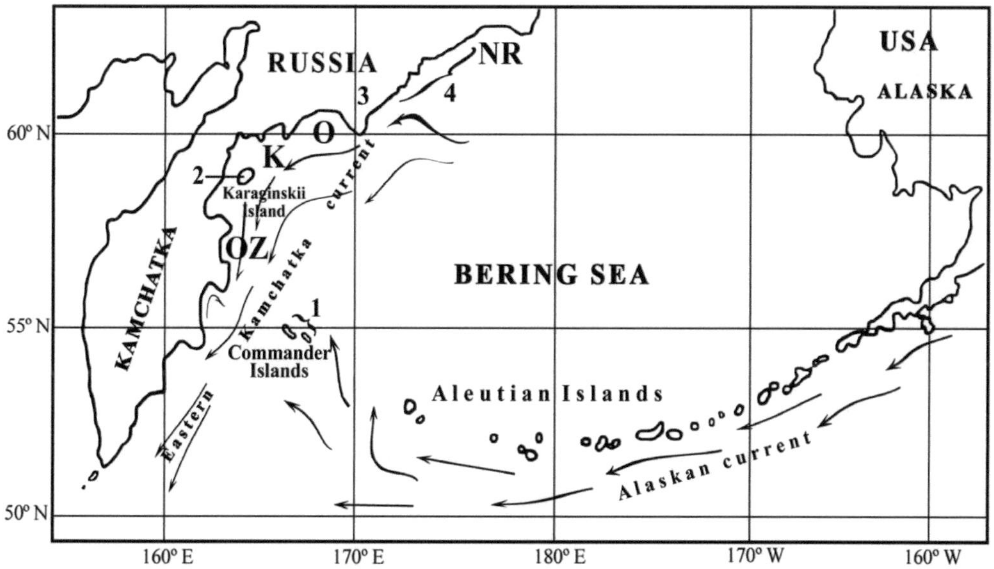

Fig. 1. A schematic map of the part of Bering Sea including Commander Islands. Arrows show the direction of the Eastern Kamchatka and Alaskan currents. Numbers correspond to the specially protected nature areas (SPNA-s) within the studied area: **1** - Commander state biosphere reserve; **2** - Reserve "Karaginskii Island"; **3** - State nature reserve of federal significance "Koryakskii"; **4** - Nature park "Beringia". Abbreviations correspond to the names of major water areas of Bering Sea where algae were collected: **NR- Navarinskii Region, O - Olyutorskii Gulf, K- Karaginskii Gulf, OZ - Ozernoi Gulf.**

3. Species composition and distribution of marine macroalgae of the upper part of the western North Pacific (continental coast of Bering Sea and Commander Islands)

Up to the present day the general list of marine benthic algae of the western coasts of Bering Sea including Commander Islands based on our own materials contains 193 species, but taking into consideration data from the literature their number totals to 209 ones: 36 species of Chlorophyta, 129 Rhodophyta and 44 Ochrophyta (class Phaeophyceae). The list is considerably enlarged as compared to the previous publications (Selivanova and Zhigadlova, 1997a, b, c; Selivanova, 2002) due to new finds of algal samples and description of new taxa.

The author's own data on the species composition of poorly studied areas of Bering Sea are supplemented with the generalized information from separate literature sources only if this information seems to be reliable. In quoting the data of the other authors I do not always share their opinion on taxonomic status of the species and their nomenclature because quite often the data cited are based on rather old literature or archieve materials, so they are outdated, not taking into account changes in the nomenclature and systematics of algae which have occurred recently. Nevertheless, they are included in the table with corrections, where it is possible, in conformity with modern taxonomic data. The original information of the other authors in the case I do not accept their interpretation of the taxa or doubt in correctness of the species identification is not included in the present list.

As it was already mentioned marine macrophytic algae of the Commander Islands are studied more thoroughly as compared to the other areas of the Russian Pacific. By the present day the species list of algae of the Islands based on our collections includes 164 species but it is still being enlarged in the process of the material processing. Recently we found there 9 species new for whole Russian Pacific coastal area (Selivanova and Zhigadlova, in press) and described a new taxon of red algae: *Gloiocladia guiryi* (Selivanova) Selivanova (2009). It belongs to the order Rhodymeniales, family Faucheaceae and appears to be the first representative of the genus and the family Faucheaceae to be reported from the Far Eastern seas of Russia though members of the genera *Fauchea* Bory et Montagne in Montagne and *Gloiocladia* J. Agardh were well known from the other areas. They were recorded from the American Pacific coasts (e.g. O'Clair, Lindstrom, 2000; Gabrielson et al., 2000, 2006), from Japan (Yoshida, 1998), Korea (Oak et al., 2005) and China (Xia, Wang, 2000). But our species still remains the only representative of the family Faucheaceae in the Russian Pacifc sector. Originally it was described as *Fauchea guiryi* Selivanova (2008c). However a taxonomic revision of the genus *Fauchea* based on morphologic and molecular-genetic data resulted in the transfer of the overwhelming majority of the species of the genus including the generitype *F. repens* (C. Agardh) Montagne et Bory de Saint-Vincent to the genus *Gloiocladia* (Rodríguez-Prieto et al., 2007). Consequently it caused the necessity of the same transfer of *F. guiryi* to *Gloiocladia*. So a new nomenclatural combination for this species was proposed by me: *Gloiocladia guiryi* (Selivanova) Selivanova (2009).

Earlier one more taxon was described from western Bering Sea (Olytorskii Gulf, Lavrov Bay) and Glubokaya Bay (Selivanova & Zhigadlova, 2003). It represented the genus *Phycodrys* (Delesseriaceae, Rhodophyta) and was named *P. valentinae* Selivanova et Zhigadlova in honour of the well-known Russian phycologist Valentina F. Przhemenetskaya (Makienko) who has been studying marine algae of the Far Eastern seas of Russia for about 30 years. Later this species was also found by us in Avacha Gulf (Starichkov Island, south-eastern Kamchatka) (Selivanova and Zhigadlova, 2009).

The list of marine macroalgae of the Russian continental part of Bering Sea (without Commander Islands) totals to 153 species. Some of them were recorded by as as new for the Far Eastern seas of Russia (e.g. *Palmaria mollis* (Setchell et Gardner) Van der Meer et Bird, *Opuntiella californica* (Farlow) Kylin, *Membranoptera setchellii* Gardner (Selivanova, 2002).

The general list of algae of the western Bering Sea is presented below in the table format (Tables 1-3) for convenience of information retrieval and with the purposes of shortening printed space. Separate tables contain data on algae of 3 major groups of the colour spectrum corresponding to higher taxa: green (Division Chlorophyta), red (Division Rhodophyta) and brown (Division Ochrophyta, class Phaeophyceae).

It should be noted that taxonomic viewpoints of Russian and western phycologists may differ, and that is the case with interpretation of brown algae. According to the first viewpoint this taxon is treated as a class Phaeophyceae within the phylum Heterokontophyta, kingdom Chromista, empire Eukaryota (Guiry and Guiry, 2011). But we prefer to follow Russian concept of Belyakova et al. (2006) and to attribute the class Phaephyceae to the division Ochrophyta, kingdom Straminopila, empire Chromalveolata. Higher taxa (orders within divisions) are arranged in the tables according to their systematic position, while families within the orders, genera within the familes and species within the genera are given in alphabetic order. The list also contains information on the distribution of algae. Abbreviations in the column "Distribution within the studied area" correspond to the names of the areas shown on schematic map (Fig. 1). In addition to the abbreviations adopted in the legend to the figure Commander Islands are abbreviated in the tables as **CI.**

3.1 Species composition and distribution of brown algae in the studied area

On the total there are 44 species of brown algae found up to now on the shelf of the western part of Bering Sea and Commander Islands. The species list is given below in the table format (Table 1).

Record number	Taxon	Distribution within the studied area
Order Desmarestiales		
Family Desmarestiaceae		
1	***Desmarestia aculeata*** (Linnaeus) Lamouroux	**CI, K, O**
2	***Desmarestia ligulata*** (Lightfoot) Lamouroux	**CI**
3	***Desmarestia viridis*** (O.F. Müller) Lamouroux	**CI**
Order Ectocarpales		
Family Pylaiellaceae		
4	***Pylaiella littoralis*** (Linnaeus) Kjellman	**CI, K,O, NR**
Family Chordariaceae		
5	***Chordaria chordaeformis*** (Kjellman) Kawai et Kim	**O, NR**
6	***Chordaria flagelliformis*** (O.F. Müller) C. Agardh	**in all places**
7	***Coilodesme bulligera*** Strömfelt	**CI**
8	***Coilodesme californica*** (Ruprecht) Kjellman	**CI**
9	***Delamarea attenuata*** (Kjellm.) Rosenv	**CI**
10	***Dictyosiphon foeniculaceus*** (Hudson) Greville	**CI, K, O, NR**
11	***Elachista tenuis*** Yamada	**CI**
12	***Halothrix lumbricalis*** (Kützing) Reinke	**K, O**
13	***Leathesia marina*** (Lyngbye) Decaisne	**CI**
14	***Leptonematella fasciculata*** (Reinke) Silva	**CI, O**
15	***Melanosiphon intestinalis*** (Saund.) Wynne	**K, O**
16	***Punctaria plantaginea*** (Roth) Greville	**K, O**
17	***Saundersella simplex*** (Saunders) Kylin	**OZ, K, O**
18	***Stictyosiphon tortilis*** (Ruprecht) Reinke	**O**
19	***Soranthera ulvoidea*** Postels et Ruprecht	**CI**
20	***Streblonema scabiosum*** Setchell et Gardner	**CI**
Family Ectocarpaceae		
21	***Ectocarpus siliculosus*** (Dillwyn) Lyngbye	**CI, K, O**
Order Scytosiphonales		
Family Scytosiphonaceae		
22	***Analipus filiformis*** (Ruprecht) Papenfuss	**K, O**
23	***Analipus japonicus*** (Harvey) Wynne	**CI, K**

24	***Petalonia fascia*** (O.F. Müller) Kuntze	**CI, K, O, NR**
25	***Scytosiphon dotyi*** Wynne	**CI, O**
26	***Scytosiphon lomentaria*** (Lyngb.) Link	**CI, K, O**
Order Laminariales		
Family Alariaceae		
27	***Alaria angusta*** Kjellman *emend.* Petrov	**CI, OZ, K, NR**
28	***Alaria marginata*** Postels et Ruprecht	**CI, K, O**
29	***Eualaria fistulosa*** (Postels et Ruprecht) Wynne	**CI, K**
Family Arthrothamnaceae		
30	***Arthrothamnus bifidus*** (Gmelin) Ruprecht	**OZ, K**
Family Chordaceae		
31	***Chorda asiatica*** Sasaki et Kawai	**K, O**
Family Costariaceae		
32	***Agarum clathratum*** Dumortier	**in all places**
33	***Thalassiophyllum clathrus*** (Gmelin) Postels et Ruprecht	**CI, K**
Family Laminariaceae		
34	***Cymathaere triplicata*** (Postels et Ruprecht) J.Agardh	**CI**
35	***Laminaria longipes*** Bory	**CI, OZ, K**
36	***Laminaria yezoensis*** Miyabe	**CI, OZ, K, O**
37	***Saccharina bongardiana*** (Postels et Ruprecht) Selivanova, Zhigadlova et G.I. Hansen	**In all places**
38	***Saccharina dentigera*** (Kjellman) Lane, Mayes, Druehl et Saunders	**CI, K**
39	***Saccharina gurjanovae*** (A. Zinova) Selivanova, Zhigadlova et G.I. Hansen	**OZ, K**
Order Fucales		
Family Fucaceae		
40	***Fucus evanescens*** C. Agardh	**In all places**
Order Ralfsiales		
Family Ralfsiaceae		
41	***Ralfsia fungiformis*** (Gunnerus) Setchell et Gardner	**CI, K, O**
42	***Ralfsia verrucosa*** (Areschoug) Areschoug	**K**
Order Sphacelariales		
Family Sphacelariaceae		
43	***Battersia arctica*** (Harvey) Draisma, Prud'homme et Kawai	**K, O**
44	***Chaetopteris plumosa*** (Lyngbye) Kützing	**K**

Table 1. Empire Chromalveolata Kingdom Straminopila Division Ochrophyta Class Phaeophyceae

3.2 Species composition and distribution of green algae in the studied area

Up now there are 36 species of green algae found on the shelf of the western part of Bering Sea and Commander Islands. The species list is given below in the table format (Table 2).

Record number	Taxon	Distribution within the studied area
	Class Bryopsidophyceae	
	Order Bryopsidales	
	Family Derbesiaceae	
1	*Derbesia marina* (Lyngbye) Solier	CI
	Family Codiaceae	
2	*Codium ritteri* Setchell et Gardner	CI
	Class Chlorophyceae	
	Order Chlorococcales	
	Family Chlorochytriaceae	
3	*Chlorochytrium inclusum* Kjellman	In all places
	Family Endosphaeraceae	
4	*Codiolum gregarium* Braun	CI
	Class Ulvophyceae	
	Order Cladophorales	
	Family Cladophoraceae	
5	*Chaetomorpha ligustica* (Kützing) Kützing	CI, K, O
6	*Chaetomorpha linum* (O.F. Müller) Kützing	CI
7	*Chaetomorpha melagonium* (Weber et Mohr) Kützing	OZ, K
8	*Cladophora speciosa* Sakai	CI, K
9	*Rhizoclonium riparium* (Roth) Harvey	CI, K, O
	Order Ulotrichales	
	Family Gomontiaceae	
10	*Monostroma crassidermum* Tokida	K, O
11	*Monostroma grevillei* (Thuret) Wittrock	CI, K, NR
	Family Ulotrichaceae	
12	*Acrosiphonia duriuscula* (Ruprecht) Yendo	CI, K, O, NR
13	*Psedothrix groenlandica* (J.Agardh) Hanic et Lindstrom	CI, O
14	*Spongomorpha arcta* (Dillwyn) Kützing	CI
15	*Spongomorpha mertensii* (Yendo) Setchell et Gardner	CI
16	*Ulothrix flacca* (Dillwyn) Thuret in Le Jolis	CI, K, O
17	*Ulothrix implexa* (Kützing) Kützing	O
18	*Urospora penicilliformis* (Roth) Areschoug	CI, K, O
19	*Urospora wormskjoldii* (Mertens ex Horneman) Rosenvinge	CI, O
	Order Ulvales	
	Family Kornmanniaceae	
20	*Blidingia chadefaudii* (Feldman) Bliding	K
21	*Blidingia minima* (Nägeli ex Kützing) Kylin	CI, K, O
22	*Blidingia subsalsa* (Kjellman) Kornman et Sahling	CI, O

23	***Kornmannia leptoderma*** (Kjellman) Bliding	**CI**
Family Ulvaceae		
24	***Percursaria percursa*** (C. Agardh) Rosenvinge	**CI, O**
25	***Ulva clathrata*** (Roth) C. Agardh	**K, O**
26	***Ulva fenestrata*** Postels et Ruprecht	**In all places**
27	***Ulva flexuosa*** Wulfen	**CI, K**
28	***Ulva linza*** Linnaeus	**CI, K, O**
29	***Ulva procera*** (Ahlner) Hayden, Blomster, Maggs, Silva, Stanhope et Waaland	**CI, O**
30	***Ulva prolifera*** O.F. Müller	**CI, K, O**
31	***Ulvaria splendens*** Ruprecht	**In all places**
Family Ulvellaceae		
32	***Acrochaete flustrae*** (Reinke) O'Kelly	**CI, O**
33	***Acrochaete geniculata*** (Gardner) O'Kelly	**CI**
34	***Acrochaete ramosa*** (Gardner) O'Kelly	**K**
35	***Acrochaete viridis*** (Reinke) R. Nielsen	**CI, K**
Class Trebouxiophyceae		
Order Prasiolales		
Family Prasiolaceae		
36	***Prasiola borealis*** Reed	**CI**

Table 2. Empire Eukaryota kingdom Plantae subkingdom Viridiplantae division Chlorophyta

3.3 Species composition and distribution of red algae in the studied area

On the total there are 129 species of red algae found up to now on the shelf of the western part of Bering Sea and Commander Islands. The species list is given below in the table format (Table 3).

Record number	Taxon	Distribution within the studied area
Class Compsopogonophyceae		
Order Erythropeltidales		
Family Erythrotrichiaceae		
1	***Erythrocladia irregularis*** Rosenvinge	**CI, K**
Class Bangiophyceae		
Order Bangiales		
Family Bangiaceae		
2	***Bangia fuscopurpurea*** (Dillwyn) Lyngbye	**CI**
3	***Porphyra abbottiae*** Krishnamurthy	**CI**
4	***Porphyra brumalis*** Mumford	**CI, O, NR**
5	***Porphyra gardneri*** (Smith et Hollenberg) Hawkes	**CI**
6	***Porphyra kurogii*** Lindstrom in Lindstrom et Cole	**CI, K, O, NR**
7	***Porphyra miniata*** (C. Agardh) C. Agardh	**CI, K, O, NR**
8	***Porphyra ochotensis*** Nagai	**CI**

9	*Porphyra pseudolinearis* Ueda	**CI, K, O**
10	*Porphyra purpurea* (Roth) C. Agardh	**CI, O, NR**
11	*Porphyra tasa* (Yendo) Ueda	**O**
12	*Porphyra torta* Krishnamurthy	**CI, NR**
13	*Porphyra variegata* (Kjellman) Kjellman in Hus	**CI, K, O, NR**
	Class Florideophyceae	
	Order Hildenbrandiales	
	Family Hildenbrandiaceae	
14	*Hildenbrandia rubra* (Sommerfelt) Meneghini	**CI, K, O**
	Order Corallinales	
	Family Corallinaceae	
15	*Bossiella frondescens* (Postels et Ruprecht) Dawson	**CI**
16	*Corallina pilulifera* Postels et Ruprecht	**CI, K,O**
17	*Pachyarthron cretaceum* (Postels et Ruprecht) Manza	**CI, K**
	Family Hapalidiaceae	
18	*Clathromorphum circumscriptum* (Strömfelt) Foslie	**CI, K,O, NR**
19	*Clathromorphum compactum* (Kjellman) Foslie	**CI, K, O**
20	*Clathromorphum loculosum* (Kjellman) Foslie	**CI, K, O**
21	*Clathromorphum nereostratum* Lebednik	**CI, O**
22	*Clathromorphum reclinatum* (Foslie) Adey	**CI**
23	*Phymatolithon lenormandii* (Areschoug) Adey	**CI, OZ, K**
24	*Lithothamnion phymatodeum* Foslie	**CI**
25	*Lithothamnion sonderi* Hauck	**CI, OZ, K**
	Colaconematales	
	Colaconemataceae	
26	*Colaconema savianum* (Meneghini) R. Nielsen	**O**
	Order Acrochaetiales	
	Family Acrochaetiaceae	
27	*Acrochaetium alariae* (Jónsson) Bornet	**O**
28	*Acrochaetium arcuatum* (Drew) Tseng	**OZ, K**
29	*Acrochaetium microscopicum* (Nägeli ex Kützing) Nägeli	**K**
30	*Acrochaetium parvulum* (Kylin) Hoyt	**K**
31	*Rhodochorton purpureum* (Lightfoot) Rosenvinge	**CI, K**
	Order Palmariales	
	Family Meiodiscaceae	
32	*Meiodiscus spetsbergensis* (Kjellman) Saunders et McLachlan	**O**
33	*Rubrointrusa membranacea* (Magnus) Clayden et Saunders	**CI**
	Family Palmariaceae	
34	*Devaleraea compressa* (Ruprecht) Selivanova et Kloczcova	**K, O, NR**
35	*Devaleraea microspora* (Ruprecht) Selivanova et Kloczcova	**K, NR**

36	***Halosaccion firmum*** (Postels et Ruprecht) Kützing	**CI, K, O**
37	***Halosaccion glandiforme*** (Gmelin) Ruprecht	**CI, K, O, NR**
38	***Halosaccion minjaii*** I.K. Lee	**CI**
39	***Palmaria callophylloides*** Hawkes et Scagel	**CI, O**
40	***Palmaria hecatensis*** Hawkes	**CI, O**
41	***Palmaria marginicrassa*** I.K. Lee	**CI**
42	***Palmaria mollis*** (Setchell et Gardner) van der Meer et Bird	**CI, O, K**
43	***Palmaria stenogona*** (Perestenko) Perestenko	**CI, O, K**
Order Ahnfeltiales		
Family Ahnfeltiaceae		
44	***Ahnfeltia fastigiata*** (Endlicher) Makienko	**CI**
45	***Ahnfeltia plicata*** (Hudson) Fries	**CI, K**
Order Bonnemaisoniales		
Family Bonnemaisoniaceae		
46	***Pleuroblepharidella japonica*** (Okamura) Wynne	**CI**
Order Cryptonemiales		
Family Crossocarpaceae		
47	***Beringia castanea*** Perestenko	**CI**
48	***Cirrulicarpus gmelinii*** (Grunow) Tokida et Masaki	**CI**
49	***Cirrulicarpus ruprechtianum*** (Sinova) Perestenko	**CI, K, O, NR**
50	***Crossocarpus lamuticus*** Ruprecht	**CI, OZ, K**
51	***Hommersandia palmatifolia*** (Tokida) Perestenko	**CI, K, O**
52	***Kallymeniopsis lacera*** (Ruprecht) Perestenko	**In all places**
53	***Velatocarpus kurilensis*** Perestenko	**K, O**
54	***Velatocarpus pustulosus*** (Postels et Ruprecht) Perestenko	**CI, OZ, K, O**
Family Dumontiaceae		
55	***Constantinea rosa-marina*** (Gmelin) Postels et Ruprecht	**CI**
56	***Constantinea subulifera*** Setchell	**CI**
57	***Dilsea socialis*** (Postels et Ruprecht.) Perestenko	**OZ, K, O**
58	***Dumontia contorta*** (Gmelin) Ruprecht	**CI, K, O**
59	***Neodilsea natashae*** Lindstrom	**CI, K**
60	***Neodilsea yendoana*** Tokida	**CI, K, O**
Family Kallymeniaceae		
61	***Callophyllis radula*** Perestenko	**CI**
62	***Callophyllis rhynchocarpa*** Ruprecht	**CI, K, O**
63	***Euthora cristata*** (C. Agardh) J. Agardh	**In all places**
Order Halymeniales		
Family Halymeniaceae		
64	***Neoabbottiella araneosa*** (Perestenko) Lindstrom	**CI, OZ, K**
Order Gigartinales		
Family Cystocloniaceae		
65	***Fimbrifolium dichotomum*** (Lepechin) Hansen	**In all places**

66	*Rhodophyllis capillaris* Tokida	CI, K
	Family Endocladiaceae	
67	*Gloiopeltis furcata* (Postels et Ruprecht) J. Agardh	CI, K, O
	Family Furcellariaceae	
68	*Opuntiella californica* (Farlow) Kylin	OZ, K
69	*Opuntiella ornata* (Postels et Ruprecht) A. Zinova	CI, OZ, K, O
70	*Turnerella mertensiana* (Postels et Ruprecht) Schmitz	In all places
71	*Turnerella pennyi* (Harvey) Schmitz in Rosenvinge	K
	Family Gigartinaceae	
72	*Mazzaella parksii* (Setchell et Gardner) Hughey, Silva et Hommersand	CI, K
73	*Mazzaella phyllocarpa* (Postels et Ruprecht) Perestenko	CI, K, O
	Family Phyllophoraceae	
74	*Coccotylus truncatus* (Pallas) Wynne et Heine	K, O
75	*Lukinia dissecta* Perestenko	CI
76	*Mastocarpus pacificus* (Kjellman) Perestenko	CI, K, O, NR
77	*Mastocarpus papilatus* (C. Agardh) Kützing	CI
	Order Nemastomatales	
	Family Schizymeniaceae	
78	*Schizymenia pacifica* (Kylin) Kylin	K, NR
	Order Rhodymeniales	
	Family Faucheaceae	
79	*Gloiocladia guiryi* (Selivanova) Selivanova	CI
	Family Rhodymeniaceae	
80	*Sparlingia pertusa* (Postels et Ruprecht) Saunders, Strachan et Kraft	CI, K
	Order Ceramiales	
	Family Ceramiaceae	
81	*Ceramium cimbricum* H. Petersen in Rosenvinge	K
82	*Ceramium deslongchampsii* Chauvin ex Duby	K
83	*Ceramium kondoi* Yendo	CI, K, O
84	*Microcladia borealis* Ruprecht	CI
85	*Scagelia breviarticulata* Perestenko	CI
86	*Scagelia pylaisaei* (Montagne) Wynne	CI, K, O
	Family Delesseriaceae	
87	*Congregatocarpus kurilensis* (Ruprecht) Wynne	CI
88	*Hideophyllum yezoense* (Yamada et Tokida) A. Zinova	CI, OZ, K, O
89	*Hymenena ruthenica* (Postels et Ruprecht) A. Zinova	CI, OZ, K, O
90	*Laingia aleutica* Wynne	CI
91	*Membranoptera spinulosa* (Ruprecht) Kuntze	In all places
92	*Membranoptera serrata* (Postels et Ruprecht) A. Zinova	K, NR
93	*Membranoptera setchellii* Gardner	OZ, K
94	*Mikamiella ruprechtiana* (A. Zinova) Wynne	CI

95	***Nienburgia prolifera*** Wynne	**CI**
96	***Pantoneura fabriciana*** (Lyngbye) Wynne	**K**
97	***Pantoneura juergensii*** (J. Agardh) Kylin	**CI, OZ, K, NR**
98	***Phycodrys fimbriata*** (Kuntze) Kylin	**CI, OZ, O, NR**
99	***Phycodrys serratiloba*** (Ruprecht) A. Zinova	**CI, OZ, K, O**
100	***Phycodrys valentinae*** Selivanova et Zhigadlova	**O, NR**
101	***Phycodrys vinogradovae*** Perestenko et Gussarova in Perestenko	**CI**
102	***Tokidadendron bullata*** (Gardner) Wynne	**CI, O**
103	***Yendonia crassifolia*** (Ruprecht) Kylin	**CI, OZ, K, O**
	Family Rhodomelaceae	
104	***Beringiella labiosa*** Wynne	**CI**
105	***Harveyella mirabilis*** (Reinsch) Schmitz et Reinke	**K, O**
106	***Neorhodomela larix*** (Turner) Masuda	**CI, K, O**
107	***Neorhodomela oregona*** (Doty) Masuda	**CI**
118	***Neosiphonia japonica*** (Harvey) M.S.Kim et I.K.Lee	**K**
109	***Odonthalia annae*** Perestenko	**CI**
110	***Odonthalia corymbifera*** (Gmelin) Greville	**CI**
111	***Odonthalia dentata*** (Linnaeus) Lyngbye	**CI, K, O, NR**
112	***Odonthalia floccosa*** (Esper) Falkenberg	**CI**
113	***Odonthalia kamtschatica*** (Ruprecht) J. Agardh	**CI, OZ, K, NR**
114	***Odonthalia ochotensis*** (Ruprecht) J. Agardh	**CI, K, O**
115	***Odonthalia setacea*** (Ruprecht) Perestenko	**CI, K, O**
116	***Polysiphonia morrowii*** Harvey	**CI**
117	***Polysiphonia stricta*** (Dillwyn) Greville	**CI, K, O**
118	***Pterosiphonia bipinnata*** (Postels et Ruprecht) Falkenberg	**In all places**
119	***Pterosiphonia hamata*** Sinova	**CI**
120	***Rhodomela pinnata*** Perestenko	**K, O**
121	***Rhodomela sibirica*** A. Zinova et Vinogradova in Vinogradova	**K, O**
122	***Rhodomela tenuissima*** (Ruprecht) Kjellman	**K, O**
123	***Tayloriella abyssalis*** Wynne	**CI**
	Family Wrangeliaceae	
124	***Neoptilota asplenioides*** (Esper) Kylin	**CI, NR**
125	***Pleonosporium kobayashii*** Okamura	**CI**
126	***Pleonosporium vancouverianum*** (J. Agardh) Setchell et Gardner	**CI**
127	***Ptilota filicina*** J. Agardh	**CI, K, O, NR**
128	***Ptilota serrata*** Kützing	**In all places**
129	***Tokidaea serrata*** (Wynne) Lindstrom et Wynne	**CI**

Table 3. Empire Eukaryota Kingdom Plantae subKingdom Biliphyta Division Rhodophyta

In conclusion of this part I would like to point out that the studies on the species composition of any group of organisms are the basis for further scientific studies in biocenology, biogeography, ecology etc. As far as marine macrophytes are among the leading components of benthic communities of the shelf of Bering Sea their inventory is very important. In spite of the incompleteness of inventory works in this region main counters of the marine algal flora are already outlined and might serve as background of development of different scientific conceptions.

For instance knowledge on contemporary algal flora may be helpful for clarification of the problems of historical phytogeography, processes of the species formation and migration. It is also important for understanding of the regularities of forming and functioning of the ecosystems and the role of seaweeds in benthic communities. Shallow water zones of the shelf of the western part of Bering Sea and Commander Islands are also of interest for economic activity. Competent exploitation of their resources is possible only on the basis of reliable scientific data on the species composition, structure and distribution of benthic communities. These problems are discussed to some extent in the sections below.

4. Brief characteristics of the algal flora of the western part of Bering Sea and Commander Islands

In Russian phycological literature the whole water area of the Far Eastern Seas is considered to be divided into 7 large floristic regions: 1) the Sea of Japan, 2) small Kurile Islands region; 3) southern Kurile Islands region; 4) northern Kurile Islands region; 5) the Sea of Okhotsk; 6) Bering Sea; 7) Commander Islands (Perestenko, 1994). We carried out our studies in the last 2 regions and came to a conclusion that the Bering Sea region that according to Perestenko (1994) includes the whole Eastern Kamchatka coasts should be subdivided into 2 subregions practically equal in geographic extension and floristic significance: a) proper Bering Sea area in its geographic borders (from Bering Strait to Ozernoi Gulf) and b) south-eastern Kamchatka (from Kamchatskii Gulf to Lopatka Cape). The latter is out of the limits of the present study. Here we discuss marine algal flora of 2 major floristic complexes of the regions located in the upper part of North Pacific, those of Commander Islands and western continental coast of Bering Sea.

From the point of view of phytogeography both floristic complexes judging from the elements forming them may be attributed to boreal (cold-temperate) type (Table 4). The number of arctic-boreal and high-boreal species in the floristic complex of the Commander Islands totals to 41 %, in the floristic complex of western continental part of Bering Sea their number comes to 44 %. If so called wide-boreal elements are taken into account these figures increase correspondingly to 92% and 90 %.

Systematic analysis of the species list revealed that the largest orders in the floristic complex of the Commander Islands are: 1) within the divion Ochrophyta, class Phaeophyceae: Ectocarpales (contains 3 families, 11 genera and 12 species) and Laminariales (5 families, 7 genera and 10 species); 2) within the division Rhodophyta: Ceramiales (4 families, 24 genera and 38 species) and Cryptonemiales (in its traditional interpretation) (5 families, 13 genera and 17 species); 3) withihn the division Chlorophyta: Ulvales (3 families, 5 genera and 14 species). These orders include above 55 % of the total species number of the flora.

The situation on the western continental coast of Bering Sea is much the same, the largest orders in this floristic complex are: 1) within the division Ochrophyta, class Phaeophyceae: Ectocarpales (contains 3 families, 10 genera and 11 species) and Laminariales (5 families, 8 genera and 12 species); 2) within the division Rhodophyta: Ceramiales (4 families, 18 genera and 31 species) and Cryptonemiales (in its traditional interpretation) (5 families, 12 genera

Phytogeographic group	Commander Islands		Continental part of Bering Sea	
	Species number	%	Species number	%
Arctic-boreal	17	10	24	16
High boreal	50	31	43	28
Wide boreal	83	51	70	46
Boreal-tropical	8	4	10	6
Cosmopolitan	6	4	6	4
Total	164		153	

Table 4. Phytogeographic composition of the macrophytic algae floras of Bering Sea and Commander Islands

and 14 species); and 3) withihn the division Chlorophyta: Ulvales (3 families, 4 genera and 14 species). These orders include about 54 % of the total species number of the flora.

The largest higher taxon of the Commander Islands' flora is the division Rhodophyta consisting of 13 orders, 25 families, 63 genera and 103 species and exceeding the number of green and brown algae in sum (Phaeophyceae + Chlorophyta) in 1.7 times. Thus, the nucleus of the Islands' flora is represented by red algae. The largest genera within Rhodophyta, consisting of 5 and more species are as follows: *Porphyra* (10 species), *Odonthalia* (7 species), *Palmaria* (5 species) and *Clathromorphum* (5 species). The largest genus among Chlorophyta is *Ulva* (6 species), however Phaeophyceae has no genera with more than 3 species, though it contains the complex of genera *Laminaria*+ *Saccharina*, but even in this case the number of the species totals to only 4.

The division Rhodophyta is also the largest one in the flora of the western continental coasts of Bering Sea consisting of 14 orders, 25 families, 56 genera and 91 species and exceeding the total number of green and brown algae in 1.5 times. So the nucles of the Bering Sea flora is also formed by red algae though their number is less than that of Commander Islands (91 species versus 103 accordingly). The largest genus within Rhodophyta in the Bering Sea flora is *Porphyra* (8 species), other large genera containing 4 species each are: *Clathromorphum, Acrochaetium, Palmaria* and *Odonthalia*. The largest taxa among division Chlorophyta and class Phaeophyceae are correspondingly the genus *Ulva* (7 species) and the complex of genera *Laminaria*+ *Saccharina* (5 species).

The figures given above make it possible to outline main characteristics of the floras. Particularly, in accordance with data of systematic analysis in general marine algal flora of both regions can be characterized as variegated in the species composition and allochthonous by origin. It contains big number of monotypic genera and families. I mean those taxa of higher rank that are represented in the studied area by only one taxon of lower rank, i.e. a family represented by one genus and a genus represented by one species. On Commanders there are: 17 genera and 5 families in Phaeophyceae; 12 genera and 7 families in Chlorophyta; 42 genera and 9 families in Rhodophyta; in Bering Sea: 18 genera and 7 families in Phaeophyceae; 7 genera and 5 families in Chlorophyta; 35 genera and 11 families in Rhodophyta. Allochthonous type of the flora is also confirmed by relatively low ratios: species/genus (1.6 for both regions); species/family (3.6 for Commander Islands and 3.4 for Bering Sea) and genus/family (2.3 for Commander Islands and 2.1 for Bering Sea). No endemic species were found. Variegated species composition shows that both floristic complexes are under strong influence of adjacent areas. Main ecologic factors of the existence of contemporary floras of algae-macrophytes in the area are formed by ocean currents.

The Bering Sea marine algal flora presumably (on the assumption of paleonthologically verified data on the other groups of organisms) was formed mostly by low-boreal elements that adapted to fall of temperature in Cainozoic period. At present this flora is being enriched due to the penetration of high-boreal and arctic species (e.g. *Dilsea socialis, Rhodomela sibirica*) via cold-water Eastern Kamchatka current. This current goes further to the south and in such a way cold-water species are spread along the south-eastern Kamchatka coasts (up to Lopatka Cape) (Fig. 1). Thus the marine algal flora of Bering Sea has a well-defined migratory character and is under strong influence of the Arctic area.

On the other hand Commander Islands are greatly influenced by American continent due to the branch of Alaska current providing penetration of American elements of flora. I suppose this process to be main tendency of the Islands' flora development. As a result of invasion of American elements marine algal flora of the Commander Islands is more rich and diverse as compared to other areas of Bering Sea. For example, elementary analysis of the data from our Tables 1-3 show that there are 57 species of marine macrophytes (9 species of green, 10 - of brown and 38 - of red algae) that are met only on Commander Islands and are not found on the continental coast of Bering Sea. Many of these species were described from the American Pacific coasts (e.g. *Streblonema scabiosum, Acrochaete geniculata, Erythrocladia irregularis, Palmaria callophylloides, P. hecatensis, Laingia aleutica, Nienburgia prolifera, Beringiella labiosa, Tayloriella abyssalis*). Of course, absence of records of some species from the continental part of Bering Sea does not in all cases mean that these algae do not grow there, possibly it is due to less careful studies in this area in contrast to Commander Islands.

It should be noted that in spite of the fact that all above-mentioned species were described from the American continent they are not necessarily American in their origin. It is quite possible that marine algal flora of the American continent is just more comprehensively studied and new species are found and described there more often in comparison with the Russian Pacific sector. And it is no wonder, because big and well-equipped groups of phycologists from Canada and USA effectively work in their sector of the Pacific Ocean while in Russia in fact only small isolated groups of enthusiasts try to study marine algae of the western Pacific coasts. In any case the presence of the species on Commander Islands that are not met in other areas of the Russian Pacific but are common with American coast (*Coilodesme californica, Pleonosporium vancouverianum, Microcladia borealis, Tokidaea serrata, Nienburgia prolifera, Laingia aleutica, Beringiella labiosa, Odonthalia floccosa* etc.) brings together their flora with the floras of Aleutian Islands and Alaska. So Commander Islands serve as a peculiar bridge uniting American and Asian algal floras.

Comparison of the species composition of the macrophytes of the Commander Islands' shelf and that of the south-eastern Kamchatka reveals certain similarity of the floras of these two regions (Jaccard's coefficient of community K_j = 0.68) and testify to belonging of the Commander Islands flora to Asian type.[1] According to my calculations similarity between

[1] The geological data are also in favour of hypothesis of Asian origin of marine algal flora of the Commander Islands. In accordance with the theory of continental drift the Pacific islands (together with their underwater foundations) are considered to be the edge chains separated from the continental blocks. During general movement of the Earth crust along the mantle oriented mostly to the west they lagged behind and remained in the east. So their original position was closer to the Asian continent in comparison with the present day (Vegener, 1984). In some sense hypothesis of Asian origin of the Commander Islands marine algal flora is confirmed by data on terrestrial flora of some well studied

the floras of south-eastern Kamchatka and western part of Bering Sea is a little less (K_j = 0.63), and it is the least between the floras of the Commander Islands and western continental coast of Bering Sea (K_j = 0.51). I explain phenomenon of such low similarity (in other words, considerable difference) between floristic complexes of geographically close water areas by peculiarities of the system of ocean currents in this sector of the North Pacific, existence of already mentioned Eastern Kamchatka and Alaska currents (Fig. 1) providing favourable conditions for penetration of invasive species of different origin from the adjacent areas, namely from the Arctic into northern part of Bering Sea and from the American continent to Commander Islands. To tell the truth, such deviations in comparative floristic coefficients may happen due to purely statictical reasons (Kafanov et al., 2004). Nevertheless my conclusions are in agreement with the data of K.L. Vinogradova & L.P. Perestenko (1978) who also showed considerable difference of marine flora of the western Bering Sea from those of the south-eastern Kamchatka and Commander Islands.

5. The problem of diversity conservation of algae in the western Bering Sea including Commander Islands

Conservation of biodiversity is a primary condition for stability of biosphere. Studies on biodiversity became a component part of National strategy of Russia more that 10 years ago. Creation of the system of the specially protected nature areas (SPNA-s) serves as the most important instrument of biodiversity conservation in our country. At present 9 SPNA-s with marine coastal area aimed at conservation of hydrobionts diversity have been created in the studied sector of the North Pacific. The northernmost of these areas - Nature Park "Beringia" is located within administrative borders of Chukchi Autinomous District and is just mentioned but not included in our discussion.

5.1 Commander State Biosphere Reserve

Commander State Biosphere Reserve is the most important among the rest 8 SPNA-s. It embraces practically the whole archipelago Commander Islands (Fig. 1). Necessity of protection of the unique biodiversity of Commander Islands was acknowledged in Russia many years ago.

In 1958 the Soviet government prohibited fishery and other economic activity in 30-miles sea area around the Islands with the purpose of conservation of marine mammals (sea otters, fur seals). In 1993 in accordance with Decision of the government of the Russian Federation the state biosphere "Komandorskii" (Commander) reserve with the area of 3648679 hectares was established, its coastal area covering 3463300 hectares. This is one of the largest reserves in the world and taking into account the size of its coastal zone it could be considered as a marine reserve. Main objectives of this state biosphere reserve are: studies and protection of the unique natural complexes of the archipelago, genofond of plants and animals, protection of large rookeries of marine mammals, population of blue fox, nesting spots of rare birds and also conservation of historical traces of Vitus Bering's expedition. Besides that Commander Islands are the place of compact residence

Pacific islands and archipelagos. For instance, aboriginal flora of Hawaiian Islands (without agricultural and invasive species) is more allied with the flora of the Old World than with that of their nearest neighbour - North America (Drude, 1890).

of indigenous small people of the North - Aleutians with their traditional mode of life and culture that also need protection. Based on these grounds the Commander Reserve was included in the worldwide net of the biosphere reservations in the frames of the UNESCO program "The man and the biosphere" (MAB).

Organization of the state reserve helps to take under protection all marine organisms inhabiting the Islands, including marine macrophytic algae. And this fact inspires optimism because the majority of rare species included in the Red Data Books of Kamchatka and Russia grow on Commander Islands. In most cases the Islands are the only area of their inhabitance within the Russian coasts (Selivanova, 2007). Conservation of these species in the nature has not only scientific importance but is an integral part of conservation of biodiversity of the coastal zone in its broad sense (i.e. not only of the species diversity but also diversity and stability of marine ecosystems).

It is expedient to note that all species included in the Red Data Book of Kamchatka are rare because of natural reasons, their number is not reduced by human activity and their conservation needs rather non-interference in natural habitats than any active measures of protection. However it is necessary to exclude anthropogenic pollution of the coastal area and poaching in the Reserve's water area. Of course, rare algae themselves cannot be the objects of harvesting but irrational catch of other hydrobionts may cause nonreversible changes in marine ecosystems and as a result - extermination of rare species of algae. In fact there are groups of algae growing on Commander Islands that should be attributed to the category of so called vulnerable species and these are commercial laminarian species (*Laminaria, Saccharina*). At present they are abundant on the Islands and there are no real threats for decrease in their diversity. In my opinion remoteness of the Islands, small number of their human population and relatively low demand for the plant marine products will prevent local community from poach harvest of seaweeds as it happened in the vicinity of Petropavlovsk-Kamchatskii. However there is a risk of their overharvest in case of large-scale uncontrolled commercial catch. I think that this perspective should be treated as a possible threat for the biodiversity conservation on the shelf of the Commander Islands.

5.2 Reserve "Karaginskii Island"

Reserve "Karaginskii Island" is another important SPNA in the Bering Sea that covers Karaginskii Island located in Karaginskii Gulf and separated from Kamchatka Peninsula by Litke Strait (Fig. 1(2). At organization of the Reserve "Karaginskii Island" in 1996 all types of commercial catch were prohibited in the water areas around the Island. It was supposed to organize here marine biosphere reserve in succeeding years but the program did not receive financial support. At present Reserve "Karaginskii Island" does not exist any more however the Island is included in the list of wetlands in accordance with Ramsar Convention. There still remains possibility of getting support for the program of organization of marine biosphere reserve on Karaginskii Island, we consider it very expedient because of high biodiversity of marine flora and fauna of the area that needs protection.

Besides the above-mentioned SPNA-s there are areas within Bering Sea with different nature protective status aimed at conservation of marine ecosystems:

5.3 State nature reserve of federal significance "Koryakskii"

State nature reserve of federal significance "Koryakskii" was founded in 1995 by the Decision of the government of the Russian Federation. The reserve covers a part of Goven

Peninsula and adjacent water area of Lavrov Bay (Olyutorskii Gulf) (Fig. 1 (3). The protected water area totals to 83000 hectares. The purpose of creation of the "Koryakskii" Reserve was to provide protection of the whole ecosystem complex of the northern Kamchatka including marine and coastal ecosystems of Bering Sea with large colonies of sea birds.

5.4 Nature park "Beringia"

Nature park "Beringia", founded in 1992, is located in Chukchi Autunomous District and includes a part of Bering Sea water area (Fig. 1 (4).

In addition there are five **Nature Monuments of regional significance** that have protected marine area within their borders: «**Verkhoturov Island**», «**Witgenstein Cape**», «**Witgenstein Rock**», «**Yuzhnaya Glubokaya Bay**» and «**Anastasia Bay**». It is planned to crerate some new marine nature parks and state nature reserves of federal significance. However all already existing SPNA-s with the exception of Commander State Biosphere Reserve are just nominal organizations without real system of nature protection service. As far as the level of knowledge on their biodiversity is concerned it is extremely low and scientific studies seem to us to be the major task for future.

However it should not preclude the possibility of seaweed harvest on the territory of SPNA-s even with high nature protective rank. For instance, Commander Reserve is thought to be perspective for plant resources exploitation in the areas with permitted economic activity (Selivanova, 2008b). It is necessary to elaborate well-grounded recommendations on permissible level of harvest that in its turn should be based on reliable estimate of algal resources. Correct estimation of resources is important for proper decisions on biodiversity conservation not destroying ecosystem integrity and providing restoration of natural seaweed resources.

6. Preliminary data on seaweed resources of the Russian sector of North Pacific

The largest and most productive algal communities in the Russian sector of North Pacific (western continental part of Bering Sea and Commander Islands) are formed by seaweeds of the order Laminariales. These algae are main objects of commercial harvest. Unfortunately there are no exact present-day quantitative data on seaweed resources in the studied area. The only information available is just preliminary, fragmentary and rather old, dated to the first dacades of the last century (Kongisser, 1933; Gail, 1936). Naturally we need more up-dated information concerning this matter but practical estimation works require considareble expeditionary expenses, equipment and efforts of many specialists. At present it is unreal to expect that these works will be carried out. So the only way out is to make calculations using so called expert method. It is a theoretical estimation based on remote sensing data on the bottom area suitable for the growth of seaweeds and their projective cover in communities. In accordance with this expert estimation general stock of seaweeds resources in the Russian sector of North Pacific comes to 10 million tons.

In spite of such rich stock there is a risk of overharvest in case of poor organization of catch. Exceeding press on commercial algae may cause abrupt decrease of their number and futher replacement by competitive species, for example, by crustose coralline algae. This may cause irreversible changes in the structure of plant benthic communities and as a result total loss of commercial species. That is why scientific recommendations on permissible level of harvest are very important.

In my opinion the most prespective region for sustainable use of seaweed resources is a shallow water area of Commander Islands in spite of the fact that their major part is included in the state biosphere reserve. The reasons for such statement are as follows: 1) this region is one of the most pristine in the Far Eastern seas of Russia, seaweeds growing there are suitable for food and pharmaceutical use not only by the local population but also for export; 2) analysis of the benthic communities of the Islands reveal that their state is good enough and the number of laminarian algae is actually excessive; 3) it will be expedient and even necessary to organize harvesting of commercial algae in the zones of permitted economic activity; 4) in the latest few years the scheme of initial zonation of the Reserve was reconsidered and the limits of economic zones were appreciably extended.

However as it was already said, competent exploitation of seaweed resourses should be based on reliable research data on the algal species composition, number and distribution within the area. This information is obviously insufficient and additional studies are necessary.

We have at our disposal only relatively old data on the stock and distribution of the laminarian algae on Commander Islands obtained during expeditions of Kamchatka Branch of the Pacific Institute of Geography in 1986, 1989-1991. In accordance with these data *Saccharina bongardiana* forms relatively small areas (20-30 hectares) covered by kelp at the depths of 1-2 meters. In sheltered bays *S. bongardiana* is met up to the depth of 12 m, forming mixed communities with other brown algae. Mean biomass of *S. bongardiana* varied in different years. The area covered by the kelp with the dominance of this species did not exceed 10 km^2, total stock came to a 1-1.5 thousand tons. In exposed habitats *S. bongardiana* was replaced by a competitive species *S. dentigera.*

S. dentigera formed a distinct belt aroung the coast of Bering Island at the depths from 3 to 10 m. Some patches are met from the depth of 0 (in exposed to waves places) to 25 m. Plants often settled on coralline alga *Clathromorphum nereostratum* and obtaining big size were cast ashore by storms together with coralline crusts on haptera. The portion of *S.dentigera* in cast ashore mass reached 70%. The biomass of *S. dentigera* varied in different years in 3-4 times. The area occupied by the kelp of this species came to 200 km^2, and the stock averaged to 2 million tons.

At present only algae of the genera *Saccharina* and *Laminaria* are the objects of harvest in Kamchatka Region though other laminarian algae (*Alaria, Eualaria*) are also suitable for food, forage and technological use. In particular a giant alga *Eualaria fistulosa* that forms dense population around Bering Island at the depths 5-15 m may be treated as a perspective commercial species. Sometimes it is met in deeper places (up to 25 m) and vice versa rises to 1-2 m in sheltered habitats. The density of populations of *E. fistulosa* varied up to 10 times from a year to year (from 0.6 to 7 kg/m^2). Kelp formed by this alga occupied the area of about 100 km^2. Floating plants of *E. fistulosa* and their parts formed rather thick fields with projective cover up to 60%. In autumn-winter period these floating fields of *E. fistulosa* were destroyed by storms but in April-May they appeared again and reached their maximal size (10-12 m) by June-July due to the fast growth of attached algae. The stock of this alga on Bering Island was estimated at its minimal amount in 1992 to be 82 thousand tons and at its maximal amount in 1990-1991 - up to 700 thousand tons. The total amount of *E. fistulosa* at both Bering and Medny Island in 1990-es averaged to 925 thousand tons.

As is well known abundance of brown algae is subjected to unpredictable changes as a result of harvesting, so their natural fluctuations should be taken into account because harvesting in unfavourable for seaweeds years may lead to abrupt decrease in their stock.

Three species we discuss (*S. bongardiana, S. dentigera* and *E. fistulosa*) have different growth peculiarities, reproduction and demographic structure of the populations, so the strategy of their harvest should be different.

For instance, *S. bongardiana* may be harvested in the volumes of 50 % of its stock in each area (for Bering Island 0.5-0.75 thousand tons). The most expedient way of this species harvesting is cutting of algae in the intertidal zone during low tides using special knife (pruner) in 20 cm above the meristem zone at basal part of the blade. This provides fast re-growth of the blade and makes it possible to get aftercrop in 1.5-2 months, i.e. to carry out multiple harvesting during a vegetation period.

On the other hand the total harvest volume of *S. dentigera* should not exceed 30 % of the total stock because algae of this species grow much slower than *S. bongardiana* but have longer lifetime (presumably, 4-5 years of *S. dentigera* versus 2-3 years of *S. bongardiana*). Inasmuch as *S. dentigera* grows at considerable depths and constitutes a major part in cast ashore algae, collection of fresh samples of *S. dentigera* from the stormy beached seaweed mass seems to be the most rational method of its harvesting.

Harvesting of *E. fistulosa* is possible from the boat by wrapping floating parts of the plants round long hook (kanza). For the sake of fast restoration of the stock it is expedient to cut them at the depth less than 1 m from water surface. Owing to high growth rate of the fronds of *E. fistulosa* it is possible to harvest phytomass several times during one vegetation season. Plants are able to reproduct successfully because the lower parts with sporophylles are not damaged. Annual estimated yield of *E. fistulosa* for Bering Island comes to 120 thousand tons.

It should be accentuated that the use of dragrope is inadmissible in seaweed harvest as it causes considerable loss of raw material, damage of substratum, death of juvenile plants and as a result - elimination of thickets of algae. Moreover torn off plants sink and form dead bottom sediments, in the process of decay they reduce hydrogene contents in the near-bottom water layers.

Recommended by us methods of brown algae harvesting at the coasts of Commander Islands satisfy the international requirements of exploitation of seaweed resources and do not conflict with the nature protection regime in the region. They are ecologically justified because produce no negative effects on the benthic marine ecosystems and populations of marine mammals. It is necessary to take into account the depths of algal vegetation and water temperature in order to fix optimal period of harvest in each specific area. According to our data the most favourable time of seaweed harvest on Bering Island is May-June. Plants growing in shallow water areas develop more rapidly and start reproduction earlier owing to faster water warming-up than algae growing in deeper areas. That is why first of all seaweeds from shallow water zones should be harvested. According to my calculations in the intertidal zone of Commander Islands it is possible to get 3-5 kg from m^2 and in the upper sublittotal zone - up to 10-12 kg of raw seaweeds.

As far as the continental part of Bering Sea is concerned the present-day information on its seaweed resources is practically absent. That is why I would like to withhold recommendations on organization of harvest in this area. Commercial catch of seaweeds without scientific grounds at contemporary technological approach may cause irreplaceable losses of seaweeds and even destruction of the benthic communities. I think that studies on biodiversity of the western part of Bering Sea and ensuring of effective work of SPNA-s should precede organization of seaweed harvesting in the area.

7. Conclusion

Inventory of marine flora of the Russian sector of North Pacific is necessary not only for clarification of scientific problems of phycology but also for solving of some important practical problems. In particular, there is a threat of decrease of biodiversity of the region and loss of the still unknown species. Already known rare species included in the Red Data Book of Kamchatka (Selivanova, 2007) are also under the threat of loss. So I consider it very important to continue studies on biodiversity of marine coastal ecosystems of Bering Sea paying special attention to their plant components that form basic environments for valuable commercial objects (herring, rock trout, sea urchins etc.). In order to prevent their loss or decrease of number it is necessary to organize valid protection of resources of the shelf of the area. I am confident that sustainable harvest will make it possible to avoid destruction of marine ecosystem integrity and provide restoration of natural seaweed resources and their long-term exploitation. This will help to supply population of Kamchatka Region with this valuable food and technological natural raw material and organize its export to other areas of Russia and abroad.

8. Acknowledgment

I am grateful to my colleague Galina G. Zhigadlova for participation in algal collection and identification.

9. References

Belyakova, G.A.; Diakov, Yu. N. & Tarasov, K.L. (2006). *Botany in 4 volumes. V. 2. Algae and fungi,* Center 'Academia" Press, ISBN 5-7695-2750-1, Moscow, Russia, 320 pp. (In Russian)

Drude, O. (1890). *Handbuch der Pflanzengeographie.* Stuttgart, Germany, 487 S.

Gabrielson, P.W.; Widdowson, T.B.; Lindstrom, S.C.; Hawkes, M.W. & Scagel, R.F. (2000). *Keys to the benthic marine algae and seagrasses of British Columbia, Southeast Alaska, Washigton and Oregon.* Phycological Contributions N 5. University of British Columbia, ISBN 0-88865-466-9, Vancouver, BC, Canada. 189 pp.

Gabrielson, P.W.; Widdowson, T.B. & Lindstrom, S.C. (2006). *Keys to the seaweeds and seagrasses of Southeast Alaska, British Columbia, Washington and Oregon.* Phycological Contributions N 7. University of British Columbia, ISBN 0-9763817-1-0, BC, Canada. 209 pp.

Guiry, M.D. & Guiry, G.M. (2011). *AlgaeBase.* World-wide electronic publication, National University of Ireland, Galway. http://www.algaebase.org [searched on February, 24, 2011].

Kafanov, A.I.; Borisovetz, Ye.E. & Volvenko, V.I. (2004). On application of cluster analysis in the biogeographic classifications. *Zhurnal obshchei biologii [Journal of general biology]*, Vol. 65, No 3, (May-June, 2004) pp. 255-270, ISSN 0044-4596 (In Russian).

Kongisser, R.A. (1933). Hydrobiological studies in Bering Sea at the northeastern coasts of Kamchatka (Preliminary information). *Issledovaniya dal'nevostochnykh morei SSSR [Investigations of the Far Eastern seas of the USSR].* Issue 2, pp. 115–124 (In Russian).

Kussakin, O.G. & Ivanova, M.B. (1978). Intertidal zone of Chukotka part of Bering Sea, In: *Litoral' Beringova morya i yugo-vostochnoi Kamchatki [Intertidal zone of Bering Sea and south-eastern Kamchatka],* O.G. Kussakin, (Ed.), 10–40, Nauka, Moscow, Russia (In Russian)

Gail, G. (1936). Laminarian algae of the Far Eastern Seas. *Vestnik DV Filiala AN SSSR [Bulletin of the Far Eastern branch of Academy of Sciences of the USSR*], issue 19, pp. 31-64. Moscow, Russia (In Russian).

Oak, J.H.; Keum, Y.-S.; Hwang, M.S. & Oh, Y.S. (2005). New records of marine algae from Korea. II. *Algae,* Vol. 20, No 3, (September, 2005), pp. 177-181, ISSN 0273-1177.

O'Clair, R.M. & Lindstrom, S.C. (2000). *North Pacific Seaweeds,* Plant Press, ISBN 0-9664245-1-4, Auke Bay, Alaska, USA, 162 pp.

Perestenko, L.P. (1988). Additions to the flora of red algae from Bering Sea, *Novitates systematicae plantarum non vascularum (News of systematics of spore plants),* Vol. 25, pp. 54-57, Leningrad , Russia, (In Russian).

Perestenko, L.P. (1994). *Krasnye vodorosli dal'nevostochnykh morei Rossii [Red algae of the Far Eastern seas of Russia].* "Olga" Publishing House, ISBN 5-86093-007-0, St. Petersburg, Russia, 331 pp. (In Russian).

Petrov, Yu.E. (1972). Systematics of some species of the genus *Laminaria* Lamour. from the Far East. *Novitates systematicae plantarum non vascularum (News of systematics of spore plants).* Vol. 9, pp. 47–52, Leningrad, Russia, (In Russian)

Petrov, Yu.E. (1973). The genus *Alaria* in the seas of the USSR. *Novitates systematicae plantarum non vascularum (News of systematics of spore plants).* Vol. 10, pp. 49–59, Leningrad, Russia, (In Russian)

Rodríguez-Prieto, C.; Freshwater, D.W. & Sánchez, N. (2007). Vegetative and reproductive morphology of *Gloiocladia repens* (C. Agardh) Sánchez et Rodríguez-Prieto, comb. nov. (Rhodymeniales, Rhodophyta) with a taxonomic re-assessment of the genera *Fauchea* and *Gloiocladia. European Journal of Phycology,* Vol. 42, No 2 (May, 2007), pp. 145-162, ISSN 0967-0262.

Selivanova, O. N. (2001a). Dynamics of the species diversity of macrophytic algae of the shelf of the Commander Islands, *Proceedings of the 2nd Intenational Conference "Conservation of biodiversity of Kamchatka and coastal waters,* pp. 91-93, ISBN 5-8440-0026-9, Petropavlovsk-Kamchatskii, Russia, April 9-10, 2001 (In Russian).

Selivanova, O.N. (2001b). Changes in the structure of marine coastal communities of the Commander Islands caused by extermination of Steller sea cow (*Hydrodamalis gigas), Proceedings of the Intenational Conference "Man in the coastal zone: Experience of centuries",* pp. 99-104, ISBN 5-89131-029-5, Petropavlovsk-Kamchatskii, Russia, September 18-20, 2001

Selivanova, O. N. (2002). Marine benthic algae of the Russian coasts of Bering Sea (from the Ozernoi Gulf to Dezhnev Bay, including Karaginskii Island). In: *Constancea,* University of California electronic publications in Botany, Vol. 83. Available from http://ucjeps.berkeley.edu/constancea/83/selivanova/Selivanova.html

Selivanova, O.N. (2007). Marine algae-macrophytes. In: *Red Data Book of Kamchatka. V. 2. Plants, mushrooms and thermophilic organisms.* O.A. Chernyagina (Ed.), 234-254, Kamchatsky Pechatny Dvor Publishing House, ISBN 5-85857-084-4, Petropavlvosk-Kamchatsky, Russia (In Russian)

Selivanova, O.N. (2008a.) Are giant kelp algae extincting from the Northern Pacific?, *Proceedings of Symposium "Asia-Pacific Network for Global Change Research: Marine biodiversity and bioresources of the North-Eastern Asia"*, pp. 160-164, Cheju, Korea, October 21-22, 2008.

Selivanova, O.N. (2008b). The present day state of the algal communities of the Commander Islands shelf and perspectives of sustainable nature use in the region. *Rybovodstvo i rybnoye khozyaistvo [Pisciculture and fish industry]*, Vol. 2, No. 7, pp. 34-38, ISSN 2074-5990, Moscow, Russia (In Russian).

Selivanova, O.N. (2008c). *Fauchea guiryi* Selivanova sp.n.: the first record of a species of the family Faucheaceae (Rhodophyta: Rhodymeniales) in the Far Eastern seas of Russia. *Russian Journal of Marine Biology,* Vol. 34, No 6, (December, 2008), pp. 351-358 (translated into English from Russian), ISSN 1063-0740.

Selivanova, O.N. (2009). *Gloiocladia guiryi* (Selivanova) comb. nov. - a new name for the first member of the family Faucheaceae (Rhodymeniales, Rhodophyta) from the Russian Pacific. *Phycologia,* Vol. 48, No 5 (September, 2009) pp. 439–440, ISSN 0031-8884.

Selivanova, O.N. & Zhigadlova, G.G. (1997a). Marine algae of the Commander Islands. Preliminary remarks on the revision of the flora. I. Chlorophyta. *Botanica Marina.* Vol. 40, No pp. 1-8.

Selivanova, O.N. & Zhigadlova, G.G. (1997b). Marine algae of the Commander Islands. Preliminary remarks on the revision of the flora. II. Phaeophyta. *Botanica Marina.* Vol. 40, No pp. 9-13.

Selivanova, O.N. & Zhigadlova, G.G. (1997c). Marine algae of the Commander Islands. Preliminary remarks on the revision of the flora. III. Rhodophyta. *Botanica Marina.* Vol. 40, No pp. 15-24.

Selivanova, O.N. & Zhigadlova, G.G. (2000). Benthic algae of the Russian coasts of Bering Sea (including Commander Islands). I. Medny Island. *In: Proceedings of Kamchatka Institute of Ecology and Nature management,* R.S. Moiseev, (Ed.), 71-108, Kamchatsky Pechatny Dvor Publishing House, ISBN 5-85857-108-5, Petropavlovsk-Kamchatskii, Russia (In Russian)

Selivanova, O.N. & Zhigadlova, G.G. (2003). Benthic algae of the Russian coasts of Bering Sea (including Commander Islands). II. Bering Island. *In: Proceedings of Kamchatka Institute of Ecology and Nature management,* R.S. Moiseev, (Ed.), 172-208, Kamchatsky Pechatny Dvor Publishing House, ISBN 5-85857-054-02, Petropavlovsk-Kamchatskii, Russia, Issue 4 (In Russian)

Selivanova, O.N. & Zhigadlova, G.G. (2009). Algofloristic studies on Starichkov Island (Eastern Kamchatka). Species composition and distribution of benthic algae. *Journal of Siberain Federal University, Biology,* Vol. 2, No 3, (September, 2009), ISSN 1997-1389, pp. 271-285 (In Russian).

Selivanova, O.N. & Zhigadlova, G.G. (2010). Diversity, systematics, distribution and resources of marine algae-macrophytes of the Western Bering Sea. *In: Current state of the ecosystem of the Western Bering Sea,* P.S. Makarevich, (Ed.), 37- 78, Southern Science Center, Russian Academy of Sciences Publishing House, ISBN 978-5-902982-78-4, Rostov-on-Don, Russia (In Russian)

Tolstikova, N.E. (1974). New data on ecology of subtidal macrophytes of Anadyrskii Gulf, Bering Sea. *Novitates systematicae plantarum non vascularum (News of systematics of spore plants).* Vol. 11, pp. 147–152, Leningrad, Russia (In Russian).

Vegener, A. (1984). Die Entstehung der Kontinente und Ozeane [*Origin of the continents and oceans*], P.N. Kropotkin, (Ed.), Nauka, Leningrad, Russia, 285 pp. (translated from German into Russian).

Vinogradova, K.L. (1973a). Species composition of algae from the intertidal and subtidal zones of the north-western part of Bering Sea. *Novitates systematicae plantarum non vascularum (News of systematics of spore plants).* Vol. 10, pp. 32–44. Leningrad, Russia (In Russian).

Vinogradova, K.L. (1973b). On new species of *Rhodomela* Ag. and *Polycerea* J.Ag. from Bering Sea. *Novitates systematicae plantarum non vascularum (News of systematics of spore plants).* Vol. 10, pp. 22–28. Leningrad , Russia (In Russian).

Vinogradova, K.L. (1974). *Ulvovye vodorosli (Chlorophyta) morei SSSR [Ulvaceous algae (Chlorophyta) of the seas of the USSR].* Nauka. Leningrad, Russia, 165 pp. (In Russian).

Vinogradova, K.L. (1978). Algae of the south-western coasts of Bering Sea. *Novitates systematicae plantarum non vascularum (News of systematics of spore plants).* Vol. 15, pp. 3-11. Leningrad, Russia. (In Russian).

Vinogradova, K.L. (1979). *Opredelitel' vodoreslei dal'nevostochnykh morei. Zelenye vodorosli [Guide-book to the algae of the Far Eastern Seas. Green Algae].* Nauka. Leningrad, Russia. 147 pp. (In Russian).

Vinogradova, K.L.; Klochkova, N.G. & Perestenko, L.P. (1978). List of intertidal algae of the Eastern Kamchatka and western coasts of Bering Sea. In: *Litoral' Beringova morya i yugo-vostochnoi Kamchatki [Intertidal zone of Bering Sea and south-eastern Kamchatka].* O.G. Kussakin (ed.), 150–155. Nauka. Moscow, Russia (In Russian).

Vinogradova, K.L. & Perestenko, L.P. (1978). Main regularities of algal distribution in the intertidal zone of the western part of Bering Sea. In: *Zakonomernosti raspredeleniya i ekologiya pribrezhnykh biotsenozov [Regularities of distribution and ecology of the coastal communities].* pp. 72–75. Nauka. Leningrad. (In Russian).

Xia, B.-M. & Wang, Y.-Q. (2000). Studies on some new red algae from China (II). *Acta Phytotaxonomica Sinica,* vol. 38, No 1, pp. 86-95, ISSN 05291526.

Yoshida, T. (1998). *Marine algae of Japan.* Uchida Rokakuho Publishing CO., ISBN 4-7536-4049-3 C3045, Tokyo, Japan, 1222 pp. (In Japanese).

Zhigadlova, G.G. (2009). Additions to the flora of macroalgae from southeast Kamchatka and Commander Islands. *Proceedings of the 10th Intenational Conference "Conservation of biodiversity of Kamchatka and coastal waters,* pp. 232-235, ISBN 978-5-9610-0133-4, Petropavlovsk-Kamchatskii, Russia, November 17-18, 2009 (In Russian).

Zhigadlova, G.G. & Selivanova, O.N. (2004). Benthic algae of the Russian coasts of Bering Sea. III. Karaginskii Gulf. In: *Proceedings of Kamchatka Branch of Pacific Institute of Geography,* R.S. Moiseev, (Ed.), 47-89, Kamchatsky Pechatny Dvor Publishing House, ISBN 5-85857-062-3, Petropavlovsk-Kamchatskii, Russia, Issue 5 (In Russian).

9

Fungal Diversity – An Overview

Sara Branco
[1]Biodiversity and Climate Research Center
[2]Mountain Research Center (CIMO), ESA – Polytechnic Institute of Braganca
[1]Germany
[2]Portugal

1. Introduction

Fungi are cryptic, understudied and hyperdiverse organisms. In this chapter I address the wonders of fungal diversity, including recent advances on the understanding of the evolution of the kingdom *Fungi*, approaches to documenting and interpreting fungal diversity, and current efforts concerning fungal conservation.

Fungi are eukaryotic organisms that cannot produce their own energy and depend on enzymatic processes to break down biopolymers that are then absorbed for nutrition. The kingdom *Fungi* encompasses tremendous biological diversity, with members spanning a wide array of lifestyles, forms, habitats, and sizes. Fungi are sister to animals (fig. 1) and include thousands of lineages, from the mushroom forming fungi, to yeasts, rusts, smuts, molds, and more or less conspicuous critters with interesting morphologies. Fungi complete indispensable ecological roles, most notably decomposition processes, but are also involved in important symbiotic associations and are known to include noteworthy parasites (Alexopoulus, 1996).

Fungi have been known and used by humans for centuries, but mycology (the scientific study of fungi) traces ist beginnings to the 18th century, with the development of the microscope (Ainsworth, 1976). While much has been discovered since then, fungi remain today a cryptic and understudied group of organisms. Recent estimates point to 1.5 million fungal species on the planet (Hawksworth, 2001) of which only ~7% have been described (Kirk et al, 2008). Furthermore, fungi assemble in very species-rich communities, making the full documentation of fungal diversity in targeted sites a particularly challenging task. Given the important roles fungi play in the maintenance and functioning of ecosystems, such documentation is often combined with functional perspectives, aimed at understanding the ecology of fungi. Advances in molecular techniques have formed the base for a boost in studies concerning fungal diversity, and the fast development of next-generation sequencing technologies promises further progress towards a more thorough understanding of fungal diversity and function.

Our current limited knowledge of fungal diversity and biology complicates an assessment of the conservation status of fungal species and has hindered the development of conservation tools and efforts. Furthermore, the absence of expedite and adequate methods to document fungal demographics has made it extremelly difficult to fit fungi into the efforts to currently established IUCN conservation categories. There have been, however, recent concerted efforts to bring fungi to conservation debates, such as the newly created Society for the Conservation of Fungi.

Fig. 1. A simplified tree of life, showing the relationships between three eukaryotic groups: fungi and animals are sister groups, with plants as their next closest relative. Taken from the tree of life web project (http://tolweb.org/tree/).

2. Fungi

2.1 The fungal tree of life

Fungi are poorly documented organisms and the phylogenetic relationships within the kingdom are not yet fully understood, but recent efforts have been shedding light on the evolutionary history of the *Fungi* (James et al., 2006, Hibbet et al., 2007). Traditionally, fungi were classified based on morphological, chemical, and anatomical characters mainly associated with spore-bearing structures (McLaughlin et al 2009). However, molecular approaches revealed the existence of repeated trait evolution and thus the prevalence of artificial groupings in these traditional classifications. Molecular data sets have permitted the development of a more natural classification and a better understanding of the fungal relationships.

In the last decades the mycological community has invested heavily in developing the field of fungal systematics. Two National Science Foundation (NSF) projects contributed to this endeavor: the Deep Hypha Research Coordination Network and the Assembling the Fungal Tree of Life project (AFTOL1). This funding allowed for the sharing of information across the mycological community and the generation of molecular data for seven loci from ~1500 species belonguing to all groups of fungi (McLaughlin et al 2009). The result was the most recent comprehensive classification of the fungal kingdom to date, based on well-supported monophyletic groups (Hibbet et al 2007, fig. 2). This fungal tree of life includes only true fungi, and does not consider non-fungal groups traditionally studied by mycologists, such as Oomycetes and slime molds. It does, however, include microsporidians (unicellular obligate endoparasitic organisms with highly reduced genomes and mithochondria (Peyretaillade et al., 2008)), several lineages of chytrids (flagelatted fungi) and zygomycetes, including the Glomeromycota (obligate symbionts of photoautotrophs that are suggested to have been crucial to the process of land colonization by plants (Pirozynski and Malloch, 1975)).

Around 98% of all described fungal species belong to the subkingdom *Dikarya* composed of *Basidiomycota* and *Ascomycota* (fig. 2). The former includes subphyla *Pucciniomycotina* (rusts, pathogens specialized in infecting plants), *Ustilagomycotina* (true smuts and some yeasts, mostly plant pathogens), and *Agaricomycotina* (including the vast majority of mushroom-forming fungi). *Ascomycota*, is also comprised of three subphyla, *Taphrinomycotina* (yeast-like

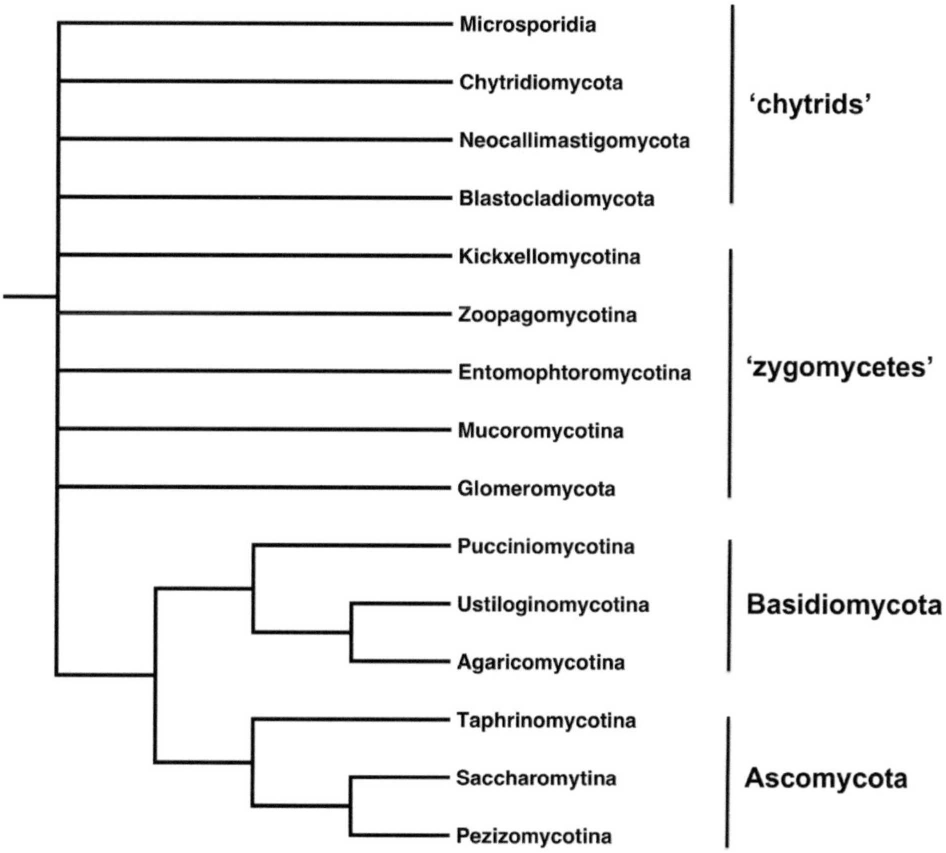

Fig. 2. The fungal tree of life (adapted from Hibbett et al., 2007 and McLaughlin et al., 2009), showing the higher level clades and the unresolved basal polytomy. The terminations -mycota refer to phyla and -mycotina to subphyla; 'chytrids' and 'zygomycetes' are informal non-monophyletic groups.

and some filamentous fungi), *Saccharomycotina* (the true yeasts), and *Pezizomycotina* (with most of the filamentous and fruit-body producing ascomycetes). There has also been extensive work to understand the arrangement taxa within these higher-level clades, a task complicated by the large numbers of fungal taxa described. As evidenced by fig. 2, the base of the tree is a large polytomy, indicating uncertatinty on the resolution of the earliest branching events.

The results of these iniciatives were a big step forward for mycological research. They provided not only a rigorous overview of the main fungal monophyletic groups, but also a framework for understanding and appreciating the evolution of fungi. Although much has been achieved, accurately reconstructing the fungal tree of life is not an easy task and much research effort must be still gathered in order to resolve the earliy branching history of this group in order to have a clear view on how different groups of fungi relate to each other. AFTOL2, an NSF funded sequel to AFTOL1, is ongoing and targetting the unresolved issues and hypotheses raised during the first phase of the project. These include resolving the basal fungal lineages, including the placement of *Microsporidia* and *Glomeromycota*, as well as resolving key lineages within the *Ascomycota* and *Basidiomycota* needed for understanding the evolution of fungal morphology and ecology (McLaughlin et al., 2009).

The availability of an accurate fungal tree of life allows for not only an appreciation of fungal diversity and evaluation of the fundamental differences across groups, but also an understanding of the evolutionary histories of different lineages that gave rise to the diversity of fungi we see today. For example, estimates point to the split between the *Ascomycota* and *Basidiomycota* having occured ~400 million years ago (Taylor and Berbee, 2006), revealing the ancient nature of the fungal phyla. Reconstructing the timing of such evolutionary events occur can be particularly interesting, allowing for comparisons with diversification patterns in other biological groups and ultimately a more thorough understanding of how life evolves.

2.2 Lessons learned from fungal phylogenetics

The use of phylogenetic approaches to reconstruct the fungal tree of life enabled a much better understanding of the evolution of fungi and made testing hypotheses on trait evolution and diversification across the kingdom possible. Two examples of such approaches are discussed below: exploring the evolution of fruit body morphology and the evolution of fungal symbioses.

2.2.1 The evolution of fruit body morphology

As mentioned above, traditional fungal classifications were based on morphology, anatomy, and biochemistry. For basidiomycete mushroom-forming fungi, the fruit body shape was traditionally one of the most important characters and as such was used for a long time as the central principle of classification. This approach gave rise to groups such as the *Hymenomycetes* (fungi with exposed hymenium, such as agaricoid species with a cap and stem, fig. 3) and the *Gasteromycetes* (fungi with gasteroid fruit bodies, that is, with internal spores and a truffle-like shape, fig. 3) that we know now are not monophyletic. Detailed anatomical investigations lead to some skepticism about these non-natural groupings, but only with the advent of early molecular studies were these suspicions confirmed: morphological dissimilar taxa could actually be very closely related. The gasteroid and agaricoid habit where shown to occur in very closely related genera, such as *Rhizopogon* and *Suillus* (Bruns et al. 1989) and *Hydnangium* and *Laccaria* (Mueller & Pine, 1994), implying that

Fig. 3 Examples of agaricoid and gasteroid fruit body morphologies. *Amanita muscaria* (upper left corner), *Armillaria* sp. (upper right corner) and *Macrolepiota* sp. (lower right corner) all agaricoid, showing exposed hymenium and *Astraeus hygrometricus* (lower left corner with internal spores. Courtesy of J. Vicente.

overall fruit body morpohology has not been a stable character across fungal evolution. Soon after this dicovery came the realization that monophyletic groups contain multiple morphologies and that these morphologies appear scattered across clades (Hibbett and Thorn, 2001), indicating that certain fruit body forms evolved multiple times independently (see Hibbett, 2007 for a review on the topic).

This phenomenon of labile fruit body morphology is not exclusive to the basidiomycetes. Another interesting example comes from a well-preserved fossil ascomycete fruit body. This flask-shaped specimen was named *Paloepyrenomycites devonicus* and classified as a pyrenomycete (*Sordariomycetes*, within the subphylum *Pezizomycotina*; Taylor et al. 1995). However, this fruit body morphology is found in several other groups within the subphylum, making it difficult to rule out the possibility that this fossil belongs to a more basal *Ascomycota* lineage, such as *Taphrinomycotina* (typically members of this clade do not fruit, however some species have open aphotecial fruit bodies), or even an earlier extinct lineage (Taylor & Berbee, 2006). These doubts make the placement of this fossil into the *Ascomycota* phylogeny difficult and impair its use for calibrating phylogenetic trees.

2.2.2 The evolution of fungal symbioses

Some fungi live in symbiotic associations with photosynthetic partners, obtaining carbohydrates from their symbionts and providing water, nutrients, or protection in return. The two most remarkable types of symbioses involving fungi are lichens and mycorrhizal associations. In lichens fungi associate with green algae or cyanobacteria (or both) to form a vegetative structure called thallus. Lichens are remarkably successful, colonizing all kinds of habitats and regions (Nash, 1996). Lutzoni et al. (2000) used a phylogenetic framework to study the evolution of lichenization in the *Ascomycota* (including most of the lichenized fungi) and found that this life-style arouse early in the evolutionary history of the phylum and that it has been easier for lineages to loose the ability to be lichenized than it is to become lichens. These results led them to conclude that many non-lichenized *Ascomycota* lineages (including important well-known fungi such as *Penicillium* and *Aspergillus*) descend from lichenized ancestors.

Mycorrhizal associations are symbioses involving fungi and plant roots. The mycorrhizal condition is the natural state for most plants under most ecological conditions. Mycorrhizas (the structure constituted by the root and the fungus) are the main organs of nutrient uptake in land plants (Smith & Read, 2008). The evolution of mycorrhizal associations had a tremendous impact on terrestrial ecosystems and is thought to have facilitated the initial colonization of land by plants (Pirozynski & Malloch, 1975). Ectomycorrhizal (ECM) fungi are one of the major groups of mycorrhizal fungi. They are mainly basidiomycetes and associate with about 30 plant families, including oaks, pines, poplars, and dipterocarps (Smith & Read, 2008). A broad phylogenetic analysis of mycorrhizal and free-living mushroom-forming fungi (*Homobasidiomycetes*, within *Agaricomycotina*) revealed that the ancestor of this group was free-living, and that ectomycorrhizal symbioses were lost and gained a number of times within the clade (Hibbett et al., 2001). This means that ectomycorrhizal fungal symbionts have evolved repeatedly from decomposer percursors and that there have been several reversals to this latter stage, with half of all Homobasiomycetes potentially deriving from ectomycorrhizal ancestors. Such findings suggest that although ECM symbioses are widespread and play relevant ecological roles in nature, they are an evolutionary unstable mutualism.

3. Fungal diversity

3.1 Documenting fungal diversity

Fungi are cryptic and hyperdiverse organisms that assemble in complex and dynamic communities. For the most part, fungi grow as a network of thin filaments on the substrate (soil, wood, insect guts, living plant parts, etc.) making them difficult to detect. Species that produce spore-bearing structures can be easier to discover, although fruiting periods can be short and fructifications ephemeral. Some species can be cultured *in vitro*, however the vast majority are not amenable to culturing, often leaving mycologists with little to work with experimentally.

Traditionally, taxonomists have been responsible for undertaking the task of uncovering new fungi. Morphological, anatomical, and sometimes chemical characters are the basis for the description of fungal species. Interestingly for new species of fungi to be formally accepted, a Latin diagnosis is still required, as recommended by the International Code of Botanical Nomenclature (McNeil et al., 2006), the code followed by mycologists to name fungi. Describing new species also requires the deposition of voucher specimens in official collections.

3.2 The rise of fungal molecular ecological studies

The last decade witnessed a substantial increase in studies focused on fungal community ecology. Conducting fungal surveys can be a tedious long-term undertaking and for a long time mycologists relied on fruit body occurrence or culturing of fungal isolates to document species occurrence and site-specific fungal diversity. Although such methods can provide important information, they tend to supply incomplete community descriptions for the reasons described in preceding sections.

The development of molecular tools to describe diversity allowed a much more straightforward, practical and rapid approach to the study of cryptic organisms such as fungi. These tools permit unveiling the communities colonizing soil (or other rich and dynamic substrates). Not only do they provide DNA-based information for identifying taxa, they also facilitate testing of ecological hypotheses, contributing for a better understanding of the structure and functioning of ecosystems. The vast majority of recent studies targeting the description of fungal communities are based on sequence data (Taylor, 2008).

In general, these molecular microbial studies target one specific short DNA region and rely on the identification of operational taxonomic units (OTUs): sequence similarity based surrogates for taxa (Sharpton et al., 2011). Although OTUs are difficult to define, they are the foundation for estimates of richness, frequency, abundance, and distributions. Most fungal environmental DNA-based diversity studies make use of the internal transcribed spacer (ITS), a nuclear ribosomal repeat unit composed of three parts, the rapidly evolving ITS1, the very conserved 5.8S, and the moderately rapid ITS2 (Horton & Bruns, 2001, Bridge et al, 2005; fig. 4).

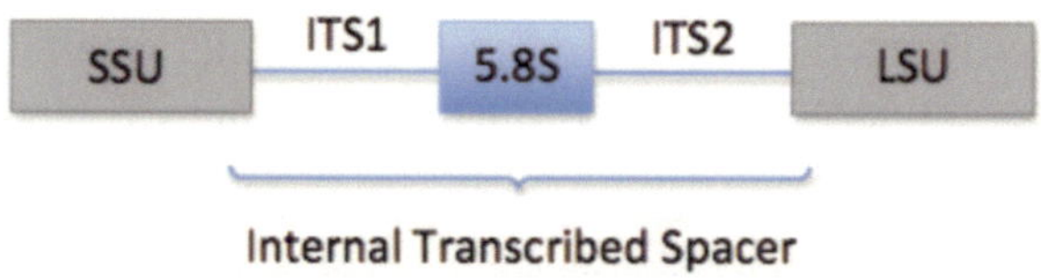

Fig. 4. Structure of the internal transcribed spacer (ITS), the nuclear ribosomal repetitive unit used to describe fungi to the species level. It is composed by the ITS1, 5.8S, and ITS2 regions, and flanked by SSU (ribosomal small subunit) and LSU (ribosomal large subunit).

ITS is used for identifying fungi at the species level. While it is far from being perfect, it offers several advantages that make it a popular that will likely be used for a long time. Genomes include numerous ribosomal DNA encoding genes distributed in tandem arrays along the same or different chromosomes (Rooney & Ward, 2005) and these copies are assumed to be extremely similar (Li, 1997). These coupled with the fact that ITS is easily amplified from low-quality samples (as opposed to single- or low-copy regions) makes it a fast and easy way to describe fungal diversity (Nilsson et al., 2008). However, there are several problems associated with using ITS to define fungal species. On the one hand, there are inherent biases associated with the use of DNA to document diversity, in particular problems with DNA extraction and amplification steps that might lead to distorted community descriptions (Avis et al., 2009). On the other hand, it is known that there is within species variability in ITS, as the different copies within a genome are not exactly identical. Furthermore, intraspecific variation differs considerably across fungal groups (Karen et al., 1997, O'Donnell & Cigelnik, 1997, Glen et al., 2001, Horton, 2002, Rooney & Ward, 2005, Pawlowska & Taylor, 2004, Avis et al., 2006, Nilsson et al., 2008). These pose challenges in determining meaningful sequence similarity cut-offs (O'Brien et al., 2005). For the most part, OTUs are defined using a 95-97% similarity cut-off with the underlying assumption that resulting units are somewhat equivalent to fungal species. However, different fungal species have been reported to have ITS similarity as high as 99% (Dettman et al., 2001, Johannesson & Stenlid, 2003), while interspecific similarity of 90% or less has been found in other species (Kuniaga et al., 1997, O'Donnell, 2000). Despite these limitations and as mentioned above, ITS is the marker of choice for fungal diversity studies and is likely to remain so in the near future.

3.2.1 Ectomycorrhizal (ECM) fungal diversity

As mentioned above, ECM fungi are one of the major functional groups of mycorrhizal fungi. They associate with plant roots by creating a sheath of fungal tissue enclosing short root tips and a net with inward hyphal growth between plant root cells (called a Hartig net). Such anatomy allows for an extensive surface area of plant-fungal contact where fungi exchange soil nutrients for plant-produced cabohydrates. For the most part, ECM fungi belong to the phylum *Basidiomycota* and associate with about 30 plant families, mainly woody perenials (Smith & Read, 2008). These fungi assemble in hyperdiverse, complex and dynamic communities and play a crucial ecological role in most temperate and some tropical habitats.

Unraveling the diversity of ECM fungi is not trivial. Although fruit body inventories provide valuable information, they by no means offer accurate estimates of ectomycorrhizal fungal diversiy. In a pioneer study Gardes & Bruns (1996) surveyed the fungi from pine forests both based on both fruitbody identification and molecular analyses of root samples. They discovered a profound disconnect between the results provided by these different types of data. In fact, the two species producing the majority of fruit bodies were not dominant at the root level, indicating fungal fruiting patterns do not reflect below ground dominances.

Root morphotyping is another approach to study ECM fungal diversity. It has been extensively developed by Agerer (1987-2002) and consists on distinguishing the different fungi based on the morphology and anatomy of ECM root tips. This is a difficult, slow and laborious method that requires extensive training.

As with other areas of mycology, molecular studies have recently revolutionized the study of ECM fungal diversity. In addition to clarifying the discrepancy between above and belowground fungal diversity, molecular surveys also revealed ECM communities as hyper-diverse (particularly when compared to plant host diversity) and composed mostly of rare species (Gehring et al., 1998, Taylor, 2002, Avis et al., 2003, Horton & Bruns, 2005, Walker et al., 2005, Avis et al., 2008, Morris et al., 2008, Branco & Ree, 2010). Figure 5 shows the typical patterns underlying ECM fungal communities: unsaturated species accumulation curves reveal the difficulty in obtaining complete community descriptions and a rank-frequency diagrams illustrate the rarity of most species. These patterns raise interesting questions, particularly from a functional perspective. The most stricking question in ectomycorrhizal ecology has been why are there so many fungal species in a given forest? What are they doing and how do they co-exist? Several explanations have been suggested, such as niche differentiation (Bruns, 1995). These could include vertical niche partitioning, where species have distinct microhabitat preferences that are distributed across a soil vertical gradient (Dickie et al., 2002), or temporal partitioning of ECM fungal communities, where species are active at different times of the year, promoting coexistence by reducing intraspecific competition (Koide et al., 2007). Although the majority of ECM fungal diversity studies are based on root tip data, fungal mycelia also live freely in soil and the community descriptions based on roots and mycelia provide different results (Koide et al., 2005), which adds another layer of complexity to the matter. Host-specificity, where different plant species associate with distinct assemblages of mycorrhizal fungi, has also been suggested as an explanation for the high ECM fungal diversity levels. In general, ECM fungi are known for not having high fidelity to their plant partners and tend to associate with a wide array of plant species. However, there is host preference, which seems to be an important factor in shaping local diversity (Dickie, 2007, Ishida et al., 2007). Inter-specific competition has been another topic of particular interest, given the high numbers of co-existing species. ECM fungi compete for access to the host, more specifically for carbon, as well as soil nutrients, and competition has recently been documented as a major player in ECM community structure (Kennedy, 2010).

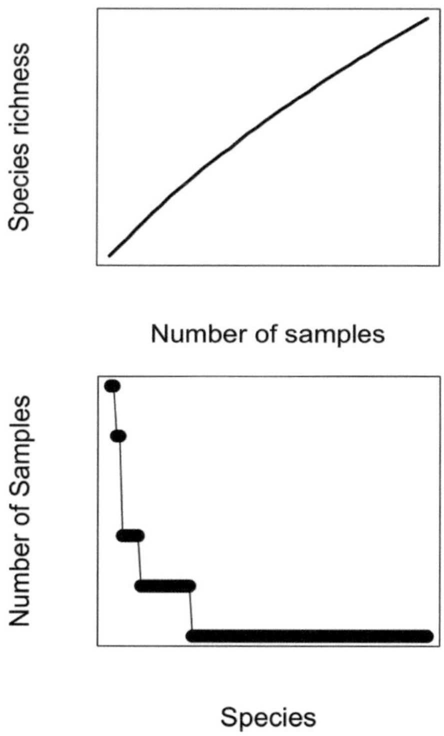

Fig. 5. Typical ECM fungal species accumulation curve (top) and species rank-frequency plot (bottom). (Adapted from Branco & Ree, 2010). As more samples are described, new species are discovered at a consistently rate. This indicates that the vast majority of species in the community are rare (see text for details).

3.3 The genomic revolution

The recent development of massively parallel DNA sequencing platforms, the so called next-generation sequencing (NGS), allowed for the democratization of genomic and metagenomic approaches due to cost reduction and wide availability (Shendure & Ji, 2008). Such technological development has been welcomed by the fungal research community, as it permits rapid studies of deeper scope than have been possible to date. Fungal communities can now be described based on millions of sequences in a very short time frame and at relatively reduced cost. Furthermore, NGS is also enabling an increase in the number of sequenced fungal genomes, providing valuable information crucial for a better understanding of fungal biology and evolution (Brockhurst et al., 2011).

Metagenomic fungal community studies have been based on massively parallel (454) pyrosequencing, a technology able to generate over a million ~500 base-pair sequences in a day (Margulis et al., 2005). 454 pyrosequencing has been preferred to other available technologies precisely because of the long sequence reads it generates, which is crucial for the OTU identification step. 454 has been used to study a wide array of fungal communities, including phyllosphere fungi (Jumpoponen & Jones, 2009, 2010), ECM fungi (Jumpponen et al., 2010, Wallender et al., 2010), AMF (Öpik et al., 2009, Lumini et al., 2010), soil fungi (Buée et al., 2009, Rousk et al., 2010), and indoor fungi (Amend et al., 2010). These studies are important contributions representing the first steps in using metagenomics to study fungal diversity. Interestingly, the 454 results published so far confirm the trends of hyperdiversity and rarity described by traditional sequencing methods (Buée et al., 2009, Jumpponen & Jones, 2009, Tedersoo et al., 2010).

As with any new technology, pyrosequencing approaches introduce many biases that are still not completely understood, such as artefactual singletons due to sequencing errors and the formation of chimeric sequences, unintentionally formed during the polymerase chain reaction step (Bellemain et al., 2009, Quince et al., 2010, Tedersoo et al., 2010). Several attempts are being made to overcome these biases, such as the development of tools like a chimera checker (Nilsson et al., 2010) and a method for extracting the variable and informative regions of the NGS generated sequences (Nilsson et al., 2010b). There have also been some discussions on the usefulness of pyrosequencing data for determining fungal abundances, with some studies advising caution when using 454 data to quantify fungal communities (Amend et al., 2010b, Unterseher et al., 2011). Undoubtedly, the technological improvements on high-throughput sequencing coupled with refinement of analytic tools will significantly increase the quality of metagenomic results in the near future, making NGS an even more powerful and informative approach.

The massive amounts of information provided by metagenomic studies are by far the most substantial source of fungal diversity data today. As mentioned above, only a small fraction of the planet's fungal diversity has been documented and it has been suggested that the sequences generated in environmental studies should be the base for describing and naming new fungal species (Hibbett et al., 2011). The authors suggest a protocol to describe fungi based on molecular sequence similarity, but stress that sequence data should be used alone only when no other sources of information are available. Although sequences from environmental sampling offer limitations for taxonomy and phylogenetics (particularly analysis of single markers), they are practical and easy to obtain, accessible through databases, good for automated approaches, and used in phylogenetic studies. Formally naming fungal species from sequence data would imply some radical changes in the procedure for species descriptions (see section 3.1 above), however it would be a very effective way to rapidly accelerate the rate of fungal discovery.

The accessibility of genomics has also enabled the possibility of a dramatic increase in the number of fungal sequenced genomes. Sequencing the genomes of ecologically and taxonomically relevant fungi is and will continue to provide information not only on those specific species, but will also permit the study of genome structure, gene evolution, metabolic and regulatory pathways and life histories (Martin et al., 2011). The sequencing and analysis of fungal genomes is ongoing, mainly through the Fungal Genomics Program (FGP; http://genome.jgi-psf.org/programs/fungi/about-program.jsf), launched by the US Department of Energy Joint Genome Institute (JGI). This program will sequence the genomes of many species, including decomposer and mycorrhizal species enabling comparative studies focused on the pathways and mechanisms involved in being a symbiont or a decomposer across the fungal tree of life. The genomes of species from lineages with no genomic information will also be sequenced, allowing further studies on fungal evolution (Martin, 2011).

4. Fungal conservation

Although fungi are cryptic and understudied organisms, there has been increasing concern regarding their conservation. As with many other organisms, fungi are affected by habitat loss, pollution, climate change, and other environmental factors. Overall fungi have no legal protection and the potential decline in fungal diversity, affecting both known and unknown species, has been a major concern among mycologists. The main reason underlying the lack of fungal conservation protocols is the challenge in gathering data on fungal populations and geographic distributions. For the most part, conservation bodies, such as the International Union for the Conservation of Nature (IUCN), rely on data describing distributions, population size and population trends to assign threat categories to species (IUCN Standards and Petitions Subcommittee, 2010). These criteria make it very difficult to apply such categories to fungi.

Nevertheless, there have been efforts to gather fungal checklists and flag species of concern with red lists, particularly among European countries. One of the most relevant initiatives has been the European Council for the Conservation of Fungi (ECCF, currently the conservation group at the European Mycological Association), created in 1985 and aimed at promoting awareness about conservation of fungi, stimulating studies and publications on fungal distributions and fungal red lists, as well as promoting international collaborations towards the compilation of a European red list of threatened fungi (http://www.wsl.ch/eccf/). In the early 2000s, ECCF submitted a list of 33 threatened fungi in Europe to be included in the Bern Convention (Dahlberg & Croneborg, 2003). This report referred to rare European macrofungal species and, for the first time, aspired to obtain continental-level legal protection for fungi. This attempt was however unsuccessful, with the Bern Convention rejecting the proposal.

More recently, the International Society for the Conservation of Fungi was established specifically with the goal of protecting fungi worldwide (Minter, 2010, Williams, 2010; http://www.fungal-conservation.org). This is the first society devoted exclusively to the conservation of fungi and aims at developing actions on four fronts: infrastructure, science, education, and politics. The political aspect is regarded as a particularly important target, as the society plans to develop and lobby for fungal conservation policies worldwide.

Hopefully the recent genomic and metagenomic developments and all the multitude of new possibilities they open for fungal research, will contribute for the development of specific protocols to describe fungal populations and distributions that can be the baseline for effective conservation strategies.

5. Conclusion

The last decades brought significant advancements to the understanding and appreciation of the kingdom *Fungi*. We have a much clearer picture of how fungi evolved, assemble and interact with each other and the environment. We also learned, however, how much we still do not know. With all the recent technological advancements, we are better poised to tackle this uncharted frontier than ever before. The use of genomic tools will enable mycology to flourish in the near future, making this a very exciting time to be a mycologist.

6. Acknowledgments

I thank R. Terry Novak, M. Keirle, P. Branco, and S. Gomes for comments on earlier drafts of this chapter.

7. References

Agerer, R. (1987-2002). *Colour Atlas of Ectomycorrhizae*. 1st-12th delivery. Einhorn Verlag. Schwäbisch Gmünd, Germany.

Ainsworth, G. C. (1976). *Introduction to the History of Mycology*. Cambrigde University Press. Cambridge, UK.

Alexopoulos, C. J. ; Mims, C. W. & Blackwell, M. 1996. *Introductory Mycology*. 4th edition. John Wiley and Sons, Inc. New York, USA.

Amend, A., Seifert, K. A., Samson, R., Bruns, T. (2010). Indoor fungal composition is geographically patterned and more diverse in temperate zones than in the tropics. *Proceedings of the National Academy of Sciences of the United States of America* 107 :13748-13753.

Amend, A., Seifert, K. & Bruns, T. (2010b). Quantifying microbial communities with 454 pyrosequencing : does read abundance count ? *Molecular Ecology* 19 :5555-565.

Avis, P. G., McLaughlin, D. J. Dentinger, B. C. & Reich, P. B. (2003) Long-term increase in nitrogen supply alters above- and below-ground ectomycorrhizal communities and increases the dominance of *Russula* spp. in a temperate oak savanna. *New Phytologist* 160:239-253.

Avis, P. G., Dickie, I. A. & Mueller, G. M. (2006). A 'dirty' business : testing the limitations of terminal restriction fragment length polymorphism (TRFLP) analysis of soil fungi. *Molecular Ecology* 15 :873-882.

Avis, P. G., Branco, S., Tang, Y., Mueller G. M. (2010). Pooled samples bias fungal community descriptions. *Molecular Ecology Resources* 1:135-141.

Avis, P. G., Mueller, G. M & Lussenhop, J. (2008) Ectomycorrhizal fungal communities in two North American oak forests respond to nitrogen addition. *New Phytologist* 179: 472-483.

Bellemain, E., Carlsen, T., Brochmann, C., Coissac, E., Taberlet, P. & Kauserud, H. (2010). ITS as an environmental DNA barcode for fungi: an *in silico* approach reveals potential PCR biases. *BMC Microbiology* 10: 189

Bridge, P. D., Spooner, B. M. & Roberts, P. J. (2005). The impact of molecular data in fungal systematics. *Advances on Botanical Research* 42 :33-67.

Branco, S. & Ree, R. H. (2010). Serpentine soils do not limit ectomycorrhizal fungal diversity. *PLoS One*:e11757.

Brockhurst, M. A. , Colegrave, N. & Rozen, D. E. (2011). Next-generation sequencing as a tool to study microbial evolution. *Molecular Ecology* 20 :972-980.

Bruns, T. D., Fogel, R., White, T. J. & Palmer, T. J. (1989). Accelerated evolution of a false truffle from a mushroom ancestor. *Nature* 339 :140-142.

Bruns, T. (1995). Thoughts on the processes that maintain local species diversity in ectomycorrhizal fungi. *Plant and Soil* 170 :63-73.

Buée, M., Reich, M., Murat, C., Morin, E., Nilsson, R. H., Uroz, S. & Martin, F. (2009). 454 pyrosequencing amalysses of forest soils reveal an unexpectedly high fungal diversity. *New Phytologist* 184 :449-456.

Dahlberg, A. & Croneborg, H. (2003). *33 threatened fungi in Europe – Complementary and revised information on candidates for listing in Appendix I of the Bern Convention*. T-PVS (2001) 34 rev 2.

Dettman, J. R., Harbinski, F. M. & Taylor, J. W. (2001). Ascospore morphology is a poor predictor of teh phylogenetic relationships of *Neurospora* and *Gelasinospora*. *Fungal Genetics and Biology* 34 :49-61.

Dickie, I. A., Xu, B. & Koide, R. T. (2002). Vertical niche differentiation of ectomycorrhizal hyphae in soil as shown by T-RFLP analysis. *New Phytologist* 156 :527-535.

Dickie, I. A. (2007). Host preference, niches and fungal diversity. *New Phytologist* 174 : 230-233.

Gardes, M. & Bruns, T. D. 1996. Community structure of ectomycorrhizal fungi in a *Pinus muricata* forest : above- and below-ground views. *Canadian Journal of Botany* 74 :1572-1583.

Gehring, C., Theimer, T., Whitman, T. & Keim, P. (1998) Ectomycorrhizal fungal community structure of pinyon pines growing in two environmental extremes. *Ecology* 79: 1562-1572.

Glen, M., Tommerup, I. C., Bougher, N. L. & Brien, P. A. (2001). Interspecific and intraspecific variation of ectomycorrhizal fungi associated with *Eucalyptus* ecosystems as revealed by ribosomal DNA PCR-RFLP. *Mycological Research* 105 :843-858.

Hawksworth, D. (2001). The magnitude of fungal diversity : the 1.5 million species estimate revisited. *Mycological Research* 12 :1422-1432.

Hibbett, D. S. & Thorn, R. G. (2001). Basidiomycota : Homobasidiomycetes. In : McLaughlin, D. J., McLaughlin, E. G. & Lemke, P. A. (eds). *The Mycota* vol. II part B. *Systematics and Evolution*. Springer-Verlag, Berlin, pp. 121-168.

Hibbett, D. S. (2007). After the gold rush, or before the flood ? Evolutonary morphology of mushroom-forming fungi (*Agaricomycetes*) in the early 21st century. *Mycological Research* 111 :1001-1008.

Hibbett, D. S., Binder, M., Bischoff, J. F., Blackwell, M., Cannon, P. F. , et al. (2007) A higher-level phylogenetic classification of the Fungi. *Mycological Research* 111 :509-547.

Hibbett, D. S., Ohman, A., Glotzer, D., Nuhn, M., Kirk, P. & Nilsson, R. H. (2011). Progress in molecular and morphological taxon discovery in Fungi and options for formal classification of environmental sequences. *Fungal Biology Reviews* 25:38-47.

Horton, T. R. & Bruns, T. D. (2001). The molecular révolution in ectomycorrhizal ecology : peeking into the black-box. *Molecular Ecology* 10 :1855-1871.

Horton. T. R. (2002). Molecular approaches to ectomycorrhizal diversity studies :variation in ITS at a local scale. *Plant and Soil* 244 :29-39.

Ishida, T. A., Nara, K. & Hgetsu, T. (2007). Host effects on ectomycorrhizal fungal communities : insight from eight host species in mixed conifer-broadleaf forests. New Phytologist 174 :430-440.

IUCN Standards and Petitions Subcommittee. (2010). Guidelines for using the IUCN red list categories and criteria. Version 8.1.

James, T. Y. , Kauff, F. , Schoch, C. L., Matheny, P. B. , Hofstetter, V. , et al. (2006). Reconstructing the early evolution of Fungi using a six-gene phylogeny. *Nature* 443 :818-822.

Johannesson, H. & Stenlid, J. (2003). Molecular markers reveal genetic isolation and phylogeography of the S and F intersterility groups of the wood-decay fungus *Heterobasidium annosum*. *Molecular Phylogenetics and Evolution* 29 :94-101.

Jumpponen, A. & Jones, K. L. (2009). Massively parallel 454 sequencing indicates hyperdiverse fungal communities in temperate *Quercus macrocarpa* phyllosphere. *New Phytologist* 184 :438-448.

Jumpponen, A. & Jones, K. L. (2010). Seasonally dynamic fungal communities in the *Quercus macrocarpa* phyllosphere differ between urban and nonurban environments. *New Phytologist* 186 :496-513.

Jumpponen, A., Jones, K. L., Mattox, J. D. & Yaege, C. (2010). Massively parallel 454 sequencing of fungal commuinities in Quercus spp. ectomycorrhizas indicates seasonal dynamics in urban and rural sites. *Molecular Ecology* 19 :41-53.

Karen, O., Hogberg, N. & Dahlberg, A. (1997). Inter- and intra-specific variation in the ITS region of rDNA of ectomycorrhizal fungi in Fennoscandia as detected by endonuclease analysis. *New Phytologist* 136 :131-325.

Kennedy, P. (2010). Ectomycorrhizal fungi and interspecific competition : species interactions, community structure, coexistence mechanisms, and future directions. *New Phytologist* 187 :895-910.

Kirk, P. M., Cannon, P. F., Minter, D. W., Stalpers, J. A. (2008). *Dictionary of the Fungi*. 10th Edition. CABI Publishing. Wallingford, UK.

Koide, R. T., Xu, B. & Sharda, J. (2005). Contrasting below-ground views of an ectomycorrhizal fungal community. *New Phytologist* 166 :251-262.

Koide, R. T., Shumway, D. L., Xu, B. & Sharda, J. (2007). On temporal partitioning of a community of ectomycorrhizal fungi. *New Phytologist* 174 :420-429.

Kuniaga, S., Natsuaki, T., Takeuchi, T, & Yokosawa, R. (1997). Sequence variation of the rDNA ITS régions within and between anastomosis groups in Rhizoctonia solani. *Current Genetics* 32 :237-243.

Li, W. H. (1997). *Molecular evolution*. Sinauer Associates. MA, USA.Lumini, E., Orgiazzi, A., Borriello, R. , Bonfante, P. & Bianciotto, V. (2010). Disclosing arbuscular mycorrhizal fungal biodiversity in soil through a land-use gradient using a pyrosequencing approach. *Environmental Microbiology* 12 :2165-2179.

Lutzoni, F., Pagel, M. & Reeb, V. 2001, Major fungal lineages are derived from lichen symbiotic ancestors. *Nature*. 411 :937-940.

Margulies, M., Egholm, M., Altman, W. E., Attiya, S., Bader, J. S., Bemben, L. A., Berka, J., Braveman, M. S., Chen, Y-J., Chen, Z., et al. (2005). Genome sequencing in microfabricated high-density picolitre reactors. *Nature* 437:376-380.

Martin, F., Cullen, D., Hibbett, D. S., Pisabarro, A., Spatafora, J. W., Baker, S. E. & Grigoriev, I. V. (2011). Sequencing the fungal tree of life. *New Phytologist.*

McLaughlin, D. J., Hibbett, D. S., Lutzoni, F. ; Spatafora, J. W. & Vilgalys, R. (2009). The search for the fungal tree of life. *Trends in Microbiology* 17 :488-497.

McNeill, J. ; Barrie, F. R. ; Brundet, H. M. ; Demoulin, V. ; Hawksworth, D. L. ; et al. (2006). *International Code of Botanical Nomenclature (Vienna Code) adopted by the Seventh International Botanical Congress, Vienna, Austria, July 2005.* Ganter Verlag. Ruggell, Liechtenstein.

Minter, D. (2010). International society for fungal conservation. *International Mycological Association Newsletter* 1 :27-29.

Morris, M., Smith, M. E., Rizzo, D. M., Rejmanek, M. & Blesdsoe, S. (2008). Contrasting ectomycorrhizal fungal communities on the roots of co-occurring oaks (*Quercus* spp.) in a California woodland. *New Phytologist* 178: 167-176.

Mueller, G. M. & Pine, E. M. (1994). DNA data provide evidence on the evolutionary relationships between mushrooms and a false truffles. *McIlvainea* 11 :61-74.

Nash, T. (1996). *Lichen Biology.* Cambridge University Press. Cambridge, UK.Nilsson, R. H., Kristiansson, E., Ryberg, M., Hallenberg, N. & Larsson, K.-H. (2008). Intraspecific ITS variability in the kingdom *Fungi* as expressed in the international sequence databases and its implications for molecular species identification. *Evolutionary Bioinformatics* 4:193-201.

Nilsson, R. H., Abarenkov, K., Veldre, V., Nylinder, S., De Wit, P., Brosché, S., Alfredssonm J. F., Ryberg, M. & Kristiansson, E. (2010). An open source chimera checker for the fungal ITS region. *Molecular Ecology Resources* 10:1076-1081.

Nilsson, R. H., Veldre, V., Hartmann, M., Unterseher, M., Amend, A., Bergsten, J., Kristiansson, E., Ryberg, M., Jumpponen, A. & Abarenkov, K. (2010b). An open source software package for automated extraction of ITS1 and ITS2 from fungal ITS sequences for use in high-throughput community assays and molecular ecology. *Fungal Ecology* 3:284-287.

O'Brien, H., Parrent, J., Jackson, J., Moncalvo, J.-M. & Vilgalys, R. (2005). Fungal community analysis by large-scale sequencing of environmental samples. *Applied Environmental Microbiology* 71:5544-5550.

O'Donnell, K. & Cigelnik, E. (1997). Two divergent intragenomic rDNA ITS2 types within a monophyletic lineage of the fungus *Fusarium* are nonorthologous. *Molecular Phylogenetics and Evolution* 7:103-116.

O'Donnell, K. (2000). Molecular phylogeny of teh *Nectria haematococca-Fusarium solani* species complex. *Mycologia* 90:919-938.

Öpik, M., Metsis, M., Daniell, T. J., Zobel, M. & Moora, M. (2009). Large-scale 454 sequencing reveals host ecological group specificity of arbuscular mycorrhizal fungi in a boreonemoral forest. *New Phytologist* 184:424-437.

Pawlowska, T. E. & Taylor, J. W. (2004). Organization of genetic variation in individuals of arbuscular mycorrhizal fungi. *Nature* 427:733-737.

Peyretaillade, E., Brousolle, V., Peyret, P., Météner, G., Gouy, M. & Vivarès, C. P. (2008). Microsporidia, amithocondrial protists, possess a 70-kDa heat shock protein gene of mithocondrial evolutionary origin. *Molecular Biology and Evolution* 15 :683-689.

Pirozynski, K. A. & Malloch D. W. 1975. The origin of land plants : a matter of mycotrophism. *Biosystems* 5 :153-164.

Rooney, A. P. & Ward, T. J. (2005). Evolution of the large ribosomal RNA multigene family in filamentous fungi : birth and death of a concerted evolution paradigm. *Proceedings of the National Academy of Sciences* 102 :5084-5089.

Rousk, J., Baath, E., Brookes, P. C., Lauber, C. L., Lozupone, C., Caporaso, J. G., Knight, R. & Fiere, N. 2010. Soil bacterial and fungal communities across a pH gradient in an arable soil. ISME Journal 4 :1340-1351.

Sharpton, T. J. ; Riesenfeld, S. J. ; Kembel, S. W. ; Ladau, J. ; O'Dwyer, J. P. ; et al. (2011). PhylOTU :a high-throughput procédure quantifies microbial community diversity and resolves novel taxa from metagenomic data. *PLoS Computational Biology* 7(1) :e1001061.

Smith, S. E. & Read, D. J. (2008). *Mycorrhizal Symbiosis.* 3rd Edition. Academic Press. Amsterdam, The Netherlands.

Taylor, A. (2002). Fungal diversity in ectomycorrhizal communities :sampling effort and species detection. *Plant and Soil* 244 :19-28.

Taylor, A. (2008). Recent advances in our understanding of fungal ecology. *Coolia* 52:197-212.Taylor, J. W. & Berbee, M. L. (2006). Dating divergences in the Fungal Tree of Life : review and new analyses. *Mycologia* 98 :838-849.

Taylor, T. N., Hass, H., Kerp, H., Krings, M. & Hanlin, R. T. (2005). Perithecial ascomycetes from the 400 million year old Rhynie chert : an example of ancestral polymorphism. *Mycologia* 97 :269-285.

Tedersoo, L., Nilsson, R. H., Abarenkov, K., Jairus, T., Sadam, A., Saar, I., Bahram, M., Bechem, Eneke, Chuyong, G. & Kõljalg. (2010). 454 pyrosequencing and Sanger sequencing of tropical mycorrhizal fungi provide similar results but reveal substancial methodological biases. *New Phytologist* 188 :291-301.

Untersher, M., Jumponnen, A., Öpik, M., Tedersoo, L., Moora, M., Dormann, C. & Schnittler, M. (2011). Species abundance distributions and richness estimations in fungal metagenomics - lessons learned from community ecology. *Molecular Ecology* 20 :275-285.

Wallander, H., Johansson, U., Sterkenburg, E., Durling, M. B., Lindahl, B. (2010). Production of ectomycorrhizal mycelium peaks during canopy closure in Norway spruce forests. *New Phytologist* 187 :1124-1134.

Walker, J., Miller O, Horton J (2005) Hyperdiversity of ectomycorrhizal fungus assemblages on oak seedlings in mixed forests in the southern Appalachian Mountains. *Molecular Ecology* 14:829-838.

Williams, N. (2010). Champions of the hidden kingdom. *Current Biology* 20:R792-R794.

10

Aquatic Fungi

Wurzbacher Christian[1], Kerr Janice[2] and Grossart Hans-Peter[1]
[1]*Leibniz-Institute of Freshwater Ecology and Inland Fisheries*
[2]*La Trobe University*
[1]*Germany*
[2]*Australia*

1. Introduction

Seventy-one percent of our planet's surface consist of water, but only 0.6% are lentic and lotic freshwater habitats. Often taken for granted, freshwaters are immensely diverse habitats and host >10% of all animal and >35% of all vertebrate species worldwide. However, no other major components of global biodiversity are declining as fast and massively as freshwater species and ecosystems. Urbanisation, economic growth, and climate change have increased pressure on freshwater resources, whilst biodiversity has given way to the increasing demands of a growing human population. The adverse impacts on aquatic ecosystems include habitat fragmentation, eutrophication, habitat loss, and invasion of pathogenic as well as toxic species. Although there is increasing evidence that freshwater fungal diversity is high, the study of the biodiversity of freshwater fungi is still in its infancy. In light of the rapid decline in freshwater biodiversity, it is timely and necessary to increase our efforts to evaluate the diversity and potential ecological function of this fascinating and diverse group of freshwater organisms.

Hyde et al (2007) have estimated that there are approximately 1.5 million fungal species on earth. Of these, only around 3000 species are known to be associated with aquatic habitats and only 465 species occur in marine waters (Shearer et al., 2007). This small proportion of aquatic fungal taxa is surprising because the aquatic environment is a potentially good habitat for many species. Based on this notion we assume that the "real" number of aquatic fungi is much larger than 3000 and includes a large variety of hitherto undescribed species with unknown ecological function.

Aquatic fungi are usually microscopic organisms, which do not produce visible fruiting bodies but grow asexually (anamorphic fungi). Their occurrence in water is rather subtle and specialised methods are needed to examine their diversity, population structure and ecological function. Water associated fungi have been known historically as "phycomycetes", a functionally defined group consisting of "true fungi" (*Eumycota*) and "analogously evolved fungus-like organisms" belonging to *Chromista* (*Oomycetes, Thraustochytridiomycetes*). Other groups formerly placed in the fungal kingdom include slime moulds (*Amobae*), *Ichthyosporae (Mesomycetozoea)* and *Actinomycetes (Bacteria),* which are now recognised as distinct taxa. While the "true fungi" are a sister group to animals, *Oomycetes* are biochemically distinct from fungi while having similar morphology, size and habitat usage (Money, 1998). Colloquially known as "water moulds", they comprise approx. 200 species inhabiting freshwater, mud and soil. Many of these are saprobes or parasites

(Czeczuga et al., 2005; Nechwatal et al., 2008). Slime moulds (*Amoebozoa*; Adl et al., 2005) are also found in freshwater habitats. Although they are relatively easy to isolate from plant detritus submerged in ponds and lakes, their ecology is little known and requires further investigation (Lindley et al., 2007).

Aquatic "true fungi" are osmoorganotrophs, absorbing nutrients across their cell wall. Most of them have a filamentous growth stage during their life cycle. This morphology enables them to invade deep into substrates and to directly digest particulate organic matter (POM) to acquire nutrients for growth and reproduction. Fungal filaments vary in length from several micrometers for the "rhizoids" of *Chytridiomycetes* to several millimetres or metres for hyphae or hyphal networks, e.g. of hyphomycetes colonising leaves, wood, and soil. However, there are always exceptions, such as unicellular yeasts, which lost filamentous growth during their evolution. Here, we will focus on diversity and function of fungi in various aquatic systems.

1.1 Characteristics of the aquatic habitats influencing fungal life

Aquatic habitats are characterised by a unique balance of allochthonous (external) and autochthonous (internal) organic matter supply, which is controlled largely by watershed characteristics, surface area and location. For example, headwater streams and small ponds receive most of their organic matter from terrestrial riparian vegetation, whereas large lakes are mainly supplied with organic matter internally from algal primary producers. Organic carbon derived from terrestrial vegetation varies substantially from that of algae. Plant remains contain large fractions of lignin, hemicelluloses and cellulose, which prolong microbial decomposition to several months, whereas algae contain much fewer recalcitrant polymers and thus are rapidly mineralised, usually within a few days. In small or shallow lentic systems submerged and emergent aquatic macrophytes often dominate the primary production, representing the most productive non-marine ecosystem worldwide. Aquatic fungi, being heterotrophs, are reliant upon photosynthetically produced organic matter. In order of decreasing biodegradability, the fungal community consumes microscopic algae, macroscopic aquatic macrophytes and terrestrial plant litter (including wood). On localised spatial scales or short-term temporal scales, carbon and nutrients from other sources may gain high importance. Resources derived from animals include fish, fish eggs, carcasses, excuviae, living zooplankton, insects, feathers and hair, while other plant-derived resources include pollen, spores, seeds and fruits (Cole et al., 1990). Interestingly, it seems to be nearly impossible to find a natural organic source that cannot be utilized by aquatic fungi (Sparrow, 1960). This notion points to either a high functional redundancy of a limited set of fungal species or to a high biodiversity of fungal specialists. Most likely, in natural systems both cases occur at the same time. Another interesting feature of aquatic habitats is the coupling of aquatic systems to terrestrial environments via animals, mainly insects, which are able to export nutrients from the aquatic ecosystem (Vander Zanden & Gratton, 2011). It will be shown later, that fungi are often closely associated with insects, which can be key organisms in aquatic freshwater systems. Although often overlooked, fungi represent a common and important component of almost every trophic level of any aquatic ecosystem.

2. Life styles of aquatic fungi

Aquatic habitats are heterogeneous in time and space and greatly differ in their physico-chemical features. Consequently, composition and abundance of aquatic fungi should differ significantly between these habitats (Wurzbacher et al., 2010). Whereas Wurzbacher et al. (2010) have recently reviewed the ecology of fungi in lake ecosystems, and present a thorough discussion on fungal communities within the different water bodies, in this book chapter we want to present a concise overview on fungal life-forms and diversity in various water bodies.

2.1 The role of fungi as decomposers, predators, endophytes, symbionts, parasites, plagues & pathogens

Aquatic fungi are heterotrophs, i.e. they *sensu stricto* depend on external organic matter, which may be dead or alive. Aquatic systems harbour a wealth of organisms that can serve as suitable hosts: algae from different phyla, cyanobacteria, protists, zooplankton, fish, birds, mussels, nematodes, crayfish, mites, insects, amphibians, mammals, plants and other fungi (Sparrow, 1960; Ellis & Ellis, 1985). Fungi are omnipresent and therefore associated with almost every organism, often as parasites, sometimes as symbionts and of course as decomposers.

Parallel to fungi in soil, aquatic fungi act as prominent decomposers of POM: foremost coarse particulate organic matter (CPOM) including plant and animal debris. Filamentous growth habit is a key feature of many aquatic fungi, and this feature is responsible for their superiority to heterotrophic bacteria as pioneer colonisers. Hyphae allow fungi to actively penetrate plant tissues and tap internal nutrients. Therefore, Gessner & Van Ryckegem (2003) describe fungal hyphae as self-extending digestive tracts that have been turned inside out growing hidden inside the substrate.

The aquatic fungi which typically decompose leaf litter and wood with a hyphal network are the polyphyletic group known as "aquatic hyphomycetes". Aquatic hyphomycetes are most common in clean, well oxygenated, flowing waters (Ingold, 1975; Bärlocher, 1992), and are characterised as anamorphic fungi with tetraradiate or sigmoid conidia (asexual reproductive structures). Taxonomically, they are mainly associated with the *Ascomycota*, and only a small percentage is affiliated with the *Basidiomycota*. In contrast, aero-aquatic hyphomycetes colonise submerged plant detritus in stagnant and slow flowing waters, such as shallow ponds and water-filled depressions. Taxonomically, most aero-aquatic fungi are classified as *Ascomycota*, although four aero-aquatic species have been classified as *Basidiomycota*, and one as *Oomycete* (Shearer et al., 2007). These fungi are adapted to habitats with fluctuating water levels subjected to periodic drying, low levels of dissolved oxygen, and elevated levels of sulfide. Therefore, they have buoyant conidia that are released at the water surface as water levels recede. Along with aquatic fungi, terrestrial fungi enter the aquatic realm as pioneer decomposers and endophytes of allochthonous plant debris. In the water, however, they are partially replaced by true aquatic hyphomycetes. After colonising the substrate and forming internal hyphal networks, the POM is macerated at least partly by the fungi themselves. This process is often accelerated by the feeding activity of macroinvertebrates, which find colonised leaves to be more palatable (compiled in Bärlocher, 1992; Gessner & Van Ryckegem, 2003). With the aid of an array of extracellular enzymes, aquatic fungi are able to degrade most of the polymeric substances in leaves (hemicelluloses, cellulose, starch, pectin and to some extent lignin; Krauss et al., 2011). Depending on leaf litter type and water chemistry, fungal leaf decomposition can extend over 1 to 6 months. The situation is slightly different for fungal decomposition of emergent macrophytes, because decomposition starts in standing shoots. Over 600 species of fungi have been recorded from the litter of *Phragmites australis* alone (Gessner & Van Ryckegem, 2003). Ninety four percent of these 600 species were members of *Ascomycota* and only 6%

belonged to *Basidiomycota*. The *Ascomycota*, in turn, comprised 30% aquatic hyphomycetes (with "naked" conidia) and 22% coelomycetous anamorphs (producing conidia inside a fruiting body). Thirty species isolated from the standing dead shoots of *Juncus effusus* (Kuehn & Suberkropp, 1998) were also mainly composed of aquatic hyphomycetes and coelomycetes. White rot *Basidiomycetes*, generally not considered being active in aquatic habitats, have also been isolated from standing dead aerial shoots in wetlands. In the case of small particles such as algae, pollen, seeds and zooplankton carcasses, decomposition is achieved by the much smaller *Oomycetes* and *Chytridiomycetes*, rather than the aquatic hyphomyctes. These organisms do not depend on macro-scale hyphal networks and are capable of very fast responses to changes in their environment. Their motile spores actively search for adequate substrates using chemotaxis. Once a suitable substrate has been reached, an appressorium is formed and the particle is invaded by tiny rhizoids tapping the internal nutrient reservoirs for production of new spores in a sporangium (either endo- or ectophytic; Sparrow, 1960; Sparrow, 1968 and references therein). Thereby, their whole life cycle can be completed in days. The short generation times and prolific spore production characterise these fungi as typical r-strategists.

Another polyphyletic group of aquatic fungi (with members of *Oomycetes, Zoopagomycotina* and *Basidiomycetes*) is specialised to hunt by using traps. These predatory fungi are often found on decomposing plant material or animal egesta. They use sticky traps, networks or slings to entrap their prey, usually amobae, rotifers, nematodes, liver flukes and small arthropods like mites. After the prey is caught, these fungi penetrate the prey's tissue and digest it from the inside. Generally, it is assumed that this behaviour supplies these fungi with additional nutrition when colonizing decomposing plant detritus. In soil, additional groups of endoparasitic fungi are found, e.g. on nematodes (Family *Hyalosporae* & *Entomophthoraceae*) which also destroy their prey from the inside (Karling, 1936; Drechsler, 1941; Peach, 1950, 1952, 1954; Sparrow, 1960; Swe et al., 2008).

An additional strategy of fungi with presumably long annual life cycles, is to grow inside living plants without affecting the host's viability. Yet, it is unclear whether the host plant benefits from these "endophytes" or if the relationship between plant and fungi is solely based on commensalism. Obviously, when the host plant enters senescence, all internal fungi have a great advantage over the secondary colonising fungi since the primary rule of "first come, first serve" is of major importance for growth and reproductive success.

An important group of endophytic fungi, which is clearly beneficial for the plant, consists of mycorrhiza forming symbionts present in the roots of several aquatic macrophytes. Many mycorrhiza forming symbionts belong to a phylum of the "lower fungi" called *Glomeromycota*. Certain orders of the *Glomeromycota* are obligate root symbionts characterized by a vesicular arbuscular mycorrhiza (VAM) that supply their hosts with nutrients. In return, the host plant provides the fungus with sugars rich in energy, amongst other things. VAM fungi were formerly believed to be purely terrestrial, but today it is known that they are particularly important in nutrient poor clear waters. For example, in oligotrophic waters, VAM fungi allow macrophytes to grow under nutrient limiting conditions by supplying the plant with solid-phase bound nutrients (Baar et al., 2011).

Freshwater algae, e.g. *Dunaliella* and the autotrophic protozoan *Euglena* can establish a mutual relationship with the fungus *Bispora* or the yeast *Cryptococcus*, respectively (Gimmler, 2001). Moreover, fungi belonging to the *Kickxellomycotina* are endosymbionts of invertebrates, especially of aquatic arthropods and - together with specialised protozoans - are summarised under the term trichomycetes (Lichtwardt, 2004; Hibbett et al., 2007).

A cornerstone of fungal lifestyle is parasitism. The life cycle of parasitic fungi is identical to that of saprophytic fungi with one exception: the host cells are still alive. Therefore, it is often impossible to separate opportunistic fungi colonizing senescent hosts from the true parasitic fungi reducing the fitness and in some cases even causing death to their previously healthy hosts. Prominent aquatic parasitic fungi belong to *Chytridiomycota* and *Oomycetes*. The host spectrum of these aquatic fungi is broad and covers every phylum including fungi itself (Sparrow, 1960; Van Donk & Bruning, 1992; Ibelings et al., 2004; Kagami et al., 2007). Encounters with fungi can be fatal to algae, particularly if their defence system is breached by the fungus. The ecological relevance of this negative interaction becomes evident when it is considered that suicide is a common defence mechanism in algae. If this controlled progress, called hypersensitivity, is initiated at the right moment during fungal infection, it results in the successful interruption of the fungal infection cycle, because the parasite's ability to reproduce via spore production is inhibited. Such "behaviour" allegorises a beneficial sanction since it protects the healthy algal population by reducing the abundance of the deadly parasite. However, if unsuccessful the parasite prevails and mass mortality of algae results. This can lead to shifts in the algal community composition.

In rare, but important cases, fungi cause severe damage to larger aquatic organisms. Some fungi, mainly but not exclusively *Oomycetes*, infect fishes or fish eggs (Noga, 1993; Chukanhom & Hatai, 2004) and thereby exert strong population pressure. This is of great importance for aquaculture since it necessitates antifungal treatments, but even in natural systems, fungi have the potential to severely harm the indigenous fish population. *Aphanomyces astaci* (*Oomycetales*) causing the crayfish plague has driven the European crayfish (family *Astacidea*) population to the edge of extinction (Reynolds, 1988). In contrast, *Coelomomyces* (*Blastocladiomycota*) effectively infecting several mosquito species (Sparrow, 1960) has been discussed as a biological control for malaria mosquitoes (Whisler et al., 1975). The most infamous fungal parasite is *Batrachochytrium dendrobatidis* (*Chytridiomycetales*), which contributes to worldwide extinction of several known and unknown amphibian species (Berger et al., 1998; Skerratt et al., 2007). Aquatic plants are also greatly affected by fungal parasites. A recently discovered plant parasite is *Pythium phragmites* (*Oomycetales*), obviously being an important causative agent of reed decline (Nechwatal et al., 2005).

Some human pathogens may also be found amongst the aquatic fungi. Common freshwater yeasts belonging to *Candida* and *Cryptococcus* are both potentially harmful to humans (e.g. *C. albicans* and *C. tropicalis*). These fungi are frequently found along bathing sites (Vogel et al., 2007). Several typical dermatophytes and keratinophylic fungi are transferred via water and can also occur in aquatic ecosystems (Ali-Shtayeh et al., 2002). *Chytridiomycetes* and "Microsporidia", however, are rarely pathogenic and only infect immune-deficient patients. Additionally, black yeasts are on occasion salt-tolerant and thus can cause problems when consuming salt preserved food (Butinar et al., 2005).

2.2 The life cycles of aquatic fungi

Life cycles of aquatic fungi cover a broad spectrum from very simple cell divisions to very complex cycles, crossing the terrestrial-water boundary. Starting with basal fungal lineages, Microsporidia are intracellular parasites with an extremely reduced genome (down to 2.3 Mbp, which is half the genome size of the common enterobacterium *Escherichia coli*). They are transmitted passively with non-motile spores, which have a size range of 1 - 50 µm. Endospores are chitinous and mature inside host-cells, where they are eventually released by an extrusion apparatus (summarised by Keeling & Fast, 2002).

Members of "Rozellida" have a similar life cycle as *Chytridiomycetes*, although diversity of *Rozella* has been so far only marginally described and is mainly based on the description of *Rozella allomyces*, a parasite living on *Allomyces sp.* The environmental clade LKM11 (van Hannen et al., 1999), the other member of Rozellida, is so far completely undescribed with scarce information about its habitat and ecology. It is known that these organisms probably have zoospores in the size range of 0.2 - 5 µm, which are most abundant above lake sediments (Mangot et al., 2009). They are also found under reduced oxygen and anoxic conditions, (Slapeta et al., 2005; Luo et al., 2005). Under anoxic conditions potential relatives of the *Neocallimastigomycota*, an obligate anaerobic symbiotic group of ruminants can be found, too (Lockhart et al., 2006; Mohamed & Martiny, 2011). However, their life cycle is similar to that of the *Chytridiomycetes*. Briefly, a zoospore is chemically attracted to its host or substrate and attaches to its surface. Then a cyst forms and tiny rhizoids (or a penetration tube) grow into the substrate to gather nutrients for (endobiotic or epibiotic) sporangium formation. Thereafter, masses of zoospores can be discharged (up to 70 000 for *Rhizophlyctis petersenii*). Sexual recombination can occur when two zoospores fuse together either in the free-swimming stage or on the host/substrate surface. Alternatively, resting spores might be formed in a prosporangium or in a zygote (Sparrow, 1960).

In principle, the life cycle of *Blastocladiomycota* is quite similar to that of the *Chytridiomycetes*, although they have hyphal growth in addition to zoospores. An important group within the *Blastocladiomycota* is comprised of members of the *Coelomomycetes*, which are often species-specific for their mosquito host. Their complete life cycle, originally described by Whisler et al. (1975), is given in figure 1.

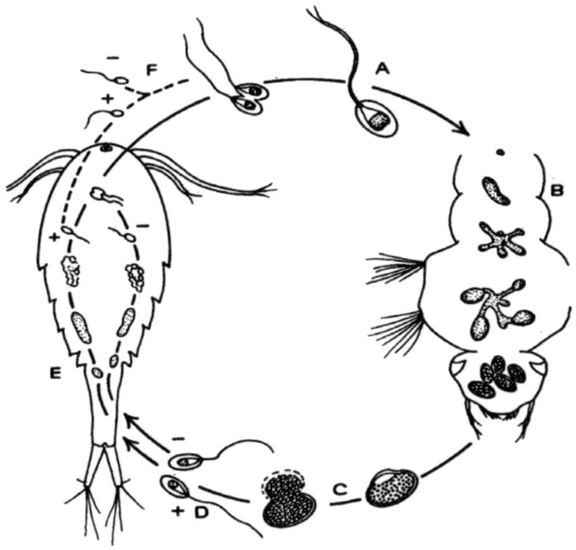

Fig. 1. Life cycle of *Coelomomyces psorophorae*. Zygote (A) infects larva of *Culiseta inornata* (B) leading to development of hyphal bodies, mycelium and, ultimately, thick-walled resistant sporangia. Under appropriate conditions these sporangia (C) release zoospores of opposite mating type (D) which infect the alternate host, *Cyclops vernalis* (E). Each zoospore develops into a thallus and, eventually, gametangia. Gametes of opposite mating type (F) fuse either in or outside of the copepod to form the mosquito-infecting zygote (Whisler et al., 1975, with permission).

Mosquitoes, like other arthropods, are potential hosts for symbiotic trichomycetes (*Harpellales*) in many lentic and lotic freshwater habitats (Lichtwardt, 2004; Koontz, 2006; Strongman & White, 2008). These gut fungi disperse via trichospores through the water column.

Parasitic members of *Entomophthorales* also use arthropods as hosts. In insects with aquatic and terrestrial life stages, these parasites are well adapted to both habitats by developing asexual conidia for dispersal in air and typical tetraradiate conidia for dispersal in water. A detailed description has been given by Hywel-Jones & Webster (1986) and is depicted in figure 2. The idea of a second host is especially inspiring, since it is known that *Entomophthorales* are also parasites of planktonic desmids (green algae; Sparrow, 1960).

Leaf decomposition is associated with high discharges of aquatic conidia of diverse shapes and sizes (e.g. Ingold, 1975), although the conidia of aquatic hyphomycetes are typically tetraradiate. Aquatic hyphomycetes reproduce asexually (figure 3), although in ca. 10 percent of all described species, teleomorphs have been found, e.g. on twigs at river margins (Webster, 1992).

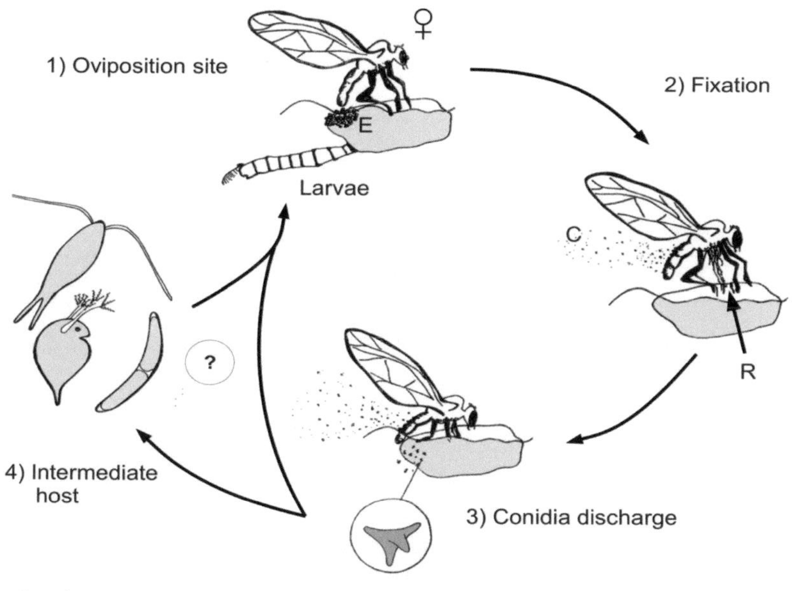

Fig. 2. Life cycle of *Erynia conica* on *Simulium sp.* (1) After oviposition (E), only infected females of *Simulium* stay at the river bank and become less active. (2) After 2-12 hrs, rhizoids (R) and pseudocystidia emerge from small swellings at the abdomen. The rhizoids anchor the animal to the ground and inhibit any further locomotion. After 15 hrs, conidiophores and primary aerial conidia emerge (C) and release. (3) After 24 hrs, when the ventral part of the fly is wetted or submerged, primary aquatic conidia are produced. Both types of conidia can be produced simultaneously in a single fly for up to 96 hrs after arrival at the oviposition site. Globose zygospores, however, stay embedded in the cadaver (teleomorphic form). (4) Aerial conidia can transform into secondary stellate aquatic conidia with typical tetraradiate symmetry upon submersion. Yet, it is not clear whether secondary hosts (zooplankton or desmids) are required for *Erynia* development because aquatic conidia were never observed to cause infection of *Simulium* (Webster 1992). Illustration drafted after descriptions of Hywel-Jones & Webster (1986).

Other filamentous fungi, such as endophytes or VAM fungi have a still more or less unknown life cycle. Though, it is similar to *Mucor* species in sediments, an interesting phenomenon occurs in this genus, which may be relevant to other fungi with yeast-like life stages. While *Mucor* usually grows in hyphal networks when oxygen is available, under certain circumstances (especially when growing anaerobically, at elevated pCO_2), growth changes to a yeast-like morphology (Orlowski, 1991). This dimorphism is known of several yeast-like species such as *Aureobasidium pullulans* or *Candida* sp. and triggers a fast adaptation to changing environmental conditions. Yeasts and yeast-like organisms have often been isolated from freshwaters, a habitat varying greatly in time and space. For example, waves, chemical gradients and currents may be highly variable over time and hence, the capability to adapt rapidly to such changes is of great advantage.

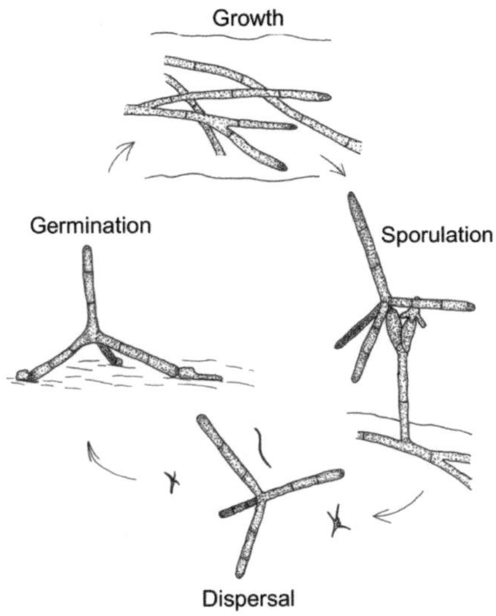

Fig. 3. Asexual life cycle of aquatic hyphomycetes. Figure reproduced from Gulis et al. (2009) with permission.

2.3 Differences in fungal morphology and ecology

Fungi can grow into the largest known organism on earth if the substrate is suitable and the environmental conditions favourable. In most cases, however, fungi remain invisible to the naked eye. Therefore, their global importance is seldom recognised even by scientists. Fungi literally tend to grow to the limit of their natural potential; the size of their cellular network is not genetically encoded, but defined by substrate and other environmental parameters. If, in the very unlikely event that a scientist attempted to prove that a whale could survive in freshwater, the whale's inevitable death would be rapidly followed by colonization of the gigantic carcass by coprophilus fungal species (as observed for various fish carcasses; Fenoglio et al., 2009). These fungi would flourish throughout the decomposition of the carcass and a single species could potentially establish an extensive network, exploiting a substantial portion of the whale's biomass. Most likely, the whale's carcass would harbour a very diverse fungal flora of several phyla and hundreds of species, supporting a whole benthic food web with nutrients and energy for years. In contrast, tiny substrates such as single celled diatoms of a few µm in diameter, only harbour a single fungal species with an evanescent low biomass. However, taking the size of a large water body and the high annual abundance of diatoms into account, the fungal biomass associated with these algae could exceed those growing within the whale carcass. Thus, substrate size is not the sole factor determining the importance of aquatic fungi in their natural habitat.

Aside from their dependence on substrate quality and quantity, fungi themselves harbour different morphologies, life stages and strategies. This is mainly due to the fact that aquatic fungi are derived from many fungal phyla comprising different cellular "blueprints" and life stages (see above). The diameter of a single fungal cell can roughly vary within an order of magnitude and there are numerous different spore morphologies extending from 1 µm small flagellated zoospores to several 100 µm large air-trapping conidia. An overview of fungal dimensions in aquatic systems is shown in figure 4.

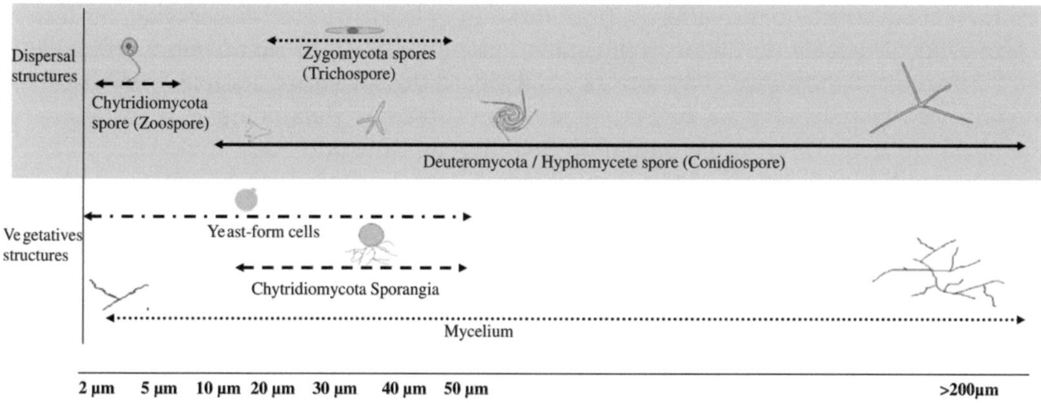

Fig. 4. Dimensions of vegetative growth forms and spores of aquatic fungi (republished from Jobard et al., 2010, with permission).

2.4 Diversity in large-scale aquatic habitats

Many aquatic fungi are saprophytes, which consume dead organic matter (Dodds, 2002), but aquatic fungi may also be parasites or symbionts. In aquatic systems, the fungal community structure greatly differs between substrates (Shearer and Webster, 1985; Findlay et al., 1990; Bärlocher & Graça, 2002; Graça et al., 2002; Mille-Lindblom et al., 2006) and with the physico-chemical properties of the respective habitats, such as flow (Pattee & Chergui, 1995; Baldy et al., 2002), dissolved oxygen concentration (Field & Webster, 1983; Medeiros et al., 2009), nutrient concentrations (Gulis & Suberkropp, 2004; Rankovic, 2005), salinity (Hyde & Lee, 1995; Roache et al., 2006), temperature (Bärlocher et al., 2008) and depth (Wurzbacher et al., 2010). Therefore, fungal communities potentially differ between streams, shallow lakes and wetlands, deep lakes, and other habitats such as salt lakes and estuaries.

2.4.1 Fungal diversity in streams

Upland stream habitats are characterised by a pool and riffle structure with relatively swift flow and high levels of dissolved oxygen. These streams are narrow and tend to be lined by overhanging riparian vegetation. These characteristics create an ideal habitat for aquatic hyphomycetes. Nikolcheva & Bärlocher (2004) have investigated the structure of fungal communities on leaves submerged in an upland stream by using molecular methods. The authors were able to resolve the diversity within the *Ascomycota, Basidiomycota, Chytridiomycota, Zygomycetes* and *Oomycetes* and found, that the leaf decomposer community was dominated by *Ascomycota*, whereas *Basidiomycota* comprised a small but consistent fraction of aquatic fungi. *Chytridiomycota* represented a substantial proportion of the fungal community in winter, while *Oomycetes* were only present in summer. *Glomeromycota*, however, were of minor importance in the stream environment. Species common in an Australian upland stream included *Tetrachaetum elegans, Lunulospora cymbiformis, Flagellospora penicillioides* and *Alatospora acuminata* (Thomas et al., 1992). These species of aquatic hyphomycetes have not yet been associated with a teleomorph, but are likely affiliated with the *Ascomycota* since they lack morhpological features characteristic of the *Basidiomycota* (Nawawi 1985).

In lowland rivers, flow remains substantial but water quality and the source of primary production are substantially different from those in upland streams. Wider channels lead to a proportional reduction in litter from riparian plants, and production from phytoplankton is of increased significance (Vannote et al., 1980). Nutrient concentrations and dissolved organic carbon may also be higher, leading to lower or fluctuating concentrations of dissolved oxygen. Thus, while aquatic hyphomycetes still dominate submerged litter in these streams (Baldy et al., 2002), fungal community composition differs from upland streams (Shearer & Webster, 1985) and biomass accumulation may be limited by competition with other microorganisms, substrate burial and lower oxygen availability (Bärlocher, 1992; Medeiros et al., 2009).

2.4.2 Fungal diversity in shallow lakes and wetlands

The dominant fungi colonising submerged plant litter in shallow, stagnant habitats common in wetlands and shallow lakes are the aero-aquatic hyphomycetes (Glen-Bott, 1951; Shearer et al., 2007). On occasion, aero-aquatic hyphomycetes may be found in streams and aquatic hyphomycetes in wetlands (Bärlocher & Kendrick, 1974; Fisher et al., 1983; Bärlocher, 1992), but aero-aquatic hyphomycetes are capable of out-competing aquatic hyphomycetes when colonising substrates in water with lower oxygen or higher nutrient concentrations (Voronin, 1997). *Oomycetes* and terrestrial fungi can also be found in wetlands (Bärlocher, 1992).

Fungal genera commonly found in wetlands include *Alternaria, Cylindrocarpon, Cladosporium, Penicillium, Fusarium, Trichoderma* and aquatic hyphomycetes (*Alatospora, Tetracladium, Helicodendron, Helicoon*; Kaushik & Hynes, 1971; Kjoller & Struwe, 1980; Ford, 1993). Aquatic lichens (a symbiotic partnership between a fungus and an alga) are potentially present in the littoral zone of wetlands, lakes and streams (McCarthy & Johnson, 1997), in particular in temperate or boreal regions (Hawksworth, 2000). There are ca. 200 species of lichenised fungi known from freshwater systems (Hawksworth, 2000). The main orders of *Oomycetes* found in aquatic environments are the *Leptomitales, Saprolegniales* and *Peronosporales*. Their requirement for dissolved oxygen varies widely among species, and many are intolerant of high salinity (Dick & Newby, 1961; Dick 1962; 1963; 1969; 1972).

2.4.3 Fungal diversity in deep lakes and reservoirs

In deep lakes and reservoirs, the abundance (as colony forming units; CFU) and diversity of aquatic fungi is greatest in both the littoral and profundal zone (Rankovic, 2005).

Considering filamentous and higher fungi, the pelagic zone only supports a few specialised fungal species, but seems to be mainly used as a medium for propagules dispersal (Wurzbacher et al., 2010). Fungi from the littoral zone, in turn, are saprobes, parasites, predators, endosymbionts or occasionally lichens. These organisms colonise substrates ranging from submerged plants and litter to the carapaces of dead micro-crustaceans (Czeczuga et al., 2002; 2004; 2007).

In the pelagic zone, fungi consist mainly of species that live parasitically on phytoplankton, zooplankton and fish. Taxonomically, the fungal community consists of *Ascomycete* and *Basidiomycete* yeasts and "zoosporic fungi" (*Chytridiomycota* and *Oomycetes*; Rankovic 2005; Lefevre et al., 2007). It has been suggested that "zoosporic fungi" and their propagules are important for pelagic food web dynamics since they are important parasites of freshwater algae and thus may be important in controlling phytoplankton blooms associated with diatoms (Kagami et al., 2004) and cyanobacteria (*Microcystis* spp.; Chen et al., 2010).

The profundal zone and lake sediments, however, mainly serve as a propagule bank, where resting spores are stored. Therefore, both aquatic and terrestrial species are frequently isolated from deep lake sediments. Moreover, it has been suggested that yeasts in lake sediments are derived from terrestrial plant litter (Kurtzman & Fell, 2004), and fungal CFU associated with the *Mucoromycotina* (*Mucor* and *Rhizopus* sp.) isolated from various Serbian reservoirs may be also of terrestrial origin (Rankovic, 2005). However, there are a few species of yeasts, *Chytridiomycetes* and *Oomycetes* that are able to grow vegetatively in lake sediments (e.g. Ali & Abdel-Raheem, 2003).

2.4.4 Other aquatic habitats

Fungi may also be found in aquatic habitats with harsh environmental conditions, such as sulfidic springs (Luo et al., 2005), acidic peat bogs and lakes (Thormann, 2006; Voronin, 2010) and volcanic lakes (Sabetta, et al., 2000). When studying fungal diversity in sediments of an estuary, Mohamed & Martiny (2011) found that community composition (at division level) did not differ substantially between fresh, brackish and seawater. However, the proportion of *Chytridiomycetes* and unknown species from basal lineages increased with salinity, and species diversity was at a maximum in the brackish zone. Although several studies have examined the fungi that can be isolated from saline lakes (Butinars et al., 2005; Zalar et al., 2005; Takishita et al., 2007) and mangroves (Suryanarayanand & Kumaresan, 2000; Kumaresan & Suryanarayanan, 2001; Ananda & Sridhar, 2002), fungal biodiversity in these systems requires further investigation.

3. Hidden biodiversity of aquatic fungi

Actual fungal biodiversity suggests that the most species-rich regions of the globe are situated in temperate rather than in tropical regions. Given that many fungal species are host or substrate specific, and that biodiversity of plants and animals is highest in tropical regions, this notion is counter-intuitive. It is very likely that sampling efforts for fungal biodiversity have been largely restricted to temperate regions, where most fungal taxonomists are situated (Shearer et al., 2007). Alternatively, seasons, cooler temperatures and moist conditions may be more amenable to fungal evolution and niche differentiation. From the above mentioned discrepancies and gaps of knowledge in diversity of aquatic fungi, it appears timely to commence co-ordinated world-wide sampling programs using consistent methodology to evaluate fungal biodiversity in various aquatic systems around the globe.

Gessner & Van Ryckegem (2003) estimated the total number of aquatic fungal species to a maximum of 20 000 different species based on the assumption that only 5% have been described so far. Whereas only a few newly described fungal species have been added in recent years, an increasing number of genetically distant environmental DNA sequences have been found (Hibbett et al., 2011). For example, biodiversity of basal fungal lineages, which bear numerous aquatic species, seems to be much higher than expected. In addition, biodiversity of these basal phyla is elevated in aquatic sediments when compared to terrestrial soil (Mohamed & Martiny, 2011). The highest estimates of global fungal diversity reach up to 5 million species (Blackwell, 2011). The above mentioned "lower fungi" belonging to *Eumycota*, excluding congruously *Oomycetes* and *Thaustrochytrids*, are listed in table 1.

Currently, the species ratio of terrestrial fungi to land plants is approximately 10.6:1. Most likely, this ratio will increase in the future since mycologists have largely increased their efforts to find new fungal species. Freshwater ecosystems can be considered as rather unexplored fungal habitats whereby the few, presently available molecular studies point to a high species diversity. Blackwell (2011) gives helpful suggestions on where to search for these hidden species and highlights insects and other animals as potential fungal habitats. For example, in a single pilot-study in 2005, Suh et al. have isolated 196 new yeast species from guts of mushroom eating beetles and thereby increased the total number of worldwide described yeast species by more than 30%. Next to fungi residing in arthropod guts, endophytes in freshwater ecosystems are another budding source of high fungal biodiversity. For example, when applying molecular tools Neubert et al. (2006) found >600 fungal operational taxonomic units (a measurement of environmental DNA sequence diversity) in single plants (*Phragmites australis*) of a single lake (Lake Constance). This remarkably high diversity of endo- and ectophytic fungi points to a so far largely hidden fungal diversity associated with higher aquatic organisms.

As already mentioned, fungal parasites in pelagic systems can greatly add to global fungal diversity, which should by far exceed even that of saprophytic fungi. This is due to the following features of parasitic fungi: (1) the presence of a specialised attack-defence co-evolution based on the red queen hypothesis and (2) a high specificity to host species of various eukaryotes. A precise estimation of their diversity is difficult since parasites can be either host strain specific (De Bruin et al., 2008) or cover a wider spectrum of hosts such as *B. dendrobatidis*. In addition to parasitic fungi, many opportunistic saprophytic fungi are host-specific (Sparrow, 1960). Nevertheless, variability in host and substrate specificity is high among aquatic fungi and it is difficult to generalise.

3.1 Hidden diversity

Several aquatic microhabitats - well studied for bacteria - have not yet been well incorporated in biodiversity studies on fungi (Wurzbacher et al., 2010). These microhabitats include biofilms (periphyton, benthic algae), floating algae, and submerged/floating macrophytes, which contribute substantially to lake primary productivity. Detrital aggregates (lake and riverine snow) are also known hotspots of bacterial activity in the pelagic zone of lakes and large rivers, but fungal contribution to these aggregates has not been evaluated. Although remineralisation processes have been well studied for bacteria, fungi have been largely excluded from these studies. The riparian/littoral zone of aquatic

Phyla	Representatives	Known Hosts	Known Substrates	Remarks
Microsporidia*	*Glugea* *Telohania* *Pleistophora*	animals (incl. protists and zooplankton)		obligate endoparasites esp. of fishes and arthropods
Rozellida*	*Rozella* *LKM11*	fungi		obligate mycoparasites, common at anoxic sites
Chytridiomycota	*Rhizophydium* *Endochytrium* *Batrachochytrium*	mycoplankton, phytoplankton, zooplankton, animals, macrophytes	phytoplankton, zooplankton, animals, plant debris, seeds, pollen, fruits, chitin, keratin, cellulose, twigs	obligate and opportunistic endoparasites & ectoparasites; saprophytes
Neocallimastigo-mycota	*Piromyces*	ruminant	cellulose	obligate anaerobe symbionts, potentially in sediments
Blastocladiomycota	*Coelomomyces* *Catenomyces*	insect larvae, eggs of liver fluke, nematodes, aquatic fungi	fruits, twigs, animal debris	endoparasites of malaria mosquito Anopheles
Glomeroycota	*Glomus*	roots of aquatic macrophytes		obligate VAM building symbionts
Subphyla of Glomeromycota				
Mucoromycotina	*Mucor*		debris	fermentative metabolism possible
Entomophthoro-mycotina	*Ancylistes* *Macrobiotophthora* *Erynia*	insects, desmids, rotifers, nematodes	vegetable debris, excrements of amphibians	endoparasites & saprophytes
Zoopagomycotina	*Zoophagus*	amoebae, rotifers, nematodes, fungi (e.g. Mucor)		endoparasites & ectoparasites or predatory fungi
Kickxellomycotina	*Harpellales*	arthropods (e.g. Chironomidae)		coprophilous species and trichomycetes (symbionts of aquatic arthropods)

Table 1. Lower fungal phyla of *Eumycota* in accordance to Hibbett et al. (2007) and Lara et al. (2010). Detailed information was obtained mainly from Sparrow (1960), Hywel-Jones & Webster (1986), Ebert (1995), Keeling & Fast (2002), Lichtwardt (2004) and Benny (2009). Asterisks mark not yet confirmed phyla.

systems is an ideal habitat for fungi and hence should be the focus of future fungal biodiversity research. Littoral food webs are very complex and a wealth of invertebrates, vertebrates and progeny suggest close interaction with a diverse community of fungi including parasitic, symbiotic and endophytic fungi. Littoral zones are highly structured by large emerged macrophytes, floating macrophytes and submerged macrophytes, which can form a dense meadow and are suitable habitats for fungal proliferation. The high diversity of algae, pelagic and benthic species, and their function as an accumulation zone for dissolved nutrients and terrigenous detritus from the catchment, renders the littoral zone an ideal fungal habitat. Littoral sediments are often well aerated by the roots of emergent and submerged macrophytes and form microenvironments with strong physico-chemical gradients frequently altered by water movement and bioturbation by invertebrates such as mussels or chironomids. Therefore, it is not surprising that Willoughby (1961) found a high diversity and activity of fungi in soils on lake margins. Monchy et al. (2011) observed a high biodiversity in littoral water, and Mohamed & Martiny (2011) found a positive relation of fungal biodiversity to abundance of macrophytes. Nevertheless, fungi are often difficult to recognize due to methodological and morphological considerations: a single observed hypha of one species is visually indistinguishable from a thousand other fungal species. Fungi are highly variable in size and many tend to grow hidden inside their substrates, all factors which make them difficult to study and easy to overlook. The recent and on-going development of modern molecular tools, however, enables ecologists to better resolve biodiversity and ecology of aquatic fungi (e.g. Neubert et al., 2006, Baar et al., 2011). Still, most aquatic plants are only superficially examined for fungi (Orlowska et al., 2004) and many unexplored aquatic microhabitats potentially serve as niches for specialists. Examples include a mutualistic relationship of a predatory *Oomycete* living inside a mussel and protecting the mussel from parasite infections, e.g. nematodes (DeVay, 1956). Another predatory fungus uses the surface structure of macroalgae and grows epiphytically on *Characea* meadows (see figure 5). The most impressive example for interspecies relationships with high impact for general fungal biodiversity considerations stems from members of *Arthropoda*. Theoretically, one single animal can simultaneously provide microhabitats for several aquatic fungi (not including saprophytic or coprophagous fungi): host muscle cells as habitat for intracellular parasites of Microsporidia (Ebert, 1995; Messick et al., 2004); in the host tissue yeasts can be found (Ebert et al., 2004); and in the haemocoel occasionally detrimental *Chytridiomycetes* occur (Johnson et al., 2006). Moreover, an obligate endoparasite of *Entomophthorales* (Sparrow, 1960) and likely a represantative of *Coelomomycetes* (Whisler et al., 1975) can be found and the animal's gut hosts yeasts and symbiotic species of *Harpellales* (trichomycetes; Strongman & White, 2008). Lastly, obligate ectoparasites belonging to an order of higher fungi called *Laboulbeniales* (*Ascomycota*) grow well on the chitinous integument. These fungi are not really aquatic, but more or less specific for arthropods, independent of habitat and are visible on their exoskeletons (Weir, 2004). Interestingly, almost all parasites and symbionts (with the exception of yeasts) are more or less host specific and *Laboulbeniales* are even sex-host specific. If we assume host specificity, a ratio of 6:1 between fungi and their arthropod host species, then a tremendous, yet hidden, fungal biodiversity is implied.

In aquatic microhabitats oxygen conditions can be extremely variable and hence it is important for fungi to be capable of survival or even growth under such conditions. Anoxic conditions are prevalent in aquatic sediments, in animal guts, in biofilms, on decomposing particles or, at a larger scale, in di- to polymictic lakes with seasonally anoxic water masses. Several fungi can withstand anoxic conditions or even grow fermentatively (Held et al., 1969). For example, archaic anoxic environments seem to be predominant habitats for lower fungi and yeasts (Stock et al., 2009; Mohamed & Martiny, 2011) but are awaiting mycologists to explore them.

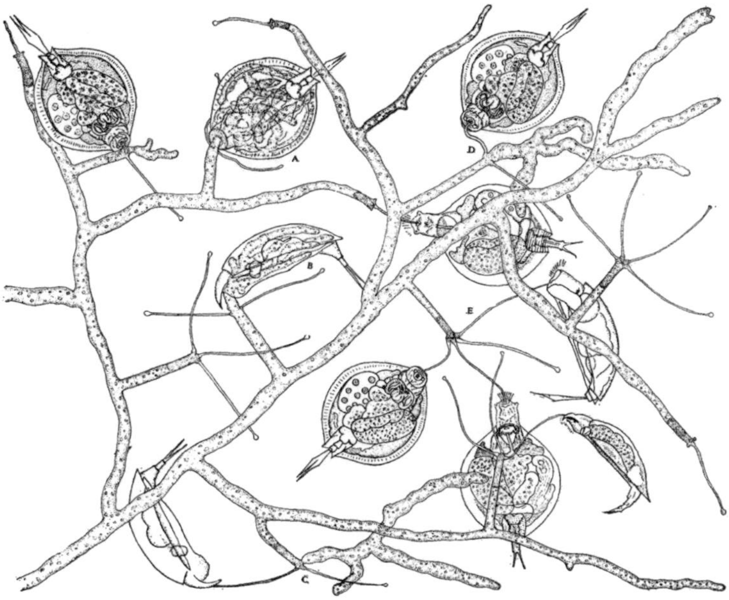

Fig. 5. *Zoophagus tentaclum* captures rotifers and grows epiphytic on *Nitella* (Figure from Karling (1936) with permission).

4. Importance of fungi for aquatic food webs

The importance of fungi as secondary producers of biomass has been well described for headwater streams with leaf litter (Suberkropp, 1992) and for reed stands in littoral zones of lakes and in marshlands. The foregut content of 109 different aquatic insects collected on submerged wood showed that in 66% of all studied insect species fungi were part of their diet (Pereira et al., 1982) and many conidia of aquatic fungi were found in faeces of fish (Sridhar & Sudheep, 2011). Furthermore, it has been shown that food web manipulations greatly alter the fungal biomass in lakes (Mancinelli et al., 2002). This suggests that saprophytic fungi transfer organic matter directly to the higher trophic levels of aquatic food webs. It is therefore likely that environmental change can have severe consequences for overall food web topology, and hence nutrient and energy cycling.

In addition, fungi can be important parasites of primary producers, e.g. phytoplankton, which fuel the aquatic food web with organic matter and energy. Lysis of aquatic organisms by fungal and protozoan parasites increases organic matter and energy cycling. These processes are often solely attributed to *Bacteria* and *Archaea*, however, aquatic fungi actively contribute as mineralisers and parasites.

4.1 Mineralisation

Aquatic systems typically lack effective herbivores meaning that most of the biomass of aquatic macrophytes and riparian plant litter enters the detrital organic matter pool and is subsequently metabolised and transformed into microbial biomass, making it available for higher trophic levels. Generally, a major fraction of carbon will be respired (as CO_2) during degradation, whereas nutrients such as phosphorus and nitrogen are efficiently recycled. Microbial mineralisation of plant litter supports a complex food-web including all kinds of microbes (*Archaea*, *Bacteria*, fungi, protozoans) and invertebrates (nematodes, trematodes, gammarids, insects, snails). As a consequence plant litter even supplies top predators such as crayfishs, amphibians, birds, fishes and bats with organic matter and energy via the microbial food web. The main basis of the microbial food web consists of fungi and bacteria growing in and on the plant debris. Microorganisms, in particular fungi, possess enzymes capable of degrading even highly polymeric substances, and filamentous fungi are capable of degrading the plant material from the inside, driving the break down of high molecular weight polymers to smaller molecules of medium molecular weight (Fischer et al., 2006). These small fragments and oligomers, e.g. sugar residues, can be readily utilized by bacteria and the so called "sugar fungi" (a sloppy term for the lower fungal phyla consisting of *Chytridiomycota, Blastocladiomycota, Mucoromycotina, Zoopagomycotina, Oomycetes*). Freshwater hyphomycetes of temperate waters are usually well adapted to lower temperatures prevailing during leaf litter input and senescence of aquatic macrophytes. During the cold season (autumn, winter and spring), filamentous fungi account for over 90 to 99% of total microbial biomass in emergent macrophytes and riparian leaf litter and their secondary production is one to two orders of magnitude higher than the bacterial production (Gulis et al., 2009). Therefore, fungal decomposition of this important POM pool seems to be of primary importance during several months in the cold season. Surprisingly, decomposition of submerged aquatic plants has not been well examined, although it is likely that filamentous fungi are of secondary importance (Mille-Lindblom et al., 2006). Thereby, other fungal taxa potentially substitute the filamentous forms, but may vary in time. For example, lower fungi are able to degrade small plant debris and particles. Foremost, *Chytridiomycetes* are suitable candidates since they are able to degrade a wide range of substrates (Sparrow, 1960). However, their saprophytic capabilities and related carbon turnover rates have not been quantified, yet. Some *Chytridiomycetes* can utilise a range of organic polymers such as glucose, starch, sucrose, cellobiose, chitin and cellulose (Gleason et al., 2011; Reisert & Fuller, 1962) whereas others possess incomplete enzymatic degradation pathways suggesting a possible complementation through other microbes. Many active *Chytridiomycetes* often occur sporadically in flooded mud of the riparian zone and submerged sediments and form a very different *Chytridiomycetes* flora compared to that of soils of the catchment area (Willoughby, 1961). This suggests that aquatic *Chytridiomycetes* include indigenous species well adapted to the prevailing environmental factors.

4.1.1 Functional redundancy of saprobes

Lawton and Brown (1993) introduced the concept of functional redundancy as a means of exploring the importance of biodiversity for ecosystem functioning. Functional redundancy is the idea that multiple species can perform the same function within an ecosystem, therefore, a reduction in number of species will not affect ecosystem functioning until all species performing a particular function are lost. However, functional redundancy is at odds with the concept of resource partitioning (Schoener, 1974), which proposes that competition between species drives them to specialise in exploiting discrete resources or ecological niches. Recent research has shown that biodiversity influences aquatic ecosystem processes such as productivity (Smith, 2007; Gustafsson & Boström, 2011) and heterotrophy (Cardinale et al., 2002), but studies of aquatic fungi show that diversity influences neither productivity (Baldy et al., 2002) nor decomposition rates (Bärlocher & Graça, 2002; Dang et al., 2005). It is likely that both functional redundancy and resource partitioning operate within aquatic ecosystems, but on different spatial and temporal scales, and with impacts at the level of individuals, populations and communities (Loreau, 2004).

In many aquatic ecosystems, saprobic fungi are important decomposer organisms. While some species show a preference for substrates derived from a particular plant species or plant tissue (i.e. leaves or wood), many fungal species are generalist saprobes (Gulis, 2001). This suggests that a large degree of functional redundancy exists among saprobic aquatic fungi at spatial scales ranging from submerged substrates to the whole ecosystem.

Aquatic fungi are microscopic organisms that interact with other species and "individuals" on a microscopic scale via enzymes and biochemical defences. Therefore, resource partitioning by fungi can be expected to occur at the molecular scale. This idea is supported by the well documented temporal succession (Garrett, 1951) that occurs as fungi colonise a submerged leaf, and the temporal partitioning of the resource that is implied. In order to exploit a substrate, fungi secrete extra-cellular enzymes that attack and degrade its chemical constituents. As separate and distinct enzymes or enzyme systems are required for the breakdown of starches, cellulose, hemicellulose, pectin, proteins, lipids and lignin, the fungal species, armed with the suite of enzymes able to efficiently degrade the most labile leaf components, become the initial colonisers. When labile resources are depleted, species able to efficiently degrade the remaining resources become dominant, and so on (Chamier, 1985). Complex plant components such as lignin may be degraded by a range of enzymes secreted by a number of fungal species (Evans et al., 1994), and this is an example of resource partitioning at the sub-molecular level (lignin moieties). It is thus likely that the biodiversity of aquatic fungi has inherent functional redundancy at larger spatial scales, but at the molecular scale, and through time there is inherent functional complementarity, competitive exclusion and resource partitioning.

4.1.2 Fungi as producers of organic matter

There are a number of studies from the past few decades that have established a strong role of fungi as important basal resources in aquatic ecosystems (Bärlocher, 1985; Albariño et al., 2008; Chung & Suberkropp, 2009), most notably in streams (Reid et al., 2008; Hladyz et al., 2009). For example, fungal biomass has been shown to be an important food source for aquatic invertebrates such as snails (McMahon et al., 1974; Newell & Bärlocher, 1993) and insect larvae (Bärlocher, 1981; 1982; 1985). Thereby, the fungal biomass is either removed from the leaf surface, or the leaf itself is consumed.

A synthesis of research from aquatic systems suggests that the functional role of aquatic heterotrophic fungi in moderating the food value of plant detritus may be more important than their role as organic matter producers (e.g. Thorp & Delong, 2002). Litter produced outside a water body may enter the water directly, as a result of abscission from riparian plants overhanging the water body, or may undergo a period of terrestrial aging before entering the water. These two pathways result in differences in litter chemistry (Baldwin, 1999) that influence their importance to the aquatic food-chain (Watkins et al., 2010). In general, fresh plant material has a higher protein content (lower C:N) and a higher proportion of readily available nutrients than aged material (Williams, 2010; Kerr et al., in prep.). However, fresh material also contains inhibitory substances such as tannins, polyphenols and aromatic oils, which function to deter microbial attack and herbivory in the living plant (Campbell & Fuchshuber, 1995; Canhoto et al., 2002; Graça et al., 2002). In contrast to fresh material, aged organic matter has a higher C:N (low C:N is correlated with higher nutritional value; Boyd & Goodyear, 1971; Hladyz et al., 2009), but a lower content of inhibitory substances.

When fungi colonise submerged plant material that has undergone terrestrial aging, the C:N ratio of the detritus declines (Bärlocher, 1985) as fungi utilise nitrogen from the water column to synthesise proteins for their own growth (Stelzer et al., 2003). They also produce lipids essential for growth (Chung & Suberkropp, 2009) and reproduction (Cargill et al., 1985) in some aquatic invertebrates. In addition to this, the activity of fungal enzymes releases sugars from structural carbohydrates (Chamier, 1985), breaks down lignins reducing leaf toughness (Leonowicz et al., 2001; Medeiros et al., 2009) and neutralises inhibitory substances such as tannins (Mahadevan & Muthukumar, 1980; Abdullah & Taj-Aldeen, 1989). Moreover, where plant detritus undergoes a period of terrestrial or standing dead aging, a more diverse consortium of fungi is able to actively degrade refractory plant components such as lignin (Bergbauer et al., 1992; Abdel-Raheem & Ali, 2004; Schulz & Thormann, 2005). Consequently, the sequential activity of terrestrial and aquatic fungi on plant detritus potentially leads to improved food value for members of the aquatic biota extending from other microorganisms to fish (Williams, 2010).

As aquatic fungi serve as a basal resource in many aquatic ecosystems, it is important to consider factors influencing their productivity. Fungal biomass increases with increasing concentration of nitrogen and phosphorus in the water column (Sridhar & Bärlocher, 1997) and decreases with lower dissolved oxygen concentrations (Medeiros et al., 2009). Thus the productivity of fungi and their importance as organic matter producers vary with climate and the availability of nutrients and organic substrates (Ferreira & Chauvet, 2010), and in some instances fungal production will not be a significant resource for the aquatic community (Bunn & Boon, 1993; Hadwen et al., 2010). Additionally, productivity will also be limited by ecological interactions such as competition (Mille-Lindblom et al., 2006) and mycotrophy (Newell & Bärlocher, 1993; Kagami et al., 2004; Lepere et al., 2007), and physical changes such as burial (Janssen & Walker, 1999; Cornut et al., 2010).

4.2 Parasites

Fungal pelagic parasites are often host-specific, but their evolution didn't stop at the species level and several fungal species developed dependencies on (1) certain algal cell types, e.g. heterocysts and akinetes (Sparrow, 1960); (2) certain cell entry sites of the host cell (Powell, 1993); and (3) certain algal strains (De Bruin et al., 2008). The latter study targets a prominent freshwater diatom called *Asterionella formosa* because it often harbours a obligatory, host-specific, very virulent fungal parasite called *Zygorhizidium planktonicum*. Infection by this fungus is often epidemic and can rapidly reach up to 90% of the host population with fatal consequences for the host. Interestingly, the authors could show that a genetically diverse host population maintains an evolutionary equilibrium between the parasite and the host population. A diminished host diversity, which is promoted, e.g. by disturbance or algal monoculture, would allow a rapid adaptive evolution of the parasite with a serious aftermath.

The occurrence of hyperparasites is really amazing since such fungi represent parasites of the algae's fungal parasites. Examples of these hyperparasites of fungal parasites on *Cyclotella, Cosmarium and Asterionella* are given by Canter-Lund & Lund (1995). Fungal hyperparasites belong to the genus *Rozella*. This genus was formerly assigned to the *Chytridiomycetes* and is now proposed to be part of the unique fungal phylum of the Rozellida (Lara et al., 2010). All members of *Rozella* are considered to be parasites of lower fungi (*Chytridiomycetes, Blastocladiomycetes, Oomycetes*). It is intriguing to think about the minimum population size of parasitic/saprobic fungi needed to sustain an obligate mycoparasitic fungal population. This suggests that a very common and stable mycoplankton population must exist in aquatic systems. Therefore, parasitism can be regarded as a key driver of food-web stability and POM transfer.

4.3 Stabilisation of ecosystems

As shown above, fungi possess multiple ecological functions in aquatic food webs. They often have a dual role which is on the one hand consumption of organic matter and on the other hand transmission of energy and genetic information (Amundsen et al., 2009; Rasconi et al., 2011). Parasitic fungi, for example, can selectively alter food web topology and thereby increase interactions and nestedness of ecosystems. Parasites including fungi, for example, interlink organisms of all trophic levels (resulting in twice as many links as without parasites) and thus increase food chain length and number of trophic levels. Amundsen et al. (2009) show that 50% of all parasites are trophically transmitted and thereby exploit different trophic levels and largely increase omnivory in the trophic web. They also show that the number of trophically transmitted parasite-host links is positively correlated with the linkage density of the host species, i.e. highly connected species have a higher rate of infection, in particular those with complex life cycles. Therefore, parasites play a prominent role in ecological networks, significantly increasing interaction strength and hence selectively changing food web links.

Parasites are ubiquitous in the aquatic environment and have subtle, sublethal or even lethal impacts. Their impacts on hosts are propagated up and down food webs and thus are manifested throughout the entire community. Environmental changes, however, greatly affect their dynamics and hence parasites can be seen as indicators of many aspects of host physiology. Parasites are uniquely situated within food webs, and following their transmission process could serve management and ecosystem conservation (Marcogliese, 2004; Lafferty et al., 2006). In general, the diversity of parasites reflects the overall diversity within the ecosystem (see Rasconi et al., 2011). In many pelagic systems, fungal parasites are 1) a driver of phytoplankton community structure, 2) crucial for organic matter and energy transfer, 3) important for food web dynamics by affecting fitness and reproduction of many aquatic organisms and 4) causes of intra-specific variability and even increased speciation. Since fungal parasites largely increase the number of trophic levels and often lower the dominance of a few species they also increase ecosystem stability and most likely even functional diversity. Fungi are also potential vectors of genetic elements and hence may also transfer genetic information between organisms of different trophic levels. In any case, they lead to a higher biodiversity by affecting key evolutionary parameters and also functional diversity, e.g. by transferring terrestrial material including leaves and pollen - otherwise unavailable for aquatic organisms - to higher trophic levels (e.g. Masclaux et al. 2011). Hence, aquatic fungi should be seen as key variables for food web structure and genetic as well functional diversity of the aquatic community rendering it less susceptible to changes in environmental variables.

5. Assessing fungal biodiversity and functionality in aquatic ecosystems

Next to classical molecular techniques for assessing *in situ* fungal communities (Bärlocher 2007), massive parallel DNA sequencing in combination with data management systems such as GenBank (Benson et al., 2008), it is now possible to fully explore fungal biodiversity. Amplicon sequencing has been used to explore fungal diversity in soils (Buée et al., 2009; Lim et al., 2010). The approach was recently used in an assessment of estuarine biodiversity (Chariton et al., 2010) and for the first time in two lakes (Monchy et al., 2011), and similar work from a lowland river floodplain has not been published to date (Colloff & Baldwin in prep; Kerr et al., in prep.).

These studies greatly extend our understanding of dimensions and structure of the fungal kingdom. However, in many cases, all we know of newly discovered species is the sequence of a small part of their genome, with no insights regarding morphology, physiology or ecology of the specimen. In the future, a combination of techniques such as transcriptomics (Bhadauria et al., 2007), proteomics (Doyle 2011), metabolomics (Tan et al., 2009) may allow us to evaluate physiological and ecological inferences based on DNA sequences. Classical culture techniques, however, will remain important for studying morphology, preserving voucher specimens, and generally expanding our knowledge of undescribed species associated with novel sequences.

At present, there are a number of biases in the representation of described species in databases such as GenBank. For practical reasons, investigations have focused on the ecology and diversity of macro-fungi and pathogens of plants and animals. In order to use DNA sequence databases to identify fungi in environmental samples, it is at first necessary to fill the database with accurate and appropriate information on fungal sequences with taxonomic descriptions. It is unlikely that this work can keep pace with the potential of current technologies to generate sequence data. Nevertheless, improved sequence analysis techniques are required to link information of the "omics" studies with those of the environment including short- and long-term changes. Although identification of fungi in the environment has been improved a lot throughout the past years, there is an obvious lack in fundamental ecological methods, e.g. methods for differentiating between the biomass of fungal species are still needed. FISH methods, which allow determining fungal biomass have only recently emerged (Mangot et al., 2009) and ergosterol measurements are only applicable to CPOM where algae are not present (e.g. *Chlorella* – a typical fresh water alga - contains ergosterol) and can't detect the presence of many species of lower fungi. To understand the importance of fungi for energy and organic matter cycling in aquatic systems, we need to greatly improve our techniques, e.g. by defining new marker molecules to measure the biomass and activity of fungi in their natural environment.

6. Concluding remarks

Present estimations suggest that global fungal diversity greatly exceeds that of any other group of microbes. As documented above, their function is of global importance for nutrient cycling and ecosystem health. For example, networks of fungal hyphae interconnect a whole habitat and trigger the transport of macro- and micronutrients over large distances to other heterotrophic microbes (Harms et al., 2011). Aquatic habitats are no exceptions in this respect and the loss of fungi severely affects aquatic food web topology and hence functioning (Lafferty et al., 2008). Hypothetical scenarios resulting from the loss of fungal diversity include: aggradation of aquatic ecosystems via the accumulation of CPOM and polymers, a decline in macroinvertebrate food sources, a reduction in the rate and range of decontamination of industrial toxins, diminished total diversity in planktonic communities and the development of fungal monocultures that would potentially impact on total biodiversity. Since fungal biodiversity is representative of ecosystem functioning and thus of ecosystem health, it is in the interests of human society to explore the fungal biodiversity present in natural environments, especially aquatic habitats.

7. Acknowledgment

We would like to thank A. Grossherr for illustrations and helpful comments and the Leibniz society and the German Science foundation (DFG GR1540/15-1) for funding. Janice Kerr would like to thank the Murray-Darling Freshwater Research Centre (Wodonga, Australia) and its staff for their support in production of the manuscript.

8. References

Abdel-Raheem, A. M. & Ali, E. H. (2004). Lignocellulolytic enzyme production by aquatic hyphomycetes species isolated from the Nile's delta region. *Mycopathologia, 157*(3), 277-286

Abdullah, S. K. & Taj-Aldeen, S. J. (1989). Extracellular enzymatic activity of aquatic and aero-aquatic conidial fungi. *Hydrobiologia, 174*, 217-223

Adl, S. M., Simpson A. G. B., et al. (2005). The new higher level classification of eukaryotes with emphasis on the taxonomy of protists. *The Journal of Eukaryotic Microbiology, 52*(5), 399-451

Ali, E. H. & Abdel-Raheem, A. (2003). Distribution of zoosporic fungi in the mud of major Egyptian lakes. *Journal of Basic Microbiology, 43*(3), 175-184

Amundsen, P.A. and Lafferty, K.D. and Knudsen, R. and Primicerio, R. and Klemetsen, A. and Kuris, A.M. (2009). Food web topology and parasites in the pelagic zone of a subarctic lake. *Journal of Animal Ecology, 78*(3), 566-572. doi: 10.1111/j.1365-2656.2008.01518.x

Ananda, K. & Sridhar, K. R. (2002). Diversity of endophytic fungi in the roots of mangrove species on the west coast of India. *Canadian Journal of Microbiology, 48*(10), 871-878.

Albariño, R., Villanueva, V. D., et al. (2008). The effect of sunlight on leaf litter quality reduces growth of the shredder *Klapopteryx kuscheli*. *Freshwater Biology, 53*(9), 1881-1889

Ali-Shtayeh, M. S., Khaleel, T. K. M., & Jamous, R. M. (2002). Ecology of dermatophytes and other keratinophilic fungi in swimming pools and polluted and unpolluted streams. *Mycopathologia, 156*(3), 193-205

Baar, J., Paradi, I., Lucassen, E. C. H. E. T., Hudson-Edwards, K. a, Redecker, D., Roelofs, J. G. M., et al. (2011). Molecular analysis of AMF diversity in aquatic macrophytes: A comparison of oligotrophic and utra-oligotrophic lakes. *Aquatic Botany, 94*(2), 53-61. doi: 10.1016/j.aquabot.2010.09.006

Baldwin, D. S. (1999). Dissolved organic matter and phosphorus leached from fresh and 'terrestrially' aged river red gum leaves: implications for assessing river-floodplain interactions. *Freshwater Biology, 41*, 675-685

Baldy, V., Chauvet, E., et al. (2002). Microbial dynamics associated with leaves decomposing in the mainstem and floodplain pond of a large river. *Aquatic Microbial Ecology, 28*(1), 25-36

Bärlocher, F. (1981). Fungi on the food and in the faeces of *Gammarus pulex. Transactions of the British Mycological Society, 76*, 160-165

Bärlocher, F. (1982). The contribution of fungal enzymes to the digestion of leaves by *Gammarus fossarum* Koch (Amphipoda). Oecologia, 52, 1-4

Bärlocher, F. (1985). The role of fungi in the nutrition of stream invertebrates. *Botanical Journal of the Linnean Society, 91*, 83-94

Bärlocher, F. (1992)(ed). *The Ecology of aquatic hyphomycetes*. Springer, Berlin. ISBN 3540544003

Bärlocher, F. (2007). Molecular approaches applied to aquatic hyphomycetes. *Fungal Biology Reviews 21*, 19-24

Bärlocher, F. & Kendrick, B. (1974). Dynamics of the fungal population on leaves in a stream. *The Journal of Ecology, 62*(3), 761-791

Bärlocher, F. & Graça, M. A. S. (2002). Exotic riparian vegetation lowers fungal diversity but not leaf decomposition in Portuguese streams. *Freshwater Biology, 47*(6), 1123-1135

Bärlocher, F., Seena, S., et al. (2008). Raised water temperature lowers diversity of hyporheic aquatic hyphomycetes. *Freshwater Biology, 53*(2), 368-379

Benny, L. G. (2009). Zygomycetes. Retrieved April 2011, from http://www.zygomycetes.org/

Benson, D. A., Karsch-Mizrachi I., et al. (2008). GenBank. *Nucleic Acids Research 36* (suppl 1), D25-D30

Bergbauer, M., Moran, M. A., et al. (1992). Decomposition of lignocellulose from a freshwater macrophyte by aero-aquatic fungi. *Microbial Ecology, 23*(2), 159-167

Berger, L., Speare, R., Daszak, P., Green, D. E., Cunningham, A. A., Goggin, C. L., et al. (1998). Chytridiomycosis causes amphibian mortality associated with population declines in the rain forests of Australia and Central America. *Proceedings of the National Academy of Sciences of the United States of America, 95*(15), 9031-6

Bhadauria, V., Popescu L., et al. (2007). Fungal transcriptomics. *Microbiological Research 162*(4), 285-298

Blackwell, M. (2011). The Fungi: 1, 2, 3 ... 5.1 million species? *American Journal of Botany, 98*(3), 426-438. doi: 10.3732/ajb.1000298

Boyd, C. E. & Goodyear, C. P. (1971). Nutritive quality of food in ecological systems. *Archiv für Hydrobiolog,e 69*, 256-270

Buée, M., Reich M., et al. (2009). 454 Pyrosequencing analyses of forest soils reveal an unexpectedly high fungal diversity. *New Phytologist 184*(2), 449-456

Bunn, S. E. & Boon, P. I. (1993). What sources of organic carbon drive food webs in billabongs? A study based on stable isotope analysis. *Oecologia, 96*(1), 85-94

Butinar, L., Santos, S., Spencer-Martins, I., Oren, A., & Gunde-Cimerman, N. (2005). Yeast diversity in hypersaline habitats. *FEMS Microbiology Letters, 244*(2), 229-234

Campbell, I. C. & Fuchshuber, L. (1995). Polyphenols, condensed tannins, and processing rates of tropical and temperate leaves in an Australian stream. *Journal of the North American Benthological Society, 14*(1), 174-182

Canhoto, C., Bärlocher, F., et al. (2002). The effects of Eucalyptus globulus oils on fungal enzymatic activity. *Archiv für Hydrobiologie, 154*(1), 121-132

Canter-Lund, H., & Lund, J. W. G. (1995). *Freshwater Algae -- Their microscopic world explored* (pp. 279-324). Biopress Ltd., Bristol

Cardinale, B. J., Palmer, M. A., et al. (2002). The influence of substrate heterogeneity on biofilm metabolism in a stream ecosystem. *Ecology, 83*(2), 412-422

Cargill, A. S., Cummins, K. W., et al. (1985). The role of lipids, fungi, and temperature in the nutrition of a shredder caddisfly, Clistoronia magnifica. *Freshwater Invertebrate Biology, 4*(2), 64-78

Chamier, A.-C. (1985). Cell-wall degrading enzymes of aquatic hyphomycetes: a review. *Botanical Journal of the Linnean Society, 91*, 67-81

Chariton, A. A., Court L. N., et al. (2010). Ecological assessment of estuarine sediments by pyrosequencing eukaryotic ribosomal DNA. *Frontiers in Ecology and the Environment* 8(5), 233-238

Chen, M., Chen, F., et al. (2010). Microbial eukaryotic community in response to Microcystis spp. bloom, as assessed by an enclosure experiment in Lake Taihu, China. *FEMS Microbiology Ecology*, 74(1), 19-31. doi: 10.1111/j.1574-6941.2010.00923.x

Chukanhom, K., & Hatai, K. (2004). Freshwater fungi isolated from eggs of the common carp (*Cyprinus carpio*) in Thailand. *Mycoscience, 45*(1), 42-48. doi: 10.1007/s10267-003-0153-9

Chung, N. & Suberkropp, K. (2009). Contribution of fungal biomass to the growth of the shredder, *Pycnopsyche gentilis* (Trichoptera: Limnephilidae). *Freshwater Biology, 54*(11), 2212-2224

Cole, J. J., Caraco, N. F., & Likens, G. E. (1990). Short-range atmospheric transport: A significant source of phosphorus to an oligotrophic lake. *Limnology and Oceanography, 35*(6), 1230-1237. doi: 10.4319/lo.1990.35.6.1230

Colloff, M. J. and Baldwin, D. S. (in prep). Diversity within the soil community in a semi-arid floodplain

Cornut, J., Elger, A., et al. (2010). Early stages of leaf decomposition are mediated by aquatic fungi in the hyporheic zone of woodland streams. *Freshwater Biology*, 55(12), 2541-2556. doi: 10.1111/j.1365-2427.2010.02483.x

Czeczuga, B., Godlewska, A., et al. (2004). Aquatic fungi growing on feathers of wild and domestic bird species in limnologically different water bodies. *Polish Journal of Environmental Studies, 13*(1), 21-31

Czeczuga, B., Kozlowska, M., et al. (2002). Zoosporic aquatic fungi growing on dead specimens of 29 freshwater crustacean species. *Limnologica, 32*(2), 180-193

Czeczuga, B., Muszyńska, E., et al. (2007). Aquatic fungi and straminipilous organisms on decomposing fragments of wetland plants. *Mycologia Balcanica, 4* (1-2), 31-44

Dang, C. K., Chauvet, E., et al. (2005). Magnitude and variability of process rates in fungal diversity-litter decomposition relationships. *Ecology Letters, 8*(11), 1129-1137

De Bruin, A., Ibelings, B. W., Kagami, M., Mooij, W. M., & Van Donk, Ellen. (2008). Adaptation of the fungal parasite *Zygorhizidium planktonicum* during 200 generations of growth on homogeneous and heterogeneous populations of its host, the diatom *Asterionella formosa*. *The Journal of Eukaryotic Microbiology, 55*(2), 69-74. doi: 10.1111/j.1550-7408.2008.00306.x

DeVay, J. E. (1956). Mutual Relationships in Fungi. *Annual review of microbiology, 10*, 115-140

Dick, M. W. (1962). The occurrence and distribution of Saprolegniaceae in certain soils of South-East England. II. Distribution within defined areas. *Journal of Ecology, 50*(1), 119-127

Dick, M. W. (1963). The occurrence and distribution of Saprolegniaceae in certain soils of South-East England: III. Distribution in relation to pH and water content. *The Journal of Ecology, 51*(1), 75-81

Dick, M. W. (1969). Morphology and taxonomy of the Oomycetes, with special reference to Saprolegniaceae, Leptomitaceae and Pythiaceae I. Sexual reproduction. *New Phytologist, 68*(3), 751-775

Dick, M. W. (1972). Morphology and taxonomy of the Oomycetes, with special reference to Saprolegniaceae, Leptomitaceae, and Pythiaceae. II. Cytogenetic systems. *New Phytologist, 71*(6), 1151-1159

Dick, M. W. and H. V. Newby (1961). The occurrence and distribution of Saprolegniaceae in certain soils of south-east England. I. Occurrence. *Journal of Ecology, 49*, 403-419

Dodds, W. K. (2002). Freshwater Ecology: Concepts and Environmental Applications. Academic Press, San Diego. ISBN 0122191358

Doyle, S. (2011). Fungal proteomics: from identification to function. *FEMS Microbiology Letters*. doi: 10.1111/j.1574-6968.2011.02292.x

Drechsler, C. (1941). Predaceous Fungi. *Biological Reviews, 16*(4), 265-290. doi: 10.1111/j.1469-185X.1941.tb01104.x

Ebert, D. (1995). The ecological interactions between a microsporidian parasite and its host *Daphnia magna*. *Journal of Animal Ecology, 64*(3), 361–369

Ellis, M. B., & Ellis, J. P. (1985). *Microfungi on Land Plants: An Identification Handbook* (1st ed.). Macmillan Pub Co

Evans, C. S., Dutton, M. V. , et al. (1994). Enzymes and small molecular mass agents involved with lignocellulose degradation. FEMS Microbiology Reviews 13(2-3), 235-239

Fenoglio, S., Bo, T., Cammarata, M., Malacarne, G., & Frate, G. (2009). Contribution of macro- and micro-consumers to the decomposition of fish carcasses in low-order streams: an experimental study. *Hydrobiologia, 637*(1), 219-228. doi: 10.1007/s10750-009-9998-z

Ferreira, V. & Chauvet, E. (2010). Synergistic effects of water temperature and dissolved nutrients on litter decomposition and associated fungi. *Global Change Biology, 17*(1), 551-564

Field, J. I. & Webster, J. (1983). Anaerobic survival of aquatic fungi. *Transactions of the British Mycological Society, 81*(2), 365-369

Findlay, S., Howe, K., et al. (1990). Comparison of detritus dynamics in two tidal freshwater wetlands. *Ecology, 71*(1), 288-295

Fischer, H., Mille-Lindblom, C., Zwirnmann, E., & Tranvik, L. J. (2006). Contribution of fungi and bacteria to the formation of dissolved organic carbon from decaying common reed (*Phragmites australis*). *Archiv für Hydrobiologie, 166*(1), 79-97. doi: 10.1127/0003-9136/2006/0166-0079

Fisher, J. P., Davey, R. A., et al. (1983). Degradation of lignin by aquatic and aeroaquatic hyphomycetes. *Transactions of the British Mycological Society, 80*(1), 166-168

Ford, T. E. (1993). *Aquatic microbiology: An ecological approach.* Blackwell Scientific Publications, Boston. ISBN 0865422257

Garrett, S. D. (1951). Ecological Groups of Soil Fungi: A Survey of Substrate Relationships *New Phytologist, 50*(2), 149-166

Gessner, M. O., & Van Ryckegem, G. (2003). Water fungi as decomposers in freshwater ecosystems. In G. Bitton (Ed.), *Encyclopaedia of Environmental Microbiology,* Wiley, New York. ISBN 9780471263395

Gimmler, H. (2001). Mutalistic Relationships Bewtween Algae and Fungi (Excluding Lichens). In K. Esser, U. Lüttge, J. W. Kadereit, & B. W. (Eds.), *Progress in Botany* (Vol. 62, pp. 194-214). Springer

Gleason, Frank H., Marano, A. V., Digby, A. L., Al-Shugairan, N., Lilje, O., Steciow, M. M., et al. (2011). Patterns of utilization of different carbon sources by *Chytridiomycota. Hydrobiologia, 659*(1), 55-64. doi: 10.1007/s10750-010-0461-y

Glen-Bott, J. I. (1951). *Helicodendron giganteum* n. sp. and other aerial-sporing Hyphomycetes of submerged dead leaves. *Transactions of the British Mycological Society, 34*(3), 275-279. doi: 10.1016/S0007-1536(51)80052-4

Graça, M. A. S., Pozo, J., et al. (2002). Effects of *Eucalyptus* plantations on detritus, decomposers, and detritivores in streams. *The Scientific World, 2,* 1173-1185

Gulis, V. (2001). Are there any substrate preferences in aquatic hyphomycetes? *Mycological Research, 105*(9), 1088-1093.

Gulis, V. & Suberkropp, K. (2004). Effects of whole-stream nutrient enrichment on the concentration and abundance of aquatic hyphomycete conidia in transport. *Mycologia, 96*(1), 57-65

Gulis, V., Kuehn, K. A., & Suberkropp, K. (2009). Fungi. In G. Likens (Ed.), *Encyclopedia of Inland Waters* (pp. 233-243). Elsevier, Oxford. doi: 10.1002/0471263397.env314

Gustafsson, C. & Boström, C. (2011). Biodiversity influences ecosystem functioning in aquatic angiosperm communities. Oikos, . doi: 10.1111/j.1600-0706.2010.19008.x

Hadwen, W. L., Fellows, C. S. et al. (2010). Longitudinal trends in river functioning: Patterns of nutrient and carbon processing in three Australian rivers. *River Research and Applications, 26*(9), 1129-1152

Hannen, E. J. van, Mooij, W., Agterveld, M. P. van, Gons, H. J., & Laanbroek, H. J. (1999). Detritus-Dependent Development of the Microbial Community in an Experimental System: Qualitative Analysis by Denaturing Gradient Gel Electrophoresis. *Applied and environmental microbiology, 65*(6), 2478-2484

Harms, H., Schlosser, D., & Wick, Y.L. (2011). Untapped potential: exploiting fungi in bioremediation of hazardous chemicals. *Nature Reviews Microbiology,* 9, 177-192. doi: 10.1038/nrmicro2519

Hawksworth, D. L. (2000). Freshwater and marine lichen-forming fungi. *Fungal Diversity, 5*, 1-7

Held, A. A., Emerson, R., Fuller, M. S., & Gleason, F H. (1969). *Blastocladia* and *Aqualinderella*: fermentative water molds with high carbon dioxide optima. *Science, 165*(3894), 706-8. doi: 10.1126/science.165.3894.706

Hibbett, D. S., Binder, M., Bischoff, J. F., Blackwell, Meredith, Cannon, P. F., Eriksson, O. E., et al. (2007). A higher-level phylogenetic classification of the Fungi. *Mycological Research, 111*(Pt 5), 509-47. doi: 10.1016/j.mycres.2007.03.004

Hibbett, D. S., Ohman, A., Glotzer, D., Nuhn, M., Kirk, P., & Nilsson, R. H. (2011). Progress in molecular and morphological taxon discovery in Fungi and options for formal classification of environmental sequences. *Fungal Biology Reviews, 25*(1), 38-47. doi: 10.1016/j.fbr.2011.01.001

Hladyz, S., Gessner, M. O. et al. (2009). Resource quality and stoichiometric constraints on stream ecosystem functioning. *Freshwater Biology, 54*(5), 957-970

Hyde, K., Bussaban, B. et al. (2007). Diversity of saprobic microfungi. *Biodiversity and Conservation, 16*(1), 7-35

Hyde, K. D. & Lee, S. Y. (1995). Ecology of mangrove fungi and their role in nutrient cycling: what gaps occur in our knowledge? *Hydrobiologia, 295*(1), 107-118

Hywel-Jones, N. L., & Webster, J. (1986). Scanning electron microscope study of external development of *Erynia conica* on *Simulium*. *Transactions of the British Mycological Society, 86*(3), 393-399. doi: 10.1016/S0007-1536(86)80183-8

Ibelings, B. W., Bruin, A. D., Kagami, M., Rijkeboer, M., Brehm, M., & Donk, E. V. (2004). Host parasite Interactions between freshwater phytoplankton and chytrid fungi (*Chytrdidiomycota*). *Journal of Phycology, 40*(3), 437-453. doi: 10.1111/j.1529-8817.2004.03117.x

Ingold, C. T. (1975). An illustrated guide to aquatic and waterborne hyphomycetes (Fungi imperfecti) with notes on their biology. *Freshwater Biological Association Scientific Publication* 30. Titus Wilson & Son Ltd, Kendal. ISBN 900386223

Janssen, M. A. & Walker, K. F. (1999). Processing of riparian and wetland plant litter in the River Murray, South Australia. *Hydrobiologia, 411*, 53-64

Jobard, M., Rasconi, S., & Sime-Ngando, T. (2010). Diversity and functions of microscopic fungi: a missing component in pelagic food webs. *Aquatic Sciences, 72*(3), 255-268. doi: 10.1007/s00027-010-0133-z

Johnson, P. T. J., Longcore, J. E., Stanton, D. E., & Carnegie, R. B. (2006). Chytrid infections of *Daphnia pulicaria*: development, ecology, pathology and phylogeny of *Polycaryum laeve*. *Freshwater Biology, 51*, 634-648. doi: 10.1111/j.1365-2427.2006.01517.x

Kagami, M., Van Donk, E., et al. (2004). *Daphnia* can protect diatoms from fungal parasitism. *Limnology and Oceanography, 49*(3), 680-685

Karling, J. (1936). A new predacious fungus. *Mycologia, 28*(4), 307–320

Kaushik, N. K. & Hynes, H. B. N. (1971). The fate of dead leaves that fall into streams. *Archiv für Hydrobiologie 68*, 645-515

Keeling, P. J., & Fast, N. M. (2002). Microsporidia: biology and evolution of highly reduced intracellular parasites. *Annual Review of Microbiology, 56*, 93-116. doi: 10.1146/annurev.micro.56.012302.160854

Kjoller, A. & Struwe, S. (1980). Microfungi of decomposing red alder leaves and their substrate utilisation. *Soil Biology & Biochemistry, 12*, 425-431

Koontz, J. A. (2006). Physiological studies on a new isolate of the gut fungus, *Smittium culisetae* (Trichomycetes: *Harpellales*), from wetland mosquito larvae, *Aedes vexans* (Diptera: Culicidae). *Transactions of the Kansas Academy of Science, 109*(3), 175–183. doi: 10.1660/0022-8443(2006)109

Krauss, G.-J., Solé, M., Krauss, G., Schlosser, D., Wesenberg, D., & Bärlocher, F. (2011). Fungi in freshwaters: ecology, physiology and biochemical potential. *FEMS Microbiology Reviews, 35, 620-651.* doi: 10.1111/j.1574-6976.2011.00266.x

Kuehn, K. A. & Suberkropp K. (1998). Diel fluctuations in rates of CO_2 evolution from standing dead leaf litter of the emergent macrophyte *Juncus effusus. Aquatic Microbial Ecology*, 14(2), 171-182. doi: :10.3354/ame014171

Kumaresan, V. & Suryanarayanan, T. S. (2001). Occurrence and distribution of endophytic fungi in a mangrove community. *Mycological Research, 105*(11), 1388-1391

Lafferty, K.D. and Dobson, A.P. and Kuris, A.M. (2006). Parasites dominate food web links. *Proceedings of the National Academy of Sciences of the United States of America,* 103(30), 11211-11216. doi: 10.1073/pnas.0604755103

Lafferty, K.D., Allesina, S., et al. (2008). Parasites in food webs: the ultimate missing links. *Ecology Letters*, 11, 533-546. doi: 10.1111/j.1461-0248.2008.01174.x

Lara, E., Moreira, D., & López-García, P. (2010). The environmental clade LKM11 and *Rozella* form the deepest branching clade of fungi. *Protist, 161*(1), 116-21. doi: 10.1016/j.protis.2009.06.005

Lawton, J. H. & Brown, V. K. (1993). Redundancy in ecosystems. In Schulze, E.-D. & Mooney, H. A. (Eds.), *Biodiversity and ecosystem function.* New York, Springer, 255-270

Leonowicz, A., Cho, N., et al. (2001). Fungal laccase: properties and activity on lignin. *Journal of Basic Microbiology, 41*(3-4), 185-227

Lepere, C., Domaizon, I., et al. (2007). Community composition of lacustrine small eukaryotes in hyper-eutrophic conditions in relation to top-down and bottom-up factors. *FEMS Microbiology Ecology, 61*(3), 483-495

Lefèvre, E., Bardot, C. et al. (2007). Unveiling fungal zooflagellates as members of freshwater picoeukaryotes: evidence from a molecular diversity study in a deep meromictic lake. *Environmental Microbiolog, 9*(1), 61-71

Lichtwardt, R. (2004). Trichomycetes: Fungi in Relationship with Insects and Other Arthropods. In J. Seckbach (Ed.), *Symbiosis* (Vol. 4, pp. 575-588). Springer-Netherlands. doi: 10.1007/0-306-48173-1_36

Lim, Y., Kim B., et al. (2010). Assessment of soil fungal communities using pyrosequencing. *The Journal of Microbiology 48*(3), 284-289

Lindley, L. A., Stephenson, S. L. , et al. (2007). Protostelids and myxomycetes isolated from aquatic habitats. *Mycologia, 99*(4), 504-509

Lockhart, R. J., Van Dyke, M. I., Beadle, I. R., Humphreys, P., & McCarthy, A. J. (2006). Molecular biological detection of anaerobic gut fungi (*Neocallimastigales*) from landfill sites. *Applied and Environmental Microbiology, 72*(8), 5659-61. doi: 10.1128/AEM.01057-06

Loreau, M. (2004). Does functional redundancy exist? Oikos 104(3), 606-611

Luo, Q., Krumholz, L. R., Najar, F. Z., Peacock, A. D., Roe, B. A., White, D. C., et al. (2005). Diversity of the Microeukaryotic Community in Sulfide-Rich Zodletone Spring (Oklahoma). *Applied and Environmental Microbiology, 71*(10), 6175-6184. doi: 10.1128/AEM.71.10.6175-6184.2005.

Mancinelli, G., Costantini, M. L., & Rossi, L. (2002). Cascading effects of predatory fish exclusion on the detritus-based food web of a lake littoral zone (Lake Vico, central Italy). *Oecologia, 133*(3), 402-411

Mangot, J.-F., Lepère, C., Bouvier, C., Debroas, D., & Domaizon, I. (2009). Community structure and dynamics of small eukaryotes targeted by new oligonucleotide probes: new insight into the lacustrine microbial food web. *Applied and Environmental Microbiology, 75*(19), 6373-81. doi: 10.1128/AEM.00607-09

Marcogliese, D.J. (2004). Parasites: Small Players with Crucial Roles in the Ecological Theater. *EcoHealth,* 1, 151-164. doi: 10.1007/s10393-004-0028-3

Mahadevan, A. & Muthukumar, G. (1980). Aquatic microbiology with reference to tannin degradation. *Hydrobiologia, 72*(1), 73-79

Masclaux, H., Bec, A., Kagami, M., Perga, M.-E., Sime-Ngando, T., Desvilettes, C.& Bourdier, G. (2011). Food quality of anemophilous plant pollen for zooplankton. *Journal Limnology and Oceanography,* 56(3), 939-946. doi: 10.4319/lo.2011.56.3.0939

McCarthy, P. M. & Johnson, P. N. (1997). *Verrucaria amnica,* a New Aquatic Lichen Species from New Zealand. *The Lichenologist, 29,* 385-388

McMahon, R., Hunter, R. D. , et al. (1974). Variation in aufwuchs at six freshwater habitats in terms of carbon biomass and of carbon: *nitrogen* ratio. *Hydrobiologia, 45*(4), 391-404

Medeiros, A. O., Pascoal, C., et al. (2009). Diversity and activity of aquatic fungi under low oxygen conditions. *Freshwater Biology, 54*(1), 142-149

Messick, G. A., Overstreet, R. M., Nalepa, T. F., & Tyler, S. (2004). Prevalence of parasites in amphipods *Diporeia* spp. from Lakes Michigan and Huron, USA. *Diseases Of Aquatic Organisms, 59,* 159-170

Mohamed, D. J., & Martiny, J. B. H. (2011). Patterns of fungal diversity and composition along a salinity gradient. *The ISME journal, 5*(3), 379-88. doi: 10.1038/ismej.2010.137

Monchy, S., Sanciu, G., Jobard, M., Rasconi, S., Gerphagnon, M., Chabé, M., et al. (2011). Exploring and quantifying fungal diversity in freshwater lake ecosystems using rDNA cloning/sequencing and SSU tag pyrosequencing. *Environmental Microbiology*. doi: 10.1111/j.1462-2920.2011.02444.x

Mille-Lindblom, C., Fischer, H., et al. (2006). Antagonism between bacteria and fungi: substrate competition and a possible tradeoff between fungal growth and tolerance towards bacteria. *Oikos, 113*(2), 233

Mille-Lindblom, C., Fischer, H., & Tranvik, L. J. (2006). Litter-associated bacteria and fungi - a comparison of biomass and communities across lakes and plant species. *Freshwater Biology, 51*(4), 730-741. doi: 10.1111/j.1365-2427.2006.01532.x

Nawawi, A. (1985). Basidiomycetes with branched, water-borne conidia. *Botanical Journal of the Linnean Society, 91*(1-2), 51-60

Nechwatal, J., Wielgloss, A., & Mendgen, Kurt. (2005). *Pythium phragmitis* sp. nov., a new species close to *P. arrhenomanes* as a pathogen of common reed *Phragmites australis*. *Mycological Research, 109*(12), 1337-1346. doi: 10.1017/S0953756205003990

Neubert, K., Mendgen, K, Brinkmann, H., & Wirsel, S. G. R. (2006). Only a few fungal species dominate highly diverse mycofloras associated with the common reed. *Applied and Environmental Microbiology, 72*(2), 1118-1128

Newell, S. Y. & Bärlocher, F. (1993). Removal of fungal and total organic matter from decaying cordgrass leaves by shredder snails. *Journal of Experimental Marine Biology and Ecology, 171*(1), 39-49

Noga, E. (1993). Water mold infections of freshwater fish: Recent advances. *Annual Review of Fish Diseases, 3*, 291-304. doi: 10.1016/0959-8030(93)90040-I

Orlowska, M., Lengiewicz, I., & Suszycka, M. (2004). Hyphomycetes Developing on Water Plants and Bulrushes in Fish Ponds. *Polish Journal of Environmental Studies, 13*(6), 703-707

Orlowski, M. (1991). *Mucor* Dimorphism. *Microbiological Reviews*, 55(2), 234-258

Pattee, E. & Chergui, H. (1995). The application of habitat templets and traits to Hyphomycete fungi in a mid-European river system. *Freshwater Biology, 33*, 525-539

Peach, M. (1950). Aquatic predacious fungi. *Transactions of the British Mycological Society, 33*(1-2), 148-153. doi: 10.1016/S0007-1536(50)80058-X

Peach, M. (1952). Aquatic predacious fungi. II. *Transactions of the British Mycological Society, 35*(1), 19-23. doi: 10.1016/S0007-1536(52)80002-6

Peach, M. (1954). Aquatic predacious fungi. III. *Transactions of the British Mycological Society, 37*(3), 240-247. doi: 10.1016/S0007-1536(54)80007-6

Pereira, C. R. D., Anderson, N. H., & Dudley, T. (1982). Gut Content Analysis of Aquatic Insects from Wood Substrates. *Melanderia* (39), 23-33

Powell, M. J. (1993). Looking at mycology with a Janus face: a glimpse at Chytridiomycetes active in the environment. *Mycologia, 85*(1), 1–20

Rankovic, B. (2005). Five Serbian reservoirs contain different fungal propagules. *Mycologia, 97*(1), 50-56

Rasconi, S., Jobard, M., Sime-Ngando, T. (2011). Parasitic fungi of phytoplankton: ecological roles and implications for microbial food webs. *Aquatic Microbial Ecology* 62(2), 123-137. doi:10.3354/ame01448

Reid, D. J., Quinn, G. P., et al. (2008). Terrestrial detritus supports the food webs in lowland intermittent streams of south-eastern Australia: a stable isotope study. *Freshwater Biology, 53*(10), 2036-2050

Reisert, P. S., & Fuller, M. S. (1962). Decomposition of Chitin by Chytriomyces Species. *Mycologia, 54*, 647-657

Reynolds, J. (1988). Crayfish extinctions and crayfish plague in central Ireland. *Biological Conservation, 45*(4), 279-285. doi: 10.1016/0006-3207(88)90059-6

Roache, M. C., Bailey, P. C., et al. (2006). Effects of salinity on the decay of the freshwater macrophyte, *Triglochin procerum*. *Aquatic Botany, 84*(1), 45-52

Sabetta, L., Costantini, M. L., Maggi, O., Persiani, A. M. & Rossi, L. (2000). Interactions between detritivores and microfungi during the leaf detritus decomposition in a volcanic lake (Lake Vico - central Italy). Hydrobiologia, 439 (1-3), 49-60

Schoener, T. W. (1974). Resource partitioning in ecological communities. *Science, 185*(4145), 27-39

Schulz, M. J. & Thormann, M. N. (2005). Functional and taxonomic diversity of saprobic filamentous fungi from Typha latifolia from central Alberta, Canada. *Wetlands, 25*(3), 675-684

Shearer, C. A. & Webster, J. (1985). Aquatic hyphomycete communities in the River Teign I. Longitudinal distribution patterns. *Transactions of the British Mycological Society, 84*(3), 489-501

Shearer, C., Descals, E., Kohlmeyer, B., Kohlmeyer, J., Marvanová, L., Padgett, D., et al. (2007). Fungal biodiversity in aquatic habitats. *Biodiversity Conserv, 16*(1), 49-67. doi: 10.1007/s10531-006-9120-z

Skerratt, L. F., Berger, Lee, Speare, Richard, Cashins, S., McDonald, Keith Raymond, Phillott, A. D., et al. (2007). Spread of Chytridiomycosis has caused the rapid global decline and extinction of frogs. *EcoHealth, 4*(2), 125-134. doi: 10.1007/s10393-007-0093-5

Sridhar, K. R., & Sudheep, N. M. (2011). Do the tropical freshwater fishes feed on aquatic fungi? *Frontiers of Agriculture in China, 5*(1), 77-86. doi: 10.1007/s11703-011-1055-9.

Slapeta, J., Moreira, D., & López-Garcia, P. (2005). The extent of protist diversity: insights from molecular ecology of freshwater eukaryotes. *Proceedings of the Royal Society B: Biological Sciences, 272*(1576), 2073-2081. doi: 10.1098/rspb.2005.3195

Smith, V. H. (2007). Microbial diversity-productivity relationships in aquatic ecosystems. *FEMS Microbiology Ecology, 62*(2), 181-186

Sparrow, F. K. (1960). *Aquatic Phycomycetes* (Second Rev.). The University of Michigan Press, Ann Arbor

Sparrow, F. K. (1968). Ecology of Freshwater Fungi. In G. C. Ainsworth & A. S. Sussman (Eds.), *The fungi - an advanced treatise, Volume III, The fungal population* (pp. 41-93). Academic Press, New York, London

Sridhar, K. R. & Bärlocher, F. (1997). Water chemistry and sporulation by aquatic hyphomycetes. *Mycological Research, 101*(5), 591-596

Stelzer, R. S., Heffernan, J., et al. (2003). The influence of dissolved nutrients and particulate organic matter quality on microbial respiration and biomass in a forest stream. Freshwater Biology 48, 1925-1937

Stock, A., Jürgens, K., Bunge, J., & Stoeck, T. (2009). Protistan diversity in suboxic and anoxic waters of the Gotland Deep (Baltic Sea) as revealed by 18S rRNA clone libraries. *Aquatic Microbial Ecology, 55,* 267-284. doi: 10.3354/ame01301

Strongman, D. B., & White, M. M. (2008). Trichomycetes from lentic and lotic aquatic habitats in Ontario, Canada. *Botany, 86*(12), 1449-1466. doi: 10.1139/B08-107

Suberkropp, K. (1992). Interactions with Invertebrates. In Bärlocher, F. (Ed.), *The Ecology of Aquatic Hyphomycetes.* Ecological Studies 94. Springer, Berlin. ISBN 3540544003

Suh, S., Mchugh, J., Pollock, D., & Blackwell, M. (2005). The beetle gut: a hyperdiverse source of novel yeasts. *Mycological Research, 109*(3), 261-265. doi: 10.1017/S0953756205002388

Suryanarayanand, T. S. & Kumaresan, V. (2000). Endophytic fungi of some halophytes from an estuarine mangrove forest. *Mycological Research, 104*(12), 1465-1467

Swe, A., Jeewon, R., Pointing, S. B., & Hyde, K. D. (2008). Diversity and abundance of nematode-trapping fungi from decaying litter in terrestrial, freshwater and mangrove habitats. *Biodiversity and Conservation, 18*(6), 1695-1714. doi: 10.1007/s10531-008-9553-7

Takishita, K., Tsuchiya, M., et al. (2007). Genetic diversity of microbial eukaryotes in anoxic sediment of the saline meromictic Lake Namako-ike (Japan): On the detection of anaerobic or anoxic-tolerant lineages of eukaryotes. *Protist, 158*(1), 51-64

Tan, K.-C., Ipcho S. V. S., et al. (2009). Assessing the impact of transcriptomics, proteomics and metabolomics on fungal phytopathology. *Molecular Plant Pathology 10*(5), 703-715

Thomas, K., Chilvers, G. A., et al. (1992). Aquatic hyphomycetes from different substrates: Substrate preference and seasonal occurrence. *Australian Journal of Marine and Freshwater Research, 43*, 491-509

Thormann, M. N. (2006). Diversity and function of fungi in peatlands: A carbon cycling perspective. *Canadian Journal of Soil Science, 86*, 281-293

Thorp, J. H. & Delong, M. D. (2002). Dominance of autochthonous autotrophic carbon in food webs of heterotrophic rivers. *Oikos, 96*(3), 543-550

Van Donk, E, & Bruning, K. (1992). Ecology of aquatic fungi in and on algae. In W. Reisser (Ed.), *Algae and symbioses* (pp. 567-592). Biopress Ltd. , Bristol. ISBN 9780948737152

Vander Zanden, M. J., & Gratton, C. (2011). Blowin' in the wind: reciprocal airborne carbon fluxes between lakes and land. *Canadian Journal of Fisheries and Aquatic Sciences, 68*(1), 170-182. doi: 10.1139/F10-157

Vannote, R. L., Minshall, G. W., et al. (1980). The river continuum concept. *Canadian Journal of Aquatic Science,s 37*, 130-137

Vogel, C., Rogerson, A., Schatz, S., Laubach, H., Tallman, A., & Fell, J. (2007). Prevalence of yeasts in beach sand at three bathing beaches in South Florida. *Water Research, 41*(9), 1915-1920

Voronin, L. V. (1997). Aquatic and aero-aquatic hyphomycetes in small lakes from Vorkuta vicinities. *Mikologiya i Fitopatologiya, 31*(2), 9-17

Voronin, L. V. (2010). The fungi of small acid lakes. *Inland Water Biology, 3*(3), 254-259. doi: 10.1134/S1995082910030089

Watkins, S. C., Quinn, G. P. , et al. (2010). Changes in organic-matter dynamics and physicochemistry, associated with riparian vegetation loss and river regulation in floodplain wetlands of the Murray River, Australia. *Marine and Freshwater Research, 61*(10), 1207-1217

Webster, J. (1992). Anamorph-Teleomorph Relationships. In Bärlocher, F. (Ed.), *The Ecology of Aquatic Hyphomycetes.* Ecological Studies 94. Springer, Berlin. ISBN 3540544003

Weir, A. (2004). The Laboulbeniales — An Enigmatic Group of Arthropod-Associated Fungi. In J. Seckbach (Ed.), *Symbiosis - Cellular Origin, Life in Extreme Habitats and Astrobiology, Vol 4* (pp. 611-620). Springer Netherlands. doi: 10.1007/0-306-48173-1_38

Whisler, H. C., Zebold, S. L., & Shemanchuk, J. A. (1975). Life history of *Coelomomyces psorophorae. Proceedings of the National Academy of Sciences of the United States of America, 72*(2), 693-696

Williams, J. L. (2010). Fungi in the floodplain wetlands of lowland rivers. Department of Environmental Management and Ecology. Albury-Wodonga, Australia, La Trobe University. Ph.D. Thesis.

Kerr, J. L., Baldwin, D. S., et al. (In prep.). Using high resolution IR micro-spectroscopy to untangle aquatic decomposition of leaves by fungi.

Kerr, J. L., Baldwin D. S., et al. (in prep.). Metagenomic analysis reveals taxonomically and functionally diverse soil fungi across a flooding-drying gradient in a semi-arid floodplain.

Willoughby, L. G. (1961). The ecology of some lower fungi at Esthwaite water. *Transactions of the British Mycological Society, 44*(3), 305-IN2. doi: 10.1016/S0007-1536(61)80024-7

Wurzbacher, C. M., Bärlocher, F. et al. (2010). Fungi in lake ecosystems. *Aquatic Microbial Ecology, 59*(2), 125-149

Zalar, P., Kocuvan, M. A., et al. (2005). Halophilic black yeasts colonize wood immersed in hypersaline water. *Botanica Marina, 48*(4), 323-326

11

Mycoflora and Biodiversity of Black Aspergilli in Vineyard Eco-Systems

Cinzia Oliveri and Vittoria Catara
Università degli Studi di Catania
Dipartimento di Scienze delle Produzioni Agrarie ed Alimentari, Catania
Italy

1. Introduction

Environmental conditions in vineyard eco-systems are of particular interest because they can influence the fungal populations associated with grapes, fungal–plant interactions, and production of secondary metabolites, including mycotoxins. Some fungal species are pathogenic to grapevines, infecting the roots, canes, leaves and fruit (Hewitt, 1988; Tournas & Katsoudas, 2005). Grape contamination by different moulds occurs during vineyard pre-harvesting, harvesting and grape processing (Magnoli et al., 2003). Moulds commonly isolated from grapes are *Alternaria*, *Cladosporium* and *B. cinerea*, the latter causing bunch rot. Pathogenic and opportunistic species of *Fusarium*, *Penicillium* and *Aspergillus* can also colonize inducing grape disease.

In heavily infected fruit, moulds alter chemical composition and mould enzymes adversely affect wine flavor and colour as well as yeast growth during alcoholic fermentation (Fleet, 1999; Fleet, 2001). Some vineyard fungal species are capable of producing toxic secondary metabolites (mycotoxins) in infected tissue, which may contaminate grapes and grape products such as wine, grape juice and dried vine fruit (reviewed in Nielsen et al., 2009). The mycotoxins of greatest significance include aflatoxins, citrinin, patulin and ochratoxin A (OTA) and recently fumonisin B_2 (FB_2) (Frisvad et al., 2007; Logrieco et al., 2010; Morgensen et al., 2010a, 2010b; Susca et al., 2010). The most important mycotoxin in grapes and the grape-wine chain is OTA first reported by Zimmerli and Dick (1996). It has nephrotoxic, carcinogenic (2B group) (IARC, 1993), teratogenic and immunotoxic effects (Abarca et al., 2001; Castegnaro & Pfohl-Leszkowicz, 2002, Da Rocha et al., 2002; Pfohl-Leszkowicz et al., 2002; Petzinger & Weidenbach, 2002; Vrabcheva et al., 2000). Thereafter, several authors reported OTA contamination in wine and the presence of OTA-producing fungi in grapes in different wine-growing areas around the world (Battilani et al., 2006; Leong et al., 2007; Medina et al., 2005; Sage et al., 2004; Tjamos et al., 2004). The contamination of grapes with OTA can occur in the field, even without visible symptoms, while the grapes are still on the vine (Serra et al. 2006).

2. Occurrence and biodiversity of black aspergilli from grapes

2.1 Role in Ocratoxin A contamination

Several surveys conducted in the Mediterranean, South America, and Australia reported fungal species belonging to *Aspergillus* section *Nigri* (also known as Black Aspergilli-BA) as the major responsible for OTA contamination in grape (reviewed in Perrone et al., 2007). Several BA have been isolated from grape or from vineyard soil/air such as *A. niger* aggregate (namely *A. niger* sensu stricto, *A. tubingensis, A. foetidus,* and *A. brasiliensis), A. carbonarius* and the uniseriate species *A. aculeatus, A. japonicus,* and *A. uvarum* (Medina et al., 2005; Perrone et al., 2008). Only a few species produce OTA among the *A. niger* aggregate, *A. carbonarius,* and *A. japonicus* (Battilani et al., 2003a). *A. ochraceus* (belonging to section *Circumdati),* although able to produce OTA, have only occasionally been isolated from grape. The most frequently occurring species are *A. niger* aggregate and *A. carbonarius,* respectively, although the highest percentage of OTA-producing strains has been detected in the latter species (Serra et al., 2005).

OTA contamination of dried vine fruit was also found to be due to black aspergilli in Europe, including Spain, Hungary and other parts of the world such as Argentina and Australia (Varga & Kozakiewicz, 2006). In spite of the higher incidence of species belonging to the *A. niger* aggregate found in vineyards, only 5-10 % of *A. niger* OTA-producing strains were detected, whereas more than 50% and, in some studies up to 100%, in *A. carbonarius* (Battilani et al. 2006; Heenan et al,. 1998 Perrone et al., 2006a; Serra et al., 2005). Other *Aspergillus* species, such as *A. helicotrix, A. ellipticus* and A. *heteromorphus, A. ochraceus* are rare (Bau et al., 2005)

2.2 Isolation and Identification tools

Black aspergilli are isolated and identified at genus and species level by morphological criteria: colour, density and colony appearance (layer colour, wrinkled, umbilical, thick or flat) and microscope observation (conidial head, conidiophore and conidia characters) in accordance with appropriate keys (Klick, 2002; Klick and Pitt, 1988; Pitt & Hocking, 1999, 2009; Samson et al., 2004, 2007). The taxonomy of *Aspergillus* section *Nigri* is widely studied but although identification at section level is quite easy, at species level it is much more complex since morphologically taxa differences are very subtle requiring taxonomic expertise.

For macromorphological observations, Czapek yeast autolysate (CYA), malt extract autolysate (MEA), Czapek yeast autolysate with 5% NaCl (CYAS) agar, yeast extract-sucrose (YES) agar, oatmeal agar (OA) and Czapek agar (CZA) are used (Samson et al., 2004). Some differential growth media e.g. DYSG Agar, Coconut Cream Agar (Heenan et al., 1998) and MEA-B (Pollastro et al., 2006) may facilitate the recognition of ochratoxigenic black aspergilli (Samson et al., 2007). Useful physiological features are very good growth and sporulation at 37 °C as well as growth and acid production on CREA agar (Samson et al., 2004). Species can be identified by micromorphological analysis of the fungal structures by light microscopy. Scanning Electron Microscopy (SEM) is helpful for vesicle observation which is necessary for distinguishing between uniseriate (i.e. *A. aculeatus, A. japonicus*) and biseriate species (ie. *A. carbonarius, A. ibericus, A. niger*) and conidia ornamentation which can distinguish between *A. niger* aggregate, *A. carbonarius* and *A. ibericus* (Serra et al., 2006; Varga et al., 2000) (Table 1).

As for other fungal species studies based on molecular sequence analysis of ribosomal and ubiquitous genes (ITS, IGS, calmodulin, ß-tubulin, elongation factor) and polymorphisms by obtained Amplified Fragment Length Polymorphism (AFLP), Random Amplified Polymorphic DNA (RAPD) and microsatellites, have been performed for Aspergilli isolated from grapes. These studies have provided useful information on the taxonomy of BA and methods for their detection, identification and monitoring (reviewed in Abarca et al., 2004; Geiser et al., 2007; Niessen et al., 2005; Perrone et al., 2007, 2009; Samson et al., 2007).

Species	Conidiophore	Conidial size (µm)	OTA production
A. japonicus/A. aculeatus	Uniseriate	4–5	Negative
A. niger aggregate	Biseriate	3–5	Positive (low %)
A. sclerotioniger	Biseriate	5–6	Positive
A. carbonarius	Biseriate	7–9	Positive (high %)
A. ibericus	Biseriate	5–7	Negative

Table 1. Some characteristics of the main black *Aspergillus* species (from Serra et al., 2006)

PCR primers for detecting BA target the generic sequences for identifying or characterizing the fungus taxonomy at the intraspecific level or the sequence of genes involved in mycotoxin biosynthesis which are not necessarily able to distinguish fungal species (Bau et al., 2005; Dao et al., 2005; Perrone et al., 2007; Sartori et al., 2006; Schmidt et al.. 2003, 2004; Serra et al., 2005).

Species-specific primers based on ITS sequence differences were developed for *A. ellipticus, A. heteromorphus, A. japonicus, A. niger* (Gonzales-Salgado et al., 2005), *A. carbonarius* and *A. ochraceus* (Patiño et al., 2005). Some others were developed for *A. carbonarius* identification based on SCAR primers (Pollastro et al., 2003; Pelegrinelli-Fungaro et al. 2004), whereas PCR-RFLP analysis is necessary to distinguish the strains of *Aspergillus niger* aggregate into two groups: *A. niger* and *A. tubingensis* (N and T type) (Accensi et al., 2001, Gonzalez-Salgado et al., 2005).

More primers have been developed for the genes encoding the polyketide synthases (PKSs) involved in OTA biosynthesis in both *Aspergillus* and *Penicillium* (Ayoub et al., 2010; Atoui et al., 2006; Dao et al., 2005; O'Callaghan et al., 2003).

Recent studies have reported the use of degenerate primers targeting the ketosynthase domain (KS) which identified a new pks gene from *A. carbonarius* (AC*pks*) (Atoui et al., 2006; Gallo et al., 2009). By screening *Aspergillus* isolates with AC*pks* specific primers, AC*pks* homologues appeared to be present in *A. sclerotioniger* and *A. ibericus* which are closely related to *A. carbonarius*.

A duplex real-time PCR assay for simultaneously detecting members of the *Aspergillus niger* aggregate and *A. carbonarius* was developed by López-Mendoza et al. (2009) and Selma et al. (2009). They targeted the beta -ketosynthase and acyl transferase domains of the poliketide synthase of *A. carbonarius* and the *A. niger* aggregate, the assay allowing preferential amplification at greater concentrations providing a fast and accurate tool to monitor, OTA-producing species in grapes in a single reaction. These approaches gave rise to molecular diagnostic assays based on expression profiling and which determine the molecular triggers controlling OTA biosynthesis in *Aspergillus* spp.

More recently real-time/quantitative PCR (qPCR) protocols have detected and quantified ochratoxigenic fungi, developed using constitutive genes (González-Salgado et al., 2009; Morello et al., 2007; Mulè et al., 2006) or genes involved in toxin biosynthesis (Atoui et al., 2007; Schmidt et al., 2004; Selma et al., 2009).

2.3 European biodiversity monitoring

DNA-based fingerprinting techniques such as AFLP, RFLP. RAPD, ap-PCR, and the sequencing of subgenomic DNA fragments have drastically improved the understanding of the occurrence and biodiversity of *Aspergillus* spp. in grapes and vineyards worldwide.

Sequencing techniques were primarily useful at the species identification level, whereas fingerprinting techniques were exploited at the intraspecific level (Ferracin et al., 2009; Dachoupakan et al., 2009; Perrone et al., 2007). Although all these studies contribute to analysing the species composition and genetic diversity of grape mycobiota, no genotypical differences could be established between OTA producers and non producers (Ilic et al., 2001, 2004; Martinez-Culebraz et al., 2009; Chiotta et al., 2011). Moreover, there was no correlation between genotype, the ability to produce OTA and geographical origin (Niessen et al., 2005). Surveys conducted in Europe during the four-year EU project 'Wine-Ochra Risk' (QLK1-CT 2001-01761) indicated a significant correlation between the incidence of grape infected by black aspergilli, as potentially OTA producer, at harvest and in climatic conditions and geography (latitude and longitude); there was increasing incidence from West to East and North to South (Battilani et al., 2006). Aspergilli in vineyards varied depending on years and geographic areas: France, Greece and Israel were the areas with the highest incidence, followed by South Italy, Spain and Portugal (Abarca et al., 2001; Battilani et al., 2006; Guzev et al., 2006; Logrieco et al., 2007; Otteneder & Majerus., 2000; Sage et al., 2002). In countries with colder temperate climates such as Germany, Northern Hungary, the Czech Republic as well as the northern parts of Portugal, France and Italy, BA has not often been isolated from grapes, although sometimes OTA has been detected in wines. The identification of OTA producing *Penicillium* species from grapes in Northen Italy and France suggests they could be responsible for contamination in these regions (Battilani et al., 2001; Rousseau, 2004).

Surveys in 107 vineyards in the Mediterranean basin have identified four main *Aspergillus* populations: *A. carbonarius, A. tubingensis, A. niger,* and a group of *Aspergillus* 'uniseriate' isolates morphologically indistinguishable from *A. japonicus* and *A. aculeatus.* The latter could be clearly distinguished by molecular tools such as AFLP, RFLP and sequence analyses (Bau et al. 2006; Perrone et al. 2006a, 2006b). Highest genetic variability was observed in the *A. niger* group due to its complexity and the difficulty of identifying it at species level by both AFLP and the sequencing of calmodulin and β-tubulin subgenomic fragments (Perrone et al., 2007). In Australian vineyards, Leong et al. (2007) documented the dominance of *A. niger* over *A. carbonarius* and *A. aculeatus.* Polyphasic studies using macro- and micromorphology, secondary metabolite profiles, partial sequences of β-tubulin, calmodulin and ITS genes, and AFLP analysis led to the description of a new *Aspergillus* species: *A. ibericus,* which is closely related to *A. carbonarius* but unable to produce OTA (Serra et al. 2006) and *A. brasiliensis* belonging to the *A. niger* aggregate (Varga et al., 2007), both isolated from grapes in the Iberian Peninsula; *A. uvarum,* morphologically very similar to *A. japonicus* and *A. aculeatus,* but clearly distinct by the molecular analysis of grapes samples in Portugal, Italy, France, Israel, Greece and Spain.

In Italy, field surveys studied the fungi associated with grapes and their ability to produce OTA in different grape-growing areas (as regards grape variety and farming methods) in the north and south of the country (Battilani et al., 2002, 2006). Analysis of these grape samples revealed that *A. niger* aggregate was the prevalent species and *A. carbonarius* was mostly found in Southern Italy and Sicily (Lucchetta et al., 2010; Oliveri, 2007). *A. carbonarius* was never dominant at different growth stages, or in different geographical areas and years, but it was confirmed as the key fungus because of the high percentage of strong OTA producing isolates in the population.

In sixteen vineyards located in 13 provinces (including Modena, Imola, Ravenna, Brindisi and the warmest places such as Trapani and Ragusa), the effect of geographic area on fungal flora was confirmed, even though a major role was played by meteorological conditions, both on fungal colonisation and the OTA content in bunches. BA were present in bunches from setting, colonising most berries at early veraison (Battilani et al., 2006). The detection of isolates belonging to the *A. niger* aggregate, A. *carbonarius* and uniseriate varied with growth stage. At setting and berry pea-sized stages, more than 50% of isolates belonged to uniseriate; starting from early veraison, the *A. niger* aggregate became dominant (about 50%) whereas A. *carbonarius* was around 20% from pea-size to harvesting. The region with the highest percentage of grapes berries colonised by the *A. niger* aggregate was Veneto, while the lowest was in central Emilia Romagna. The highest incidence of *A. carbonarius* was detected in Puglia and the lowest in Emilia Romagna and Veneto. The number of OTA-producing strains among BA, isolated in each vineyard at different growth stages, was generally very limited (an encouraging result) (Battilani et al., 2006).

Molecular techniques to investigate strain variation in toxigenic and non-toxigenic black *Aspergillus* spp. showed that isolates of A. *carbonarius* and *A. niger* clustered into species groups, however, within species, strains displaying similar degrees of toxigenicity did not cluster together when characterized by RAPD techniques (Ilic et al., 2001, 2004). The profile of the *Aspergillus terreus* species isolated from dried grapes, analysed by RAPD, indicated great genomic diversity (Narasimhan & Asokan, 2010). Martínez-Culebras et al. (2009) recently carried out a study on ochratoxigenic mycobiota in grapes by ap-PCR sequence analysis of the ITS and IGS regions and their ability to produce OTA. Based on ap-PCR profiles, derived from two microsatellite primers, three main groups were obtained by UPGMA cluster analysis corresponding to *A. carbonarius, A. niger* and *A. tubingensis*. The cophenetic correlation values corresponding to ap-PCR UPGMA analysis showed higher genetic variability in *A. niger* and *A. tubingensis* than in *A. carbonarius*. In addition, no genotypical differences could be established between OTA producers and non-producers in all the species analysed. Regarding uniseriate black aspergilli, low divergence was found between *A. aculetus* and *A. uvarum*. OTA-production seems to be strain related since it was found in different clusters, with either ap-PCR or IGS-ITS phylogenetic analysis.

2.4 Epidemiology in vineyard

Black aspergilli are affected by several factors in the vine environment, i.e., grape status , the number of damaged grape berries, meteorological conditions, vineyard location, the cropping system as well as chemical treatments (Battilani et al., 2003b, 2006; Belli et al., 2005; 2007a; Blesa et al., 2006; Clouvel et al., 2008; Hocking et al., 2007; Leong et al., 2006). Generally fungi have been detected in vineyards and on grapes from setting. However, grape aspergilli increase gradually, reaching their maximum values at the beginning of veraison and ripening (Battilani et al., 2002). As *Aspergillus* species are not considered primary pathogens, various grape damage, such as attack by other fungi or mechanical injury, dramatically increases the risk of fungal infection by these species and OTA contamination (Serra et al., 2006; Belli et al., 2007b). However, grape damage due to insects, birds or other fungal infections, is the primary factor affecting the development of the disease and OTA accumulation in grapes (Cozzi et al., 2006).

Some Australian studies have demonstrated that vineyard soil at a depth of 0–5 cm beneath the vines is the primary reservoir of black aspergilli (Clarke et al., 2003; Kazi et al. 2004; Leong et al. 2006). Concentrations were also higher in the soil directly beneath the vines compared to the inter-row area. It is postulated that air movement deposits spores from the soil onto the grapes berry surfaces, because BA spores in air samples were higher closer to the soil (Kazi et al., 2003a). Soil temperature could also affect the incidence of *A. carbonarius* in the soil; the optimal temperature for spore survival was around 25 °C, with counts decreasing at 15 °C and 35°C. Survival at 40 °C was poor (Kazi et al., 2003b, 2004).

Agronomic practices and biological and chemical treatments have been found to reduce BA colonization and OTA levels in grapes and grape-derived products (reviewed in Varga & Kozakiewicz, 2006).

3. Biodiversity in a restricted geographical area: A case study from the Mount Etna Slopes

3.1 Vineyard mycobiota

Most studies address the biodiversity of grape mycoflora and above all of BA in countries and in wine producing areas. Recently we performed a two-year survey in a restricted geographical area to assess the population density and biodiversity of the mycoflora associated with the grapes, air and soil of the vineyards on the slopes of Mount Etna (eastern Sicily, Italy) where there is a long tradition of grape cultivation. This area is characterized by a temperate Mediterranean climate, with an average annual rainfall of 800 mm and high day/night temperature fluctuations. Moreover, the area is characterized by old and authoctonous wine-grape cultivars, i.e. Etna rosso DOC (Nerello mascalese, Nerello Cappuccio) and Etna Bianco (Carricante, Catarratto).

Special emphasis was made on toxigenic fungal species, e.g. *Aspergillus* Sect. *Nigri* spp. and *Penicillium* spp. (Oliveri, 2006; Oliveri et al., 2008). It is postulated that air movement deposits soil fungal spores onto grapes, because their incidence in air samples increases closer to the soil. So, healthy grape soil beneath the vines was ecologically monitored and its air was sampled at two different heights at the pea stage, early veraison and ripening using plating methods. Spore-producing filamentous fungi were detected, identified at the genus level, and then the *Aspergillus* and *Penicillium* strains were isolated and identified at the species level (Fig. 1).

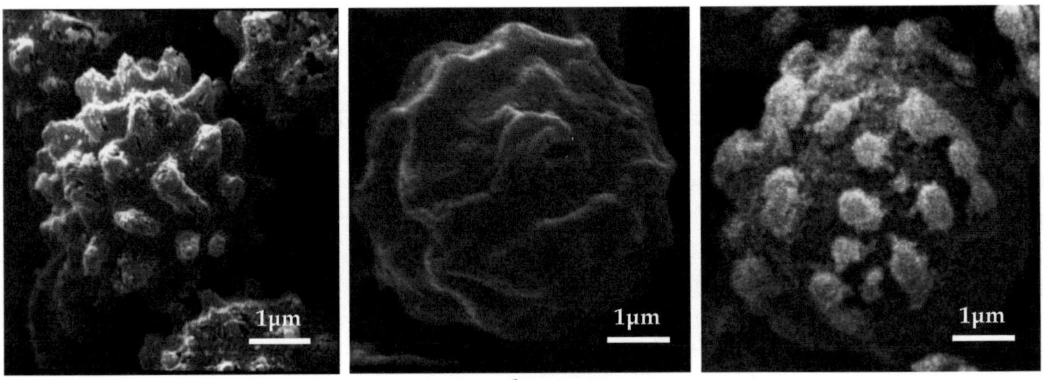

a b c

Fig. 1. Scanning electron microscopy pictures of a) *A. carbonarius*, *A. niger* b) and c) *A. japonicus* spores, where spore ornamentation differences are clearly seen (bar = 1 μm) (Zeiss, DSM 940).

The most frequent genera isolated from air, soil and grapes by increasing order were *Aspergillus, Penicillium, Cladosporium* and *Rhizopus*.

Fungi belonging to *Aspergillus* spp. were present in all the sampled matrices from the pea stage and they were predominant to *Penicillium* spp. from early veraison to ripening. Fungi from 6 genera were isolated on grapes at the ripening stage in eight different vineyards (Fig. 2).

Population densities of *Aspergillus* spp. in grape wash water ranged between 4.5 x 10^3 and 1.2 x 10^4 cfu ml^{-1}.

The most frequent *Penicillium* species isolated from the vineyard eco-system were *P. chrysogenum*, *P. expansum* and *P.olsonii* (Tab. 2). *P. verrucosum* was isolated from only one soil sample. Among the *Aspergillus* species, the most frequent were from section *Nigri*; the level of contamination by *A. ochraceus* and *A. flavus* was low.

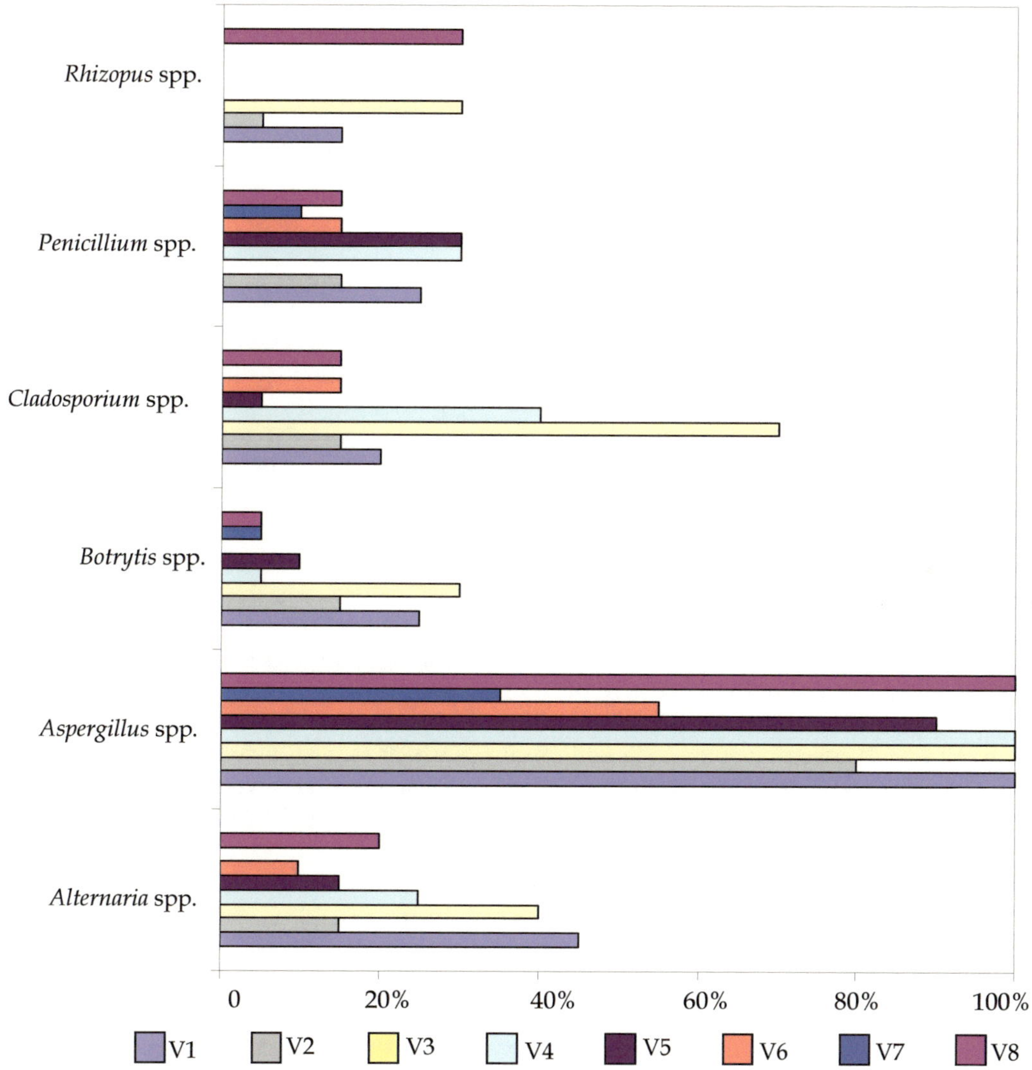

Fig. 2. Population composition (%) detected in 8 vineyards (V1-V8) from 20 grape samples, during the ripening stage and over a 2-year survey.

Genus	Species	n. samples [a]		
		air (224)	Soil (112)	Grapes (160)
Aspergillus	*niger* aggregate	125	94	128
	carbonarius	43	49	38
	flavus	18	21	1
	ochraceus	8	20	1
Penicillium	*aurantiogriseum*	4	3	n.d.
	chrysogenum	50	55	n.d.
	expansum	63	22	15
	italicum	29	19	14
	olsonii	n.d.	7	n.d.
	verrucosum	n.d.	1	n.d.

n.d.= not detected
[a] in parenthesis the total number of analyzed samples for each source

Table 2. *Aspergillus* and *Penicillium* spp. isolated from air, soil and grape samples over a two year survey.

3.2 Black Aspergilli and OTA producers

Black aspergilli appeared in all the tested samples, their incidence being higher at early veraison and ripening (Oliveri et al., 2006; Oliveri, 2006) (Fig.3 a-c). According to macro- and micromorphological characteristics, they were identified and classified into two main groups, *A. niger* aggregate and *A. carbonarius* (Fig.3 g-n). *A. ochraceus* has occasionally been detected in grape samples (Fig.3 d-f).

A subset of 66 strains was selected for further analysis. PCR assays supported the morphological identification. *A. niger, A. carbonarius* and *A. japonicus* were identified by target sequences for each species according to assays described by González-Salgado et al. (2005) and Patiño et al. (2005). In order to characterise the species in *A. niger* aggregate, i.e. *A. niger* and *A. tubingensis,* very difficult to differentiate by classical morphological criteria, the RFLP analysis with *Rsa*I was performed. This differentiation is very important to as to avoid overestimating toxicological contamination and related risks. A primer annealing site or restriction nuclease cleavage site was further confirmed by ITS sequencing which also confirmed the identity of the isolates (Oliveri et al., 2008).

The OTA production of isolates belonging to *A. carbonarius, A. niger, A. tubingensis* and *A. japonicus* was assessed by enzyme-linked immunosorbent assay (Oliveri et al., 2006b, 2008). 56% of strains were shown to produce OTA. *A. carbonarius* isolates were the strongest OTA producers with some of them producing high concentrations of OTA (>40 ppb).

3.3 Intraspecific variability

A fAFLP protocol was used to assess specific and intraspecific variability (Oliveri et al., 2008). In agreement with other studies (Perrone et al., 2006a, 2006b), the AFLP technique generated enough polymorphism to differentiate between and within the species of black aspergilli. *A. niger, A. tubingensis, A. carbonarius* and *A. japonicus* strains were clearly differentiated, although *A. niger* strains clustered into two different groups. Intraspecific variability didn't correlate with the isolate origin. In fact isolates from different vineyards either of grape or the environment could also cluster in the same or in different clusters. Perrone et al. (2006a, 2006b) analyzed representative strains from the main wine producing

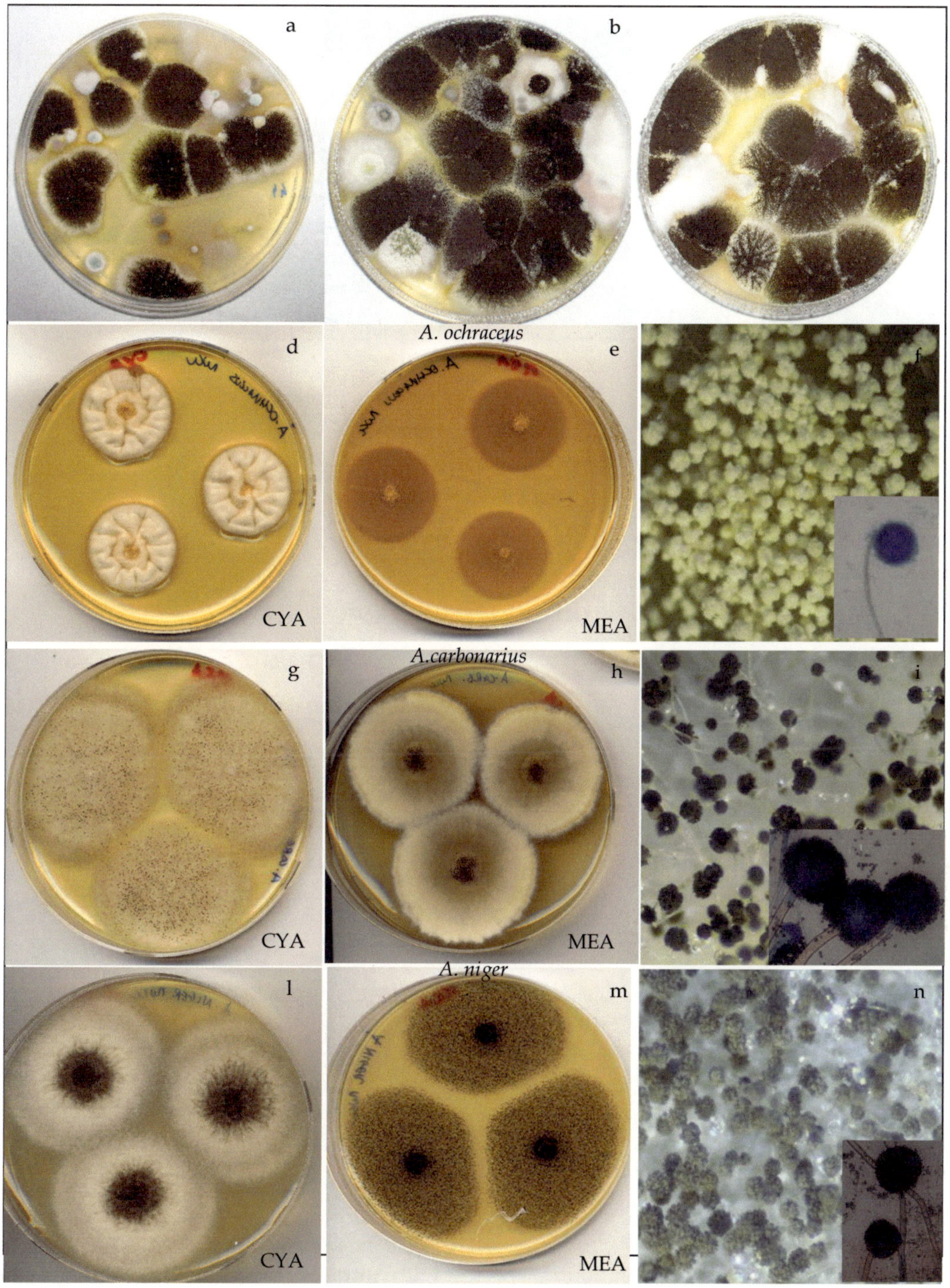

Fig. 3. Fungal colonies isolated from grape (a) , soil (b) and air (c) samples from vineyard. Colony morphologies and light microscopy pictures of conidiphores and conidia of representative isolates belonging to *Aspergillus ochraceus* (d–f), *A. carbonarius* (g - i), *A. niger* (l - n).

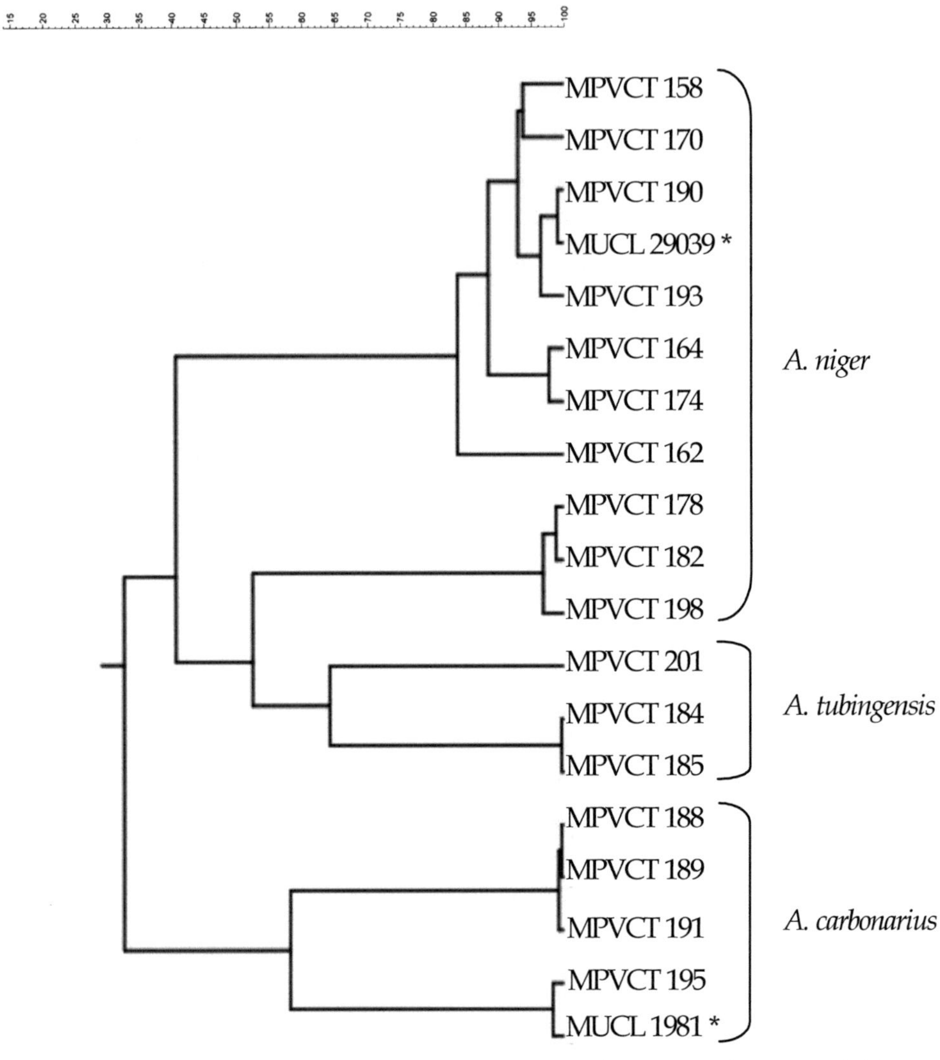

MPVCT: Micothèque of Institute of Plant Pathology, University of Catania, Italy; USA; MUCL: Micothèque de L'Universitè Catholique de Lovain, Belgium
* Reference strains

Fig. 4. UPGMA dendrogram obtained from fAFLP analysis with the selective primer pair E-AT (Cy5-labelled) and M-CT on 17 isolates (10 *A. niger*, 3 *A. tubingensis*, 4 *A. carbonarius*) and 2 reference strains isolated from grape samples in the same vineyard. Fragments between 50 and 600 bp were analysed with PHYLIP® v. 3.66 software

European countries (Italy, France, Spain, Portugal, Greece and Israel) and the four main groups were obtained by AFLP clustering analysis of the strains, three of them showing a well-defined homogeneous population/species with intraspecific homology higher than 48%: *A.carbonarius*, *A. tubingensis* and *Aspergillus* 'uniseriate'. The fourth cluster, called *A. niger* 'like', showed low homology with *A. niger* 'type strain' and high internal heterogeneity. The intra-population variability of *Aspergillus* Section *Nigri* strains isolated from grape samples within the same vineyard proved that mixed populations of *A. niger* and *A. carbonarius* were present and most of them were OTA producers (Fig. 4). As for larger population studies, no correlation was found between genotypes and mycotoxin production (Martinez-Culebraz et al., 2009).

4. Conclusion

This chapter summarizes data on grape mycoflora, toxigenic fungi and mycotoxin contamination at the pre-harvesting, harvesting and processing stages. Grape rotting and spoilage can be caused by a variety of fungal species, including *Botrytis cinerea, Penicillium, Aspergillus, Alternaria* and *Cladosporium.* In recent years, black *Aspergillus* species (Section *Nigri*) and in particular *A. carbonarius* and *A. niger* aggregate have been described as the main source of grape contamination with the mycotoxin ochratoxin A. In this chapter, we highlighted how *Aspergillus* species distribution on European grapes may occur and vary in relation to meteorological conditions and geographical areas and several studies have shown an increase in the amount of OTA in warmer climates. The literature on various molecular methods used for species identification is reviewed and a critical evaluation of the usefulness of various techniques and genomic loci for the species identification of black aspergilli is presented. Reports of the occurrence of black aspergilli in vineyards and their potential toxigenicity must be reconsidered on the basis of the wide molecular biodiversity found within morphologically undistinguishable strains of this section. Mycotoxin production is a characteristic of the species, so by studying the species intraspecific biodiversity can predict potential mycotoxin hazards. Different isolates belonging to the black aspergilli species showed varying abilities to produce OTA so it becomes utmost importance to guarantee a quality control of the grapes and grape derived products, through accurate contaminant mycoflora identification.

5. References

Abarca, M.L., Accensi, F., Bragulat, M.R. & Cabañes, F.J. (2001). Current importance of ochratoxin A-producing *Aspergillus* spp. *Journal of Food Protection,* Vol. 64, pp. 903–906, ISSN 0362-028X

Abarca, M.L., Accens,i F., Cano, J., Cabañes, F.J. (2004). Taxonomy and significance of black aspergilli. *Antonie Van Leeuwenhoek,* Vol. 86, pp. 33–49, ISSN 0003-6072

Accensi, F., Cano, J., Figuera, L., Abarca, M.L. & Cabañes, F.J. (1999). New PCR method to differentiate species in the *Aspergillus niger* aggregate. *FEMS Microbiology Letters,* Vol. 180, No.2, pp. 191–196, ISSN 0378-1097

Atoui, A., Dao, P., Mathieu, F. & Lebrihi, A. (2006). Amplification and diversity analysis of ketosynthase domains of putative polyketide synthase genes in *Aspergillus ochraceus* and *Aspergillus carbonarius. producers of ochratoxin A. Molecular Nutrition & Food Research,* Vol. 50, pp. 448-493, ISSN 1613-4133

Atoui, A., Mathieu, F. & Lebrihi, A. (2007). Targeting a polyketide synthase gene for *Aspergillus carbonarius* quantification and ochratoxin A assessment in grapes using real-time PCR. *International Journal of Food Microbiology,* Vol. 115, pp. 313–318, ISNN 0168-1605

Ayoub, F., Reverberi, M., Ricelli, A., D'Onghia, A. M. & Yaseen, T. (2010). Early detection of *Aspergillus carbonarius* and *A. niger* on table grapes: a tool for quality improvement. *Food Additives and Contaminants,* Vol. 27, No. 9, pp. 1285-1293, ISNN 0265-203X

Battilani, P., Giorni, P. & Pietri, A. (2001). Role of cultural factors on the content of ochratoxin A in grape. *Journal of Plant Pathology,* Vol. 83,p. 231, ISSN 0929-1873

Battilani, P., & Pietri, A. (2002). Ochratoxin A in grapes and wine. *European Journal of Plant Pathology,* Vol. 108, pp.639-643, ISSN 0929-1873

Battilani, P., Pietri, A., Bertuzzi, T., Languasco, L., Giorni, P. & Kozakiewicz, Z. (2003). Occurrence of ochratoxin A-producing fungi in grapes grown in Italy. *Journal of Food Protection,* Vol. 66, pp. 633–636, ISSN 0362-028X

Battilani, P., Giorni, P. & Pietri, A. (2003b). Epidemiology and toxin-producing fungi and ochratoxin A. *European Journal of Plant Pathology,* Vol. 109, pp. 723–730, ISSN 0929-1873

Battilani, P., Barbano, C., Marin, S., Sanchis, V., Kozakiewicz, Z. & Magan, N. (2006). Mapping of *Aspergillus* Section *Nigri* in Southern Europe and Israel based on geostatistical analysis. *International Journal of Food Microbiology* ,Vol. 111, pp. S72-S82, ISNN 0168-1605

Bau, M., Castellà, G., Bragulat, M.R. & Cabañes F.J. (2005). DNA-based characterization of ochratoxin-A.producing and non-producing *Aspergillus carbonarius* strains from grapes. *Research in Microbiology,* Vol. 156, pp. 375-381, ISSN 0923-2508

Bellí, N., Mitchell, D., Marín, S., Alegre, I., Ramos, A. J., Magan, N. & Sanchis, V.,(2005). Ochratoxin A-producing fungi in Spanish wine grapes and their relationship with meteorological conditions. *European Journal of Plant Pathology,* Vol. 113, pp. 233-239, ISSN 0929-1873

Bellí, N., Marín, S., Argilés, E., Ramos, A.J. & Sanchis, V. (2007a). Effect of chemical treatments on ochratogenic fungi and common mycobiota of grapes (*Vitis vinifera*). *Journal of Food Protection,* Vol. 70, No.1, pp. 157-163, ISSN 0362-028X

Bellí, N., Marín, S., Coronas, I., Sanchis, V. & Ramos, A.J. (2007b). Skin damage, high temperature and relative humidity as detrimental factors for *Aspergillus carbonarius* infection and ochratoxin A production in grapes. *Food Control,* Vol. 18, pp. 1343-1349, ISNN 0956-7135

Blesa, J., Soriano, J. M., Moltó, J. C. & Mañes, J. (2006). Factors affecting the presence of ochratoxin A in wines. *Critical Reviews in Food Science and Nutrition,* Vol. 46, pp. 473-478, ISSN 1040-8398.

Castegnaro, M. & Pfohl-Leszkowicz, A. (2002). Les mycotoxines: Contaminations omnipresentes dans l'alimentation animales et humaines. La sécurité alimentaire du consommateur, Lavoisier, Tec & doc., Moll &Moll (Eds.), Paris.

Chiotta, M.L., Reynoso, M.M., Torres, A.M., Combina, M. & Chulze, S.N. (2011). Molecular characterization and toxigenic profile of *Aspergillus* section *Nigri* populations isolated from the main grape-growing regions in Argentina. *Journal of Applied Microbiology,* Vol. 110, No.2, pp. 445-454, ISSN 1364-5072

Clarke, K., Emmett, R.W., Kazi, B.A., Nancarrow, N. & Leong, S.,(2003). Susceptibility of dried grape varieties to berry splitting and infection by *Aspergillus carbonarius Proceedings of 8th International Congress of Plant Pathology,* vol. 2, p. 140, Christchurch, New Zealand

Clouvel, P., Bonvarlet, L., Martinez, A., Lagouarde, P., Dieng, I. & Martin, P. (2008). Wine contamination by ochratoxin A in relation to vine environment. *International Journal of Food Microbiology,* Vol. 123, pp. 74-80, ISNN 0168-1605

Cozzi, G., Pascale, M. ,Perrone, G. ,Visconti, A. & Logrieco, A. (2006) Effect of *Lobesia botrana* damages on black aspergilli rot and ochratoxin A content in grapes. *International Journal of Food Microbiology* Vol. 111, No.S1, pp. S88-S92

Da Rocha R., Palacios, V. ,Combina, M. ,Fraga, M.E., De Oliveira Rekson, A., Magnoli, C.E. & Dalcero, A.M. (2002) Potential ochratoxin A producers from wine grapes in Argentina and Brazil. *Food Additives and Contaminants* 19, pp. 408-414.

Dachoupakan, C., Ratomahenina, R., Martinez, V., Guiraud, J.P., Baccou, J.C. & Schorr-Galindo, S. (2009). Study of the phenotypic and genotypic biodiversity of potentially ochratoxigenic black aspergilli isolated from grapes. *International Journal of Food Microbiology,* Vol. 132, No. 1, pp. 14-23, ISNN 0168-1605

Dao, H.P., Mathieu, F. & Lebrihi, A. (2005). Two primer pairs to detect OTA producers by PCR method. *International Journal of Food Microbiology,* Vol. 104, pp. 61-67, ISNN 0168-1605

Ferracin, L.M., Frisvad, J.C., Taniwaki, M.H., Iamanaka, B.T, Sartori, D., Schapovaloff, M.E. & Pelegrinelli Fungaro, MH. (2009). Genetic relationships among strains of the *Aspergillus niger* Aggregate. *Brazilian Archives of Biology and Technology,* Vol. 52, pp. 241-248, ISSN 1516-8913

Fleet, G.H. (1999) Microorganisms in food ecosystems. *International Journal of Food Microbiology,* Vol. 50, pp. 101–117, ISNN 0168-1605

Fleet, G.H. (2001) Wine. In: *Food Microbiology Fundamentals and Frontiers,* 2nd edn, Doyle, M.P., Beuchat, L.R. and Montville, T.J. (Ed.), 747–772. Washington, DC: ASM Press.

Frisvad, J.C., Smedsgaard. J., Samson. R.A., Larsen, T.O. & Thrane, U. (2007). Fumonisin B2 production by *Aspergillus niger. Journal of Agricultural and Food Chemistry,* Vol. 14;55, No.23, pp. 9727-32, ISSN 0021-8561

Gallo, A., Perrone, G., Solfrizzo, M., Epifani, F., Abbas, A., Dobson, A.D.W. & Mulè, G. (2009). Characterisation of a pks gene which is expressed during ochratoxin A production by *Aspergillus carbonarius. International Journal of Food Microbiology,* Vol. 129, (October 2009) pp. 8–15, ISNN 0168-1605

Geiser, D.M., Klich, M.A., Frisvad, J.C., Peterson, S.W., Varga, J. & Samson R.A. (2007). The current status of species recognition and identification in *Aspergillus. Studies in Mycology,* Vol. 59, pp. 1-10, ISSN 0166-0616

Gonzáles-Salgado, A., Patiño, B., Vázquez, C. & Gonzáles-Jaén, M.T. (2005). Discrimination of *Aspergillus niger* and other *Aspergillus* species belonging to section *Nigri* by PCR assays. *FEMS Microbiology Letters,* Vol. 245, pp. 353–361, ISSN 0378-1097

González-Salgado, A., Patino, B., Gil-Serna, J., Vazquez, C. & González -Jaen, M.T. (2009). Specific detection of *Aspergillus carbonarius* by SYBRs Green and TaqMans quantitative PCR assays based on the multicopy ITS2 region of the rRNAgene. *FEMS Microbiology Letters,* Vol. 295, pp. 57–66, ISSN 0378-1097

Guzev, L., Danshin, A., Ziv, S.& Lichter, A. (2006). Occurrence of ochratoxin A producing fungi in wine and table grapes in Israel. *International Journal of Food Microbiology,* Vol. 111, pp. S67–S71, ISNN 0168-1605

Heenan, C.N., Shaw, K.J. & Pitt, J.I. (1998) Ochratoxin A production by *Aspergillus carbonarius* and *A. niger* isolates and detection using coconut cream agar. *Journal of Food Mycology*, Vol. 1, pp. 67–72.

Hewitt, W.B. (1988). Berry rots and raisin moulds. In: Pearson, R.G., Goheen, A.C. (Eds.), Compendium of Grape Diseases. The American Phytopathological Society, pp. 26–28,APS Press, ISBN 92-9043-153-9. St. Paul, Minnesota.

Hocking, A.D., Su-lin, L. Leong, S.L., Kazi, B.A., Emmett, R.W., Eileen, S. & Scott E.S. (2007). Fungi and mycotoxins in vineyards and grape products. *International Journal of Food Microbiology*, Vol. 119, No.1-2, (October 2007), pp. 84-88, ISNN 0168-1605

IARC (1993) IARC Monographs on the Evaluation of Carcinogenic Risks to Humans (1993). Ochratoxin A. In: Some Naturally Occurring Substances: Food Items and Constituents, Heterocyclic Aromatic Amines and Mycotoxins. IARC Press, Lyon, France Vol. 56, pp. 489–521.

Ilic, Z., Pitt, J.I. & Carter, D.A. (2001). Identifying genes associated with ochratoxinA production in *Aspergillus carbonarius* and *Aspergillus niger* . *Proceedings of 21st Fungal Genetics Conference*, Pacific Grove, USA, March 13–18, 2001

Ilic, Z., Pitt, J.I., & Carter, D.A., (2004). Genetic diversity and ochratoxin production in the black Aspergilli. Australian Mycological Society, joint meeting with the Australian Society for Microbiology Annual Scientific Meeting, Sydney, Australia, September 27–28, 2004

Kazi, B.A., Emmett, R.W., Clarke, K. & Nancarrow, N., (2003a). Black *Aspergillus* mould in Australia. *Proceedings of 8th International Congress of Plant Pathology*, Vol. 2, p. 119, Christchurch, New Zealand

Kazi, B.A., Emmett, R.W., Nancarrow, N. & Clarke, K. (2003b). Effects of temperature, moisture and/or irrigation on the survival of *Aspergillus carbonarius* in soil. *Proceedings of 8th International Congress of Plant Pathology*, Vol. 2, p. 140, Christchurch, New Zealand

Kazi, B.A., Emmett, R.W., Nancarrow, N. & Clarke, K. (2004). Incidence of *Aspergillus carbonarius* in Australian vineyards. In: Ophel-Keller, K., Hall, B. (Eds.), 75–76, 3rd Australasian Soil-borne Disease Symposium, South Australian Research and Development Institute, Adelaide, Australia

Klich, M.A.& Pitt, J.I. (1988). A Laboratory Guide to Common *Aspergillus* Species and Their Teleomorphs. CSIRO Division of Food Processing, North Ryde, New South Wales, Australia.

Klich, M.A (2002). Identification of common *Aspergillus* Species. Centraalbureau Voor Schimmelcultures, Utrech, The Netherlands.

Leong, S.L., Hocking, A.D., Pitt, J.I., Kazi, B.A., Emmett, R.W. & Scott, E.S. (2006). Australian research on ochratoxigenic fungi and ochratoxin A. *International Journal of Food Microbiology*, Vol. 111, pp. S10-S17, ISNN 0168-1605

Leong, S.L., Hocking, A.D. & Scott, E.S. (2007). *Aspergillus* producing ochratoxin A: isolation from vineyard soils and infection of Semillon bunches in Australia. *Journal of Applied Microbiology*, Vol. 102, pp. 124–133, ISSN 1364-5072.

Logrieco, A., Moretti, A., Perrone, G. & Mulè, G. (2007). Biodiversity of complexes of mycotoxigenic fungal species associated with *Fusarium* ear rot of maize and *Aspergillus* rot of grape. *International Journal of Food Microbiology*, Vol. 119, pp. 11–16, ISNN 0168-1605

Logrieco, A., Ferracane, R., Visconti, A. & Ritieni, A. (2010). Natural occurrence of fumonisin B_2 in red wine from Italy. *Food Additives and Contaminants*, Vol. 27, No. 8, pp. 1136-1141, ISNN 0265-203X

López-Mendoza, M.C., Crespo-Sempere, A. & Martínez-Culebras P.V. (2009). Identification of *Aspergillus tubingensis* strains responsible for OTA contamination in grapes and wine based on the acyl transferase domain of a polyketide synthase gene. *International Journal of Food Science & Technology* , Vol. 44,No.11, (November 2009), pp. 2147–2152, ISSN 0950-5423

Lucchetta, G., Bazzo, I. Cortivo, G. dal Stringher, L., Bellotto, D., Borgo, M. & Angelini, E. (2010). Occurrence of black aspergilli and ochratoxin A on grapes in Italy. *Toxins*, Vol. 2, No.4, pp. 840-855, ISNN 2072-6651

Martínez-Culebras, P.V., Crespo-Sempere, A., Sánchez-Hervás, M., Elizaquivel, P., Aznar, R., Ramón, D. (2009). Molecular characterization of the black *Aspergillus* isolates responsible for ochratoxin A contamination in grapes and wine in relation to taxonomy of *Aspergillus* section *Nigri*. *International Journal of Food Microbiology*, Vol. 132, No.1, (June 2009), pp. 33-41, ISNN 0168-1605

Magnoli, C.E., Violante, M., Combina, M., Palacio, G. & Dalcero, A.M. (2003). Mycoflora and ochratoxin-producing strains of *Aspergillus* section *Nigri* in wine grapes in Argentina. *Letters in Applied Microbiology*, Vol. 37, pp. 179–184, ISSN 0266-8254

Medina, A., Mateo, R., López-Ocaña, L., Valle-Algarra, F.M. & Jiménez, M. (2005). Study of Spanish grape mycobiota and ochratoxin A production by isolates of *Aspergillus tubingensis* and other members of *Aspergillus* Section *Nigri*. *Applied and Environmental Microbiology* 71, pp. 4696–4702, ISSN 0099- 2240

Morello, L.G., Sartori, D., De Oliveira Martinez, A.L., Vieira, M.L.C., Taniwaki, M.H. & Pelegrinelli Fungaro, M.H. (2007). Detection and quantification of *Aspergillus westerdijkiae* in coffee beans based on selective amplification of β-tubulin gene by using real-time PCR. *International Journal of Food Microbiology*, Vol. 119, pp. 270–276, ISNN 0168-1605

Mogensen, J. M., Frisvad, J. C. , Thrane, U. & Nielsen, K.F. (2010a). Widespread occurrence of the mycotoxin fumonisin B_2 in wine. *Journal of Agricultural and Food Chemistry*, Vol. 58, No.8, pp. 4853-4857, ISSN 0021-8561

Mogensen, J. M. Larsen, T. O. & Nielsen, K. F. (2010b). Production of fumonisin B_2 and B_4 by *Aspergillus niger* on grapes and raisins. *Journal of Agricultural and Food Chemistry*, Vol. 58, No.2, pp. 954-958, ISSN 0021-8561

Mulè, G., Susca, A., Logrieco, A., Stea, G. & Visconti, A. (2006). Development of a real-time PCR assay for the detection of *Aspergillus carbonarius* in grapes. *International Journal of Food Microbiology*, Vol. 111, Supplement 1, (September 2006), pp. S28–S34, ISNN 0168-1605

Narasimhan B. & Asokan M. (2010). Genetic variability of *Aspergillus terreus* from dried grapes using RAPD-PCR. *Advances in Bioscience and Biotechnology*, Vol. 1, No.4, (October 2010) pp. 345-353, ISSN 2156-8456

Niessen, L., Schmidt, H., Mühlencoert, E., Farber, P. & Karolewiez, A. (2005). Advances in the molecular diagnosis of ochratoxin A-producing fungi. *Food Additives and Contaminants*, Vol. 22, No. 4, pp. 324–334, ISNN 0265-203X

Nielsen, K.F., Mogensen, J.M., Johansen, M., Larsen, T.O. & Frisvad, J.C. (2009). Review of secondary metabolites and mycotoxins from the *Aspergillus niger* group. *Analytical and Bioanalytical Chemistry*, Vol. 395, No.5, pp. 1225-1242, ISNN 1618-2642

O'Callaghan, J., Caddick, M.X. & Dobson, A.D. (2003). A polyketide synthase gene required for ochratoxin A biosynthesis in *Aspergillus ochraceus*. *Microbiology*, Vol. 149,pp. 3485-3491, ISSN 1350-0872

Oliveri, C. (2006). Caratterizzazione morfologica e molecolare di specie ocratossigene di *Aspergillus* sez. *Nigri* isolate in vigneti dell'Etna ed esperienze di lotta biologica. PhD thesis, University of Catania, 179 pp.

Oliveri, C., Fardella, M., Grimaldi, V. & Catara, V. (2006a). Characterization of OTA-producing *Aspergillus* spp., strains isolated from sicilian vineyards. *Journal of Plant Pathology*, Vol. 88, No.3, pp. S51-S52, ISSN 1125-4653

Oliveri, C., Fardella, M., Grimaldi, V. & Catara, V. (2006b). Characterization of OTA-producing *Aspergillus* spp. strains isolated from sicilian vineyards *Journal of Plant Pathology*, Vol. 88, No.3, p. S52.

Oliveri, C., Torta, L. & Catara, V. (2008). A polyphasic approach to the identification of ochratoxin A-producing black *Aspergillus* isolates from vineyards in Sicily. *International Journal of Food Microbiology*, Vol. 27, (June 2008), pp. 147-154, ISNN 0168-1605

Otteneder, H. & Majerus, P. (2000). Occurrence of ochratoxin A (OTA) in wines: influence of the type of wine and its geographical origin. *Food Additives and Contaminants*, Vol. 17, pp. 793–798, ISNN 0265-203X

Patiño, B., Gonzáles-Salgado, A., Gonzáles-Jaén, M.T.& Vázquez, C. (2005). PCR detection assays for the ochratoxigen-producing *Aspergillus carbonarius* and *A. ochraceus* species. *International Journal of Food Microbiology*, Vol. 104, pp. 207–214, ISNN 0168-1605

Pelegrinelli-Fungaro, M.H.., Vissotto, P.C., Sartori, D., Vilas-Boas, L.A., Furlaneto, M C.& Taniwaki, M.H. (2004), A molecular method for detection of*Aspergillus carbonarius* in coffee beans. *Current Microbiology*, Vol. 49, pp. 123-127, ISSN 0343-8651

Perrone, G., Mulè, G., Susca, A., Battilani, P., Pietri, A. & Logrieco, A. (2006a). Ochratoxin A production and AFLP analysis of *Aspergillus carbonarius*, *Aspergillus tubingensis*, and *Aspergillus niger* strains isolated from grapes in Italy. *Applied and Environmental Microbiology*, Vol. 72, pp. 680–685, ISSN 0099- 2240

Perrone, G., Susca, A., Epifani, F. & Mulè, G. (2006b). AFLP characterization of Southern Europe population of *Aspergillus* Section *Nigri* from grapes. *International Journal of Food Microbiology*, Vol. 111, pp. S22–S27, ISNN 0168-1605

Perrone, G., Susca, A., Cozzi, G., Ehrlich, K., Varga, J., Frisvad, J.C., Meijer, M., Noonim, P., Mahakarnchanakul, W. & Samson, R.A. (2007). Biodiversity of *Aspergillus* species in some important agricultural products. *Studies in Mycology* 59, 53–66, ISSN 0166-0616

Perrone, G., Varga, J., Susca, A., Frisvad, J.C., Stea, G., Kocsubé, S., Tóth, B., Kozakiewicz, Z.& Samson, R.A. (2008). *Aspergillus uvarum* sp. nov., an uniseriate black *Aspergillus* species isolated from grapes in Europe. *International Journal of Systematic and Evolutionary Microbiology*, Vol.58, (2008), pp. 1032–1039, ISSN 14665026

Petzinger, E.& Weidenbach, A. (2002), Mycotoxins in the food chain: the role of ochratoxins. *Livestock Production Science*, Vol. *76*, (2002), pp. 245-250, ISSN 0301-6226.

Pfohl-Leszkowicz, A., Petkova-Bocharova, T., Chernozemsky, I.N. & Castegnaro, M. (2002). Balkan endemic nephropathy and associated urinary tract tumours: a review on aetiological causes and the potential role of mycotoxins. *Food additives and Contaminants*, Vol. 19, No. 3, (2002), pp.282-302, ISNN 0265-203X

Pitt, J.I. & Hocking, A.D. (1999). Fungi and Food Spoilage, 2nd edn. Aspen Publishers, Inc., Gaithersburg, Md.

Pitt, J.I. & Hocking, A.D. (2009). Fungi and Food Spoilage, 3rd edn. Springer-Verlag New York Inc., 04.09.2009, pp.540, ISBN 0387922067

Pollastro, S., De Miccolis Angelini, R., Abbatecola,A., De Guido, M.A. & Faretra, F. (2005). Real-Time PCR for quantitative detection of *Aspergillus carbonarius* in grapes and musts. *Proceedings of International Workshop: Ochratoxin A in grapes and wine: prevention and control*, p. 43, Marsala (TP), Italy, October 20-21, 2005

Pollastro, S., De Miccolis, R.M. & Faretra, F. (2006). A new semi-selective medium for the ochratoxigenic fungus *Aspergillus carbonarius*. *Journal of Plant Pathology*, Vol. 88, pp. 107–112, ISSN 1125-4653

Rousseau, J., 2004. Ochratoxin A in wines: Current knowledge. Retrieved from Vinidea.net, Wine Internet Technical Journal.

Sartori, D., Furlaneto, M.C., Martins, M.K., Ferreira de Paula, M.R., Pizzirani-Kleiner, A.A., Taniwaki, M.H. & Pelegrinelli-Fungaro, M.H. (2007). PCR method for the detection of potential ochratoxin-producing *Aspergillus* species in coffee beans. *Research in Microbiology*, Vol.157, 350–354, ISSN 0923-2508.

Sartori, D., Taniwaki, M.T., Iamanaka, B., & Pelegrinelli Fungaro M.H. (2010). Molecular Diagnosis of Ochratoxigenic Fungi. (2010) Part 1, pp. 195-212

Sage, L., Krivobok, S., Delbos, E., Seigle-Murandi, F.& Creppy, E.E. (2004). Fungal microflora and ochratoxin A production in grapes and musts from France. *Agricultural and Food Chemistry*, Vol. 50, pp. 1306-1311, ISSN 0021-8561

Sage, L., Garon, D., Seigle-Murandi, F. (2004). Fungal microflora and ochratoxin A risk in French vineyards. *Journal of Agricultural and Food Chemistry*, Vol. 52, pp. 5764–5768, ISSN 0021-8561

Samson, R.A., Hoekstra, E.S. & Frisvad, J.C., 2004. In: Samson, R.A., Hoekstra, E.S. (Eds.), Introduction to Food- and Airborne Fungi, 7th ed. Centraalbureau Voor Schimmelcultures, Utrech, The Netherlands. 389 pp.

Samson, R.A., Noonim, P., Meijer, M., Houbraken, J., Frisvad, J.C. & Varga, J. (2007). Diagnostic tools to identify black aspergilli. *Studies in Mycology*, Vol. 59, pp. 129–145, ISSN 0166-0616

Schmidt, H., Ehrmann, M., Vogel, R. F., Taniwaki, M. H. & Niessen, L. (2003). Molecular Typing of *Aspergillus ochraceus* and construction of species specific SCAR-primers based on AFLP. *Systematic & Applied Microbiology*, Vol. 26, pp.138-146, ISSN 0723-2020

Schmidt, H., Bannier, M., Vogel, R.F. & Niessen, L. (2004). Detection and quantification of *Aspergillus ochraceus* in green coffee by PCR. *Letters in Applied Microbiology*, Vol. 38, pp. 464–469, ISSN 0266-8254

Selma, M.V., Martinez-Culebras, P.V., Elizaquivel, P. & Aznar, R. (2009). Simultaneous detection of the main black aspergilli responsible for ochratoxin A (OTA) contamination in grapes by multiplex real-time polymerase chain reaction. *Food Additives and Contaminants*, Vol. 26, No.2, pp. 180-188, ISNN 0265-203X

Serra, R., Braga, A. & Venâncio, A. (2005). Mycotoxin-producing and other fungi isolated from grapes for wine production, with particular emphasis on ochratoxin A. *Research in Microbiology,* Vol. 156, pp. 515–521, ISSN 0923-2508

Serra, R., Cabañes, J., Perrone, G., Kozakiewicz, Z., Castellá, G., Venâncio, A. & Mulè, G. (2006). *Aspergillus ibericus*: a new species of the Section *Nigri* isolated from grapes. *Mycologia,* Vol. 98, No.2, pp. 295–306, ISSN 0027-5514

Serra, R., Mendonca, C. & Venâncio, A. (2006). Fungi and ochratoxin A detected in healthy grapes for wine production. *Letters in Applied Microbiology,* Vol. 42, pp. 42–47, ISSN 0266-8254

Susca, A., Proctor, R.H., Mulè, G., Stea, G., Ritieni, A., Logrieco, A. & Moretti, A. (2010). Correlation of Mycotoxin Fumonisin B_2 Production and Presence of the Fumonisin Biosynthetic Gene *fum8* in *Aspergillus niger* from Grape. *Journal of Agricultural and Food Chemistry,* Vol. 58, No.6, pp. 9266-9272, ISSN 0021-8561

Tournas, V.H. & Katsoudas, E. (2005). Mould and yeast flora in fresh berries, grapes and citrus fruits. *International Journal of Food Microbiology* Vol. 105, pp.11–17, ISNN 0168-1605

Tjamos, S.E., Antoniou, P.P., Kazantzidou, A., Antonopoulos, D.F., Papageorgiou, I, & Tjamos, E.C. (2004). *Aspergillus niger* and *Aspegillus carbonarius* in Corinth raisin and wine-producing vineyards in Greece: population composition, ochratoxin A production and chemical control. *Journal of Phytopathology, Vol.* 152:, pp. 250–255, ISSN 0931-1785

Tjamos, S.E., Antioniou, P.P & Tjamos, E.C. (2006). *Aspergillus* spp., distribution, population composition an ochratoxin A production in wine producing vineyards in Greece. *International Journal of Food Microbiology,* Vol. 111, pp. S61–S66, ISNN 0168-1605

Varga, J., Kevei F., Hamari, Z., Toth, B., Teren, J., Croft, J.H. & Kozakiewicz, Z. (2000). Genotypic and phenotypic variability among black aspergilli. In: Samson RA, Pitt JI, eds. Integration of modern taxonomic methods for *Penicillium* and *Aspergillus* classification. Amsterdam, the Netherlands: Harwood Academic Publishers. pp. 397–411.

Varga J & Kozakiewicz Z. (2006). Ochratoxin A in grapes and grape-derived products. *Trends in Food Science & Technology,* Vol. 17, pp. 72-81. ISSN 0924-2244

Varga, J., Kocsubé, S., Tóth, B., Frisvad, J.C., Perrone, G., Susca, A., Meijer, M. & Samson, R.A. (2007). *Aspergillus brasiliensis* sp. nov., a biseriate black *Aspergillus* species with world-wide distribution. *International Journal of Systematic and Evolutionary Microbiology,* Vol. 7, pp. 1925-1932, ISSN 1466-5026

Vrabcheva, T., Usleber, E., Dietrich, R.& Märtlbauer, E. (2000). Co-occurrence of ochratoxin A and citrinin in cereals from Bulgarian villages with a history of Balkan endemic nepropathy. *Journal of Agricultural and Food Chemistry,* Vol. 48, (2000), pp. 2483–2488, ISSN 0021-8561

Zimmerli, B. & Dick, R. (1996). Ochratoxin A in table wine and grape-juice: occurrence and risk assessment. *Food Additives and Contaminants,* Vol. 13, pp. 655–668, ISNN 0265-203X

12

Biodiversity of Yeasts in the Gastrointestinal Ecosystem with Emphasis on Its Importance for the Host

Vladimir Urubschurov and Pawel Janczyk
Chair for Nutrition Physiology and Animal Nutrition, University of Rostock
Germany

1. Introduction

Thinking of the diversity of the microbial world most readers will focus their attention to the bacteria and archea. However, among most of the ecosystems present on Earth, such as soil or intestine of animals, another microbial group has established: the yeasts. Their biodiversity has been hardly investigated although they possess probably as much adaptation potential as bacteria, considering the enormous differences between the habitats and the challenges the different ecosystems must face.

In the chapter the authors would like to provide to the reader the state of the art in the field of intestinal yeast research, with focus on the diversity of the yeasts in the gastrointestinal tract of animals - insects and mammals. Up to date there are about 1,500 yeast species known, belonging to two phyla *Ascomycota* (Suh et al., 2006a) and *Basidiomycota* (Fell et al., 2000; Scorzetti et al., 2002) of the Dikarya subkingdom (Kurtzman & Fell, 2006). These unicellular fungi are considered as ubiquitous microorganisms, which can be found in a vast variety of different ecological systems associated with terrestrial and underwater flora and fauna (Rosa & Peter, 2006). Nevertheless, based on the currently researches it could be suggested that only 1% of the diversity of yeast species has been described yet (Kurtzman & Fell, 2006).

The gastrointestinal tract (GIT) of animals remains a largely unexplored habitat. Most of the yeasts were isolated from the GIT of beetles and other insects. The current knowledge about yeasts' diversity in the digestive tract of vertebrates, especially of farm animals, is still based generally on the findings from 50's and 70's of the XXth Century. Furthermore, the taxonomy of yeasts undergoes continuous revision, e.g., variety of yeasts has double names or even many synonyms. This came off due to the fact that sometimes the same yeasts have been described by different scientists (Kurtzman & Fell, 1998) or several yeasts were invalidly classified, e.g., species assigned to genus *Torulopsis* were reclassified to the genus *Candida* (Yarrow & Meyer, 1978). Moreover, it transpires frequently when yeast species previously described based on its phenotypic characteristics has been later phylogenetically analysed and on that basis reclassified into another genus, consequently obtained a new name (Kurtzman & Fell, 2006). Therefore, few yeast species will be mentioned in the further sections with a double name.

Furthermore, in this review we will provide some consideration to the importance of the yeasts for the host. Advantages and disadvantages of the contemporary methods used for diversity studies will also be pointed.

2. Biodiversity of yeasts in the gastrointestinal ecosystem

Microorganisms live in the diverse habitats of the world. In course of evolution, some microbes adapted to the extreme environment prevailing in the gastrointestinal ecosystem of human and animals. Gastrointestinal tracts of mammals (Hooper & Gordon, 2001; Bauer et al., 2006; Ley et al., 2008) and insects (Dillon & Dillon, 2004; Hongoh, 2010; Grunwald et al., 2010) harbour vast bacterial communities which undoubtedly play an important role for the maturation and proper function of mucosal and systemic immune systems, nutrient metabolism and host health. In contrast, the knowledge of yeasts which naturally occur in intestine and thereby belong to the intestinal microbiota still remains deficient.

2.1 Yeasts' diversity in the alimentary tract of insects

Insects are among the most diverse group of animals that has been found worldwide (Chapman, 2009) and they unavoidably come into contact with yeasts widespread in various habitats like soil (Botha, 2006), plants (Fonseca & Inàcio, 2006) and fresh and marine water (Nagahama, 2006). This has been confirmed by the fact that yeasts can be found on body surface as well as in the entrails of insects: beetles, bees, flies, lacewings, termites, and mosquitoes; and their larvae. Table 1 summarises the yeast species that were recently discovered in the GIT of insects and that have particularly been identified using molecular methods.

2.1.1 Yeasts associated with flowers and gut of bees

In general, insects can be considered as either as a vector carrying yeasts on the body surface or the consumers of the yeasts (for review see Phaff & Starmer, 1987; Ganter, 2006). For instance, since yeasts regularly occur in flowers they are considered as autochthonous for this environment, and so they are closely associated with flower-visiting insects (Lachance et al., 2001; Brysch-Herzberg, 2004). In floristic nectar, ascomycetous yeasts belonging to the genera *Metschnikowia, Kodamaea, Wickerhamiella* have been found in higher abundance, whereas basidiomycetous yeasts (e.g. *Cryptococcus* spp., *Rhodotorula* spp., *Pseudozyma* spp.) were rarely isolated (Lachance et al., 2001; Brysch-Herzberg, 2004). Brysch-Herzberg (2004) counted an astonishing number of yeast cells (up to 16,000 cells/µl nectar) from the nectar samples of *Digitalis purpurea*. In this study, yeasts were isolated from nectar, plant materials, honey and from body of bumblebees, but unfortunately not from the GIT. In another study (Batra et al., 1973), the same yeasts (species of *Candida, Endomycopsis, Oidium, Hansenula, Rhodotorula, Saccharomyces, Schizosaccharomyces, Pichia,* and *Zygosaccharomyces*) were found in nectar and in the crops of bees, however, in the last niche yeasts were determined to be in higher, about 10 to 100 fold, concentration. Gilliam et al. (1974) summarized the data of yeasts isolated from the digestive tract of honey bees and also reported about their own investigations on yeasts in intestines of 388 adult worker honey bees.
Seven species were observed in the study, but three: *Candida* (*Torulopsis*) *magnoliae, Candida parapsilosis,* and *Candida* (*Torulopsis*) *glabrata* were found most frequently providing evidence for their dominance in this environment.

2.1.2 Yeasts associated with digestive tract of ants and termites

Ants, belonging like bees to the order *Hymenoptera*, have been also associated closely with variety of yeasts harbouring their nests (Carreiro et al., 1997; Rodrigues et al., 2009; Pagnocca et al., 2010). Some yeasts can pass into the infrabuccal pocket, a pouch in the ants'

Species	Host organism	References
Ambrosiozyma monospora	owlfly (*Ascalaphidae*)	(Nguyen et al., 2007)
Aureobasidium pullulans	leaf beetles (*Chrysomelidae*)	(Molnar et al., 2008)
Candida aglyptinia sp. nov.	round fungus beetle (*Leiodidae*)	(Suh et al., 2006b)
Candida alai sp. nov.	click beetle (*Elateridae*)	(Suh et al., 2008)
Candida ambrosiae	pleasing fungus beetle (*Erotylidae*); darkling beetle (*Tenebrionidae*); sap-feeding beetle (*Nitidulidae*)	(Suh et al., 2004b)
Candida amphixiae sp. nov.	handsome fungus beetle (*Endomychidae*)	(Suh et al., 2005b)
Candida anneliseae sp. nov.	pleasing fungus beetles (*Erotylidae*); darkling beetles (*Tenebrionidae*); rove beetle (*Staphylinidae*); false darkling beetles (*Melandryidae*); clown beetle (*Histeridae*); minute tree-fungus beetle (*Ciidae*)	(Suh et al., 2004b)
Candida arabinofermentans	bark beetles (*Scolytinae*)	(Rivera et al., 2009)
Candida ascalaphidarum sp. nov.	owlfly (*Ascalaphidae*); largus bug (*Largidae*); fungus weevil (*Anthribidae*)	(Nguyen et al., 2007)
Candida atakaporum sp. nov.	pleasing fungus beetle (*Erotylidae*)	(Suh et al., 2004b)
Candida atbi sp. nov.	sap-feeding beetles (*Nitidulidae*)	(Suh et al., 2006b)
Candida athensensis sp. nov.	beetles: cucujoid, curculionid, *Megalodacne fasciata* (*Erotylidae*)	(Suh & Blackwell, 2004)
Candida barrocoloradensis sp. nov.	sap-feeding beetles (*Nitidulidae*)	(Suh et al., 2006b)
Candida blattae sp. nov.	cockroach (*Blattidae*); dobsonfliy (*Corydalidae*)	(Nguyen et al., 2007)
Candida blattariae sp. nov.	cockroach (*Blattaria*)	(Suh et al., 2005b)
Candida bohiensis sp. nov.	click beetles (*Elateridae*); leaf beetle (*Chrysomelidae*)	(Suh et al., 2008)
Candida bokatorum sp. nov.	pleasing fungus beetles (*Erotylidae*); ground beetle (*Carabidae*); sap-feeding beetle (*Nitidulidae*); false darkling beetle (*Melandryidae*); darkling beetle (*Tenebrionidae*)	(Suh et al., 2004b)
Candida bolitotheri sp. nov.	darkling beetles (*Tenebrionidae*); pleasing fungus beetle (*Erotylidae*)	(Suh et al., 2004b)

Species	Host organism	References
Candida bribrorum sp. nov.	pleasing fungus beetles (*Erotylidae*); darkling beetles (*Tenebrionidae*)	(Suh et al., 2004b)
Candida buenavistaensis sp. nov.	longhorned beetle (*Cerambycidae*); scarab beetle (*Scarabaeidae*)	(Suh et al., 2008)
Candida cf *neerlandica*	owlfllies (*Ascalaphidae*); earwig (*Labiduridae*); cricket (*Gryllidae*)	(Nguyen et al., 2007; Suh et al., 2008)
Candida chickasaworum sp. nov.	pleasing fungus beetles (*Erotylidae*); minute tree-fungus beetle (*Ciidae*)	(Suh et al., 2004b)
Candida choctaworum sp. nov.	darkling beetles (*Tenebrionidae*); minute tree-fungus beetles (*Ciidae*); fungus weevil (*Anthribidae*)	(Suh et al., 2004b)
Candida chrysomelidarum sp. nov.	leaf beetles (*Chrysomelidae*)	(Nguyen et al., 2006)
Candida corydali sp. nov.	dobsonflies and fishflies (*Corydalidae*)	(Nguyen et al., 2007)
Candida derodonti sp. nov.	tooth-necked fungus beetles (*Derodontidae*)	(Suh & Blackwell, 2005)
Candida dosseyi sp. nov.	dobsonflies (*Corydalidae*)	(Nguyen et al., 2007)
Candida elateridarum sp. nov.	click beetle (*Elateridae*)	(Suh & Blackwell, 2004)
Candida emberorum sp. nov.	pleasing fungus beetles (*Erotylidae*); handsome fungus beetle (*Endomychidae*)	(Suh et al., 2004b)
Candida endomychidarum sp. nov.	handsome fungus beetle (*Endomychidae*)	(Suh et al., 2005b)
Candida ernobii	bark beetles (*Scolytinae*)	(Rivera et al., 2009)
Candida fermentati	sap-feeding beetle (*Nitidulidae*); scarab beetle (*Scarabaeidae*)	(Suh & Blackwell, 2004)
	fishfly (*Corydalidae*)	(Nguyen et al., 2007)
	mosquitoes (*Culicidae*)	(Gusmão et al., 2010)
Candida frijolesensis sp. nov.	handsome fungus beetle (*Endomychidae*); elephant beetle (*Scarabaeidae*)	(Suh et al., 2008)
Candida gatunensis sp. nov.	sap-feeding beetles (*Nitidulidae*)	(Suh et al., 2006b)
Candida gigantensis sp. nov.	click beetles (*Elateridae*)	(Suh et al., 2008)
Candida guaymorum sp. nov.	pleasing fungus beetles (*Erotylidae*); scarab beetle (*Scarabaeidae*)	(Suh et al., 2004b)
Candida intermedia	mosquitoes (*Culicidae*)	(Ricci et al., 2011a)

Species	Host organism	References
Candida kruisii	sap-feeding beetles (*Nitidulidae*)	(Suh et al., 2006b)
Candida kunorum sp. nov.	sap-feeding beetle (*Nitidulidae*)	(Suh et al., 2004b)
Candida labiduridarum sp. nov.	earwig (*Labiduridae*); cricket (*Gryllidae*); owlflies (*Ascalaphidae*)	(Suh et al., 2008)
Candida lessepsii sp. nov.	unidentified beetle	(Suh et al., 2005b)
	bark beetles (*Scolytinae*)	(Rivera et al., 2009)
Candida lycoperdinae sp. nov.	sap-feeding beetle (*Nitidulidae*); scarab beetle (*Scarabaeidae*)	(Suh et al., 2006b)
Candida maltosa	click beetles (*Elateridae*); bess beetles (*Passalidae*); scarab beetle (*Scarabaeidae*)	(Suh et al., 2008)
Candida maxii sp. nov.	darkling beetle (*Tenebrionidae*)	(Suh et al., 2004b)
Candida membranifaciens	cranefly (*Tipulidae*); dobsonfly (*Corydalidae*); green lacewings (*Chrysopidae*)	(Suh et al., 2005b; Nguyen et al., 2007)
Candida michaelii sp. nov.	handsome fungus beetle (*Endomychidae*)	(Suh et al., 2005b)
Candida nodaensis	mosquitoes (*Culicidae*)	(Gusmão et al., 2010)
Candida oregonensis	bark beetles (*Scolytinae*)	(Rivera et al., 2009)
Candida pallodes sp. nov.	sap-feeding beetles (*Nitidulidae*)	(Suh et al., 2006b)
Candida panamensis sp. nov.	sap-feeding beetles (*Nitidulidae*); darkling beetle (*Tenebrionidae*)	(Suh et al., 2006b)
Candida panamericana sp. nov.	pleasing fungus beetle (*Erotylidae*); rove beetle (*Staphylinidae*); darkling beetle (*Tenebrionidae*)	(Suh et al., 2004b)
Candida picachoensis sp. nov.	green lacewings (*Chrysopidae*)	(Suh et al., 2004a)
	leaf beetles (*Chrysomelidae*)	(Nguyen et al., 2006)
Candida piceae	bark beetles (*Scolytinae*)	(Rivera et al., 2009)
Candida pimensis sp. nov.	green lacewings (*Chrysopidae*)	(Suh et al., 2004a)
		(Nguyen et al., 2006)
Candida plutei sp. nov.	rove beetle (*Staphylinidae*)	(Suh & Blackwell, 2005)
Candida pseudorhagii	click beetle (*Elateridae*)	(Suh et al., 2008)
Candida quercitrusa	dobsonfly (*Corydalidae*)	(Nguyen et al., 2007)
	cotton bollworm (*Noctuidae*)	(Molnar et al., 2008)

Species	Host organism	References
Candida sake	crambid snout moths (*Crambidae*); cotton bollworm (*Noctuidae*)	(Molnar et al., 2008)
Candida sinolaborantium	handsome fungus beetle (*Endomychidae*); cerambycid larvae (*Cerambycidae*)	(Suh et al., 2005b)
Candida smithsonii sp. nov.	endomychid larva (*Endomychidae*); Iphiclus beetle (*Erotylidae*)	(Suh & Blackwell, 2004)
Candida stri sp. nov.	sap-feeding beetles (*Nitidulidae*)	(Suh et al., 2006b)
Candida taliae sp. nov.	darkling beetle (*Tenebrionidae*)	(Suh et al., 2004b)
Candida temnochilae sp. nov.	bark-gnawing beetle (*Trogossitidae*); bess beetle (*Passalidae*)	(Suh et al., 2005b)
Candida tenuis	minute tree-fungus beetle (*Ciidae*)	(Suh et al., 2005b)
Candida terraborum sp. nov.	pleasing fungus beetle (*Erotylidae*)	(Suh et al., 2004b)
Candida tetrigidarum sp. nov.	elephant beetle (*Scarabaeidae*); pygmy grasshopper (*Tetrigidae*)	(Suh et al., 2008)
Candida tritomae sp. nov.	pleasing fungus beetles (*Erotylidae*); scarab beetle (*Scarabaeidae*)	(Suh et al., 2006b)
Candida tropicalis	owlfly (*Ascalaphidae*); bess beetles (*Passalidae*); ichneumon wasps (*Ichneumonidae*); dobsonfly and fishfly (*Corydalidae*); roach (*Blattidae*)	(Nguyen et al., 2007; Suh et al., 2008)
Candida wounanorum sp. nov.	pleasing fungus beetle (*Erotylidae*)	(Suh et al., 2004b)
Candida xestobii	crambid snout moths (*Crambidae*); leaf beetle (*Chrysomelidae*)	(Molnar et al., 2008)
Candida yuchorum sp. nov.	pleasing fungus beetle (*Erotylidae*)	(Suh et al., 2004b)
Clavispora lustansiae	blister beetle (*Meloidae*)	(Rao et al., 2007)
Cryptococcus flavescens	crambid snout moth (*Crambidae*); cotton bollworm (*Noctuidae*); leaf beetles (*Chrysomelidae*)	(Molnar et al., 2008)
Cryptococcus luteolus	crambid snout moths (*Crambidae*); leaf beetles (*Chrysomelidae*)	(Molnar et al., 2008)
	green lacewings (*Chrysopidae*)	(Woolfolk & Inglis, 2004)
Cryptococcus oeirensis	leaf beetle (*Chrysomelidae*)	(Molnar et al., 2008)
Cryptococcus victoriae	crambid snout moth (*Crambidae*)	(Molnar et al., 2008)
	green lacewings (*Chrysopidae*)	(Woolfolk & Inglis, 2004)

Species	Host organism	References
Cryptococcus zeae	crambid snout moths (*Crambidae*); leaf beetle (*Chrysomelidae*)	(Molnar et al., 2008)
Geotrichum carabidarum sp. nov.	ground beetle (*Carabidae*); geometrid larva (*Geometridae*); pleasing fungus beetle (*Erotylidae*)	(Suh & Blackwell, 2006)
Geotrichum cucujoidarum sp. nov.	minute tree-fungus beetle (*Ciidae*); cucujoid beetle (*Cucujoidae*); hining fungus beetle (*Scaphidiinae*)	(Suh & Blackwell, 2006)
Geotrichum histeridarum sp. nov.	clown beetles (*Histeridae*); pleasing fungus beetle (*Erotylidae*); geometrid larvae (*Geometridae*); tiger moths (*Arctiidae*)	(Suh & Blackwell, 2006)
Hanseniaspora uvarum	crambid snout moths (*Crambidae*)	(Molnar et al., 2008)
	mosquitoes (*Culicidae*)	(Ricci et al., 2011a)
Hanseniaspora vineae	dobsonflies and fishflies (*Corydalidae*)	(Nguyen et al., 2007)
Issatchenkia orientalis	scarab beetle (*Scarabaeidae*)	(Rao et al., 2007)
Kodamaea laetipori sp. nov.	darkling beetles (*Tenebrionidae*); scarab beetles (*Scarabaeidae*)	(Suh & Blackwell, 2005)
Kodamaea ohmeri	dobsonfly (*Corydalidae*); sap-feeding beetle (*Nitidulidae*); pleasing fungus beetle (*Erotylidae*)	(Suh & Blackwell, 2005; Nguyen et al., 2007)
	mosquitoes (*Culicidae*)	(Gusmão et al., 2010)
Kuraishia capsulata	bark beetles (*Scolytinae*)	(Rivera et al., 2009)
Kuraishia cf. molischiana	bark beetles (*Scolytinae*)	(Rivera et al., 2009)
Lachancea fermentati	dobsonflies and fishflies (*Corydalidae*)	(Nguyen et al., 2007)
Lachancea thermotolerans	dobsonflies and fishflies (*Corydalidae*)	(Nguyen et al., 2007)
Lodderomyces elongisporus	bark and ambrosia beetle (*Scolytinae*)	(Suh et al., 2008)
Metschnikowia andauensis sp. nov.	cotton bollworm (*Noctuidae*)	(Molnar & Prillinger, 2005)
Metschnikowia chrysoperlae sp. nov.	green lacewings (*Chrysopidae*)	(Suh et al., 2004a)
	dobsonfly (*Corydalidae*)	(Nguyen et al., 2007)
Metschnikowia corniflorae sp. nov.	soldier beetles (*Cantharidae*)	(Nguyen et al., 2006)
Metschnikowia fructicola	crambid snout moths (*Crambidae*); cotton bollworms (*Noctuidae*)	(Molnar & Prillinger, 2005)
Metschnikowia noctiluminum sp. nov.	green lacewings (*Chrysopidae*)	(Nguyen et al., 2006)
Metschnikowia pulcherrima	dobsonfly (*Corydalidae*)	(Nguyen et al., 2007)
	green lacewings (*Chrysopidae*)	(Woolfolk & Inglis, 2004)

Species	Host organism	References
Pichia americana	bark beetles (*Scolytinae*)	(Rivera et al., 2009)
Pichia canadensis	bark beetles (*Scolytinae*)	(Rivera et al., 2009)
Pichia caribbica	pleasing fungus beetle (*Erotylinae*)	(Rao et al., 2007)
	mosquitoes (*Culicidae*)	(Gusmão et al., 2007)
Pichia glucozyma	bark beetles (Scolytinae)	(Rivera et al., 2009)
Pichia guilliermondii	fishfly and dobsonfly (*Corydalidae*); owlfly (*Ascalaphidae*)	(Nguyen et al., 2007)
	bark beetle (*Scolytinae*)	(Rivera et al., 2009)
	crambid snout moths (*Crambidae*); leaf beetle (*Chrysomelidae*)	(Molnar et al., 2008)
	mosquitoes (*Culicidae*)	(Gusmão et al., 2010)
Pichia mexicana	bark beetles (*Scolytinae*)	(Rivera et al., 2009)
Pichia nakazawae var. akitaensis	handsome fungus beetle (*Endomychidae*)	(Suh et al., 2005b)
Pichia scolyti	bark beetles (*Scolytinae*)	(Rivera et al., 2009)
Pseudozyma apsidi	cotton bollworms (*Noctuidae*); leaf beetles (*Chrysomelidae*)	(Molnar et al., 2008)
Pseudozyma flocculosa	cotton bollworms (*Noctuidae*)	(Molnar et al., 2008)
Pseudozyma prolifica	cotton bollworms (*Noctuidae*)	(Molnar et al., 2008)
Rhodotorula aurantiaca	leaf beetle (*Chrysomelidae*)	(Molnar et al., 2008)
Rhodotorula glutinis	crambid snout moths (*Crambidae*); leaf beetle (*Chrysomelidae*)	(Molnar et al., 2008)
Saccharomyces cariocanus	dobsonflies (*Corydalidae*)	(Nguyen et al., 2007)
Saccharomyces cerevisiae	dobsonflies and fishflies (*Corydalidae*)	(Nguyen et al., 2007)
Saccharomyces fermentans	fishfly (*Corydalidae*)	(Nguyen et al., 2007)
Saprochete gigas	dobsonfly and fishfly (*Corydalidae*)	(Nguyen et al., 2007)
Sporobolomyces coprosmae	leaf beetles (*Chrysomelidae*)	(Molnar et al., 2008)
Tilletiopsis washingtonensis	leaf beetle (*Chrysomelidae*)	(Molnar et al., 2008)
Torulaspora delbrueckii	fishfly (*Corydalidae*)	(Nguyen et al., 2007)
Trichosporon insectorum sp. nov.	bess beetles (*Passalidae*); scarab beetles (*Scarabaeidae*)	(Fuentefria et al., 2008)
Trichosporon mycotoxinivorans sp. nov.	lower termite (*Mastotermitidae*)	(Molnar et al., 2004)
Trichosporon xylopini sp. nov.	darkling beetles (*Tenebrionidae*); bess beetles (*Passalidae*)	(Gujjari et al., 2010)
Wickerhamomyces anomalus (*Pichia anomala*)	mosquitoes (*Culicidae*)	(Ricci et al., 2011a)

Table 1. Yeasts detected in the alimentary tract of insects

oral cavity (Hansen & Klotz, 2005). Based on physiological characteristics using a microbial identification system BIOLOG, Mankowski & Morrell (2004) identified 19 species from 155 yeast isolates collected from nest, surrounding soil and frass as well as from exoskeleton and infrabuccal pockets of carpenter ants. From 17 isolates found in the infrabuccal pockets, ten were identified as *Debaryomyces polymorphus* and other species (*Pichia guilliermondii, Candida ergatensis, Candida edax, Bulleromyces* spp. and *Cryptococcus laurentii*) occurred only once or twice. Besides of soil samples, *D. polymorphus* was the most often isolated yeast from the all analysed materials. Further social insects, such as termites may harbour high yeasts' numbers (10^7-10^8 cells/ml) in their gut (Schäfer et al., 1996). Schäfer et al. (1996) cultured 35 yeast isolates from the intestinal contents of termite species, *Zootermopsis nevadensis* and *Neotermes castaneus*, but the authors reported presence of only 15 yeast strains, as their enzymatic activity were significant to the study. These phenotypes were related to the genera *Candida, Sporothrix, Pichia* and *Debaryomyces*. In another study, *Debaryomyces hansenii* and *Sporothrix albicans* as well as species of *Trichosporon* and *Rhodosporidium* could also be found in the hindgut of the termites from families: *Mastotermitidae, Hodotermitidae, Kalotermitidae* and *Rhinotermitidae* and roaches (Prillinger et al., 1996).

2.1.3 Yeasts associated with the gut of some pests

In the gut of some maize' pests (*Diabrotica virgifera, Helicoverpa armigera* and *Ostrina nubialalis*), Molnar et al. (2008) isolated 97 yeast strains; furthermore they detected yeasts as well as other fungi of the genera: *Acremonium, Aspergillus, Cladosporium* and *Fusarium* by means of cloning and denaturing gradient gel electrophoresis (DGGE). The occurence of clones was given in percents. All methods reveald that *Metschnikowia* spp., closely related to *Metschnikowia pulcherrima, Cryptococcus* spp. (*Cr. luteolus, Cr. zeae* and *Cr. flavescens*) as well as *Candida* spp., bearing close similarity to *C. xestobii* or *C. sake,* and *Pseudozyma* spp. were the most frequently identified yeasts. *Pichia guiliermondii* and *Rhodotorula* species were less common. Some of occassionaly found yeasts e.g. *Aureobasidium pullulans, Candida quercitrusa, Hanseniaspora uvarum, Sprobolomyces coprosmae, Tilletiopsis washingtonensis* were detected however only via culturing. There are some publications reporting presence of the yeasts in the gut of mosquitos (*Diptera: Culicidae*), which are known to be vectors of many diseases in humans. Gusmão et al. (2007; 2010) identified *Pichia caribbica, Pichia guilliermondii, Pichia* (syn. *Kodamaea*) *ohmeri, Candida fermentati* and *Candida nodaensis* in the diverticulum of *Aedes aegypti*. Ricci et al. (2011a) investigated yeasts in the gut of *Anopheles stephensi* using molecular and cultivation-dependent methods. Forty six clones that expressed fragments of the 18S rRNA gene retrieved from the gut samples of 6 adults were sequenced. Eleven clones were identified as *Wickerhamomyces anomalus,* known also as *Pichia anomala,* while others could be assigned either to genus *Candida* or *Pichia* or to unidentified fungus. Moreover, 100 colonies were cultured from 10 insect speciemens, classified based on their morphology and identified as *Candida intermedia, Hanseniaspora uvarum* and *W. anomalus* (77%, 15% and 8% respectively) by sequencing analysis of 18S and 26S rRNA genes and ITS fragments. *W. anomalus* was detectable using both approaches. Furthermore, Ricci et al. (2011a; 2011b) observed the presence of *W. anomalus* in the midgut of different mosquitos species *Anopheles stephensi, Anopheles gambiae, Aedes albopictus* and *Aedes aegypti* of both sexes as well as on larvae, pupae and gonads, thereby supposed close relationship between this yeast species and mosquitos.

2.1.4 Yeasts in the digestive tract of lacewings

Lacewings (*Neuroptera*: *Chrysopidae*) are one of the predators admitted as biological control agents of pests. During the scanning and transmission electron microscopical studies, a large numbers of yeast cells were observed within lacewings' alimentary tract (Woolfolk et al., 2004; Woolfolk & Inglis, 2004; Chen et al., 2006). Woolfolk & Inglis (2004) investigated yeasts in the different parts: diverticulum, foregut, midgut, and hindgut of digestive tract of 24 lacewing adults (*Chrysoperla rufilabris*) collected at two field locations in Mississippi. With the exception of 7 insects that were yeasts-free, lacewings harboured a high concentration ($\approx 10^3$ colony forming units; CFU) of yeasts distributed in the all analysed gut sections; however the highest ($5.4 x 10^5$ CFU/g) density was in diverticulum. In total 752 yeasts were isolated in the study and arranged in five groups based on their phenotypic properties; some specimens were randomly chosen from each group for further genotyping analysis. Interestingly, 89% of the isolates were identified as *Metschnikowia pulcherrima* and the remaining 11% involved either *Cryptococcus victoriae* or *Cryptococcus luteolus* or strains that could not be assigned by the authors to any known species. Sometimes, closely related yeast species could be separated only according to the genotypic characterization, while they were showing similar physiological properties (Kurtzman & Fell, 2006) as it was the case in the study of Suh et al. (2004a). These authors isolated 14 yeasts from digestive tract of *Chrysoperla* spp. which were closely related to *M. pulcherrima*, however sufficiently variable in the D1&D2 domains of the 26S rRNA gene of the large subunit (LSU) to represent three new species: *Metschnikowia chrysoperlae, Candida picachoensis* and *Candida pimensis*. Recently, several new yeasts of *Metschnikowia* and *Candida* (see table 1) were discovered in the gut of other members of the *Neuroptera*, too (Nguyen et al., 2006; 2007).

2.1.5 Yeasts in the digestive tract of beetles

At the present time, the most yeasts were isolated from the digestive tract of beetles (*Coleoptera*). Shifrine & Phaff (1956) collected bark beetles (*Dendroctonus* and *Ips*) and their larvae from the various coniferus trees in Northern California. After sterilization of the outer surface, the beetles were dissected and yeasts were isolated from internal parts of the insects. Total of 169 yeast strains could be assigned to 13 species. *Candida silvicola* (41.4%; teleomorph *Hansenula holstii*), *Hansenula capsulata* (21.3%), *Pichia pinus* (18.9%) and *Candida curvata* (8.9%) were frequently found; other species (e.g. *C. parapsilosis, C. mycoderma, C. rugosa, Cryptococcus diffiuens*) were rarely (from 0.6 to 1.8%) isolated. Some yeasts as *Candida* (*Torulopsis*) *nitratophila, C.* (*T.*) *melibiosum, Rhodotorula crocea* and *C. silvicola* were described by these authors as new species. Recently, Rivera et al. (2009) provided an account of yeasts associated with alimentary tract of *Dendroctonus* beetles. Yeasts (403 strains) were isolated from different parts of intestine, the midgut (anterior & posterior) and the hindgut, as well as from the ovaries, eggs and frass of the beetles collected from pine trees at 34 locations in Mexico, Cuatemala and the USA. Based on the sequence analysis of several DNA regions (18S, 26S rRNA genes and ITS1) and phenotypical characteristics, the yeasts were reletated to three genera: *Candida* spp. (*C. ernobii, C. piceae, C. membranifaciens, C. lessepsii, C. arabinofermentans* and *C. oregonensis*), *Pichia* spp. (*P. americana, P. guilliermondii, P. scolyti, P. mexicana, P. glucozyma* and *P. canadensis*) and *Kurashia* spp. (*K. capsulata* and *K.* cf. *molischiana*). The exact numbers of the yeast strains isolated from the different gut sections have not been provided by the autors, however, they indicated that yeasts were present in eggs, ovaries and frass to much lesser extent than in the guts. For instance, *P. americana, C.*

ernobii and the strains related to the one *Candida* sp. were prevalent in all parts of gut and frass and *P. guilliermondii* and *C. ernobii* were cultured most frequently from the posterior midgut.

In relation to high number of the yeast isolates (richness) described above, comparatively low yeast diversity was found in the assemblage of *Dendroctonus* beetles. It thereby underlined the impact of the host and/or environmental factors on the yeasts diversity. Nevertheless, examination of yeasts harbouring the GIT of beetles from 27 families reviled a huge variety of yeasts (Suh et al., 2005a).

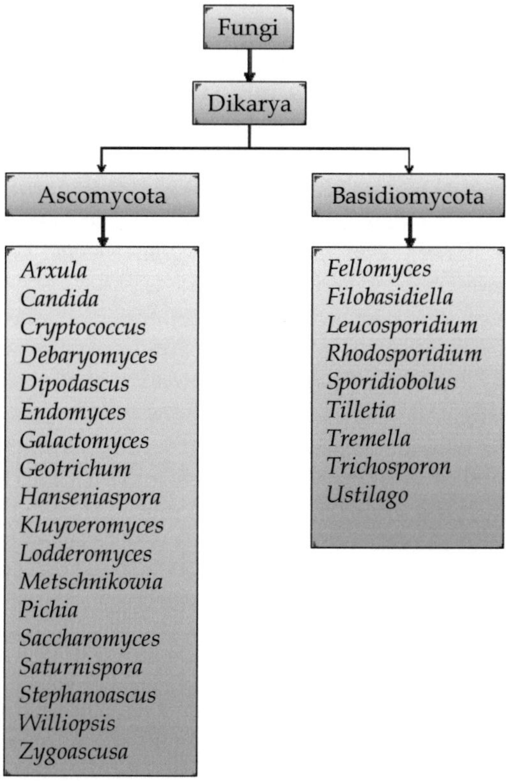

Fig. 1. Schematic representation of the different yeast taxa belonging to the two phyla *Ascomycota* and *Basidiomycota* of the Dikarya subkingdom isolated from the intestinal tract of insects during the study of Suh et al. (2005a).

During three-years-period, Suh et al. (2005a) isolated about 650 yeasts from the gut of diverse beetles collected from the south-eastern USA and Panama. Sequence analyses of the D1&D2 domains of LSU rRNA gene revealed 290 single species belonging to at least 27 taxa (Fig. 1.); the great majority of which were ascomycetous and some basidiomycetous yeasts.

It is noteworthy that nearly 200 yeasts determined throughout the study were considered by the authors to represent new, not yet described species. In the meantime, some of them (table 1) were characterized by Suh, Nguyen, Blackwell and their co-workers. Based on their observation Suh et al. (2005a) suggested that almost each beetle species may be a host for at least one unknown yeast species. In the last decades, describing of many novel species of yeasts isolated from the gut of insects corroborates this supposition.

Nowadays, there are over one million of accepted insect species; however, their number has yearly increased and is still largely undiscovered (Chapman, 2009). Thus, it can be supposed that the number of yeasts would tremendously rise, even if only the intestinal tracts of the currently known insects were explored.

2.2 Yeasts' diversity in the GIT of vertebrates with focus on farm animals

The more intensive investigations of the yeast population present in the GIT of vertebrates, based on various cultivation procedures, began in the fifties of the XXth Century. Van Uden et al. (1958) and Van Uden & Carmo Sousa (1957b) examined yeasts in the caecal samples of large number of animals: 252 cattle, 252 horses, 503 sheep, 250 goats and 250 pigs. Yeasts were also studied by Parle (1957) in the digestive tract of cows, rabbits, sheep, guinea pigs, opossums, monkeys, cats, dogs, hedgehogs, mice, pigs and rats. Lund (1974) explored yeasts and moulds in the bovine rumen. Lately, yeasts were also described in the intestinal tract of reptiles (Kostka et al., 1997), birds (Cafarchia et al., 2006; 2008; Brilhante et al., 2010; Costa et al., 2010), mice (Scupham et al., 2006), dogs (Brito et al., 2009) and fish (Gatesoupe, 2007). In these studies, the scientists have detected various ascomycetous and basidiomycetous yeasts chiefly representing the genera *Candida, Trichosporon, Pichia, Rhodotorula, Debaryomyces, Kluyveromyces* and *Saccharomyces*.

In general, the diversity of the yeast population depended on the host; but many species occurred at diverse, also not intestinal ecosystems; and several exhibited direct relationship to the individual animal. It should be noticed, however, that yeasts could not be always isolated from the investigated GIT and often they were present in small numbers. Nevertheless, taking into consideration the scarce information existing on yeasts in the gastrointestinal ecosystems of vertebrates, it is well known that relatively high variety as well as quantity of yeasts can be found in the GIT of pigs.

Here, the yeasts diversity in the GIT of farm animals representing diverse nutritional types: omnivores (pig), monogastric herbivores (horse) and ruminants will stay in focus and will be compared.

2.2.1 Yeasts in the GIT of pigs

Comparatively to all animals investigated in the study of Van Uden et al. (1958), the most frequent occurrence (88.8%) of yeasts was detected in the caecum of pigs (horses 52.4%, cattle 46.8%, sheep 6.8%, and goats 6.4%). The yeasts *Candida slooffiae, Candida krusei, Saccharomyces telluris, Candida albicans, Candida* (*Torulopsis*) *glabrata*, were commonly found in the porcine gut. However, *C. slooffiae* was isolated most frequently (48.4%). A few other yeasts such as *Saccharomyces* spp., *Pichia membranifaciens, Pichia farinose* and *Candida mycoderma* could also be identified. Roughly the same situation has been confirmed in the following studies (Van Uden & Carmo-Sousa, 1962; Mehnert & Koch, 1963), where the scientists isolated almost the same variety of yeasts from the different parts of porcine GIT. After investigation of digesta samples collected from stomach, three parts of small intestine as well as caecum and rectum of healthy 57 pigs, Van Uden & Carmo Sousa (1962) reported high animal-individual qualitative and quantitative variability if the yeast occupation. In total 15 yeast species were identified; while *C. slooffiae* was present in 27 pigs, many other species mentioned above occurred only sporadically. Moreover, *C. slooffiae* was highly abundant, from 10^2 to 10^3 CFU/g of chyme in the stomach and up to 10^6 CFU/g intestine contents in the rectum. A still higher occurrence of yeasts in the gut of pigs was detected by Mehnert & Koch (1963), up to 10^7 CFU/g in rectum. They isolated 292 yeasts from 200 digesta samples collected from stomach and rectum from 98 (of 100 examined) pigs. Apart from the *C. slooffiae* which was detectable in 75% of pigs, yeast species such as *C. krusei, S. telluris, C. albicans, C. glabrata, C. tropicalis, C. parapsilosis* and *C. pintolopesii* (60%, 26%, 9%, 4%, 3%, 3% and 2% respectively) were isolated, too. Also in this study the appearance of yeasts was variable within a part of the GIT and among examined animals. Thus, stomach was generally colonized by yeasts at lesser intense than rectum. In most animals, *C. slooffiae* and *C. krusei* were detected both in stomach and rectum, while just in a few cases the yeasts could be found only in a single part of the GIT. *C. slooffiae* and its closely related species: *S. telluris* and *C. pintolopesii* have been newly molecularly investigated and based on multigene sequence analyses they were assigned to the teleomorphic genus *Kazachstania* (Kurtzman et al., 2005).

Recently, Urubschurov et al. (2008) described yeasts' diversity in the gut of piglets around weaning which were reared at two facilities: at experimental farm (EF) with improved husbandry conditions than at commercial farm (CF). Most piglets, 33 at CF and 35 at EF, were weaned at 28 days (d) of age and fed with the same diet until 39 d in both farms. A number of piglets, namely 18 at CF and 9 at EF, were left by the sows without additional feeding. All piglets were sacrificed at 39 d of age and digesta samples from GIT were collected. D1&D2 domains of 26S rRNA gene from 173 yeast isolates obtained from 95 piglets were sequenced. The alignment to known sequences revealed close relationship to 17 species, of which the most dominated are presented in figure 2. Urubschurov et al. (2008) observed distinction of yeasts variety between both facilities that were proven by calculation of different similarity and diversity indices. In piglets from CF *Galactomyces geotrichum, Kazachstania slooffiae* and *Candida catenulata* were the most abundant ones and the other were present only at low abundances. Unlike at CF, at EF two species, namely *K. slooffiae* and *C. glabrata* were found to be the most dominating ones and the others were rarely isolated. Some of the other species could be found in piglets either only at the EF (*P. fermentans, C. tropicalis, C. oleophila, C. parapsilosis, P. guilliermondii, Rh. mucillaginosa, T. montevideense*) or at the CF (*C. silvae* and *P. farinose*). This study provided evidence for association of *K. slooffiae* with the porcine GIT. *K* (*C.*) *slooffiae* was found for the first time in 6 of 252 examined horses (Van Uden & Carmo-Sousa, 1957a), however, due to frequent occurrence and high concentration in the porcine digestive tract it can be considered to be specific for pigs.

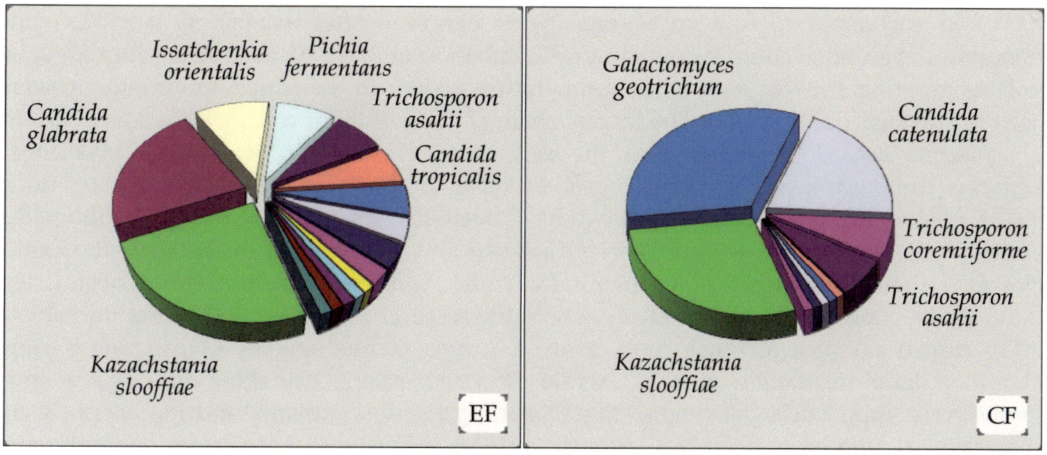

Fig. 2. Dominated yeasts isolated from the gut of 39 d old piglets, kept at experimental (EF) and commercial farm (CF), in the study of Urubschurov at al. (2008).

Furthermore, compared to other yeasts occurring in the porcine GIT, which can survive also in other ecological niches outside the animals, *K* (*C.*) *slooffiae* seems to be well adapted to the porcine gastrointestinal habitat, as this species is one of those that need high temperature to grow, comparable to the temperature of animal body, being characterized as thermophilic or psychrophobic (Travassos & Cury, 1971).

2.2.2 Yeasts in the equine GIT

Several investigators focused on the effect of yeast, *Saccharomyces cerevisiae,* on intestinal microbiota of horses and on the digestibility of different diets (e.g. Medina et al., 2002; Jouany et al., 2008; 2009). But little is known about the yeasts naturally occurring in the equine gut. Van Uden et al., (1958) studied yeasts in the caecal contents of 252 horses, and these authors revealed presence of yeasts in over half (52.4%) of the investigated animals. With occurrence of 21.8%, *Trichosporon cutaneum* was the most frequently isolated yeast, whereas in porcine intestine this species was found only one time. However, other yeasts: *C. krusei* (8.3%) as well as *C. tropicalis* (4.4%), *C. albicans* (4.4%), *C. parapsilosis* (3.6%), *C. slooffiae* (2.4%), *C. glabrata* (1.2%) and *S. telluris* (1.2%) detected in the GIT of horses were also commonly recorded in pigs.

2.2.3 Yeasts in the GIT of ruminants

As reported by Lund (1974), a different number of yeasts has been observed in the ruminal contents from cows and sheep depending on the culture conditions and incubation temperature. After at 39°C incubation of the rumen contents collected at different times from five cows, Clarke & Menna (1961) quantified yeast colonies rated from 80 to 13,000 per gram of samples; 134 colonies were isolated for further characterization. Yeasts from *Candida* spp. (*C. krusei, C. albicans, C. tropicalis* and *C. rugosa*), *Trichosporon* spp. (*T. cutaneum* and *T. sericeum*) and *Rhodotorula* spp. (*Rh. rubra/mucilaginosa, Rh. glutinis* and *Rh. macerans*) were identified; however, the *Rhodotorula* spp. could be cultured only at the temperature of 19°C. Lund (1974) examined fungal microbiota in rumen liquid of 10 fistulated and 2 non fistulated cows fed different diets. Forty nine collected samples were plated and incubated simultaneously at 25°C and at 39°C. A considerably larger number of yeast colonies, up to 1000 fold and about 20 fold on average, were observed after incubation at 25°C, while sometimes even none could be obtain after incubation at 39°C. Nevertheless, only 67 yeast isolates growing at 39°C, as it is the temperature proper to the rumen environment, were used for further identification. The largest share (77.6%) of them was identified as *C. krusei, T. cutaneum* and *T. capitatum* and the rest were *C. valida, C. ingens, C. pintolepesii, Klyveromyces bulgaricus, Saccharomycopsis lipolytica* and *Hansenula fabianii.* Other fungi (molds) belonging to the order *Mucorales* have been also found in the study. Additionally, Lund (1974) observed two yeast species *C. krusei* and *T. capitatum* in faeces of the cattle. However, their counts were lower than in the rumen samples of corresponding host. Later, Lund (1980) conducted a similar study where the researcher investigated yeasts microbiota in 16 rumen samples of musk oxen. Only 6 strains of one species, *C. parapsilosis* were identified after incubation at 37°C, while 41 yeast strains belonging to *Candida* spp., *Cryptococcus* spp., *Trichosporon* spp., *Rhodotorula* spp., *Torulopsis* spp. and *Pichia* spp. were characterized after growing at 25°C. But, the authors indicated that the rumen contents were kept frozen for a long period (more than 7 weeks), what could have had considerable effects on yeast colonization. As mentioned above, Van Uden et al. (1958) cultivated caecal samples

from a much higher number of cattle (252) as well as sheep (503) and goats (250). Among the investigated animals, cattle showed the highest (46.8%) occurrence of yeasts, whereas just a few yeasts could be found in sheep and goats, 6.8% and 6.4% of the animals, respectively. The most frequently isolated yeasts were *C. tropicalis* and *C. krusei* in cattle, and *C. albicans* in sheep. These species were also isolated from the goats, but just two times each; and *C. glabrata* four times. A few other yeasts identified as members of *Saccharomyces* spp., *Candida* spp. and *Pichia* spp. have been found only occasionally.
Quite similar results regarding yeast colonization have been obtained in the cultivation dependent studies (Clarke & Menna 1961; Lund 1974; 1980; Van Uden et al. 1958) from different geographical regions. Shin et al. (2004) explored different rumen samples (fluid, solid and epithelium) from one cow, examined for yeasts population using molecular approaches. Shin et al. (2004) have succeeded to obtain 97 clones containing 26S rRNA gene fragments from the three types of samples and to assign them to the different phylogenetic groups. Compared to 4 phylotypes from the rumen epithelium showing the closest relatedness to *Geotrichum silvicola, Acremonium alternatum, Pseudozyma rugulosa* (up to 99%) and *Galactomyces* sp. (97%), and 2 phylotypes (*Geotrichum silvicola*, 99% and *Galactomyces* sp., 97%) from the rumen solid, the highest yeast' diversity was observed in the samples of rumen fluid revealing presence of 15 various phylotypes. Only 5 (*Setosphaeria monoceras, Raciborskiomyces longisetosum, Magnaporthe grisea, Ustilago affinis* and *Pseudozyma rugulosa*) of the 15 phylotypes showed 99% identity with the sequences deposited at the NCBI GenBank. The identification rate of the others belonging also to the classes *Pezizomycotina, Urediniomycetes, Saccharomycotina* and *Hymenomycetes* ranged from 91 to 98%. These phylotypes could represent new species, because in yeasts more than 1% of the nucleotide divergence in D1&D2 domains of the 26S rRNA gene may represent a separate species (Kurtzman & Fell, 2006). In spite of the lack of inter-individual comparison, this study showed a potential existence of the other yeasts that have not been discovered yet.

3. Methods for investigating biodiversity of the yeasts from GIT

From the cited references it is obvious, the biodiversity studies depend very much on the applied method. However, this is beyond the scope of this chapter to provide very detailed description of all possible methods that could be used for studies on yeasts' diversity. Nor calculation of the different biodiversity indices is in the focus of the paragraph. This paragraph is meant to provide short discussion on the existing possibilities, their limitations and advantages, and provide the reader with some input for consideration which methods he or she would choose for his/her studies.
Any application of either method mentioned below requires correct sampling of the material. Studying the biodiversity of the yeasts harbouring the GIT the dominating yeasts are in focus of most studies, as well as their abundance and changes of the abundance in time and in relation to the diet. For these purposes faecal or digesta samples have been collected from large animals (Urubschurov et al., 2008; 2011) or whole intestines from e.g. insects have been dissected (Suh et al., 2004b; 2005a; Nguyen et al., 2007). Whereas rain worms, termites or other small animals can provide the whole GIT for the studies, only part of contents of wall of the GIT can be studied in large animals. Therefore the choice of sampling is the first bottle neck in the studies on yeasts biodiversity in the GIT. Following proceedings such as homogenization, concentration or dilution of the samples must be hereby additionally considered.

Among the methods applied for investigating the biodiversity of yeasts harbouring the GIT of animals, cultivation and morphological and/or biochemical identification have been the most often used for more than 150 years. However, these methods bear limitations such as the choice of the right cultivation medium, pH, temperature and moisture. Furthermore the yeast species that need more time for growth and are at lower abundance in the community cannot be identified in this way. It has been accepted that every ecosystem consists, next to cultivable organisms, also of viable but non-cultivable (VBNC) microorganisms, that contemporarily cannot be cultivated in laboratory because of nutrient limitation or lack of optimal living conditions (Edwards, 2000). This is why only approximately 1% of yeast species could be described so far (Kurtzman & Fell, 2006). Sabouraud agar is the medium most commonly used for cultivation of yeasts from clinical or ecological samples (Odds, 1991), however many others have been used for industrial purposes, providing alternatives for cultivation of more demanding species (King et al., 1986; Jarvis & Williams, 1987; Fleet, 1990; Deak, 1991). It is to remember, that various species can give similar colonies, and the same species can grow in a different way under different conditions. Cultivated species can be however observed under microscope, what helps for identification of the isolates. Spectrophotometric methods such as MALDI-TOF could also provide fast and good tool for identification of the isolates. Molecular methods can be also applied for identification of isolates, e.g. pyrosequencing of target genes (Borman et al., 2009; 2010). Further development of modified media and combinations of temperature, pH, aerobic/anaerobic conditions and moisture would probably increase the number of isolated yeasts, it is however laborious and very time consuming.

Cultivation-independent methods which have been used for the last two decades provided the researches with fast and specific tools for the biodiversity studies. Polymerase chain reaction (PCR), DNA-DNA hybridization or fluorescence in situ hybridization (FISH) applying probes targeting the RNA allow in theory detection of 1 single colony present in a sample population. Further separation of the specific DNA fragments performing denaturing or temperature gradient gel electrophoresis (DGGE / TGGE) allows studying the diversity of the complex community (Cocolin et al., 2002; Prakitchaiwattana et al., 2004; Molnar et al., 2008). Other molecular methods could be also applied for identification of members of a community, e.g. terminal restriction fragment length polymorphism (T-RFLP), amplified fragment length polymorphism (AFLP), multiple-locus variable number tandem repeat analysis (MLVA) (e.g. Tiedje et al., 1999; Gemmer et al., 2002). These methods are very specific, allowing targeting of specified species and thus quantification of the yeasts and calculating the biodiversity. The largest limitation for methods based on PCR is the low sensitivity, as the practice shows only 1-2% of the community can be detected in this way (Macnaughton et al., 1999). Furthermore, fingerprint methods have the bias combined to the fact that amplicons form different species with sequences of similar energetic profile may migrate to the same positions; multiple gene copies with slight sequence differences may give multiple bands for one strain or species; finally some species are phylogeneticaly very similar (Lachance et al., 2003; Janczyk et al., 2006; Borman et al., 2010). The design of probes for direct targeting needs knowledge on the sequence of the target gene and differences between species.

Pyrosequencing and other high-throughput methods provide a fast and very efficient tool for identification of the members of the complex populations. Metagenome analyses targeting the D1/D2 domain of the 26S rRNA gene or the internal transcribed regions (ITSs) allow distinction of the yeasts (Kurtzman & Fell, 2006) and seem to be very suitable methods for studying the yeast biodiversity in the GIT of animals. Pyrosequencing is a rapid method providing up to several thousands of sequences per sample in just few days. Followed by bioinformatics processing, alignment to known species is performed resulting not only in a phylogenetic tree but also in description of the species diversity. Unknown species can be also detected in this way. The high cost provides the limitation for the wide application of this method; however it is to expect that in the near future the high-throughput sequencing will be as expensive as the other commonly used molecular tools.

A microarray has been recently developed allowing characterization of pig GIT bacterial community, targeting over 800 phylotypes (Pérez Gutiérrez, 2010). Microarrays for yeasts would need to be developed to provide further molecular tool for studying the biodiversity and its changes caused by different extrinsic factors.

4. Role of yeasts in GIT

Studying diversity of yeasts harbouring the GIT of animals would be incomplete without consideration of the role that these microorganisms play for the host. For years the yeasts harbouring the GIT of animals and humans have been considered rather as harmful to the host's health. Indeed, there are some species belonging to *Candida, Cryptococcus, Malassezia, Trichosporon* and *Geotrichum* that could be pathogenic to members of the Animal kingdom (Fidel et al., 1999; Girmenia et al., 2005; Cabañes, 2010). Furthermore, many researchers have evaluated yeasts in association with various diseases and if they found representatives of this group they acted against them applying medical treatment (Schulze & Sonnenborn, 2009). However, there is just as little known about yeasts harbouring the GIT of healthy animals to understand their importance there, and growing evidences appear for their role in the proper function and survival of the host. In fact, the current knowledge about yeasts in the digestive tract of vertebrates is still based on the findings from 50's and 70's of the XXth Century; therefore there is a great demand for the scientific evaluation in this field. As enough reports exist concerning pathogenic yeasts, in this paragraph a possible positive impact of yeasts on the gut ecology and host health will be discussed.

There are nice reviews (e.g. Phaff & Starmer, 1987; Ganter, 2006) pronouncing a yeast-insect relationship. Gatesoupe (2007) gave an insight into the ecology of yeasts naturally occurring in the intestinal tract of fish, and thereby emphasized a possible importance of yeasts to the host.

Similarily to the probiotic strains of *Saccharomyces cerevisiae* (Buts, 2009), the cells of some intestinal yeasts could have a trophic effects since they provide a source of B vitamins, proteins, trace minerals and essential amino acids. Besides, the major portion (> 90%) of the yeast cell walls comprise of polysaccharides such as β-glucans, mannans and chitin, which composition and structure are specific for individual yeast (Latgé, 2007). In many human and animals studies, β-glucans and mannans have been comprehensively investigated; they may play important diverse roles for the host immune system and exhibit antimicrobial activity against bacteria thereby influencing the establishment of the intestinal microbiota and promising to promote host's health. Therefore, several studies concentrated on use of the live or dead yeast cells in human and animal nutrition as supplements or as a remedy for acute diarrhoea in humans (Bekatorou et al., 2006; Buts & De Keyser, 2006; Fleet, 2007; Buts, 2009). Furthermore, due to production of several enzymes, some yeast species, e.g. found in the gut of termites (Schäfer et al., 1996; Molnar et al., 2004) and beetles (Suh et al., 2003), are able to degrade hemicelluloses that are being the main carbohydrates of herbivorous diet, and also detoxify toxins that can appear in the feed. The possibility cannot be excluded that some yeasts harbouring GIT of herbivorous animals may produce extracellular enzymes (e.g. exohemicellulases, exocellulases) or show endocellulolytic activity, and thereby contribute to their digestion by braking down complex, indigestible fibre into simple carbohydrates.

It is still a prevalent opinion, that yeasts harbouring the digestive tract of animals have only minor importance for the host. The main scientific argument up to date is the negligible quantity of yeasts. Nevertheless, yeasts may be of physiological relevance, even though they are present to a much lesser extent than bacteria. In fact, yeasts could provide a relevant biomass, as their have a cell volume 30- to 100-fold higher than bacteria (Gatesoupe, 2007). Commensal yeasts may interact with intestinal bacteria and due to this interplay affect microbial diversity and host organism. An example of such yeasts-bacteria interrelationship provides the study of Urubschurov et al. (2011) who examined changes of yeasts and major bacterial groups (lactobacilli, enterobacteria and enterococci) in the faeces of piglets after weaning. They observed that the increase of yeasts number, where the dominating species was *Kazachstania slooffiae,* significantly correlated with the increase of lactobacilli and decrease of enterobacteria numbers. Other studies hypothesized that specific yeasts frequently occurred in high quantity at the digestive tract of lacewings (Woolfolk & Inglis, 2004; Woolfolk et al., 2004) and mosquitoes (Ricci et al., 2011a; 2011b) and were symbiotically related to the host.

These first indications need further confirmation but they already show that the yeasts cannot be considered negligible any more.

5. Conclusions

Yeasts belong to gastrointestinal microbiota even though they are not as frequent as the bacteria or archea. However, it does not disclude their importance for the host and for the members of the complex microbial community. Despite long time of research, whereas our knowledge on bacterial intestinal communities has increased dramatically during last decade, still only little is known on the intestinal yeasts. This review provides an overview on what has been done in the field of intestinal yeast research up till now, and the reader surely agrees that much more work needs to be done. Not only the diversity of the intestinal yeasts and its changes depending on different conditions shall be further uncovered. The importance of yeasts for the host and the interplay between yeasts and other members of the intestinal milieu is also waiting to be explored. New cultivation techniques; cultivation combined with molecular techniques will need to be further developed to overcome the existing limitations.

6. References

Batra, L.R., Batra, S.W.T. & Bohart, G.E. (1973). Mycoflora of Domesticated and Wild Bees (Apoidea). *Mycopathol Mycol Appl,* Vol.49, No.1 (August 2005 in SpringerLink), pp. 13-44, ISSN: 0027-5530.

Bauer, E., Williams, B.A., Smidt, H., Verstegen, M.W., & Mosenthin, R. (2006). Influence of the gastrointestinal microbiota on development of the immune system in young animals. *Curr Issues Intest Microbiol,* Vol.7, No.2, pp. 35-51.

Bekatorou, A., Psarianos, C. & Koutinas, A.A. (2006). Production of food grade yeasts. *Food Technol. Biotech.* Vol.44, No.3, pp. 407-415, ISSN: 1330-9862.

Borman, A.M., Linton, C.J., Oliver, D., Palmer, M.D., Szekely, A., Odds, F.C. & Johnson, E.M. (2009). Pyrosequencing analysis of 20 nucleotides of internal transcribed spacer 2 discriminates *Candida parapsilosis, Candida metapsilosis,* and *Candida orthopsilosis. J Clin. Microbiol.,* Vol.47, No.7 (July 2009), pp. 2307-2310, ISSN: 0095-1137.

Borman, A.M., Linton, C.J., Oliver, D., Palmer, M.D., Szekely, A. & Johnson, E.M. (2010). Rapid molecular identification of pathogenic yeasts by pyrosequencing analysis of 35 nucleotides of internal transcribed spacer 2. *J Clin. Microbiol.,* Vol.48, No.10 (October 2010), pp. 3648-3653, ISSN: 1098-660X.

Botha, A. (2006). Yeasts in Soil. In: *The Yeast Handbook: Biodiversity and Ecophysiology of Yeasts,* C.A. Rosa & G. Peter (Ed.), pp. 221-240, Springer-Verlag, ISBN-10: 3-540-26100-1, Berlin Heidelberg.

Brilhante, R.S., Castelo-Branco, D. S, Soares, G.D., Astete-Medrano, D.J., Monteiro, A.J., Cordeiro, R.A., Sidrim, J.J. & Rocha, M.F. (2010). Characterization of the gastrointestinal yeast microbiota of cockatiels (*Nymphicus hollandicus*): a potential hazard to human health. *J Med. Microbiol.,* Vol.59, No.Pt6 (June 2010), pp. 718-723, ISSN: 0022-2615.

Brito, E.H., Fontenelle, R.O., Brilhante, R.S., Cordeiro, R.A., Monteiro, A.J., Sidrim, J.J. & Rocha, M.F. (2009). The anatomical distribution and antimicrobial susceptibility of yeast species isolated from healthy dogs. *Vet. J,* Vol.182 No.2, (November 2009), pp. 320-326, ISSN: 1090-0233.

Brysch-Herzberg, M. (2004). Ecology of yeasts in plant-bumblebee mutualism in Central Europe. *FEMS Microbiol. Ecol.,* Vol.50, No.2 (November 2004), pp. 87-100, ISSN: 1574-6941.

Buts, J.P. & De Keyser, N. (2006). Effects of *Saccharomyces boulardii* on intestinal mucosa. *Dig. Dis. Sci.,* Vol.51, No.8, pp. 1485-1492, ISSN: 0163-2116.

Buts, J.P. 2009. Twenty-five years of research on *Saccharomyces boulardii* trophic effects: updates and perspectives. *Dig. Dis. Sci.,* Vol.54, No.1, pp. 15-18, ISSN: 0163-2116.

Cabañes, F.J. (2010). Yeast pathogens of domestic animals. In: *Pathogenic Yeasts, The Yeast Handbook,* H.R. Ashbeee & E.M. Bingell (Ed), pp. 253-279 Springer-Verlag, ISBN 978-3-642-03149-6, Berlin Heidelberg.

Cafarchia, C., Camarda, A., Romito, D., Campolo, M., Quaglia, N.C., Tullio, D. & Otranto, D. (2006). Occurrence of yeasts in cloacae of migratory birds. *Mycopathologia,* Vol.161, No.4 (March 2006), pp. 229-234, ISSN 0301-486X.

Cafarchia, C., Romito, D., Coccioli, C., Camarda, A. & Otranto, D. (2008). Phospholipase activity of yeasts from wild birds and possible implications for human disease. *Med Mycol,* Vol.46, No.5, pp. 429-434, ISSN 1369-3786.

Carreiro, S.C., Pagnocca, F.C., Bueno, O.C., Bacci, M.B., Hebling, M.J.A. & da Silva, O.A. (1997). Yeasts associated with nests of the leaf-cutting ant *Atta sexdens rubropilosa* Forel, 1908. *Anton. Leeuw. Int. J. G.,* Vol.71, No.3 (October 2004 in SpringerLink), pp. 243-248, ISSN 0003-6072.

Chapman, A.D. (2009). Numbers of Living Species in Australia and the World. 2nd ed. *Australian Biological Resources Study,* Canberra, pp. 66.

Chen, T.Y., Chu, C.C., Hu, C., Mu, J.Y., & Henneberry, T.J. (2006). Observations on midgut structure and content of *Chrysoperla carnea* (Neuroptera: Chrysopidae). *Ann. Entomol. Soc. Am.,* Vol.99, No.5 (September 2006), pp. 917-919, ISSN: 0013-8746.

Clarke, R.T. & Menna, M.E.D. (1961). Yeasts from Bovine Rumen. *J. Gen. Microbiol.*, Vol.25, No.1, pp. 113-117, ISSN: 0022-1287..

Cocolin, L., Aggio, D., Manzano, M., Cantoni, C. & Comi, G. (2002). An application of PCR-DGGE analysis to profile the yeast populations in raw milk. *Int. Dairy J.*, Vol.12, No.5, pp. 407-411, ISSN: 0958-6946.

Costa, A.K.F., Sidrim, J.J.C., Cordeiro, R.A., Brilhante, R.S.N., Monteiro, A.J. & Rocha, M.F.G. (2010). Urban Pigeons (*Columba livia*) as a Potential Source of Pathogenic Yeasts: A Focus on Antifungal Susceptibility of *Cryptococcus* Strains in Northeast Brazil. *Mycopathologia*, Vol.169, No.3 (October 2009 in SpringerLink) pp. 207-213, ISSN: 0301-486X.

Deak, T. (1991). Foodborne Yeasts. In: *Advances in Applied Microbiology*, Vol.36, pp. 179-278, Academic Press, Inc., ISBN: 0-12-002636-8, USA.

Dillon, R.J. & Dillon, V.M. (2004). The gut bacteria of insects: Nonpathogenic interactions. *Annu. Rev. Entomol.*, Vol.49, No.1 (January 2004), pp. 71-92, ISSN: 0066-4170.

Edwards, C. (2000). Problems posed by natural environments for monitoring microorganisms. *Mol. Biotechnol.*, Vol.15, No.3 (August 2007 in SpringerLink), pp. 211-223, ISSN: 1073-6085.

Fell, J.W., Boekhout, T., Fonseca, A., Scorzetti, G. & Statzell-Tallman, A. (2000). Biodiversity and systematics of basidiomycetous yeasts as determined by large-subunit rDNA D1/D2 domain sequence analysis. *Int. J. Syst. Evol. Microbiol.*, Vol.50, No.3, pp. 1351-1371, ISSN: 1466-5026.

Fidel, P.L.Jr., Vazquez, J.A. & Sobel, J.D. (1999). *Candida glabrata*: review of epidemiology, pathogenesis, and clinical disease with comparison to *C. albicans*. *Clin. Microbiol. Rev.*, Vol.12, No.1, pp. 80-96, ISSN: 0893-8512.

Fleet, G.H. (1990). Food spoilage yeasts. In: *Yeast technology*. J.F.T. Spencer & D.M. Spencer (Ed), pp. 124-166, Springer, Berlin Heidelberg, ISBN: 3-540-50689-6, New York.

Fleet, G.H. (2007). Yeasts in foods and beverages: impact on product quality and safety. *Curr. Opin. Biotechnol.*, Vol.18, No.2 (April 2007), pp. 170-175, ISSN: 0958-1669.

Fonseca, A. & Inàcio, J. (2006). Phylloplane Yeasts. In: *The Yeast Handbook: Biodiversity and Ecophysiology of Yeasts.* C.A. Rosa & G. Peter (Ed), pp. 263-301, Springer-Verlag, ISBN: 978-3-540-26100-1, Berlin Heidelberg.

Fuentefria, A.M., Suh, S.O., Landell, M.F., Faganello, J., Schrank, A., Vainstein, M.H., Blackwell, M. & Valente, P. (2008). *Trichosporon insectorum* sp. nov., a new anamorphic basidiomycetous killer yeast. *Mycol. Res*, Vol.112, No.1 (January 2008) pp. 93-99, ISSN: 0953-7562.

Ganter, P.F. (2006). Yeast and Invertebrate Associations. In: *The Yeast Handbook: Biodiversity and Ecophysiology of Yeasts.*, C.A. Rosa & G. Peter (Ed), pp. 303-370, Springer-Verlag, ISBN: 978-3-540-26100-1, Berlin Heidelberg.

Gatesoupe, F.J. (2007). Live yeasts in the gut: Natural occurrence, dietary introduction, and their effects on fish health and development. *Aquaculture*, Vol.267, No.1-4 (July 2007), pp. 20-30, ISSN: 0044-8486.

Gemmer, C.M., DeAngelis, Y.M., Theelen, B., Boekhout, T. & Dawson, T.L.J.Jr. (2002). Fast, noninvasive method for molecular detection and differentiation of *Malassezia* yeast species on human skin and application of the method to dandruff microbiology. *J Clin. Microbiol.*, Vol.40, No.9 (September 2002), pp. 3350-3357, ISSN: 0095-1137.

Gilliam, M., Wickerham, L.J., Morton, H.L., & Martin, R.D. (1974). Yeasts Isolated from Honey Bees, Apis-Mellifera, Fed 2,4-D and Antibiotics. *J. Invertebr. Pathol.*, Vol.24, No.3 (November 1974), pp. 349-356, ISSN: 0022-2011.

Girmenia, C., Pagano, L., Martino, B., D'Antonio, D., Fanci, R., Specchia, G., Mei, L., Buelli, M., Pizzarelli, G., Venditti, M. & Martino, P. (2005). Invasive infections caused by *Trichosporon* species and *Geotrichum capitatum* in patients with hematological malignancies: a retrospective multicenter study from Italy and review of the literature. *J. Clin. Microbiol.* Vol.43, No.4 (April 2005), pp. 1818-1828, ISSN: 0095-1137.

Grunwald, S., Pilhofer, M. & Holl, W. (2010). Microbial associations in gut systems of wood- and bark-inhabiting longhorned beetles [Coleoptera: Cerambycidae]. *Syst. Appl. Microbiol.*, Vol.33, No.1 (January 2010), pp. 25-34, ISSN: 0723-2020.

Gujjari, P., Suh, S.O., Lee, C.F. & Zhou, J.J. (2010). *Trichosporon xylopini* sp. nov., a hemicellulose-degrading yeast isolated from wood-inhabiting beetle *Xylopinus saperdioides*. *Int J Syst. Evol. Microbiol.*, IJSEM Papers in Press. Published October 29, 2010 as doi:10.1099/ijs.0.028860-0

Gusmão, D.S., Santos, A.V., Marini, D.C., Russo, E.D., Peixoto, A.M.D., Bacci, M., Berbert-Molina, M.A. & Lemos, F.J.A. (2007). First isolation of microorganisms from the gut diverticulum of *Aedes aegypti* (Diptera : Culicidae): new perspectives for an insect-bacteria association. *Mem I Oswaldo Cruz*, Vol.102, No.8 (December 2007), pp. 919-924, ISSN: 0074-0276.

Gusmão, D.S., Santos, A.V., Marini, D.C., Bacci, M., Berbert-Molina, M.A. & Lemos, F.J.A. (2010). Culture-dependent and culture-independent characterization of microorganisms associated with *Aedes aegypti* (Diptera: Culicidae) (L.) and dynamics of bacterial colonization in the midgut. *Acta Tropica*, Vol.115, No.3 (September 2010), pp. 275-281, ISSN: 0001-706X.

Hansen, L.D. & Klotz, J.H. (2005). *Carpenter ants of the United States and Canada.* Cornell University Press, ISBN 0-8014-4262-1, Ithaca, New York.

Hongoh, Y. (2010). Diversity and Genomes of Uncultured Microbial Symbionts in the Termite Gut. *Biosci. Biotechnol. Biochem.*, Vol.74, No.6 (June 2010), pp. 1145-1151, ISSN: 1347-6947.

Hooper, L.V. & Gordon, J.I. (2001). Commensal host-bacterial relationships in the gut. *Science*, Vol.292, No.5519 (May 2001), pp. 1115-1118.

Janczyk, P., Pieper, R. & Souffrant, W.B. (2006) 16S rDNA polymerase chain reaction - denaturing gradient gel electrophoresis used to study the pig intestinal microflora. Abstracts of International Conference "Sustainable Animal Health through Eubiosis - Relevance for Man"., Ascona, Switzerland, 8-13 Oct. 2006 p.16

Jarvis, B. & Williams, A.P. (1987). Methods for detecting fungi in foods and beverages. In: *Food and beverage mycology.* L.R. Beuchat (Ed), pp. 599-636 Van Rostrand, ISBN: 0-442-21084-1, New York, USA.

Jouany, J.P., Gobert, J., Medina, B., Bertin, G. & Julliand, V. (2008). Effect of live yeast culture supplementation on apparent digestibility and rate of passage in horses fed a high-fiber or high-starch diet. *J. Anim Sci.*, Vol.86, No.2, pp. 339-347.

Jouany, J.P., Medina, B., Bertin, G. & Julliand, V. (2009). Effect of live yeast culture supplementation on hindgut microbial communities and their polysaccharidase and glycoside hydrolase activities in horses fed a high-fiber or high-starch diet. *J. Anim Sci.*, Vol.87, No.9 (May 2009), pp. 2844-2852.

King, A.D., Pitt, J.I., Beuchat, L.R. & Corry, J.E.L. (1986). *Methods for the mycological examination of food.* Plenum, ISBN-10: 0306424797, New York, USA.

Kostka, V.M., Hoffmann, L., Balks, E., Eskens, U. & Wimmershof, N. (1997). Review of the literature and investigations on the prevalence and consequences of yeasts in reptiles. *Vet. Rec.* Vol.140, No.11, pp. 282-287.

Kurtzman, C.P. & Fell, J.W. (1998). Definition, Classification and Nomenclature of the Yeasts. In: *The Yeasts, a Taxonomic Study*, 4th ed, Kurtzman, C.P. & J.W. Fell, (Ed), pp. 3 -5, Elsevier, ISBN: 0 444 81312 8, Amsterdam.

Kurtzman, C.P., Robnett, C.J., Ward, J.M., Brayton, C., Gorelick, P. & Walsh, T.J. (2005). Multigene phylogenetic analysis of pathogenic candida species in the *Kazachstania* (*Arxiozyma*) *telluris* complex and description of their ascosporic states as *Kazachstania bovina* sp. nov., *K. heterogenica* sp. nov., *K. pintolopesii* sp. nov., and *K. slooffiae* sp. nov. *J. Clin. Microbiol.*, Vol.43, No.1 (January 2005), pp. 101-111, ISSN: 1098-660X.

Kurtzman, C.P. & Fell, J.W. (2006). Yeast systematics and phylogeny - implications of molecular identification methods for studies in ecology. In: *The Yeast Handbook: Biodiversity and Ecophysiology of Yeasts*, Rosa, C.A. & G. Peter, (Ed), pp. 11-30, Springer-Verlag, ISBN: 978-3-540-26100-1, Berlin Heidelberg.

Lachance, M.A., Starmer, W.T., Rosa, C.A., Bowles, J.M., Barker, J. S. & Janzen, D.H. (2001). Biogeography of the yeasts of ephemeral flowers and their insects. *FEMS Yeast Res*, Vol.1, No.1 (April 2001), pp. 1-8, ISSN: 1567-1356.

Lachance, M.A., Daniel, H.M., Meyer, W., Prasad, G.S., Gautam, S. P. & Boundy-Mills, K. (2003). The D1/D2 domain of the large-subunit rDNA of the yeast species *Clavispora lusitaniae* is unusually polymorphic. *FEMS Yeast Res.*, Vol.4, No.3 (December 2003), pp. 253-258, ISSN: 1567-1356.

Latgé, J.P. (2007). The cell wall: a carbohydrate armour for the fungal cell. *Molecular Microbiology*, Vol.66, No.2 (October 2007), pp. 279-290.

Ley, R.E., Hamady, M., Lozupone, C., Turnbaugh, P.J., Ramey, R.R., Bircher, J.S., Schlegel, M.L., Tucker, T.A., Schrenzel, M.D., Knight, R. & Gordon, J.I. (2008). Evolution of mammals and their gut microbes. *Science*, Vol.320, No.5883 (June 2008), pp. 1647-1651.

Lund, A. (1974). Yeasts and Molds in Bovine Rumen. *J. Gen. Microbiol.*, Vol.81, No.2 (April 1974), pp. 453-462, ISSN: 0022-1287.

Lund, A. (1980). Yeasts in the rumen contents of musk oxen. *J Gen. Microbiol.*, Vol.121, No.1, pp. 273-276, ISSN: 0022-1287.

Macnaughton, S.J., Stephen, J.R., Venosa, A.D., Davis, G.A., Chang, Y.J. & White, D.C. (1999). Microbial population changes during bioremediation of an experimental oil spill. *Appl Environ. Microbiol.*, Vol.65, No.8 (August 1999), pp. 3566-3574, ISSN: 0099-2240.

Mankowski, M.E. & Morrell, J.J. (2004). Yeasts associated with the infrabuccal pocket and colonies of the carpenter ant *Camponotus vicinus*. *Mycologia*, Vol.96, No.2 (March 2004), pp. 226-231, ISSN: 0027 5514.

Medina, B., Girard, I.D., Jacotot, E. & Julliand, V. (2002). Effect of a preparation of *Saccharomyces cerevisiae* on microbial profiles and fermentation patterns in the large intestine of horses fed a high fiber or a high starch diet. *J. Anim Sci.*, Vol.80, No.10, (October 2002), pp. 2600-2609, ISSN: 00218812.

Mehnert, B. & Koch, U. (1963). Über das Vorkommen von im Verdauungstrakt des Schweines vermehrungsfähiger Hefen. *Zbl. Bakter. I. Orig.*, Vol.188, pp. 103-120.

Molnar, O., Schatzmayr, G., Fuchs, E. & Prillinger, H. (2004). *Trichosporon mycotoxinivorans* sp nov., a new yeast species useful in biological detoxification of various mycotoxins. *Syst. Appl. Microbiol.*, Vol.27, No.6 (December 2004), pp. 661-671, ISSN: 0723-2020.

Molnar, O. & Prillinger, H. (2005). Analysis of yeast isolates related to *Metschnikowia pukherrima* using the partial sequences of the large subunit rDNA and the actin gene; description of *Metschnikowia andauensis* sp nov. *Syst. Appl. Microbiol.*, Vol.28, No.8 (October 2005), pp. 717-726, ISSN: 0723-2020.

Molnar, O., Wuczkowski, M. & Prillinger, H. (2008). Yeast biodiversity in the guts of several pests on maize; comparison of three methods: classical isolation, cloning and DGGE. *Mycol. Prog.*, Vol.7, No.2 (April 2008 in SpringerLink), pp. 111-123, ISSN: 1861-8952.

Nagahama, T. (2006). Yeast Biodiversity in Freshwater, Marine and Deep-Sea Environments. In: *The Yeast Handbook: Biodiversity and Ecophysiology of Yeasts,* C.A. Rosa & G. Peter (Ed), pp. 241-262, Springer-Verlag, ISBN: 978-3-540-26100-1, Berlin Heidelberg.

Nguyen, N.H., Suh, S.O., Erbil, C.K. & Blackwell, M. (2006). *Metschnikowia noctiluminum* sp. nov., *Metschnikowia corniflorae* sp. nov., and *Candida chrysomelidarum* sp. nov., isolated from green lacewings and beetles. *Mycol. Res*, Vol.110, No.Pt3 (March 2006), pp. 346-356, ISSN: 0953-7562.

Nguyen, N.H., Suh, S.O. & Blackwell, M. (2007). Five novel *Candida* species in insect-associated yeast clades isolated from Neuroptera and other insects. *Mycologia,* Vol.99, No.6, pp. 842-858.

Odds, F.C. (1991). Sabouraud('s) agar. *J Med Vet. Mycol*, Vol.29, No.6, pp. 355-359.

Pagnocca, F.C., Legaspe, M.F., Rodrigues, A., Ruivo, C.C., Nagamoto, N.S., Bacci, M.Jr. & Forti, L.C. (2010). Yeasts isolated from a fungus-growing ant nest, including the description of *Trichosporon chiarellii* sp. nov., an anamorphic basidiomycetous yeast. *Int J Syst. Evol. Microbiol.*, Vol.60, No.Pt6 (June 2010), pp. 1454-1459.

Parle, J.N. (1957). Yeasts Isolated from the Mammalian Alimentary Tract. *J. Gen. Microbiol.*, Vol.17, No.2 (October 1957), pp. 363-367, ISSN: 0022-1287.

Pérez Gutiérrez, O. (2010). Unraveling Piglet Gut Microbiota Dynamics in Response to Feed Additives microbiota. PhD Thesis, Wageningen University, Wageningen, The Netherlands, ISBN 978-90-8585-684-9.

Phaff, H.J. & Starmer, W.T. (1987). Yeasts associated with plants, insects and soil. In: *The yeasts*, A.H. Rose & J. S. Hartison (Ed), pp. 123-180. Academic Press Inc., ISBN 0-12-596411-0, London.

Prakitchaiwattana, C.J., Fleet, G.H. & Heard, G.M. (2004). Application and evaluation of denaturing gradient gel electrophoresis to analyse the yeast ecology of wine grapes. *FEMS Yeast Res.*, Vol.4, No.8 (September 2004), pp. 865-877, ISSN: 1567-1356.

Prillinger, H., Messner, R., Konig, H., Bauer, R., Lopandic, K., Molnar, O., Dangel, P., Weigang, F., Kirisits, T., Nakase, T. & Sigler, L. (1996). Yeasts associated with termites: A phenotypic and genotypic characterization and use of coevolution for dating evolutionary radiations in asco- and basidiomycetes. *Syst. Appl. Microbiol.*, Vol.19, No.2, pp. 265-283, ISSN: 0723-2020.

Rao, R.S., Bhadra, B. & Shivaji, S. (2007). Isolation and characterization of xylitol-producing yeasts from the gut of colleopteran insects. *Curr. Microbiol.*, Vol.55, No.5 (August 21, 2007), pp. 441-446, ISSN: 0343-8651.

Ricci, I., Damiani, C., Scuppa, P., Mosca, M., Crotti, E., Rossi, P., Rizzi, A., Capone, A., Gonella, E., Ballarini, P., Chouaia, B., Sagnon, N.F., Esposito, F., Alma, A., Mandrioli, M., Sacchi, L., Bandi, C., Daffonchio, D. & Favia, G. (2011a). The yeast *Wickerhamomyces anomalus* (*Pichia anomala*) inhabits the midgut and reproductive system of the Asian malaria vector *Anopheles stephensi*. *Environ. Microbiol.*, Vol.13, No.4 (April 2011), pp. 911-921, ISSN: 1462-2912

Ricci, I., Mosca, M., Valzano, M., Damiani, C., Scuppa, P., Rossi, P., Crotti, E., Cappelli, A., Ulissi, U., Capone, A., Esposito, F., Alma, A., Mandrioli, M., Sacchi, L., Bandi, C., Daffonchio, D. & Favia, G. (2011b). Different mosquito species host *Wickerhamomyces anomalus* (*Pichia anomala*): perspectives on vector-borne diseases symbiotic control. *Anton. Leeuw. Int. J. G.*, Vol.99, No.1 (Januar 2011), pp. 43-50, ISSN: 0003-6072.

Rivera, F.N., Gonzalez, E., Gomez, Z., Lopez, N., Hernandez-Rodriguez, C., Berkov, A. & Zuniga, G. (2009). Gut-associated yeast in bark beetles of the genus *Dendroctonus Erichson* (Coleoptera: Curculionidae: Scolytinae). *Biol. J. Linn. Soc.*, Vol.98, No.2 (October 2009), pp. 325-342, ISSN: 0024-4066.

Rodrigues, A., Cable, R.N., Mueller, U.G., Bacci, M.Jr. & Pagnocca, F.C. (2009). Antagonistic interactions between garden yeasts and microfungal garden pathogens of leaf-cutting ants. *Antonie Van Leeuwenhoek*, Vol.96, No.3 (May 2009 in SpringerLink) , pp. 331-342, ISSN: 0003-6072.

Rosa, C.A. & Peter, G. (2006). *The Yeast Handbook: Biodiversity and Ecophysiology of Yeasts*. Germany: Springer-Verlag, ISBN-10: 3-540-26100-1, Berlin Heidelberg.

Schäfer, A., Konrad, R., Kuhnigk, T., Kampfer, P., Hertel, H. & Konig, H. (1996). Hemicellulose-degrading bacteria and yeasts from the termite gut. *J Appl. Bacteriol.*, Vol.80, No.5 (May 1996), pp. 471-478, ISSN: 0021-8847.

Schulze, J. & Sonnenborn, U. (2009). Yeasts in the gut: from commensals to infectious agents. *Dtsch. Arztebl. Int.*, Vol.106, No.51-52 (December 2009), pp. 837-842, ISSN: 1866-0452.

Scorzetti, G., Fell, J.W., Fonseca, A. & Statzell-Tallman, A. (2002). Systematics of basidiomycetous yeasts: a comparison of large subunit D1/D2 and internal transcribed spacer rDNA regions. *FEMS Yeast Res.*, Vol.2, No.4 (December 2002), pp. 495-517, ISSN: 1567-1356.

Scupham, A.J., Presley, L.L., Wei, B., Bent, E., Griffith, N., McPherson, M., Zhu, F., Oluwadara, O., Rao, N., Braun, J. & Borneman, J. (2006). Abundant and diverse fungal microbiota in the murine intestine. *Appl. Environ. Microbiol.*, Vol.72, No.1 (January 2006), pp. 793-801, ISSN: 0099-2240.

Shifrine, M. & Phaff, H. J. (1956). The Association of Yeasts with Certain Bark Beetles. *Mycologia*, Vol.48, No. 1, (January- February 1956), pp. 41-55, ISSN: 00275514.

Shin, E.C., Kim, Y.K., Lim, W.J., Hong, S.Y., An, C.L., Kim, E.J., Cho, K.M., Choi, B.R., An, J.M., Kang, J.M., Jeong, Y.J., Kwon, E.J., Kim, H. & Yun, H. D. (2004). Phylogenetic analysis of yeast in the rumen contents of cattle based on the 26S rDNA sequence. *J Agric Sci*, Vol.142, No.05 (March 2005), pp. 603-611, ISSN: 00218596.

Suh, S.O., Marshall, C.J., Mchugh, J. V. & Blackwell, M. (2003). Wood ingestion by passalid beetles in the presence of xylose-fermenting gut yeasts. *Mol. Ecol.*, Vol.12, No.11 (November 2003) , pp. 3137-3145, ISSN: 0962-1083.

Suh, S.O. & Blackwell, M. (2004). Three new beetle-associated yeast species in the *Pichia guilliermondii* clade. *FEMS Yeast Res*, Vol.5, No.1 (October 2004), pp. 87-95, ISSN: 1567-1356.

Suh, S.O., Gibson, C.M. & Blackwell, M. (2004a). *Metschnikowia chrysoperlae* sp. nov., *Candida picachoensis* sp. nov. and *Candida pimensis* sp. nov., isolated from the green lacewings *Chrysoperla comanche* and *Chrysoperla carnea* (Neuroptera: Chrysopidae). *Int J Syst. Evol. Microbiol.*, Vol.54, No.Pt5 (September 2004), pp. 1883-1890, ISSN: 1466-5026.

Suh, S.O., Mchugh, J.V. & Blackwell, M. (2004b). Expansion of the *Candida tanzawaensis* yeast clade: 16 novel *Candida* species from basidiocarp-feeding beetles. *Int J Syst. Evol. Microbiol.*, Vol.54, No.Pt6 (November 2004), pp. 2409-2429, ISSN: 1466-5026.

Suh, S.O. & Blackwell, M. (2005). Four new yeasts in the *Candida mesenterica* clade associated with basidiocarp-feeding beetles. *Mycologia*, Vol.97, No.1 (January- February 2005), pp. 167-177, ISSN: 0027-5514.

Suh, S.O., Mchugh, J.V., Pollock, D.D. & Blackwell, M. (2005a). The beetle gut: a hyperdiverse source of novel yeasts. *Mycol. Res.*, Vol.109, No.Pt3 (March 2005), pp. 261-265, ISSN: 0953-7562.

Suh, S.O., Nguyen, N.H. & Blackwell, M. (2005b). Nine new *Candida* species near *C. membranifaciens* isolated from insects. *Mycol. Res*, Vol.109, No.Pt9 (September 2005), pp. 1045-1056, ISSN: 0027-5514.

Suh, S.O. & Blackwell, M. (2006). Three new asexual arthroconidial yeasts, *Geotrichum carabidarum* sp. nov., *Geotrichum histeridarum* sp. nov., and *Geotrichum cucujoidarum* sp. nov., isolated from the gut of insects. *Mycol. Res*, Vol.110, No.Pt2 (Februar 2006), pp. 220-228, ISSN: 0953-7562.

Suh, S.O., Blackwell, M., Kurtzman, C.P. & Lachance, M.A. (2006a). Phylogenetics of *Saccharomycetales*, the ascomycete yeasts. *Mycologia*, Vol.98, No.6 (November-December 2006), pp. 1006-1017, ISSN: 0027-5514.

Suh, S.O., Nguyen, N.H. & Blackwell, M. (2006b). A yeast clade near *Candida kruisii* uncovered: nine novel *Candida* species associated with basidioma-feeding beetles. Mycol. Res Vol.110, No.Pt2 (December 2006), pp. 1379-1394, ISSN: 0953-7562.

Suh, S.O., Nguyen, N.H. & Blackwell, M. (2008). Yeasts isolated from plant-associated beetles and other insects: seven novel *Candida* species near *Candida albicans*. *FEMS Yeast Res.*, Vol.8, No.1 (Februar 2008), pp. 88-102, ISSN: 1567-1356.

Tiedje, J.M., Asuming-Brempong, S., Nüsslein, K., Marsh, T.L. & Flynn, S.J. (1999). Opening the black box of soil microbial diversity. *Appl. Soil Ecol.*, Vol.13, No.2 (October 1999), pp. 109-122, ISSN: 0929-1393.

Travassos, L.R.R.G. & Cury, A. (1971). Thermophilic Enteric Yeasts. *Annu. Rev. Microbiol.*, Vol.25, No.1 (October 1971), pp. 49-74, ISSN: 0066-4227.

Urubschurov, V., Janczyk, P., Pieper, R. & Souffrant, W.B. (2008). Biological diversity of yeasts in the gastrointestinal tract of weaned piglets kept under different farm conditions. *FEMS Yeast Res.*, Vol.8, No.8 (December 2008), pp. 1349-1356, ISSN: 1567-1356.

Urubschurov, V., Janczyk, P., Souffrant, W.B., Freyer, G. & Zeyner, A. (2011). Establishment of intestinal microbiota with focus on yeasts of unweaned and weaned piglets kept under different farm conditions. *FEMS Microbiol.Ecol.*, in Press

Van Uden, N. & do. Carmo-Sousa, L. (1957a). *Candida slooffii* nov.sp., a thermophilic and vitamin deficient yeast from the equine intestinal tract. *Portug. Acta Biol*, Vol.V, pp. 7-17, ISSN 0874-9035.

Van Uden, N. & do. Carmo-Sousa, L. (1957b). Yeasts from the bovine caecum. *J. Gen. Microbiol.*, Vol.16, No.2 (April 1957), pp. 385-395, ISSN: 1350-0872.

Van Uden, N. & do. Carmo-Sousa, L. (1962). Quantitative aspects of the intestinal yeast flora of swine. *J. Gen. Microbiol.*, Vol.27, No.1 (January 1962), pp. 35-40, ISSN: 1350-0872.

Van Uden, N., do. Carmo-Sousa, L., & Farinha, M. (1958). On the intestinal yeast flora of horses, sheep, goats and swine. *J. Gen. Microbiol.*, Vol.19, No.3 (December 1958), pp. 435-445, ISSN: 1350-0872.

Woolfolk, S.W., Cohen, A.C. & Inglis, G.D. (2004). Morphology of the alimentary canal of *Chrysoperla rufilabris* (Neuroptera : Chrysopidae) adults in relation to microbial symbionts. *Ann Entomol Soc Am* Vol.97, No.4, pp. 796-808, ISSN: 0013-8746.

Woolfolk, S.W. & Inglis, G.D. (2004). Microorganisms associated with field-collected *Chrysoperla rufilabris* (Neuroptera: Chrysopidae) adults with emphasis on yeast symbionts. *Biol Control*, Vol.29, No.2 (February 2004), pp. 155-168, ISSN: 1049-9644.

Yarrow, D. & Meyer, S.A. (1978). Proposal for Amendment of Diagnosis of Genus *Candida* Berkhout Nom Cons. *Int. J. Syst. Bacteriol.*, Vol.28, No.4, pp. 611-615, ISSN: 1466-5026.

13

Biodiversity of *Trichoderma* in Neotropics

Lilliana Hoyos-Carvajal[1] and John Bissett[2]
[1]*Universidad Nacional de Colombia, Sede Bogotá*
[2]*Agriculture and Agri-Food Canada, Eastern Cereal and Oilseed Research Centre, Ottawa*
[1]*Colombia*
[2]*Canada*

1. Introduction

Trichoderma species frequently are predominant over wide geographic regions in all climatic zones, where they are significant decomposers of woody and herbaceous materials. They are characterized by rapid growth, an ability to assimilate a diverse array of substrates, and by their production of an range of antimicrobials. Strains have been exploited for production of enzymes and antibiotics, bioremediation of xenobiotic substances, and as biological control agents against plant pathogenic fungi and nematodes. The main use of *Trichoderma* in global trade is derived from its high production of enzymes. *Trichoderma reesei* (teleomorph: *Hypocrea jecorina*) is the most widely employed cellulolytic organism in the world, although high levels of cellulase production are also seen in other species of this genus (Baig et al., 2003, Watanabe et al., 2006). Worldwide sales of enzymes had reached the figure of $ 1.6 billion by the year 2000 (Demain 2000, cited by Karmakar and Ray, 2011), with an annual growth of 6.5 to 10% not including pharmaceutical enzymes (Stagehands, 2008). Of these, cellulases comprise approximately 20% of the enzymes marketed worldwide (Tramoy et al., 2009). Cellulases of microbial origin are used to process food and animal feed, biofuel production, baking, textiles, detergents, paper pulp, agriculture and research areas at all levels (Karmakar and Ray, 2011). Most cellulases are derived from *Trichoderma* (section Longibrachiatum in particular) and *Aspergillus* (Begum et al., 2009). *Trichoderma* is also an efficient degrader of heteropolysaccharides such as xylan, and xylanases and mannanases are of importance in the production of fine paper (Watanabe et al., 2006). In addition, some strains of *Trichoderma* are agents of bioremediation, capable of assimilating heavy metals (Akhtar et al., 2009; Guillermina et al., 2002) and of degrading cyanide (Ezzi and Lynch, 2005) and pesticides with high persistence in the environment (Cross, 1999, Tang et al., 2009). The genus *Trichoderma* includes strains altogether producing an extremely wide range of metabolites, including compounds with antifungal activities (phenolic compounds, 6-α-pentyl-pyrone, viridofunginas, harzianopiridona), antibiotics (anthraquinone, harzianodiona, gliotoxin), plant growth regulators (ciclonerodiol, α-harzianopiridona-pentyl-pyrone), antimicrobial peptides including more than 200 peptaibols, and even viridiol phytotoxic compounds with potential pharmaceutical uses as anti-tumor and immunomodulatory compounds (harzianodiona and gliotoxin). These and other metabolites that are unclassified inhibitors and anti-virus agents expand the prospects of industrial, pharmaceutical or other commercial uses of this organism (Sivasithanparam and Ghisalberti, 1998, Supothina et al., 2007, Vinal et al., 2006, Xiao-Yan et al., 2006).

Many species of *Trichoderma* are closely associated with plant roots and specific strains may form endophytic associations with their plant host (Bailey et al., 2006, Evans et al., 2003, Hoyos-Carvajal et al., 2009b; Manesh et al., 2006, Sette et al., 2006, Viterbo & Chet 2006, Yedidia et al., 2000). As endophytes they are particularly effective biological control agents of fungi in the rhizosphere, producing antimicrobials, activating plant defence mechanisms, and stimulating plant growth and vigour by solubilizing minerals and providing other nutrients and growth regulating compounds (Alfano et al., 2007; Altomare et al. 1999; Sharon et al., 2001; Vinale et al., 2006, Woo et al., 2006, Yedidia et al., 2000). The multiple roles of *Trichoderma* in biotrophic decomposition, parasitism and endophytic associations are of particular importance to the sustainability of agricultural and natural ecosystems (Harman et al. 2004). However, one of the great impediments to the study of *Trichoderma* has been the incorrect and confusing application of species names, making comparisons and generalizations from many published studies unreliable (Kopchinskiy et al., 2005). In addition, different isolates of *Trichoderma* species may exhibit as much variation in metabolic activity as observed among species, making careful study of their biodiversity essential to fully exploit the potential of these fungi.

2. *Trichoderma* taxonomy, a tool to assess diversity

There remain many difficulties in the morphological identification of *Trichoderma* due to the homoplasy of morphological and phenetic characters, particularly among the *Trichoderma* anamorph forms (Chaverri & Samuels, 2003; Druzhinina *et al.*, 2006). For very many years since the genus was first described by Persoon in 1794, and connected to its sexual state by Tulasne and Tulasne in 1865, the genus continued without additions and was commonly assumed to comprise a single species, *T. viride.* This concept resulted in misleading species identifications which are still evident today. The type species, *T. viride sensu stricto,* is a relatively rare species more or less restricted in its distribution to regions in Europe and North America, and yet it is frequently cited as a native biological control agent in other regions (Jacklist *et al.*, 2006). Similarly, the widely reported *T. aureoviride*, for example, appears to have a limited distribution in northern Europe (Lieckfeldt *et al.*, 2001).

Rifai (1969) made the initial approach to understand the diversity of *Trichoderma*, introducing the concept of "species aggregates" in *Trichoderma* and featuring Nine of them, but clarifying that these aggregate species could include multiple species not distinguishable by morphological characters. Later revisions of Bissett (1984, 1991 a, b, c, 1992) and Gams & Bissett (1998), increased the number of species based on morphological distinctions and made connections between anamorph and teleomorph states to include also some species previously placed in the genus *Gliocladium*. Studies on *Hypocrea* demonstrated the overlapping morphological characteristics among species in the anamorph genus *Trichoderma* (e.g. Chaverri and Samuels, 2003; Jaklitsch, 2009), definitively showing that morphological distinctions were not reliable indicators of the degree of genetic divergence between species, and that morphological observations alone were insufficient for accurate identification of species of *Trichoderma*.

To compensate for the lack of reliable morphological characters, research in *Trichoderma* biodiversity over the past 20 years has concentrated on the development of a variety of molecular markers to differentiate species, including isozymes, RAPDs, RFLP, AFLP and, most recently, the nucleotide sequences of various gene loci. The introduction of molecular tools resulted in greatly expanding the number of species recognized in *Trichoderma*. 104

species of *Trichoderma* are listed on the website of the International Commission on the Taxonomy of Fungi subcommision on *Trichoderma* and *Hypocrea* (www.isth.info, Druzhinina & Kopchinskiy 2008), and 193 named taxa are represented to date by sequences deposited in Genbank (www.ncbi.nlm.nih.gov/Taxonomy/Browser/wwwtax.cgi?id=29859).

2.1 Species concepts

In the past, species in *Trichoderma* were defined primarily by the application of the concept Morphological Species Recognition (MSR), sometimes in combination with other phenetic characters. However, morphological identifications are highly prone to error due to the lack of definitive morphological characteristics and variations in culture. Consequently, perhaps 50% or more of the *Trichoderma* isolates deposited in culture collections may be incorrectly named based on morphological identifications. Furthermore, *Trichoderma* strains evidently cannot consistently be crossed to apply the Biological Species Recognition (BSR) concept based on their reproductive behavior. Therefore, Genealogical Concordance Phylogenetic Species Recognition (GCPSR) (Taylor et al., 2000), based on the concordance of multiple gene phylogenies, is an attractive alternative to apply the Phylogenetic Species Concept (PSC) in recognizing species of *Trichoderma*.

2.1.1 Cryptic species or phylogenetic species

Complementary methodologies have been applied to differentiate and characterize *cryptic species* or *phylogenetic species* in a fungi, correlating morphological, biogeographic, biochemical, ecological and, most recently, phylogenetic traits (e.g. refs). Applying the PSC proposed by Taylor et al. (2000), Chaverri et al. (2003) examined the internal transcribed spacer regions of rDNA (ITS1 and ITS2), the large intron of the transcription elongation factor 1-α (*tef1a)*, and short fragments of the actin (*act*1) and calmodulin (*cal1*) exon sequences in *H. lixii/T. harzianum*, to determine seven phylogenetic lineages in *T. harzianum*. However, they declined to recognize the lineages as phylogenetic species since they could not be reliably distinguished morphologically. Similarly applying GCPSR, Samuels et al. (2006) found that the *T. koningii* species aggregate includes three well-separated main lineages defined by phenotypic characters, and further recognized twelve taxonomic species and one variety within the three lineages: (1) T. *koningii, T. ovalisporum* and the new taxa *T. caribbaeum* var. *caribbaeum, T. caribbaeum* var. *aequatoriale, T. dorotheae, T. dingleyae, T. intricatum, T. koningiopsis, T. petersenii and T. taiwanense*; (2) the new species *T. rogersonii* and *T. austrokoningii*, and (3) the new anamorph species *T. stilbohypoxyli*. Druzhinina et al (2010b) recently revisited the genetic diversity in *T. harzianum,* examining three unlinked gene loci for 93 strains isolated worldwide. Their data illustrated clearly the complex history of speciation in the *H. lixii-T. harzianum* species group, rejecting the anamorph/teleomorph combination in favour of separate species status for *H. lixii* and *T. harzianum,* with the phylogenetic position of most isolates not resolved and attributed to a diverse network of recombining strains lacking strict genetic borders. In a similar study employing multiple gene phylogenies and multiple methods of phenotype profiling, Druzhina et al. (2010a) demonstrated that isolates previously identified as *H. jecorina* comprised four phylogenetic species, including *H. jecorina/T. reesei sensu stricto* containing most of the teleomorph isolates and the wild-type strain of *T. reesei* (QM6a) that has subsequently been genetically modified and employed in biofuel production. Conversely, all of the strains isolated as anamorphs from soil were referred to *T. parareesei.* It becomes clear from these studies that phylogenetic structure within these complex species groupings must be taken into account in selecting potential isolates to use in industrial applications. For example, although the name "*T. harzianum*" has been uniformly applied to the biological control agent in the past, there is now increasing evidence that several, genetically diverse species are used in biocontrol (Druzhinina and Kubicek, 2005).

2.2 Methods to identify species in *Trichoderma*

2.2.1 Morphological analysis

Different laboratories have used a variety of media culture for morphological observations in *Trichoderma*. In general, a relatively simple media, such as malt extract agar 2% (MEA), is useful for the production of conidia and the observation of complex branching conidiophores (macronematous). A rich culture medium such potato dextrose agar (PDA) is useful for observing pigment production and harvesting mycelium to isolate DNA. Conidiophore structure and morphology is observed from conidiophores taken from the edge of conidiogenous pustules or fascicles when conidia were maturing (usually after 4-7 days of incubation). The morphology and size of conidia should be observed at maturity after approximately 14 days of growth (Bissett 1984, 1991a, b, c, 1992; Gams and Bissett, 1998). The preliminary identification of species or species aggregates based on characteristic morphologies can be attempted using keys and descriptions in the available taxonomic literature (e.g. Gams & Bissett, 1998).

2.2.2 Molecular analysis

Initially, it was presumed that sequences of the ITS regions of rDNA were sufficient for identification of most fungal species (e.g. Lieckfeldt & Seifert, 2000). However, the popular application of BLAST (in Genbank for example) to identify species based on ITS sequence homologies can provide misleading identifications. Kopchinskiy et al. (2005) found numerous identification errors among sequences deposited in Genbank, which from time to time also does not include all species of the genus. It is now apparent that ITS alone is not sufficiently informative to resolve closely related species in *Trichoderma,* illustrated by the more recent studies employing GCPSR based on multiple gene phylogenies to resolve cryptic species. In addition, paralogous copies of RNA coding genes have been found in some genera of Hypocreales which can result in misleading identifications based on ITS alone (O'Donnell, 2000; Lieckfeldt & Seifert, 2000, Chaverry et al., 2003b, Hoyos et al., 2009a). Numerous genes have now been investigated for the application of GCPSR to resolve species limits in *Trichoderma,* with genes such as translation elongation factor 1-α (TEF), RNA polymerase II subunit B2 (RPB2), chitinase 18-5 (ECH42), calmodulin 1 (CALM1), actin, β-tubulin2 (TUBB2), LAS1 nuclear protein and ATP citrate lyase subunit A (ACLA) providing informative loci for multi-gene studies.

Other molecular tools have been developed for identification of *Trichoderma*. Druzhinina et al. (2005) presented a unique 'bar-code' system for *Trichoderma,* based on 'hallmark' regions in sequences of ITS 1 and 2, and using several of these oligonucleotide regions as genetic fingerprints. These are stored in a MySQL database and integrated with their TrichOKey program for identification (www.isth.info). This program can be used to supplement traditional identification methods. For other gene loci they have developed the program TrichoBLAST (Kopchinskiy et al., 2005), which allows alignment and comparison of sequences of ITS 1 and 2 and fragments of *tef*1α and RPB2. Complementing these methods they developed TrichoMARK to detect one or more sequence fragments of these genes as phylogenetic markers. The latter program is capable of distinguishing the five groups of species haplotypes that have identical ITS 1 and 2 sequences, *viz.*: *T. tomentosum / T. cerinum, T. longipile / T. crassum, T. koningii / T. ovalisporum / T. muroiana, H. lutea / H. melanomagna,* and *T. longibrachiatum / H. orientalis / H. cerebriformis.* In the case of *H. lixii / T. harzianum,* the program detects intraspecific differences accurately in this cluster, which contains several putative phylogenetic species. The ISTH website (www.isth.info) also provides the primer sequences and protocols necessary for sequencing the genes used for identification.

2.2.3 Metabolic tests

These may be based on the profiles of particular enzyme classes such as chitinases or cellulases, although other metabolic profiling techniques have been developed to validate new species which can also potentially provide data on the ecological roles of species (e.g. Kubicek et al. 2003, Hoyos-Carvajal et al., 2009a). The latter studies employed Biolog FF ® microplates (Biolog Inc., Hayward CA) comprising 96 cells containing different carbon sources and redox reagents sensitive to the activity of the enzyme succinate dehydrogenase in the citric acid cycle. Photometry at 590 nm and 750 nm provide quantitative measurements of assimilation and growth (measuring mycelial density) and respiratory activity on the different substrates. The metabolic profiles obtained may be characteristic of species, and the assimilation of specific substrates may allow hypotheses on the ecological role of species. For example, the assimilation of polyols such as maltitol and adonitol could indicate activity of dehydrogenases relevant to survive droughty conditions.

3. *Trichoderma:* distribution and biogeography

Broad studies on the taxonomy and biodiversity of *Trichoderma* have been carried out in North America and some regions of Europe (*e.g.* Bissett 1991a,b,c, 1992), where the distribution of species is now reasonably well known, particularly for specific taxa or groups (Lieckefeldt et al., 2001). Some regions have been studied in detail, e.g. Wuczkowski et al., 2003, investigated the genetic diversity of a European river-floodplain landscape near Vienna, and Migheli et al., 2009 studied the biodiversity of *Trichoderma* in Sardinia, a Mediterranean hot spot of biodiversity, analyzing the influence of abiotic factors on the distribution of species *Trichoderma.* In the latter study, 482 strains of *Hypocrea/Trichoderma* were identified from undisturbed and disturbed environments (forest, shrub lands and undisturbed or extensively grazed grass steppes), with the finding that most of the strains were pan-European and/or pan-global species. Meinke et al., 2010 described the *Trichoderma* communities in rhizosphere of four varieties and transgenic lines of potato in Germany. They observed a heterogeneous distribution and varying diversity of *Trichoderma* dependent on soil characteristics, climate and management practices, in this case not related to the crop variety.

Studies in previously uninvestigated regions or habitats have frequently led to the discovery of new taxa. Kullning et al. (2000) examined 76 isolates from Russia, Nepal and North India, reporting seven species (*T. asperellum, T.atroviride, T. ghanense, T. hamatum, T. harzianum, T. virens* and *T. oblongisporum*) and five new taxa. They also found T. *harzianum* the most genetically diverse speices, with the *T. harzianum* complex representing the majority of isolates. A similar study was conducted by Kubicek et al. (2003) in Southeast Asia, where they reported *T. asperellum, T.atroviride, T. ghanense, T. hamatum, T. harzianum, T. koningii, T. spirale, T. virens, T. viride* and *H. jecorina* (anam: *T. reesei*), along with seven new species among 96 isolates tested (Bissett et al., 2003). The *T. harzianum* complex was equally prevalent, exhibiting high metabolic and morphological variability that may explain the wide distribution of this species aggregate over different habitats (Kubicek et al., 2003).

Sadfi-Zouaoui et al., 2009, in a study encompassing four different bioclimatic zones in Tunisia, assessed the genetic diversity of endemic species of *Trichoderma* and their association to bioclimatic zones. *T. harzianum*, divided into six clades, was the prevalent species complex. *T. harzianum* and *T. longibrachiatum* were predominant in forest soils in north Tunisia; *T. harzianum, T. saturnisporum* and *Trichoderma* sp. indet. were isolated from forest soils in central Tunisia; *T. atroviride* and *T. hamatum* were found in cultivated fields in northeast Tunisia; and *T. harzianum* and *T. hamatum* were present in oasis soils in south Tunisia. Zhang et al. (2005) assessed the biodiversity and biogeography of *Trichoderma* in China, sampling four disparate regions: north (Hebei province), south-east (Zhejiang province), west (Himalayan, Tibet) and south-west (Yunnan province). *T. asperellum, T. koningii, T. atroviride, T. viride, T. velutinum, T. cerinum, T. virens, T. harzianum, T. sinensis, T. citrinoviride,* and *T. longibrachiatum* were identified along with two putative new species. This study revealed a north-south gradient in species distribution in eastern Asia. Tsurumi et al. (2010) explored the biodiversity of *Trichoderma* in Mongolia, Japan, Vietnam and Indonesia, isolating 332 strains and finding T. *harzianum, T. hamatum, T. virens* and *T. crassum* in most habitats. *T. koningiopsis, T. atroviridae* and *T. stramineum* also were frequently isolated, except in cool sites where they were replaced by *T. polysporum* and *T. viridescens.* In tropical areas *T. ghanense, T.brevicompactum* and *T. erinaceum* were prevalent. In addition they discovered five unidentifiable isolate groups and 26 singular unidentified strains.

The most comprehensive survey over any one biogeographic region was performed by Jaklitsch (2009, 2011). He employed three genetic markers to identify 620 specimens of *Hypocrea* occurring in 14 countries in temperate regions of Europe, identifying 75 species including 29 previously undiscovered, thus greatly expanding the number of species known in that region. His observations suggest that the biodiversity of *Hypocrea/Trichoderma* above soil exceeds the diversity residing in soil. He also speculated that the majority of species may be nectrotrophic on other fungi colonizing wood and bark. It now appears that the majority of *Trichoderma* species are capable of sexual recombination and form a teleomorph, and a comparatively smaller number may be clonally propagating agamospecies.

As a result of these recent discoveries, generalizations on the distribution of *Trichoderma* have become increasingly problematic. Their occurrence will be modulated by microclimatic components, substrate availability, rhizosphere associations, soil chemistry, complex ecological interactions and many other factors. The introduction of invasive species, biocontrol agents, and agricultural perturbations result in changes in specific patterns of distribution that cannot be clearly identified, as suggested by Migheli et al. (2009) in finding the colonization of pan-European pan-global *Hypocrea/Trichoderma* species on the island of Sardinia, which may or may not involve the displacement of native strains.

4. Diversity of *Trichoderma* in tropical America

Comparatively few comprehensive studies have been undertaken to assess the diversity of *Trichoderma* in neotropical regions. Since agriculture is a vital segment of local economies in the neotropics, most research in this region on *Trichoderma* has been directed to their biological control activities against phytopathogens. Studies have focused on biological control of plant pathogens with economical importance in cacao plantations, orchards, coffee, beans, cotton, flowers and rubber tree plantations, (Castro, 1996, Carsolio *et al.*, 1994, Hebbar *et al.*, 1999; Hoyos-Carvajal et al., 2008; Rivas & Pavone, 2010; Samuels *et al*, 2000, Samuels *et al.*, 2006,), to control the symbiotic fungus of the leaf-cutting ant *Atta cephalotes* (López & Orduz, 2003), as well as to study the ability of *Trichoderma* to improve plant vigour and stimulate crop growth (Bae et al., 2009, Hoyos-Carvajal et al., 2009b, Resende et al., 2004).

Our knowledge of the distribution of *Trichoderma* species is constantly evolving in the context of current advances toward resolving the taxonomy of the genus. As a consequence, we can anticipate in future years to better understand the biogeography of *Trichoderma* species as research is pursued in new regions and to resolve complex species aggregates. For example, Samuels et al. (2006) determined that the species commonly cited in literature, *Trichoderma koningii,* in the strict sense is a relatively uncommon species restricted to temperate Europe and North America. From within the *T. koningii* aggregate he erected numerous new species, describing *T. caribbaeum* var. *aequatoriale, T. koningiopsis,* and *T. ovalisporum* as endophytes of *Theobroma* species in tropical America, and *T. ovalisporum* also from the woody liana *Banisteropsis caapi* in Ecuador. *T. koningiopsis* (previously identified as *T. koningii*) was determined to be common in tropical America, occurring also on natural substrata in East Africa, Europe and Canada, from ascospores in eastern North America, and as an endophyte in *Theobroma. T. stilbohypoxyli,* described as a parasite of *Stilbohypoxylon* species in Puerto Rico, was found to be more common in the tropics. Samuels et al. (1998) reported on the diversity of *Trichoderma* section *Longibrachiatum,* revealing diversity in neotropical areas resulting in the description of new species in this section. Jaklitsch et al. (2006), in revising the *T. viride* species complex, reported *T. viridescens* as a species found in Peru at high elevation, and *T. neokoningii* in a tropical region in Peru. He also described, as new species, *T. scalesiae* isolated as an endophyte from the trunk of daisy tree (*Scalesia pedunculata*) in the Galapagos Islands of Ecuador, *T. paucisporum* as a mycoparasite of *Moniliophthora roreri* on pods of *Theobroma cacao* in Ecuador, and *T. gamsii,* an apparently cosmopolitan species that has been found in Italy, Rwanda, South Africa, and Romania as well as Guatemala. Recent studies undertaken to find biocontrol agents in specific crops such as cocoa also has resulted in the determination of other new species in neotropical regions (Jaklitsch et al., 2006, Samuels et al., 2000, Samuels et al., 2006).

4.1 Can we generalize on the soil-inhabiting species of *Trichoderma* occurring in the neotropics?

Hoyos-Carvajal et al. (2009a) carry out a systematic survey of *Trichoderma* species in seven countries: Mexico, Guatemala, Panama, Peru, Ecuador, Brazil and Colombia, isolating primarily from soil and employing complementary observations on morphology, metabolism and sequences of ITS 1 and 2 and the 5′ region of *tef*-1a encompassing four introns. They identified 182 *Trichoderma* isolates finding a wide diversity of species over this region of the neotropics - *T. asperellum* (26 isolates), *T. asperelloides* (34 isolates, as *T. asperellum* 'B') *T. atroviride* (3), *T. brevicompactum* (5), *T. crassum* (3), *T. erinaceum* (3), *T. gamsii* (2), *T. hamatum* (2), *T. harzianum* (49), *T. koningiopsis* (6), *T. longibrachiatum* (3), *T. ovalisporum* (1), *T. pubescens* (2), *T. rossicum* (4), *T. spirale* (1), *T. tomentosum* (3), *T. virens* (8), *T. viridescens* (7), *T. parareesei* (3, as *H. jecorina*), along with 11 presumptive new species that have not yet been described. Analyses of variance were performed on metabolic data for the Colombian isolates. Highly significant differences ($P < 0.0001$) in assimilation were observed for 42 substrates among the 12 species isolated from Colombia (*T. asperellum, T. atroviride, T.*

brevicompactum, T. erinaceum, T. hamatum, T. harzianum, T. koningiopsis, T. longibrachiatum, T. virens, T. viridescens, T. parareesei and *Trichoderma* sp. 210). The highest growth rates were observed on 23 substrates for *T. viridescens* isolated from mostly from rhizosphere of *Impatiens*, for *T. asperellum* obtained from a broad range of substrates on five substrates, and for *T. harzianum* from varied habitats and *T. parareesei* from African palm on three substrates. Seven substrates for which *T. viridescens* had the fastest growth rate were substrates not or scarcely assimilated by any other species (sedoheptulosan, glucuronamide, 2-aminoethanol, D-lactic acid methyl ester, putrescine, L-alaninamide, γ-hydroxyphenylacetic acid), perhaps indicating an ability to grow on recalcitrant substrates, although a similar pattern has been observed in other studies, contrasting isolates from undisturbed forests habitats, capable of growing on recalcitrant substrates, with isolates from agricultural habitats (Bissett unpublished). *Trichoderma viridescens* and *T. harzianum* showed positive growth on the largest number of significant substrates (41 and 34 respectively), indicating possible adaptation to a relatively broad range of habitats or niches and reflected in their wide distibutions. Slowest growth rates were observed for *T. erinaceum* from maize rhizosphere on 15 substrates, and for *Trichoderma* sp. 210 from river sand on 11 substrates. *T. longibrachiatum* and *Trichoderma* sp. 210 in section Longibrachiatum, along with *T. erinaceum* assimilated the fewest substrates (19-25 substrates).

The considerable biodiversity of *Trichoderma* in neotropical regions was evident in the study by Hoyos-Carvajal et al. (2009a). Nineteen species were identified from 182 isolations, and eleven so far undescribed species were discovered from rainforest soils and other specific habitats such as river sand, humus and wood in Peru, Mexico, Guatemala and Colombia. In a study of *Trichoderma* in Venezuelan soils the most abundant species was *T. harzianum,* followed by *T. virens, T. brevicompactum, T. theobromicola, T. koningiopsis, T. ovalisporum, T. asperellum, T. pleurotum* and *T. koningiopsis* (Rivas & Pavone, 2010). These observations are added to new species of *Trichoderma* from neotropics described in recent years, mainly as endophytes in plants (Jaklitsch et al., 2006, Samuels et al., 2006), and are evidence of the significant biodiversity of *Trichoderma* in the tropical region (table 1).

Unlike the studies conducted by Kullning et al. (2000) and Kubicek et al. (2002), assessing *Trichoderma* biodiversity of specific geographic areas, where the most common species was *T. harzianum,* in contrast in neotropical areas studied by Hoyos-Carvajal et al. (2009a) was *T. asperellum* (33% of strains) and *T. harzianum* the second most common (27%). Genetic variation was evident for both species, and two (*T. asperellum*) or three (*T. harzianum*) distinct genotypes were evident in the analysis of *tef* sequences and metabolic profiles. *T. asperellum,* which is often isolated from tropical regions (Druzhinina et al., pers. comm.), could be divided into two groups (A and B), which more recently have been described as separate species, *T. asperellum* and *T. asperelloides* respectively (Samuels et al., 2009, 2010). *T. asperellum* includes isolates from Brazil, Peru and Colombia originating in soils with poorly degraded materials such as fallen leaves or crop residuals in colder climate zones. T. *asperelloides* includes strains collected in Colombia and Ecuador that exhibit a preference for soils and substrates with high organic content, and often adapted to the rhizosphere of crops in Andean zones. These two species could not be differentiated by morphological characters or by growth rates, suggesting the development of ecologically or geographically isolated lineages as has been reported for *T. harzianum* (Chaverri et al., 2003) and *T. koningii* (Samuels et al., 2006). Metabolic differences were apparent between the closely related species, with *T. asperellum* better able to assimilate a wider range of C-substrates including some organic acids, polyols and amino acids, although growth rates on readily assimilated substrates such as glucose and N-acetyl-D-glucosamine were essentially the same. *T. asperelloides* had faster growth only on the disaccharide gentobiose. All differences between the two species were a matter of rate of growth, rather than growth/no growth. Notably, *T. asperellum* had significantly higher growth on substrates that are usually not at all assimilated by fungi, such as D-psicose, sedoheptulosan and γ-hydroxybutyric acid. *T. asperellum* was isolated from soils with poorly incorporated litter or crop residuals in colder climate zones (15ºC), growing on a wider variety of poorly metabolized substrates, and concomitantly under more difficult nutritional conditions in forest soils or debris. The pattern of growth on recalcitrant substrates for *T. asperellum* may be correlated with occurrence in relatively undisturbed, forested soils or other natural habitats. *T. asperelloides* was associated with agricultural soils crops, and in Colombia displayed a pattern of affinity for readily available substrates such as sugars, from comparatively warmer climates (23–28 ºC), and the rhizosphere soils where this strain was collected are associated with tropical crops such as African palm (*Elaeis guineensis*), coffee (*Coffea arabiga*), black mulberry (*Morus* sp.), avocado (*Persea americana*) and some grasses, in areas with a comparatively high organic matter contents. Despite the apparent different habitat preferences, *T. asperellum* and *T. asperelloides* did not exhibit differences in growth rates over the range 5-40°C, both species with temperature optima near 28˚C.

In the neotropics, the second most prevalent species from neotropical soils studied by Hoyos-Carvajal et al. (2009a) was *T. harzianum*, commonly associated with the rhizosphere of cultivated plants and frequently employed as a biocontrol agent against soil-borne phytopathogens. The predominance of *T. harzianum* in many different environments might be explained by its ability to assimilate a comparatively wide array of carbon substrates. The concept of *T. harzianum* as a genetically variable complex, comprising reproductively isolated biological species, recent agamospecies and phylogenetically unresolved relict lineages as determined by Druzhinina et al. (2010) is coherent with the adaptive range of this taxon. In the study of neotropical *Trichoderma* by Hoyos-Carvajal et al. (2009a), *T. harzianum* A was characterized by growth on poorly metabolized substrates, and strains from Colombia were isolated from a variety of environments, but commonly Andean soils associated with *Impatiens* sp. Group A also included strains from Mexico, consistent with the distribution found by Chaverri et al. (2003) for *H. lixii* lineage 1, and also includes isolates from Panama and Peru. The second clade, *T. harzianum* B, comprised mostly strains from the rhizosphere of tobacco, and sequences are coincident with lineage 5 of Chaverri et al. (2003) that included strains from Japan, Mexico and Cameroon, and with lineage 6 from Europe. *T. harzianum* C comprised nine strains from Colombia together with lineage 3 identified by Chaverri et al. (2003) from the United States, lineage 7 from Japan and Mexico, lineage 2 from Europe, and lineage 4 from Cameroon. *T. harzianum* A had fastest growth on L serine, i-erythriol, glycerol, D-sorbitol, D-ribose, α-D-lactose, L-threonine, L-proline, D-mannitol, and L-sorbose; however, there was no significant difference among the three genotypes on glucose, and all three genotypes had similar linear growth rates in culture on PDA over the temperatures range 5-40ºC. Significantly higher growth rates for *T. harzianum* A on the monosaccharide polyols i-erythritol, D-sorbitol, and D-mannitol, and on the fatty acid glycerol could indicate the presence of dehydrogenases allowing an adaptation to relatively dry habitats. *T. harzianum* B was the only genotype able to assimilate the disaccharide sucrose, although it exhibited relatively poor growth on the disaccharide lactose which was readily assimilated by genotypes A and C.

In the study carried out by Hoyos-Carvajal et al. (2009a), Colombia becomes the most intensively surveyed neotropic region for *Trichoderma* biodiversity to date, comprising 116 isolates, representing 11 described species and one new taxon. As was mentioned, the most commonly isolated species from Colombia were *T. asperellum* (inclusive of more recently distinguished *T. asperelloides)* and *T. harzianum*. The prevalence of these species in Colombia may be explained by their genetic variability, seen in the several distinct genotypes found for each species, and their corresponding ability to grow on a wide variety of carbon substrates. However, the isolation methods used in this study, which are commonly used to isolate from the soil and rhizosphere, would be selective for soil-inhabiting species such as *T. asperellum* and *T. harzianum,* which are fast growing and sporulate early, allowing them to be recognized ahead of slower developing species. The majority of new species in this study

Species	**Reference**
T. asperellum	Samuels *et al.*, 1999
T. asperelloides	Samuels *et al.*, 2010
T. atroviride	Bissett, 1992.
T. brevicompactum	Krauss *et al.*, 2004.
T. caribbaeum	Samuels *et al.* 2006
T. caribbaeum var. aequatoriale	Samuels *et al.* 2006
T. crassum	Bissett, 1991 a
T. erinaceum	Bissett *et al.*, 2003
T. evansii	Samuels & Ismaiel, 2009
T. gamsii	Jaklitsch *et al.*, 2006
T. hamatum	Gams y Bissett, 1998
T. harzianum	Chaverry *et al.*, 2003; Gams & Bissett, 1998
T. koningiopsis	Samuels *et al.*, 2006
T. lieckfeldtiae	Samuels & Ismaiel, 2009
T. longibrachiatum	Gams y Bissett
T. neokoningii	Jaklitsch *et al.*, 2006
T. ovalisporum	Holmes *et al.*, 2004
T. parareesei	Atanasova *et al.*, 2010
T. paucisporum	Samuels *et al.* 2006b
T. pleurotum	Komoń-Zelazowska et al., 2007
T. pubescens	Bissett 1991a
T. reesei (H. jecorina)	Gams and Bissett, 1998
T. rossicum	Bissett *et al.*, 2003.
T. scalesiae	Jaklitsch *et al.*, 2006
T. spirale	Bissett 1991a
T. stilbohypoxyli	Samuels *et al.* 2006a
T. theobromicola	Samuels *et al.* 2006b
T. tomentosum	Bissett 1991a
T. virens	Bissett 1991[a]
T. viridescens	Jaklitsch *et al.*, 2006

Table 1. *Trichoderma* species currently identified from the tropics with references to morphological descriptions.

were isolated from other neotropical countries, notably Peru (6 new species) and Guatemala (3 new species) from which there were far fewer isolates. However, sampling in these countries was selective for unusual substrates above ground, resulting in the high proportion of novel strains. Therefore, the study by Hojos-Carvajal et al. (2009) would not account for the above ground biodiversity of species in Colombia on account of a relatively selective (but typical) sampling regime, although it is indicative of the wide distribution of *T. asperellum* and *T. harzianun* in soils, as reported in previous studies for other regions (Kullnig et al., 2000, Kubicek et al., 2003, Chaverry et al., 2003).
The various studies of *Trichoderma* in the neotropics have expanded the known biogeographical and ecological distribution of many *Trichoderma* species. For example, *T. virens* (rain forest in Perú; rotten wood, rhizosphere of rice, tobacco and grassland in Colombia), *T. pubescens* (rain forest soil in Perú), *T. strigosum* (Perú rain forest soil), and *T. tomentosum* (cloud forest soil, Guatemala), were originally described from North America and Europe where they are relatively uncommon (Bissett, 1991 b). *T. ovalisporum*, previously reported from Ecuador as an endophyte in *Banisteriopsis caapi* and *Theobroma* sp. (Samuels et al., 2006), was isolated as an apparent saprophyte from soil in Panama. The infrequent isolation of these species also from neotropical soils suggests that these species may be restricted to specific ecozones, habitats or niches (Hoyos-Carvajal et al., 2009a). Samuels et al. (1998) reported *H. jecorina* (anam.: *T. reesei*) to be common in the pantropical region, and it is an important producer of cellulase enzymes. Hoyos-Carvajal et al., 2009a reported the species in typically warm soils cultivated with African palm in Colombia, but these strains did not assimilate sucrose, which had been reported for isolations of *H. jecorina* from the eastern Pacific (Samuels et al., 1998). We now know that the species reported by Hoyos-Carvajal et al. (2009) was in fact *T. parareesei*, recently differentiated from the sympatric species *H. jecorina* by Druzhinina et al. (2010a).
Eleven neotropical clades were differentiated from known *Trichoderma* species by Hoyos-Carvajal et al. (2009a) based on morphologic, metabolic and molecular differences and these remained undescribed. These are presumed to represent new taxa in *Trichoderma* and are the subject of ongoing investigations. The high proportion of apparently new species in this study is an indication that we have only begun to explore the biodiversity of *Trichoderma* in the neotropic regions.

5. Conclusions

Trichoderma species represent a major component of soil biodiversity with an important role in maintaining soil and plant health. The numbers, diversity, roles, and interactions of *Trichoderma* species in the environment are only now being discovered as tools are developed to distinguish the anamorph forms most commonly encountered. Significant and novel biodiversity of *Trichoderma* in the neotropics has been demonstrated, although we have only begun to explore the diversity of regions, habitats and substrates that exist in the region. We are now able to account taxonomically for a significant component of their biological diversity, to begin to predict biological activities, and to communicate results through the use of accurately determined names. The identification of *Trichoderma* species, as for species in other economically important and species-rich genera, is increasingly reliant on molecular data as the limits of phenotypic characters to distinguish species are reached. Many new species of *Trichoderma* will undoubtedly be distinguished as molecular tools are developed for ecological and metagenomic studies. Agriculture is the main economic activity in neotropical regions, and *Trichoderma* is the most important biocontrol agent against soil-borne phytopathogens. Consequently neotropical investigations have concentrated on the application of *Trichoderma* to control crop diseases, and discovering new metabolites and mechanisms of action. We can now appreciate the importance in preserving the biodiversity of delicate ecosystems such as rain forest and Andean forest, as reservoirs of metabolites and diverse and unique ecological niches for habitation by animals, plants and microorganisms. Conservation is facilitated as we increase our knowledge of the fundamental role of *Trichoderma* in nutrient cycling and in the complex interactions within the soil biota.

6. References

Akhtar, K., Khalid, A., Akhtar, M., & Ghauri, M. (2009). Removal and recovery of uranium from aqueous solutions by Ca-alginate immobilized *Trichoderma harzianum*. *Bioresource Technology*, Vol. 100, No. 20, (May 2009), pp. 4551–4558, ISSN 09608524.

Alfano, G., Lewis Ivey, M., Cakir, C., Bos, J., Miller, S., Madden, L., Kamoun, S., & Hoitink, H. (2007). Systemic modulation of gene expression in tomato by *Trichoderma hamatum* 382. *Phytopathology*, Vol. 97, No. 4, (November 2006), pp. 429–437, ISSN 0031-949X.

Altomare, C., Norwell, W., Björkman, T., & Harman, G. (1999). Solubilization of phosphates and micronutrients by the plant-growth-promoting and biocontrol fungus *Trichoderma harzianum* Rifai 1295-22. *Applied and Environmental Microbiology*. Vol.65, No. 7, (April 1999), pp. 2926-2933, ISSN 0099-2240.

Atanasova, L., Jaklitsch, W.M., Komoń-Zelazowska, M., Kubicek, C.P. and Druzhinina, L.S. 2010. The clonal species *Trichoderma parareesei sp. nov.*, likely resembles the ancestor of the cellulase producer *Hypocrea jecorina/T. reesei*. Appl. Environ. Microbiol. doi:10.1128/AEM.01184-10.

Bae, H., Sicher, R., Kim, M., Kim, S., Strem, M., Melnick, R., & Bailey, B. (2009).The beneficial endophyte *Trichoderma hamatum* isolate DIS 219b promotes growth and delays the onset of the drought response in Theobroma cacao. *Journal of Experimental Botany*. Vol. 60, No. 11, (February 2009), pp. 3279–3295, ISSN 1460-2431.

Baig, M., Mane, V., More, D., Shinde, L., & Baig, M. (2003). Utilization of banana agricultural waste: production of cellulases by soil fungi. *Jornal of Environmental Biology*. Vol.24, No.2, (April 2003), pp. 173-176, ISSN 0254-8704.

Bailey, B., Bae, H., Strem, M., Roberts, D., Thomas, S., Crozier, J., Samuels, G., Choi, I., & Holmes, K. (2006). Fungal and plant gene expression during the colonization of cacao seedlings by endophytic isolates of four *Trichoderma* species. *Planta*. Vol. 224, No.6, (April 2006), pp.1449-1464, ISSN 1432-2048.

Begum, F., Absar, N., & Shah, M. (2009). Purification and Characterization of Extracellular Cellulase from *A. oryzae* ITCC-4857.01. *Journal of Applied Sciences Research*. Vol. 5, No. 10, (October 2009), pp. 1645-1651, ISSN 1819-544X.

Bissett, J. (1984). *A* revision of the genus *Trichoderma*. I. Section Longibrachiatum sect. nov. *Canadian Journal of Botany*. Vol. 62, No. 5, (April 1983), pp. 924-931, ISSN 1480-3305.

Bissett, J. (1991a). A revision of the genus *Trichoderma*. II. Infrageneric classification. *Canadian Journal of Botany*. Vol.69, No.11, (February 1991), pp. 2357-2372, ISSN 1480-3305.

Bissett, J. (1991b). A revision of the genus *Trichoderma*. III. Sect. Pachybasium. *Canadian Journal of Botany*. Vol. 69, No. 11, (February 1991), pp. 2373–2417, ISSN 1480-3305.

Bissett, J. (1991c). A revision of the genus *Trichoderma*. IV. Additional notes on section Longibrachiatum. *Canadian Journal of Botany*. Vol. 69, No. 11, (February 1991), pp. 2418 - 2420, ISSN 1480-3305.

Bissett, J. (1992). *Trichoderma atroviride*. *Canadian Journal of Botany*. Vol. 70, No. 3, (June 1991), pp. 639–641, ISSN 1480-3305.

Bissett, J., Szacaks, G., Nolan, C., Druzinina, I., Grandiger, C., & Kubicek, C. (2003). New species of *Trichoderma* from Asia. *Canadian Journal of Botany*. Vol. 81, No. 6, (December 2002), pp. 570-586, ISSN 1480-3305.

Carsolio, C., Gutierrez, A., Jiménez, B., Van Montagu, M., & Herrera-Estrella, A. (1994). Characterization of ech-42, a *Trichoderma harzianum* Endochitinase gene expressed during mycoparasitism. *Proceedings of the National Academy of Sciences*. Vol.91, No. 23, (June 1994), pp. 10903–10907, ISSN 1091-6490.

Castro, B. (1996). Antagonismo de algunos aislamientos de *Trichoderma koningii* originados en suelo colombiano contra *Rosellinia bunodes, Sclerotina sclerotiorum* y *Pythium ultimum*. *Fitopatología Colombiana*. Vol. 19, No. 2, (Diciembre 1995), pp. 7-18, ISSN.

Chaverri, P., & Samuels, G. (2003). *Hypocrea/Trichoderma* (Ascomycota, Hypocreales, Hypocreaceae): species with green ascospores. *Studies in Mycology*. Vol. 48, No. 1, (2003), pp. 1-116, ISSN 0166-0616.

Chaverri, P., Castlebury, L., Samuels, G., & Geiser, D. (2003b). Multilocus phylogenetic structure within the *Trichoderma harzianum*/Hypocrea lixii complex. *Molecular Phylogenetics and Evolution*. Vol. 27, No. 2, (September 2002), pp. 302-313, ISSN 1055-7903.

Cross, D. (1999). Determination of limiting factors in fungal bioremediation: 1999 Progress Report. In: *National Center for Environmental Research*. Available from: http://es.epa.gov/ncer/fellow/progress/96/crossde99.html

Druzhinina, I., & Kopchinskiy, A. (2008). International Subcommission on *Trichoderma* and *Hypocrea* Taxonomy. *ISTH*. Available from: http://isth.info/biodiversity/index.php (accessed 22-Jun-2008)

Druzhinina, I., Kopchinskiy, A., & Kubicek, C. (2006). The first 100 *Trichoderma* species characterized by molecular data. *Mycoscience*. Vol.47 No. 2, (June 2006), pp. 55–64, ISSN 1618-2545.

Druzhinina, I., Kopchinskiy, A., Komon, M., Bissett, J., Szakacs, G., & Kubicek, C. (2005). An oligonucleotide barcode for species identification in *Trichoderma* and *Hypocrea*. *Fungal Genetics and Biology*. Vol. 42, (June 2005), pp. 813-828, ISSN 1087-1845.

Druzhinina, I.; Kubicek, C. (2005). Species concepts and biodiversity in *Trichoderma* and *Hypocrea*: from aggregate species to species clusters?. *Journal of Zhejiang University SCIENCE*. Vol. 6B, No. 2, (October 2004), pp. 100-112, ISSN 1009-3095.

Druzhinina, I.S., M. Komoń-Zelazowska, L. Atanasova, V. Seidl and C.P. Kubicek. 2010a. Evolution and ecophysiology of the industrial producer *Hypocrea jecorina* (anamorph *Trichoderma reesei*) and a new sympatric agamospecies related to it. PLoS ONE 5(2)e9191, 15pp.

Druzhinina, I.S., C.P. Kubicek, M. Komoń-Zelazowska, T. B. Mulaw and J. Bissett. (2010b). The *Trichoderma harzianum* demon: complex specieation history resulting in coexistence of hypothetical biological species, recent agamospecies and numerous relict lineages. BMC Evolutionary Biology 10: 94, 14pp.

Evans, C., Holmes, K., & Thomas, S. (2003). Endophytes and mycoparasites associated with an indigenous forest tree, *Theobroma gileri*, in Ecuador and a preliminary assessment of their potential as biocontrol agents of cocoa diseases. *Mycological Progress*. Vol. 2, No. 2, (May 2003), pp. 149-160, ISSN 1861-8952.

Ezzi, M., & Lynch, J. (2005). Biodegradation of cyanide by *Trichoderma* spp. and *Fusarium* spp. *Enzyme Microbiology and Technology*. Vol. 36, No. 7, (March 2004), pp. 849–854, ISSN 0141-0229.

Gams, W., & Bissett, J. (1998). Morphology and identification of *Trichoderma*. In: *Trichoderma and Gliocladium. Basic biology, taxonomy and genetics,* Kubicek, C.P. & Harman, G.E, pp. 3-34, Taylor & Francis Ltd., ISBN 978-0-7484-0572-5, London, UK.

Guillermina, M., Romero, M., Cazau, M., & Bucsinszky, A. (2002). Cadmium removal capacities of filamentous soil fungi isolated from industrially polluted sediments, in La Plata (Argentina). *World Journal of Microbiology and Biotechnology*. Vol. 18, No. 9, (July 2002), pp. 817–820, ISSN 1573-0972.

Harman, G., Howell, C., Viterbo, A., Chet, I., & Lorito, M. (2004). *Trichoderma* species-oportunistic, avirulent plant symbionts. *Nature Reviews Microbiology*. Vol. 2, No. 1, (January 2004), pp. 43 - 56, ISSN 1740-1526.

Hebbar, P., Lumsden, R., Krauss, U., Soberanis, W., Lambert, S., Machado, R., Dessomoni, C., & Aitken, M. (1999). Biocontrol of cocoa diseases in Latin America - status of field trials. In: *Workshop Manual - Research Methodology for the Biological Control of Plant Diseases with Special Reference to Fungal Diseases of Cocoa,* Hebbar, P., Krauss, U, CATIE, Turrialba, Costa Rica.

Hoyos, L., Chaparro, P., Abramsky, M., Chet, I., & Orduz, S. (2008). Evaluation of *Trichoderma* spp. isolates against *Rhizoctonia solani* and *Sclerotium rolfsii* under *in vitro* and greenhouse conditions. *Agronomía Colombiana.* Vol. 26, No. 3, (Noviembre 2008), pp. 451-458, ISSN 0120-9965.

Hoyos-Carvajal, L., Orduz, S., & Bissett, J. (2009a). Genetic and metabolic biodiversity of *Trichoderma* from Colombia and adjacent neotropic regions. *Fungal Genetics and Biology*. Vol. 46. No. 9, (April 2009), pp. 615–631, ISSN 1087-1845.

Hoyos-Carvajal, L., Orduz, S., & Bissett, J. (2009b). Growth stimulation in bean (*Phaseolus vulgaris* L.) by *Trichoderma*. *Biological Control*. Vol. 51, No. 3, (July 2009), pp. 409–416, ISSN 1049-9644.

Jaklitsch, W., Samuels, G., Dodd, S., Lu, B., & Drizhinina, I. (2006). *Hypocrea rufa/ Trichoderma viride*: a reassessment, and description of five closely related species with and without warted conidia. *Studies in Mycology*. Vol. 56, No. 1, (2006), pp. 135–177, ISSN 0166-0616.

Jaklitsch, W.M. 2009. European species of *Hypocrea* Part I. The green-spored species. Studies in Mycology 63:1-91. ISSN 0166-0616.

Jaklitsch, W.M. 2011. European species of *Hypocrea* part II: species with hyaline ascospores. Fungal Diversity 48:1-250.

Karmakar, M. & Ray, R. (2011). Current trends in research and application of microbial cellulases. *Research Journal of Microbiology.* Vol. 6, No.1, (2011), pp. 41-53. ISSN 186-4935.

Koenig, D., Bruce, R., Mishra, S., Barta, D., & Pierson, D. (1994). Microbiological characterization of a regenerative life support system. *Advances in Space Research.* Vol. 14, No. 11, (November 1994), pp. 377-382, ISSN 0273-1177.

Komoń-Zelazowska, M. J. Bissett, D. Zafari, L. Hatvani, L. Manczinger, S. Woo, M. Lorito, L. Kredics, C. P. Kubicek and I.S. Druzhinina. 2007. Genetically closely related but phenotypically divergent *Trichoderma* species cause green mold disease in oyster mushroom farms worldwide. Appl. Environ. Microbiol. 2007 November; 73(22): 7415–7426. ISSN 0099-2240.

Kopchinskiy, A., Komon, M., Kubicek, C., & Druzhinina, I. (2005). Mycological research news. *Mycological Research.* Vol. 109, No. 6, (June 2005), pp. 657-660, ISSN 1469-8102.

Kubicek, C., Bissett, J., Druzhinina, I., Kulling-Grandiger, C., & Szakacs, G. (2003). Genetic and metabolic diversity of *Trichoderma*: a case study on South East Asian isolates. *Fungal genetics and biology.* Vol. 38, No. 3, (April 2003), pp. 310-319, ISSN 1087-1845.

Kulling, C., Szakacs, G., & Kubicek, C. (2000). Molecular identification of *Trichoderma* species from Russia, Siberia and the Himalaya. *Mycological Research.* Vol. 104, No. 9, (December 1999), pp. 1117-1125, ISSN 1469-8102.

Lieckfeldt, E., & Seifert, K. (2000). An evaluation of the use of ITS sequences in the taxonomy of the Hypocreales. *Studies in Mycology.* Vol. 45, No. 1, (May 2000), pp. 35-44, ISSN 0166-0616.

Lieckfeldt, E., Kullnig, C., Kubicekb, C., & Samuels, G. (2001). *Trichoderma aureoviride*: phylogenetic position and characterization. *Mycological Research.* Vol. 105, No. 3, (August 2000), pp. 313-322, ISSN.

Lopez, E., & Orduz, S. (2003). *Metarhizium anisopliae* and *Trichoderma viride* for control nests of the fungus-growing and *Atta cephalotes. Biological Control.* Vol. 27, No. 2, (December 2002), pp. 194-200, ISSN 1049-9644.

Migheli, Q., Balmas, V., Komoñ-Zelazowska, M., Scherm, B., Fiori, S., Kopchinskiy, A., Kubicek, C., & Druzhinina, I. (2009). Soils of a Mediterranean hot spot of biodiversity and endemism (Sardinia, Tyrrhenian Islands) are inhabited by pan-European, invasive species of *Hypocrea/Trichoderma. Environmental Microbiology.* Vol. 11, No. 1, (June 2008), pp.35-46, ISSN 1462-2920.

O'Donnell, K., Kistler, H., Tacke, B., & Casper, H. (2000). Gene genealogies reveal global phylogeographic structure and reproductive isolation among lineages of *Fusarium graminearum*, the fungus causing wheat scab. *Proeedings of National Academy of Sciences of the United States of America.* Vol. 97, No. 14, (January 2000), pp. 7905–7910, ISSN 1091-6490.

Resende, M., Oliveira, J., Guimarães, R., Von Pinho, R., & Vieira, A. (2004). Inoculação de sementes de milho utilizando o *Trichoderma harzianum* como promotor de crescimento. *Ciência e Agrotecnologia.* Vol. 28, No. 4, (Marzo 2004), pp. 793-798, ISSN 1413-7054.

Rifai, M. (1969). *A revision of the genus Trichoderma* (1 edition). Commonwealth Mycological Institute. ISBN 0851990002, Kew, Surrey. England.

Rivas, M., & Pavone, D. (2010). Diversidad de *Trichoderma* spp. en plantaciones de *Theobroma cacao* l. del estado Carabobo, Venezuela, y su capacidad biocontroladora sobre *Crinipellis pernicosa* (Stahel) Singer. *Interciencia: Revista de ciencia y tecnología de América.* Vol. 35, No. 10, (Septiembre 2010), pp. 777-783, ISSN 0378-1844.

Sadfi-Zouaoui, N., Hannachi, I., Rouaissi, M., Hajlaoui, M., Rubio, M., Monte, E., Boudabous, A., & Hermosa, M. (2009). Biodiversity of *Trichoderma* strains in Tunisia. *Canadian Journal of Microbiology.* Vol. 55, No. 2, (October 2008), pp. 154-162, ISSN 0008-4166.

Sadfi-Zouaoui, N.; Hannachi, I.; Rouaissi, M.; Hajlaoui, M. R.; Rubio, M. B.; Monte, E.; Boudabous, A.; Hermosa, M. R. 2009. Biodiversity of *Trichoderma* strains in Tunisia. *Canadian Journal of Microbiolog.*, Vol. 55, No 2, (February 2009), pp. 154-162(9) ISSN 0008-4166

Samuels, G. (2006a). *Trichoderma*: Systematics, the sexual state, and ecology. *Phytopathology.* Vol. 96, No. 2, (October 2005), pp. 195–206, ISSN 0031-949X.

Samuels, G., Suarez, C., Solis, K., Holmes, K., Thomas, S., Ismaiel, A., & Evans, H. (2006b). *Trichoderma theobromicola* and *T. paucisporum*: two new species isolated from cacao in South America. *Mycological Research.* Vol. 110, No. 4, (November 2005), pp. 381–392, ISSN 0953-7562.

Samuels, G., & Ismaiel, A. (2009). *Trichoderma evansii* and *T. lieckfeldtiae* two new *T. hamaturn*-like species. *Mycologia.* Vol. 101, No. 1, (October 2008), pp. 142-156, ISSN 1557-2536.

Samuels, G., Ismaiel, A., Bon, M., De-Repinis, S., & Petrini, O. (2010). *Trichoderma asperellum sensu lato* consist of two cryptic species. *Mycologia.* Vol., 102 No. 4, (June 2010), pp. 944-966, ISSN 1557-2536.

Samuels, G., Pardo-Schulteiss, R., Hebbar, K., Lumsden, R., Bastos, C., Costa, J., & Bezerra, J. (2000). *Trichoderma stromaticum* sp. nov., a parasite of the cacao witches broom pathogen. *Mycology Research.* Vol. 104, No. 6, (August 1999), pp.760-764, ISSN 0953-7562.

Samuels, G., Petrini, O., Kulhs, J., Lieckfeldt, E., & Kubicek, C. (1998). The *Hypocrea schweinitzii* complex and *Trichoderma* sect. Longibrachiatum. *Studies in Mycology.* Vol. 41, No. 1, (1998), pp. 2-54, ISSN 0166-0616.

Sette, L., Passarini, M., Delarmelina, C., Salati, F., & Duarte, M. (2006). Molecular characterization and antimicrobial activity of endophytic fungi from coffee plants. *World Journal of Microbiology and Biotechnology.* Vol. 22, No. 11, (March 2006), pp. 1185-1195, ISSN 1573-0972.

Sharon, E., Bar-Eyal, M., Chet, I., Herrera-Estrella, A., Kleifeld, O., & Spiegel, Y. (2001). Biological control of the root-knot nematode *Meloidogyne javanica* by *Trichoderma harzianum*. *Phytopathology.* Vol. 91, No. 7, (March 2001), pp.687-693, ISSN 0031-949X.

Shoukouhi, P., & Bissett, J. (2008). Preferred primers for sequencing the 5′ end of the translation elongation factor 1-alpha gene (eEF1a1). *ISTH* Available from: http://www.isth.info/methods.

Supothina S., Isaka, M., & Wongsa, P. (2007). Optimization of culture conditions for production of the anti-tubercular alkaloid hirsutellone A by *Trichoderma gelatinosum* BCC 7579. *Letters in Applied Microbiology*. Vol. 44, No. 5, (November 2006), pp. 531-537, ISSN 0266-8254.

Tang, J., Liu, L., Hua, S., Chen, Y., & Chen, J. (2009). Improved degradation of organophosphate dichlorvos by *Trichoderma atroviride* transformants generated by restriction enzyme-mediated integration (REMI). *Bioresource Technology* Vol. 100, No. 1, (May 2008), pp. 480–483, ISSN 0960-8524.

Taylor, J., Jacobson, D., Kroken, S., Kasuga, T., Geiser, D., Hibbett, D., & Fisher, M. (2000). Phylogenetic Species Recognition and Species Concepts in Fungi. *Fugal Genetics and Biology*. Vol. 31, No. 1, (September 2000), pp. 21-32, ISSN 1087-1845.

Tramoy, P. (2008). Review on The Enzyme Market. *Lifescience online*. 1 noviembre de 2010, Available from: http://www.lifescience-online.com/articles.html.

Tsurumi, Y., Inaba, S., Susuki, S., Kamijo, S., Widyastuti, Y., Hop, D., Balijinova, T., Sukarno, N., Nakagiri, A., Susuki, K., & Ando, K. (2010). Distribution of *Trichoderma* species in four countries of Asia. *9th International Mycological Congress*. Edimburg, Scotland, August 2010.

Vinale, F., Marra, R., Scala, F., Ghisalberti, E., Lorito, M., & Sivasithamparam, K. (2006). Major secondary metabolites produced by two commercial *Trichoderma* strains active against different phytopathogens. *Letters in Applied Microbiology*. Vol. 43, No. 2, (March 2006), pp. 143–148, ISSN 0266-8254.

Viterbo, A., & Chet, I. (2006). TasHyd1, a new hydrophobin gene from the biocontrol agent *Trichoderma asperellum*, is involved in plant root colonization. *Molecular Plant Pathology*. Vol. 7, No. 4, (July 2006), pp. 249–258, ISSN 1464-6722.

Watanabe, N., Akiba, T., Kanai, R., & Harata, K. (2006). Structure of an orthorhombic form of xylanase II from *Trichoderma reesei* and analysis of thermal displacement. *Acta Crystallographica Section D: Biological Crystallography*. Vol. 62, No.7, (July 2006), pp. 784–792, ISSN 1600-5724.

Woo, S., Scala, F., Ruocco, M., & Lorito, M. (2006). The molecular biology of the interactions between *Trichoderma* spp., phytopathogenic fungi, and plants. *Phytopathology*. Vol. 96, No. 2, (June 2005), pp.181-185, ISSN 0031-949X.

Wuczkowski, M., Druzhinina, I., Gherbawy, Y., Klug, B., Prillinger, H., & Kubicek, C. (2003). Species pattern and genetic diversity of *Trichoderma* in a mid-European, primeval floodplain-forest. *Microbiological Research*. Vol. 158, No. 2, (December 2002), pp.125-133, ISSN 0944-5013.

Xiao-Yan, S., Qing-Tao, S., Shu-Tao, X., Xiu-Lan, C., Cai-Yun, S., & Yu-Zhong, Z. (2006). Broad-spectrum antimicrobial activity and high stability of Trichokonins from *Trichoderma koningii* SMF2 against plant pathogens. *FEMS Microbiology Letters*. Vol. 260, No. 1, (April 2006), pp. 119–125, ISSN 1574-6968.

Yedidia, I., Benhamou, N., Kapulnik, Y., & Chet, I. (2000). Induction and accumulation of PR proteins activity during early stages of root colonization by the mycoparasite

Trichoderma harzianum strain T-203. *Plant Physiology and Biochemistry*. Vol. 38, No. 11, (August 2000), pp. 863-873, ISSN 0981-9428.

Zhang, C., Druzhinina, I., Kubicek, C., & Xu, T. (2005). *Trichoderma* biodiversity in China: evidence for a North to South distribution of species in East Asia. *FEMS Microbiology Letters*. Vol. 251, No. 2, (August 2005), pp.251-257, ISSN 1574-6968.

14

Genetic Diversity and Population Differentiation of Main Species of *Dendrolimus* (Lepidoptera) in China and Influence of Environmental Factors on Them

Gao Baojia[1,2], Nangong Ziyan[2] and Gao Lijie[2]
[1]*Hebei North University, Zhangjiakou*
[2]*Agricultural University of Hebei, Baoding*
China

1. Introduction

As the major forestry pest insects in China, *Dendrolimus* included *Dendrolimus punctatus* Walker, *D. punctatus tabulaeformis* Tsai *et* liu, *D. punctatus spectabilis* Butler, *D. superans* Butler, *D. houi* Lajonquiere and *D. kikuchii* Matsumura. During sequential outbreaks, economical damage can be so extensive for forest appears to be burned and unable to withstand such a long period of defoliation. Gene diversity and genetic structure among the main species of *Dendrolimus* were assessed using morphological diversities, allozyme, Random Amplified Polymorphism DNA (RAPD), Amplified Fragment Length Polymorphism (AFLP), Inter-Simple Sequence Repeat (ISSR), mitochondrial DNA and Simple Sequence Repeat (SSR), which provide the powerful tool for investigation of genetic variation. The influence of ecological factors on the genetic diversity is also discussed by the correlation analysis.

2. Genetic diversities among the main species and geographical populations of *Dendrolimus*

2.1 Morphological diversities among 4 species of *Dendrolimus*

For investigating the morphological diversity, many pupae of *D. punctatus* Walker, *D. superans* Butler, *D. houi* Lajonquiere and *D. kikuchii* Matsumura were collected in 2006(table1).

Species	Location	Code	Latitude and Longitude	
D.punctatus Walker:	Yujiang , Jiangxi	YJM	28° 11'N	116°54'E
D.superans Butler:	Zhangjiakou, Hebei	ZJKL	40° 50'N	114°53'E
D.houi Lajonquiere:	Guangyuan, Sichuan	GYYN	31° 53'N	105°16'E
D.kikuchii Matsumura:	Liangping, Chongqing	LPS	30° 25'N	107°24'E

Table 1. Origin of 4 different species of the tested *Dendrolimus* materials

Morphological diversities among 4 species of *Dendrolimus* by analyzing characters such as pupae weight, pupae length, female weight, female wing length, female length, male weight, male wing length and male length. Analysis of variance for all characteristics was significantly different among species and among individuals within populations. Coefficient of variation of morphological traits showed that pupae weight is significantly higher than other traits. Pupae length, female weight and female wing length were the main traits account for the phenotypic variations according to the PCA and discriminate analysis. Based on the UPGMA cluster, four species of *Dendrolimus* may be divided into two groups: the first branch in the dendrogram included *D. punctatus* Walker and *D. superans* Butler, and *D. houi* Lajonquiere and *D. kikuchii* Matsumura were grouped together into second branch(table2-3,figure1)

Code	(PW)	(PL)	(FW)	(ML)	(FL)	(MW)	(MWL)	(FWL)	Mean
YJM	54.15	14.13	20.15	4.38	6.01	27.30	4.53	7.87	17.32
GYYN	46.62	12.89	12.85	5.20	3.72	23.09	3.53	2.59	13.81
LPS	58.14	5.21	26.56	8.29	6.00	32.13	4.65	7.58	18.57
ZJKL	49.04	9.68	33.52	4.03	8.57	50.87	9.15	9.21	21.76
Total	51.99	10.48	23.27	5.48	6.08	33.35	5.47	6.81	17.87

Note:PW:pupae weight, PL: pupae length, FW: female weight, FWL: female wing length, FL: female length, MW: male weight, MWL: male wing length, ML: male length.

Table 2. Coefficient of variation of morphological traits of 4 different species of the tested *Dendrolimu* materials

code	Eigenvalue	Contribution rate %	Cumulative contribution rate %
PW	6.0824	76.03	76.03
PL	1.0501	13.13	89.16
FW	0.3807	4.76	93.92
FWL	0.2334	2.92	96.83
FL	0.0886	1.11	97.94
MW	0.0734	0.92	98.86
MWL	0.0515	0.64	99.50
ML	0.0400	0.50	100.00

Note: PW: pupae weight, PL: pupae length, FW: female weight, FWL: female wing length, FL: female length, MW: male weight, MWL: male wing length, ML: male length.

Table 3. Eigenvalue and principle proportion of 8 character of 4 different species of the tested *Dendrolimus* materials

2.2 Genetic diversities among 4 species of *Dendrolimus* by allozyme

Gene diversity and genetic structure among 4 species of *Dendrolimus* were assessed by allozyme, which provides a powerful tool for investigation of genetic variation (table4,figure2).

A total of 12 presumed loci and 22 alleles were scored in the sampled populations by allozyme method, which 6 loci were polymorphic. At population level, the mean number of alleles per locus (*A*) was 1.2708, the percentage age of polymorphic loci (*P*) was 25.00%, and the mean expected heterozygosity per locus (*He*) was 0.0828. At species level, *A*=1.8333,

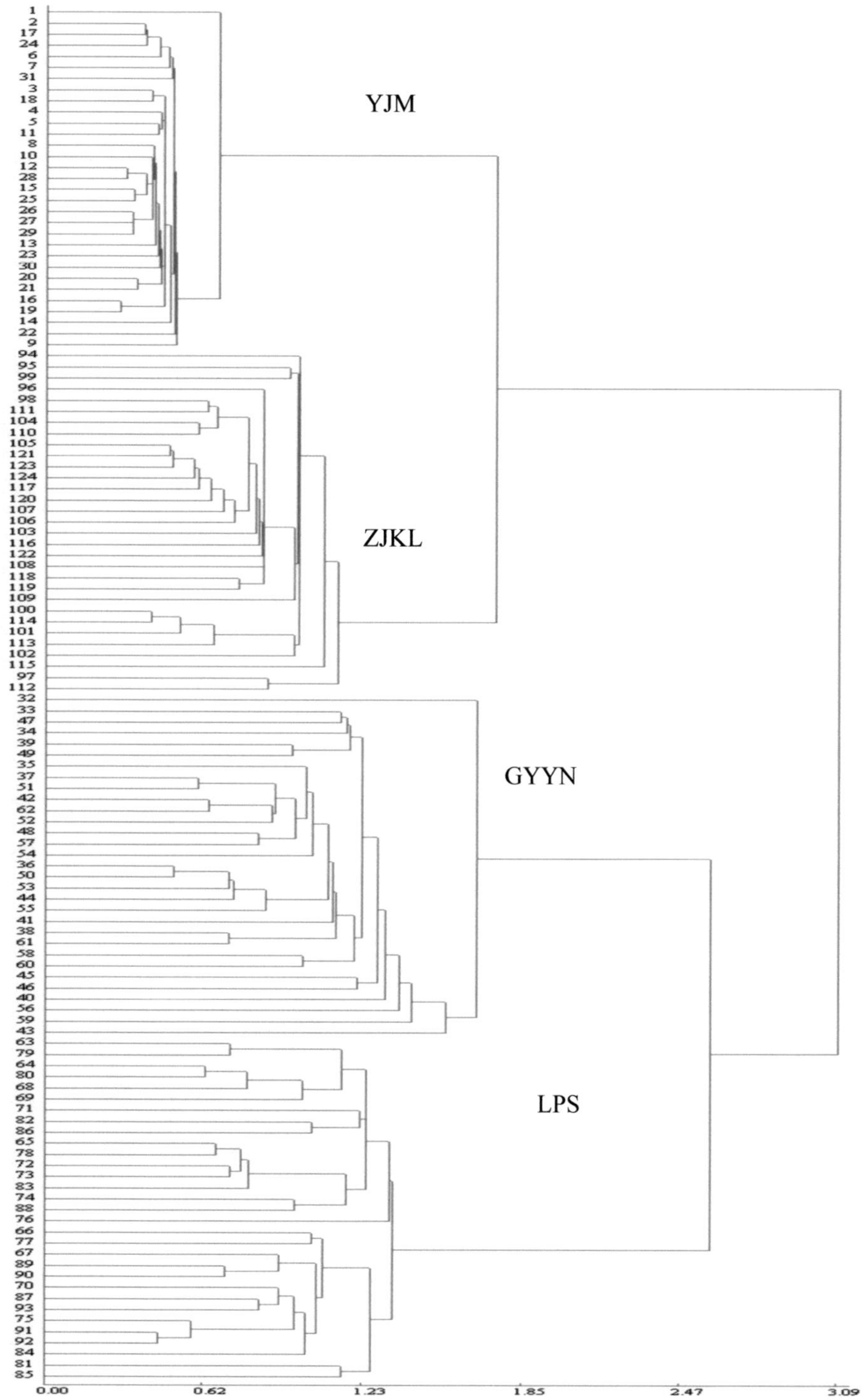

Fig. 1. UPGMA cluster based on morphological traits of 4 different species of the tested *Dendrolimus* materials

Enzyme	Abbreviation	E.C. code
Lactate dehydrogenase	LDH	E.C. 1. 1. 1. 27
Malate dehydrogenase	MDH	E.C. 1. 1. 1. 37
Malic enzyme	ME	E.C. 1. 1. 1. 40
Alcohol dehydrogenase	ADH	E.C. 1. 1. 1. 1
Formate dehydrogenase	FDH	E.C. 1.2. 1. 2
Glutamate dehydrogenas	GDH	E.C. 1.4. 1.2
Catalase	CAT	E.C. 1. 11. 1. 6
Peroxidase	POD	E.C. 1. 1. 1. 1

Table 4. The kinds of allozymes be choosed

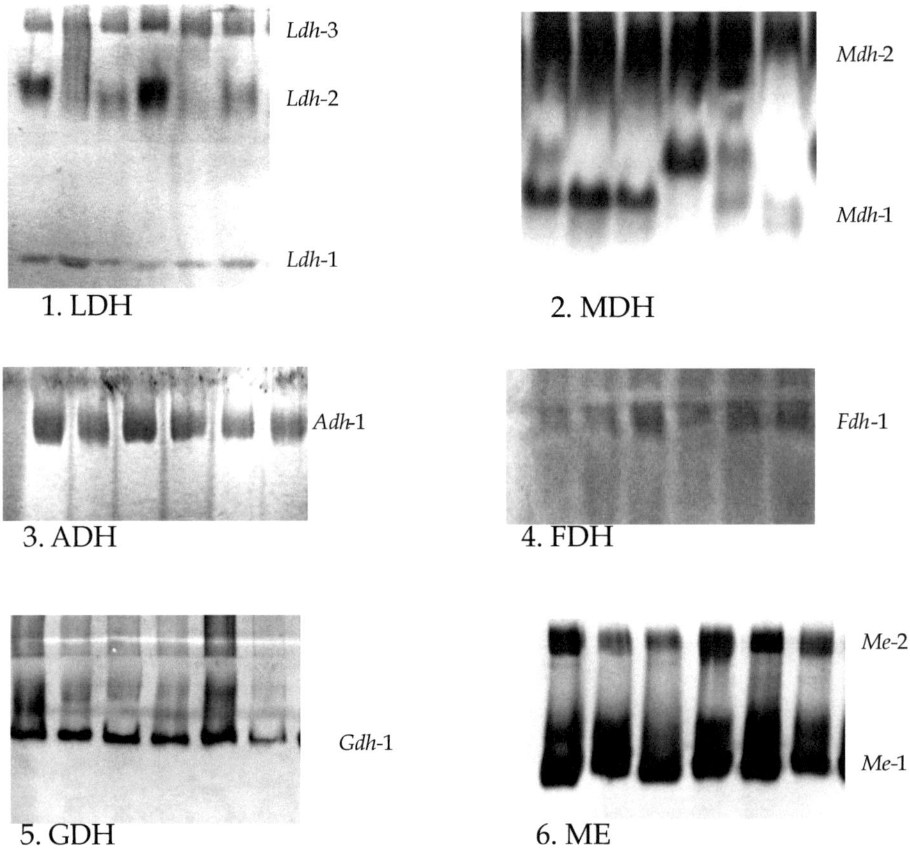

Fig. 2. Electrophoretic forms of LDH, ADH, FDH, GDH, MDH, ME
From left to right: SYC, QDC, JZC, HTM, TDM, YJM

P= 50.00%, *He* = 0.2323. So, the genetic diversity at species level was rather high than that at population level (table5,6). A UPGMA dendrogram was generated based on *Nei's* genetic distance showed 4 species of *Dendrolimus* were clustered into two groups: the first branch in the dendrogram included *D. punctatus* Walker and *D. superans* Butler, and the second branch included *D. houi* Lajonquiere and *D. kikuchii* Matsumura (figure3).

population	*A*	*Ae*	I	*P(%)*	*Ho*	*He*
YJM	1.3333	1.1462	0.1403	33.33	0.0944	0.0921
ZJKL	1.1667	1.1569	0.1129	16.67	0.1458	0.0828
GYYN	1.3333	1.1325	0.1207	25.00	0.0877	0.0771
LPS	1.2500	1.1424	0.1258	25.00	0.0972	0.0856
Mean	1.2708	1.1445	0.1249	25.00	0.1063	0.0828
Overall loci	1.8333	1.5564	0.3795	50.00	0.1076	0.2323

Note: A: Number of alleles per locus; Ae: Effective number of alleles per locus; I: The Shannon information index;
Ho: Observed heterozygosity; He: Expected heterozygosity; P: Percentage of polyporphic loci

Table 5. Genetic diversity parameters of 4 different species of the tested *Dendrolimus* materials

population	YJM	ZJKL	GYYN	LPS
YJM	****	0.9142	0.8319	0.7157
ZJKL	0.0897	****	0.7845	0.6747
GYYN	0.1840	0.2427	****	0.7967
LPS	0.3345	0.3935	0.2272	****

Note: Nei's genetic identity (above diagonal) and genetic distance (below diagonal).

Table 6. Nei's genetic identity and genetic distance of 4 different species of the tested *Dendrolimus* materials

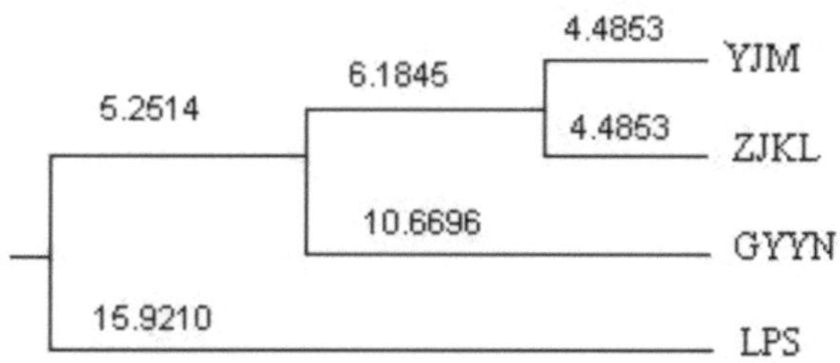

Fig. 3. UPGMA dendrograms based on Nei's genetic distance of 4 different species of the tested *Dendrolimus* materials

2.3 Genetic diversities among 3 species of *Dendrolimus* by ISSR

A set of optimized response system of ISSR has been developed, which provides a powerful tool for investigation of genetic variation of 9 natural populations(*D. punctatus spectabilis; D. punctatus tabulaeformis; D. superans*) (figure4-8,table7). Shannon index shows highest genetic diversity in Weichang population(*D. superans*) and lowest genetic diversity in Shenyang population(*D. spectabilis*), which is identical with Nei's index(table8-10).

Different geographical groups in the same populations has a genetic divergence in some extent. Genetic similarities and cluster analysis shows that inter-species heredity difference is obviously bigger than the homogeneous between populations in the heredity difference, which reflects the heredity difference degree is consistent from the DNA level with the phenotype level. *D. punctatus spectabilis* has nearer genetic distance with *D. punctatus tabulaeformis* than *D. superans*, which is identical with that researched in sexual information of pine caterpillars(figure9).

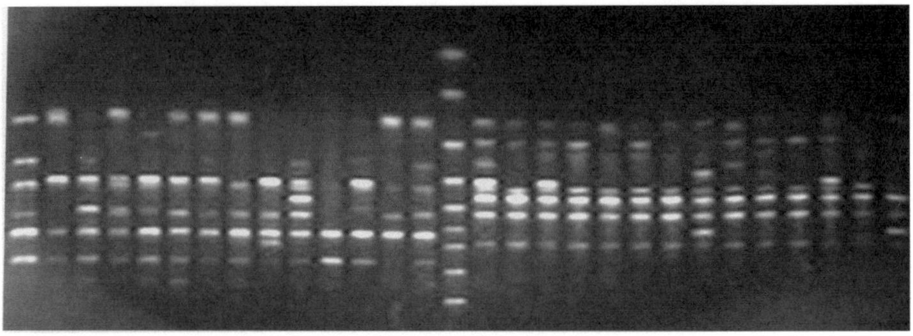

Sequence (from left to right): WL1~WL14,Marker, SC1~SC14

Fig. 4. ISSR-PCR fingerprints of 14 samples of *Dendrolimus superans* from WL and *Dendrolimus punctatus spectabilis* from SC populations using primer 106

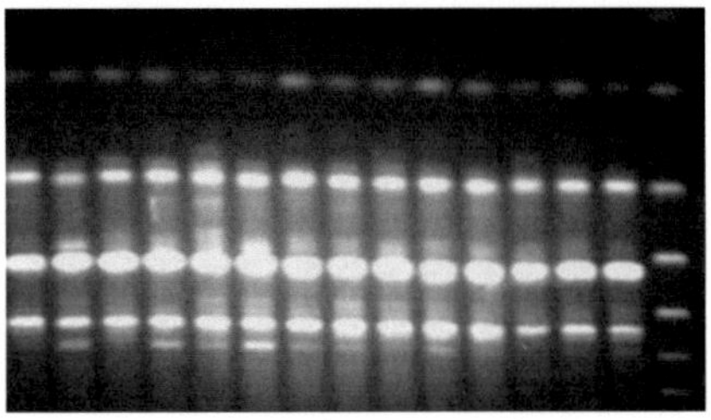

1~14: Qingdao population of D. punctatus Spectabilis M: molecular marker(DL3000bp DNA marker)

Fig. 5. ISSR profile amplified by primer 30

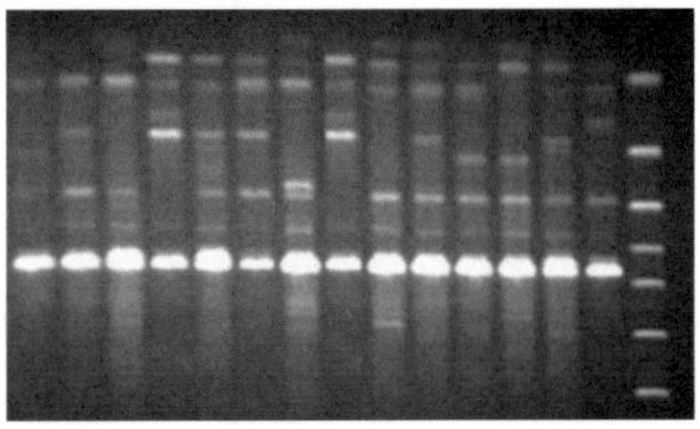

1 ~ 14: Qingdao population of D. punctatus spectabilis M: molecular marker(DL3000bp DNA marker)

Fig. 6. ISSR profile amplified by primer 32

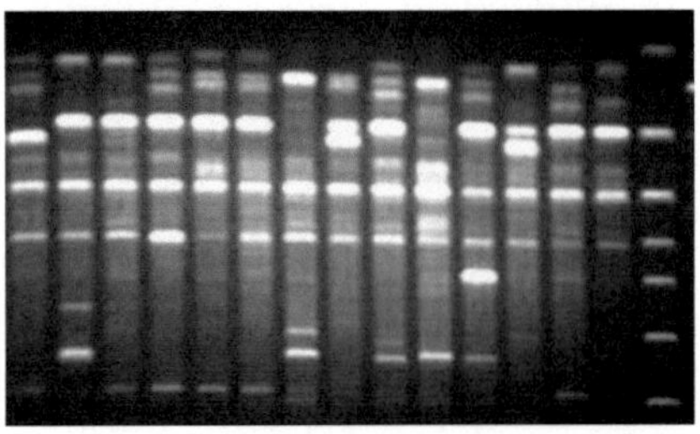

1~14: Weichang population of D.superans M:molecular marker(DL3000bp DNA marker)

Fig. 7. ISSR profile amplified by primer 33

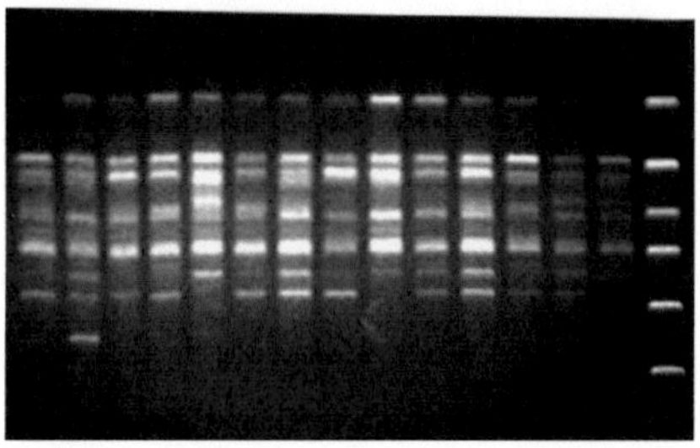

1~14: Shenyang population of D. Punctatus spectabilis M: molecular marker(DL3000bp DNA marker)

Fig. 8. ISSR profile amplified by primer 35

Primer order	Sequence ofprimer(5'-3')	Total bands	Polymorphic bands	Percentage of Polymorphic bands
30	AGAGAGAGAGAGAGAGG	11	9	81.82
32	GAGAGAGAGAGAGAGAC	17	16	94.12
33	ACCACCACCACCACCACC	14	13	92.86
35	AGCAGCAGCAGCAGCAGC	17	16	94.12
102	CACACACACACACACAG	11	11	100.00
104	ATGATGATGATGATGATG	16	14	87.50
105	GGATGGATGGATGGAT	14	13	92.86
106	VDVCTCTCTCTCTCTCT	14	14	100.00
108	CACACACACACAAC	16	15	93.75
112	AGAGAGAGAGAGAGAGTA	15	15	100.00
118	GTCGTCGTCGTCGTCGTC	17	16	94.12
120	GAGGAGGAGGAGGC	18	18	100.00
121	GTGCGTGCGTGCGTGC	15	14	93.33
	all locus	195	184	94.36%

Note: V= (G/A/C)

Table 7. Result of ISSR on adults of 3 species of *Dendrolimus* with 13 primers

population	*Na*	*Ne*	*h*	*I*	*PPB*
TC	1.1949±0.3971	1.0983±0.2359	0.0605±0.1371	0.0927±0.2035	19.49
JC	1.2256±0.4191	1.1383±0.2989	0.0794±0.1624	0.1182±0.2345	22.56
SC	1.2051±0.4048	1.0913±0.2293	0.0565±0.1310	0.0883±0.1944	20.51
QC	1.2103±0.4085	1.1158±0.2647	0.0691±0.1484	0.1047±0.2176	21.03
ZL	1.3846±0.4878	1.2055±0.3245	0.1230±0.1791	0.1873±0.2608	38.46
WL	1.4308±0.4965	1.2605±0.3588	0.1518±0.1959	0.2264±0.2824	43.08
CY	1.2256±0.4191	1.1287±0.2805	0.0759±0.1557	0.1143±0.2271	22.56
PY	1.1846±0.3890	1.1159±0.2805	0.0662±0.1517	0.0984±0.2190	18.46
SY	1.2205±0.4157	1.1561±0.3244	0.0869±0.1736	0.1268±0.2482	22.05
Average	1.2535±0.4264	1.1456±0.2886	0.0855±0.1594	0.1286±0.2319	25.35
All locus	1.9436±0.2313	1.4926±0.3278	0.2955±0.1610	0.4495±0.2127	94.36

Note: CY: *D. tabulaeformis* (Chengde population) ; PY:*D.tabulaeformis* (Pingquan population) ; SY: *D.tabulaeformis* (Shenyang population) QC:*D.spectabilis* (Qingdao population) ; TC :*D.spectabilis* (Taian population) ; SC: *D.spectabilis* (Shenyang population) ; JC: *D.spectabilis* (Jinzhou population) ; ZL: *D.superans* (Zhangjiakou population) ; WL: *D. superans* (Weichang population)

Table 8. The genetic variation among 9 populations of *Dendrolimus*

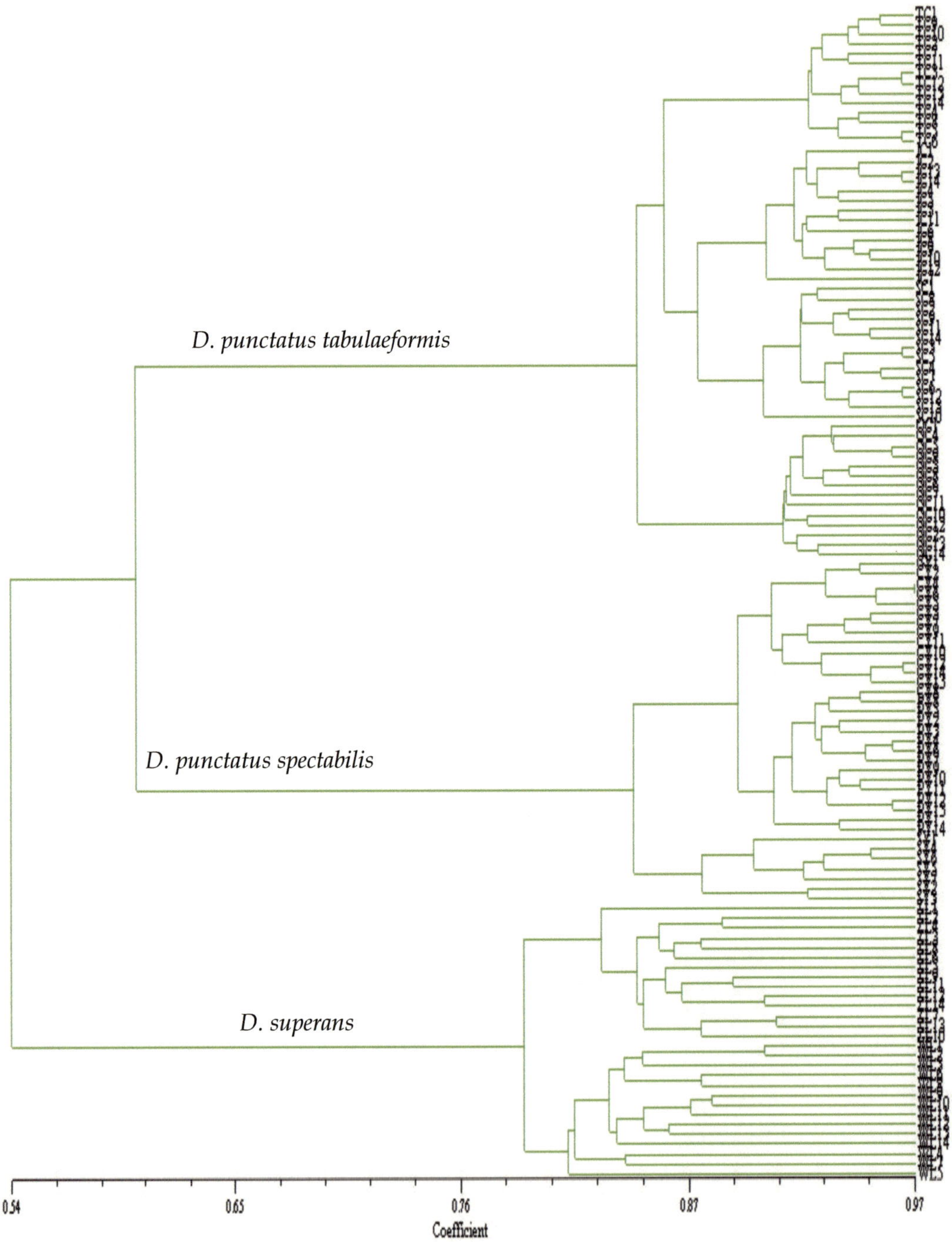

Fig. 9. Dendrogram of all D. punctatus spectabilis, D. punctatus tabulaeformis and D.superans clustered by NTSYS

species	Total gene diversity, *Ht*	Within population gene diversity, *Hs*	Genetic differentiation among populations, *Gst*	*Nm*
D.punctatus spectabilis	0.1143	0.0664	0.4192	0.6927
D.punctatus tabulaeformis	0.1132	0.0763	0.3257	1.0351
D.superans	0.1626	0.1374	0.1550	2.7265

Table 9. Genetic differentiations among 3 species of *Dendrolimus*

populations	TC	JC	SC	QC	ZL	WL	CY	PY	SY
TC	****	0.9492	0.9293	0.9305	0.6429	0.6434	0.6729	0.6735	0.6501
JC	0.0521	****	0.9512	0.9219	0.6104	0.6239	0.6833	0.6839	0.6654
SC	0.0733	0.0500	****	0.9224	0.6446	0.6535	0.7072	0.7071	0.6967
QC	0.0720	0.0813	0.0808	****	0.6154	0.6209	0.6648	0.6605	0.6510
ZL	0.4418	0.4936	0.4392	0.4855	****	0.9473	0.6970	0.6872	0.6784
WL	0.4410	0.4717	0.4254	0.4766	0.0541	****	0.6969	0.6954	0.7076
CY	0.3961	0.3808	0.3464	0.4083	0.3609	0.3612	****	0.9772	0.9259
PY	0.3952	0.3799	0.3466	0.4147	0.3751	0.3632	0.0230	****	0.9292
SY	0.4306	0.4074	0.3614	0.4292	0.3880	0.3459	0.0769	0.0735	****

Note: Nei's genetic identity (above diagonal) and genetic distance (below diagonal)

Table 10. Genetic identity and genetic distance among 9 populations of *Dendrolimus*

2.4 Genetic diversities among 3 species of *Dendrolimus* by AFLP

The genetic structure and diversity among the 3 natural populations of the *D. punctatus tabulaeformis* Tsai *et* Liu, *D. punctatus spectabilis* Butler and *D. punctatus* Walker were tested with AFLP technique(table11). At population level, gene diversity with in populations (Hs) was 0.0895; coefficient of population differentiation (Gst) was 0.7623. Genetic variation among populations accounted for the genetic diversity of the total population of 76.23%, few portions of variation exist in the population (23.77%). Gene flow between populations Nm (0.1559) on the exchange of genes between populations is not strong, can not be effectively offset by genetic drift caused by population differentiation(table12,13). Clustering results show that *D. punctatus tabulaeformis* Tsai and *D. punctatus* Walker together with a category, with the *D. punctatus spectabilis* Butler together (figure10).

Primer combination	Total bands	Polymorphic bands	Polymorphism(%)
P1-T4	46	42	91.30
P2-T3	39	33	84.62
P2-T4	62	59	95.16
P2-T5	56	52	92.86
P7-T9	33	33	100.00
P8-T1	40	39	97.50
P8-T3	40	38	95.00
P8-T4	34	31	91.18
P8-T5	34	31	91.18
P8-T6	34	33	97.06

Table 11. List of AFLP primer labels,primer sequences and amplification results of 3 different species of the tested *Dendrolimus*

population	*Na*	*Ne*	*h*	*I*	*PPB*
PYC	1.6746	1.5184	0.2870	0.4140	67.46%
SC	1.6555	1.4951	0.2739	0.3961	65.55%
HM	1.7010	1.5452	0.3003	0.4326	70.10%
Average	1.6770	1.5196	0.2871	0.4142	67.70%
All locus	1.9354	1.6707	0.3765	0.5487	93.54%

Note: PYC: *D. punctatus tabulaeformis* (Pingquan population), SC: *D.punctatus spectabilis* (Shenyang population), HM: *D.punctatus* (Walker) (Hunan population)

Table 12. The genetic diversity of 3 different species of the tested *Dendrolimus*

Pop	PYC	SC	HM
PYC	****	0.8097	0.8271
SC	0.2111	****	0.8190
HM	0.1898	0.1997	****

Note: Nei's genetic identity (above diagonal) and genetic distance (below diagonal).

Table 13. Nei's genetic identity and genetic distance of 3 different species of the tested *Dendrolimus* materials

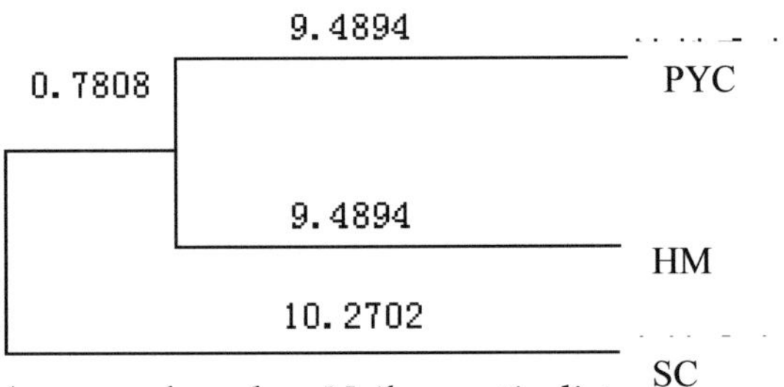

Fig. 10. UPGMA dendrograms based on Nei's genetic dist[illegible]he tested *Dendrolimus* materials

2.5 The genetic polymorphism and genetic differentiation in mitochondrial DNA of 3 species of *Dendrolimus*

In order to provide the scientific basis for prevention and treatment of *Dendrolimus*, the genetic structure of 3 species of *Dendrolimus* were probed by mitochondrial DNA. Cyt b exhibits an A/T (73.6%) bias across all sites which was the most prominent at the third position of codon with the highest content of 86.5%. A/T bias was no significant difference among the populations. Nucleotide substitution occurred mostly at the third position (table14,15). Transitions were greater than transversions, while substitutions of intraspecific populations were higher than interspecific populations.

condon site	*T*	*C*	*A*	*G*	*A+T*	*Ts*	*Tv*	*Ts/Tv*
First	23	10.0	46.9	20.0	69.9	0	1	0
Second	31	17.5	34.0	17.8	65	1	1	1.33
Third	47	1.1	39.5	12.8	86.5	6	2	3.04
total	33.5	9.5	40.1	16.9	73.6	6	4	1.45

Table 14. Nucleotide Frequency and Substitution of Coden in Vary Site

condon site		*populations*					
		GYM	*SYC*	*HBY*	*HBY*		
					HTLZYC	*PQYC*	*HTLZYH*
Avg.	*T*	*33.2*	*33.4*	*33.5*	*33.5*	*33.5*	*33.5*
	C	*9.6*	*9.1*	*9.6*	*9.7*	*9.6*	*9.5*
	A	*39.8*	*40.4*	*40.1*	*40.1*	*40.1*	*40.2*
	G	*17.5*	*17.1*	*16.7*	*16.7*	*16.7*	*16.8*
	A+T	*73*	*73.8*	*7.6*	*73.6*	*73.6*	*73.7*
First	*T*	*23*	*24*	*23*	*23*	*23*	*23*
	C	*10.0*	*9.6*	*10.0*	*10.0*	*10.0*	*10.0*
	A	*46.9*	*46.5*	*46.9*	*46.9*	*46.9*	*46.9*
	G	*20.0*	*20.0*	*20.0*	*20.0*	*20.0*	*20.0*
	A+T	*69.9*	*70.5*	*69.9*	*69.9*	*69.9*	*69.9*
Second	*T*	*31*	*30*	*31*	*31*	*31*	*31*
	C	*17.7*	*16.5*	*17.6*	*17.7*	*17.7*	*17.5*
	A	*33.3*	*35.0*	*33.9*	*33.8*	*33.8*	*34.2*
	G	*18.5*	*18.1*	*17.7*	*17.7*	*17.7*	*17.7*
	A+T	*64.3*	*65.0*	*64.9*	*64.8*	*64.8*	*64.8*
Third	*T*	*46*	*46*	*47*	*47*	*47*	*47*
	C	*1.0*	*1.2*	*1.1*	*1.2*	*1.2*	*0.9*
	A	*39.0*	*39.5*	*39.5*	*39.5*	*39.5*	*39.5*
	G	*14.0*	*13.2*	*12.5*	*12.4*	*12.4*	*12.7*
	A+T	*85.0*	*85.5*	*86.5*	*86.5*	*86.5*	*86.5*

Table 15. Nucleotide Frequencey of Cyt b gene of different populations of *Dendrolimus*

Thirty nine nucleotide sites showed mutation in this sequence fragment and the sequence variability was 10.1%, the measurement sequence encoding 129 amino acids, of which 11 mutations, accounting for 8.5% (table16,17). Nucleotide sequences and amino acid sequences in genetic distance was 0.000-0.100 and 0. 000-0.086, indicating a low genetic variation(table18).

Average nucleotide differences of interspecies and intraspecific are 8.968 and 3.934, genetic difference mainly exist in interspecies(table19).

Defined eight different haplotypes based on the mtDNA Cyt b of *Dendrolimus* populations with one shared. Four haplotypes were identified in three populations of *D. punctatus tabulaeformis* Tsai *et* Liu, accounting for 50%, and with two shared. Analysis of the haplotype and its distribution based on the mtDNA Cyt b, in interspecies has more exclusive haplotypes but the exclusive haplotypes equal the shared haplotypes. *Fst* value and gene flow showed that genetic differentiation mostly existed in interspecies, and among the intraspecific not only the gene flow occurred, but also the genetic differences did(table20,21).

codon site	*ii*	*si*	*sv*	*R*	*TT*	*TC*	*TA*	*TG*	*CC*	*CA*	*CG*	*AA*	*AG*	*GG*
Avg.	*380*	*6*	*4*	*1.45*	*139*	*1*	*3*	*0*	*37*	*1*	*0*	*152*	*5*	*63*
First	*130*	*0*	*1*	*0*	*30*	*0*	*1*	*0*	*13*	*0*	*0*	*61*	*0*	*26*
Second	*129*	*1*	*1*	*1.33*	*39*	*0*	*0*	*0*	*23*	*0*	*0*	*44*	*0*	*13*
Third	*122*	*6*	*2*	*3.04*	*59*	*1*	*2*	*0*	*1*	*0*	*0*	*48*	*5*	*14*

Table 16. The base frequency and substitution of nucleotide codes each sites

code	*site*	*ii*	*si*	*sv*	*R*	*TT*	*TC*	*TA*	*TG*	*CC*	*CA*	*CG*	*AA*	*AG*	*GG*
GYM	*1st*	130	0	0	-	30	0	0	0	13	0	0	61	0	26
	2nd	129	0	1	0	39	0	1	0	23	0	0	43	0	24
	3rd	128	1	1	1	59	1	1	0	1	0	0	50	0	18
	Avg.	387	1	1	0.5	128	1	1	0	37	0	0	154	0	68
SYC	*1st*	127	0	3	0	30	0	2	0	12	1	0	59	0	26
	2nd	126	2	2	1	39	0	0	1	21	1	0	44	2	22
	3rd	126	1	2	0.5	59	0	1	0	1	0	1	50	1	16
	Avg.	379	3	7	0.43	128	0	3	1	34	2	1	153	3	64
HBY	*1st*	130	0	0	-	30	0	0	0	13	0	0	61	0	26
	2nd	130	0	0	0.54	40	0	0	0	23	0	0	44	0	23
	3rd	126	3	0	5.3	60	1	0	0	1	0	0	50	2	15
	Avg.	385	3	1	3.78	130	1	1	0	37	0	0	155	2	64
HTLZYC (*HBY*)	*1st*	130	0	0	-	30	0	0	0	13	0	0	61	0	26
	2nd	130	0	0	-	40	0	0	0	23	0	0	44	0	23
	3rd	128	1	0	-	60	1	0	0	1	0	0	51	0	16
	Avg.	388	1	0	-	130	1	0	0	37	0	0	156	0	65
PQYC (*HBY*)	*1st*	130	1	0	-	30	0	0	0	13	0	0	61	0	26
	2nd	130	0	0	-	40	0	0	0	23	0	0	44	0	23
	3rd	128	1	0	-	60	1	0	0	1	0	0	51	0	16
	Avg.	388	1	0	-	130	1	0	0	37	0	0	156	0	65
HTLZYH (*HBY*)	*1st*	130	0	0	-	30	0	0	0	13	0	0	61	0	26
	2nd	129	0	1	0.67	39	0	1	0	23	0	0	44	0	23
	3rd	120	7	2	4.5	59	0	1	0	1	0	0	47	7	13
	Avg.	379	8	2	3.45	129	1	2	0	37	0	0	152	7	62

Note: GYM: *D. punctatus punctatus* Walker(Guiyang population); SYC:*D. punctatus spectabilis* Butler (Shenyang population); HTLZYC, PQYC, HTLZYH: *D.punctatus tabulaeformis* (Pingquan population)

Table 17. The base substitution of nucleotide of different populations of *Dendrolimus*

Code	1	2	3	4	5
GYM	***	0.086	0.016	0.016	0.016
SYC	0.100	***	0.070	0.070	0.070
HTLZYC	0.013	0.094	***	0.000	0.000
PQYC	0.013	0.094	0.000	***	0.000
HTLZYH	0.010	0.091	0.003	0.003	***

Note: nucleotides, lower triangle; amino acid, up triangle

Table 18. Pairwise distance of Cyt b gene sequence of different populations of *Dendrolimus*

	Pop	*n*	*S*	*H*	*Hd*	*Pi*	*K*
	GYM	3	3	3	1.000	0.00514	2.000
	SYC	2	10	2	1.000	0.02571	10.000
	HBY	15	24	4	0.670	0.01011	3.934
HBY	HTLZYC	9	1	2	0.556	0.00143	0.556
	PQYC	2	1	2	1.000	0.00257	1.000
	HTLZYH	4	23	4	0.833	0.02999	11.667
	total	20	39	8	0.805	0.02306	8.968

Note:n-number of individuals; *H*-Haploid; *Hd*-Haploid diversity index; *S*-sites of diversity; *K*—Average nucleotide differences; *Pi*—Polymorphic index

Table 19. Genetic diversity between different populations of *Dendrolimus*

pop	GYM	SYC	HBY
GYM	***	0.12	0.36
SYC	0.808	***	0.17
HBY	0.583	0.744	***

Note: Fst, lower triangle; Nm, up triangle

Table 20. Genetic differentiation and gene flow between any specie of *Dendrolimus*

pop	HTLAYC	PQYC	HTLZYH
HTLZYC	***	19.00	6.88
PQYC	-0.555	***	-1.40
HTLZYH	0.025	0.067	***

Note:Fst, lower triangle; Nm, up triangle

Table 21. Genetic differentiation and gene flow between any population of *D. punctatus tabulaeformis*

Cluster analysis showed that the genetic distance between *D. punctatus* Walker and sub-species of *D. punctatus tabulaeformis* is relatively close, genetic differentiation exists between *D. punctatus* Walker and sub-species of *D. punctatus spectabilis* Butler and the population genetic differentiation was related to ecological environment(figure11).

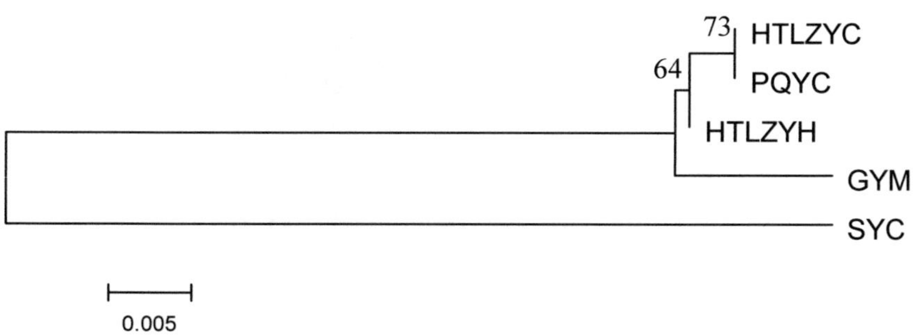

Fig. 11. NJ dendrogram based on the Cyt b gene sequence of *Dendrolimus*

2.6 The genetic diversity analysis of 6 populations of *D. punctatus* Walker by RAPD

Rrandom amplified polymorphic DNA (RAPD) were used to evaluate genetic diversity of 6 populations of the masson moth *D. punctatus* Walker. Three populations of *D. punctatus* Walker and three populations of geogrophic subspecies of *D. punctatus* Walker - *D. punctatus spectabilis* Bulter were examined and compared(table22).

Populations	Location	Code	Latitude and Longitude
D. punctatus punctatus Walker:	Huitong , Hunan	HTM	26° 40'N 109°26'E
	Tongdao, Hunan	TDM	26° 07'N 109°46'E
	Yujiang , Jiangxi	YJM	28° 11'N 116°54'E
D. punctatus spectabilis Butler:	Shenyang , Liaoning	SYC	41° 11' N 119°23' E
	Jinzhou, Liaoning	JZC	40° 27' N 119°51' E
	Qingdao, Shandong	QDC	41° 07'N 121°05' E

Table 22. Origin of the tested *D. punctatus* Walker materials

In total, 88 bands whose size ranged between 350 to 1500 bp were produced using 8 primers in RAPD analysis method (figure12-13,table23-24). Out of 88 loci, 73 bands were polymorphic at the species level. The percentages of polymorphic loci (*P*) was 25.19 %, the mean number of alleles per locus (*A*) was 1.252, the mean *Nei's* gene diversity (*h*) was 0.052, the mean Shannon's information index (*I*) was 0.090. For the species level, *P*= 82.95 %, *A*= 1.830, *h*= 0.234(table25). The coefficient of genetic differentiation between populations based on Shannon information index was 0.7490, which revealed a very high level of genetic differentiation among populations, the number of migrants per generation among populations (*Nm*) was 0.168, which revealed a very low gene flow among populations. The coefficient of genetic differentiation between populations based on *Nei's* genetic diversity was 0.7780 which revealed a very high level of genetic differentiation among populations, the number of migrants per generation among populations (*Nm*) was 0.143(table26,27).

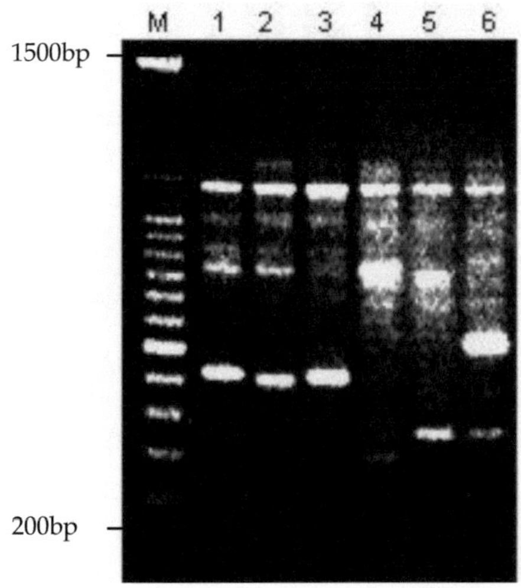

1~6: SYC, JZC, QDC, HTM, TDM, YJM. M: molecular marker (100bp DNA ladder)

Fig. 12. RAPD profile amplified by primer OP03

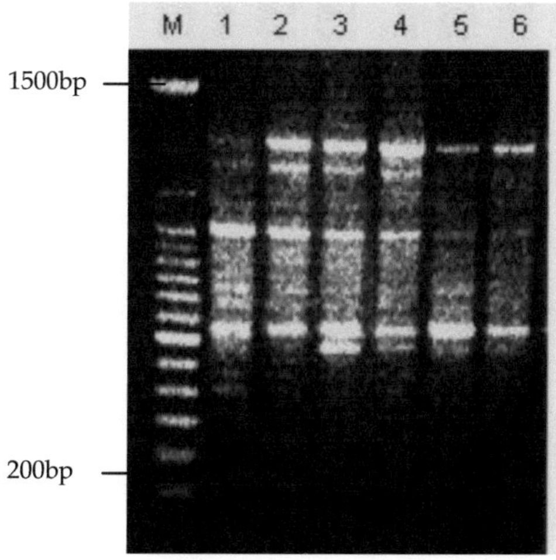

1~6: SYC, JZC, QDC, HTM, TDM, YJM. M: molecular marker (100bp DNA ladder)

Fig. 13. RAPD profile amplified by primer OP05

Primer	Sequences (5′-3′)	Primer	Sequences (5′-3′)
OP03	CTGAGACGGA	OP12	ACGACCGACA
OP05	CTGACGTCAC	OP13	GTCAGGGCAA
OP06	AGGGCCGTCT	SB01	TTCGAGCCAG
OP07	TGCCCGTCGT	SB09	TGTCATCCCC
OP08	CTCTCCGCCA	P01	GGTCCCTGAC

Table 23. RAPD Primers of the tested *D. punctatus* Walker materials

Primer order	Sequence of primer (5′-3′)	Total bands	Polymorphic bands	Percentage of polymorphic bands
OP03	CTGAGACGGA	18	12	66.67
OP05	CTGACGTCAC	20	14	70.00
OP06	AGGGCCGTCT	25	19	76.00
OP07	TGCCCGTCGT	22	22	100.00
OP08	CTCTCCGCCA	30	30	100.00
OP12	ACGACCGACA	29	13	44.83
OP13	GTCAGGGCAA	28	28	100.00
SB01	TTCGAGCCAG	29	23	79.31
SB09	TGTCATCCCC	29	23	79.31
P01	GGTCCCTGAC	28	22	78.57
Total		88	73	82.95

Table 24. Result of RAPD on adults of 6 populations of *D. punctatus* Walker materials with 10 primers

population	*A*	*Ae*	*I*	*h*	*P*
SYC	1.239	1.058	0.081	0.046	23.86
JZC	1.193	1.056	0.071	0.042	19.32
QDC	1.273	1.059	0.087	0.048	27.27
HTM	1.273	1.076	0.099	0.058	27.27
TDM	1.250	1.070	0.089	0.052	25.00
YJM	1.284	1.096	0.114	0.069	28.41
At population level	1.252	1.069	0.090	0.052	25.19
At species level	1.830	1.398	0.359	0.234	82.95

Note: *A:* Number of alleles per locus; *Ae:* Effective number of alleles per locus;
I: The Shannon information index; *h*: Nei's genetic diversity;
P: Percentage of polymorphic loci

Table 25. The genetic variation statistic among populations of *D. punctatus* Walker materials

species	*D. punctatus* Walker	
Shannon's information index	*Hsp*	0.090
	Hpop	0.359
	(*Hpop-Hsp*) / *Hpop*	0.749
	N m	0.168
Nei's genetic index	*Hs*	0.052
	Ht	0.234
	Gst	0.778
	Nm	0.143

Table 26. Genetic differentiations among populations of *D. punctatus* Walker materials

population	SYC	JZC	QDC	HTM	TDM	YJM
SYC	****	0.870	0.724	0.725	0.691	0.610
JZC	0.121	****	0.747	0.732	0.687	0.623
QDC	0.108	0.139	****	0.880	0.871	0.711
HTM	0.351	0.323	0.292	****	0.820	0.637
TDM	0.351	0.322	0.313	0.128	****	0.677
YJM	0.448	0.370	0.376	0.138	0.199	****

Table 27. RAPD estimates of Nei's unbiased genetic distance among 6 populations of *D. punctatus* Walker

Three populations of *D. punctatus* Walker and three populations of *D. punctatus spectabilis* were clustered into one branch independently: pop1 and pop3 grouped firstly in the first branch, and the pop 4 and pop 5 were first grouped together in the second branch (figure14).

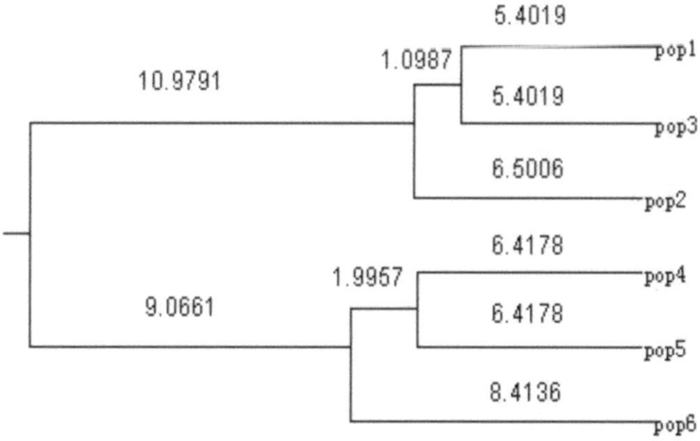

From pop1 to pop6: SYC, JZC, QDC, HTM, TDM, YJM, See table 22 for abbreviations of population codes.

Fig. 14. UPGMA dendrograms based on Nei's genetic distance of RAPD markers

2.7 Genetic diversities among 5 sub-populations of *D. punctatus tabulaeformis* Tsai *et* liu

The pupae of 5 sub-populations of *D. punctatus tabulaeformis* were collected in Pingquan, Hebei Province(table28). And the gene diversity and genetic structure of them were assessed by SSR (table29).

Sub-Population Name	Forest type	Mixed tree	Direction	Vegetation under the forest	Insect prevention
pop1	Artificial pure *Pinus tabulaeformis* forest	none	North	Few vegetation, pine needle leaf cover	Not control
pop2	Artificial mixed *P. tabulaeformis* forest	*Larix gmelinii* (Ruprecht) Kuzeneva	West	Few herb, pine needle leaf cover	
pop3	Artificial mixed *P. tabulaeformis* forest	*Quercus mongolicus* Fisch.	Northeast	Ferns and bryophyta, pine needle, broad-leaf leaf cover.	Not control
pop4	Natural pure *P. tabulaeformis* forest	None	South	Few herb, pine needle leaf cover	Matrine aircraft control district
pop5	Artificial mixed *P. tabulaeformis* forest	*Pinus armandii* Franch	Southeast	Herb, pine needle leaf cover	Matrine aircraft control district

Table 28. Origin of the tested *D. punctatus tabulaeformis* materials in Chengde City, Pingquan county

Primer	Sequences (5′—3′)	Number of allele	Size range (bp)	Tm(°C)
SSR4	TCATCCCGAGTCCCACTCA ATTGCTCTTCCTATCTGGCTA	3	78~188	52
SSR5	GTTCTCGGTCGTGGTTTTAG AACCGCTTCCGCCGATTAC	1	161	50
SSR6	CTGGCACCCCGCCGATTAC AACAAAACAATTATAAACTCTTAC	2	155~160	50
SSR7	ACCAACTTCGACACCTTCT CACTGCCCCGAACCTATAC	3	217~333	50
SSR8	ACTTCTACTGCGTGTGAACT GTCCCTTTGTCCGATAATATG	4	193~207	51
SSR9	GGAGCACCAATGAAGAATGT GTTTCTACCTCATGGGATCTTTTAGCTC	3	234~430	52
SSR10	ACGTAAAACTAATCAA CTGTCCAAAGCAAACTATC	3	187~220	52
SSR11	CTGCTAGAGCTTTCTGTGTT AAGAATTTCAATTTAAGACTGAC	6	158~241	50

Table 29. SSR primers of the tested *D. punctatus tabulaeformis* materials in this study

For genetic diversities, among 5 sub-populations of *D. punctatus tabulaeformis*, in total, polymorphic bands were produced using 8 primers in SSR analysis method (table29, figure15-17). The percentage of polymorphic loci (*P*) was 80.00%, the mean number of alleles per locus (*A*) was 2.6250, and the mean expected heterozygosity per locus (*He*) was 0.3765. For the species level, *P*=87.50%, *A*=3.1250, *He*=0.4747(table30). From the summary of F-statistics at polymorphic loci of 6 populations, we could found the *Fst* was 0.2159, which mean there show high genetic diversity among populations.The number of migrants per generation among populations (*Nm*) was 0.9081, suggesting the occurrence of rather low

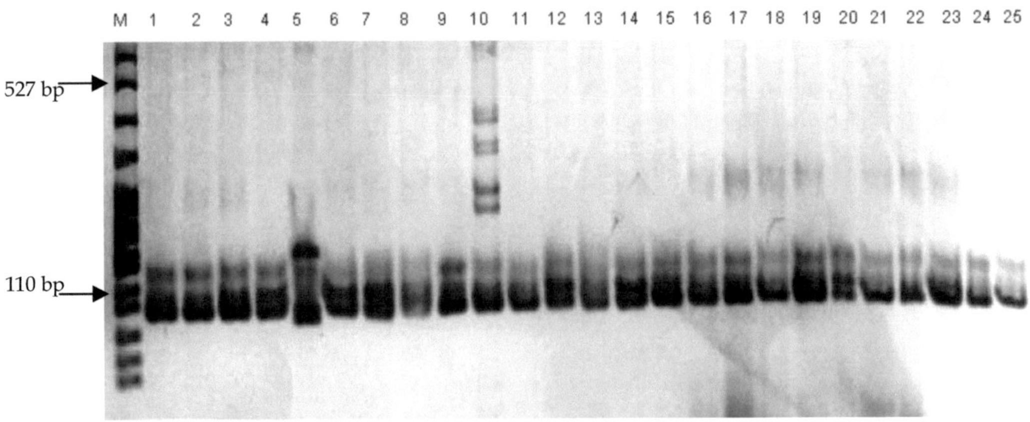

1~5,6~10,11~15,16~20,21~25 are pop1, pop2, pop3, pop4, pop5 of *D. punctatus tabulaeformis*;
M: molecular marker (PBR322/Msp I marker)

Fig. 15. SSR profile amplified by primer SSR4

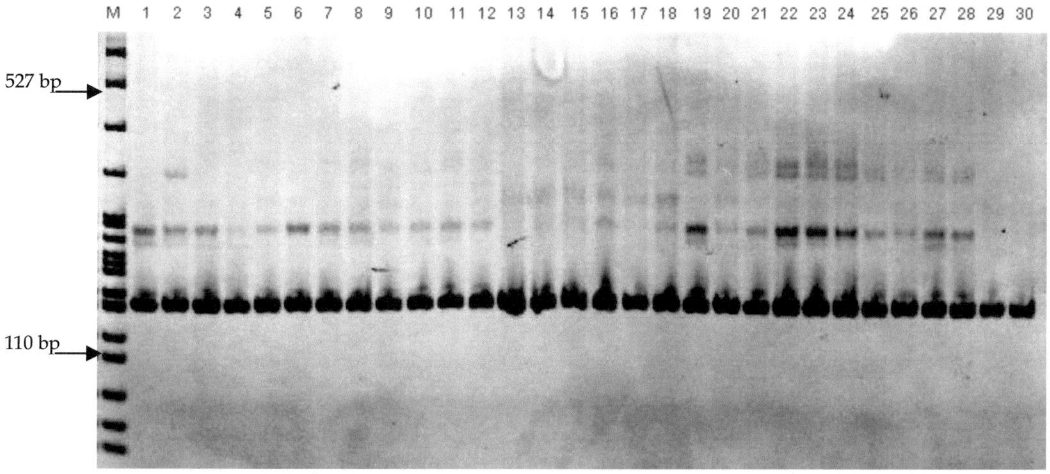

1~6, 7~12, 13~18, 19~24, 25~30 are pop1, pop2, pop3, pop4, pop5 of *D. punctatus tabulaeformis;*
M: molecular marker(PBR322 / Msp I marker)

Fig. 16. SSR profile amplified by primer SSR5

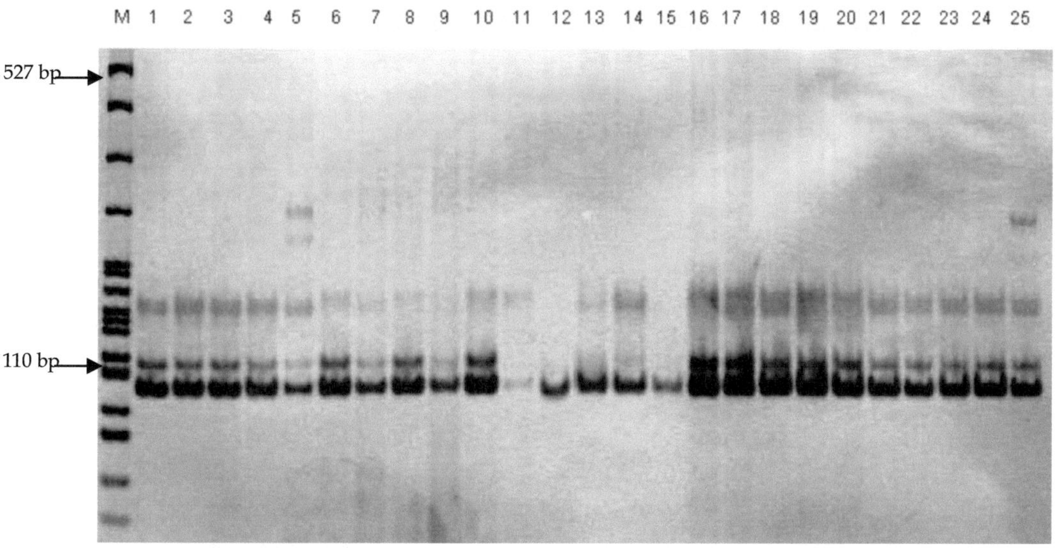

1~5,6~10,11~15,16~20,21~25 are pop1, pop2, pop3, pop4, pop5 of *D. punctatus tabulaeformis;*
M: molecular marker(PBR322 / Msp I marker)

Fig. 17. SSR profile amplified by primer SSR6

gene flow among sub-populations. A UPGMA dendrogram based on *Nei's* genetic distance showed 5 sub-populations of *D. punctatus tabulaeformis* Tsai *et* liu were clustered into two groups: the first branch in the dendrogram included pop1 and pop2, and the remaining three populations were grouped together into second branch (table31,figure18).

population	*A*	*Ae*	*I*	*Ho*	*He*	*h*	*P*
pop1	2.5000	1.8045	0.6501	0.4903	0.4109	0.4063	87.50
pop2	2.5000	1.4960	0.4981	0.3587	0.2811	0.2781	75.00
pop3	2.6500	1.7922	0.6245	0.4417	0.3690	0.3649	75.00
pop4	3.0000	2.2521	0.8020	0.4054	0.4543	0.4481	87.50
pop5	2.5000	1.8122	0.6180	0.4472	0.3674	0.3634	75.00
Mean	2.6250	1.8314	0.6385	0.4287	0.3765	0.3722	80.00
Overall	3.1250	2.1201	0.8091	0.4302	0.4747	0.4736	87.50

Note: *A:* Number of alleles per locus; *Ae:* Effective number of alleles per locus; *I:* The Shannon information index;
Ho: Observed heterozygosity; *He:* Expected heterozygosity; *P:* Percentage of polyporphic loci

Table 30. Genetic diversity parameters of 5 sub-populations in *D. punctatus tabulaeformis*

POP	pop1	pop2	pop3	pop4	pop5
pop1	****	0.9322	0.7555	0.8826	0.7714
pop2	0.0703	****	0.6573	0.8262	0.6953
pop3	0.2804	0.4197	****	0.8568	0.7905
pop4	0.1249	0.1909	0.1546	****	0.8775
pop5	0.2596	0.3635	0.2350	0.1306	****

Note: See table 3-1 for abbreviations of population codes.

Table 31. Nei's genetic identity and genetic distance of 5 sub-populations in *D. punctatus tabulaeformis*

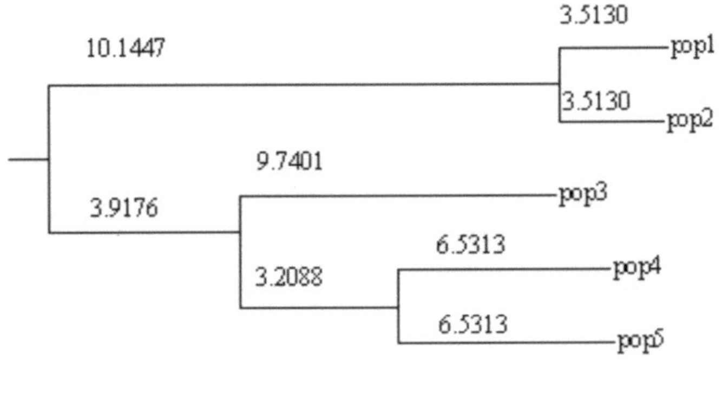

Fig. 18. Phylogenetic relationship of 5 sub-populations of *D. punctatus tabulaeformis* based on Nei's distance of SSR markers and clustered using UPGMA

3. The effect of environmental conditions on the genetic diversity of *Dendrolimus*

3.1 The effect of environmental conditions on the genetic diversity of *D. punctatus tabulaeformis* populations

The genetic diversity among 3 natural populations of the Chinese pine caterpillar (*Dendrolimus punctatus tabulaeformis*) were tested by AFLP method in one *Pinus tabulaeformis-Dahurian larch* mixed forest and two *Pinus tabulaeformis* pure forests in Pingquan county. Besides, investigations on plant species diversity, forest crown density, incidence extent, gradient and exposure of three forest communities were taken, while dilution heat method and semi-micro Macro Kjeldahl method were used to determine the content of organic matter and nitrogen in soil(table32-36).

The result of principle component analysis shows that the growth status is main factor which influent genetic diversity of *D. punctatus tabulaeformis* populations. Besides, site conditions had some effects on it. An integrated effect was produced by all site conditions, so main factor could not be judged(table37).

Mixed forest has a great influence on gene flow among different populations of *D. punctatus tabulaeformis*, because gene flow between populations in mixed forest and in pure forests was lower than that of two pure forests. Gene flow between populations in pure forest with larger species abundance and in mixed forest was higher than between population in pure forest with lower species abundance and in mixed forest, which showed that the correlation between gene flow among different populations and species abundance of pine forests is negative.

Community	Tree height	Diameter at breast height	Crown density	Exposure	Gradient	Incidence extent
I	7.9	11.6	0.6	Ubac	31.7°	1
II	7.8	10.9	0.5	Adret	34.3°	2
III	8.5	13.5	0.5	Half adret	36.0°	3

Note: I,II: pure forests; III: mixed forest.

Table 32. Statistical results of stand condition investigation

Community	Sample plot	Species number	Individual number	Simpson s diversity index	Shannon-weaver diversity index
	1	20	285	0.7962	2.9628
I	2	16	327	0.8400	2.9936
	3	17	307	0.8619	3.1770
	4	25	499	0.8722	3.3170
II	5	15	728	0.8387	2.9772
	6	20	392	0.8184	2.9746
	7	29	1021	0.8363	3.3473
III	8	31	868	0.8627	3.4626
	9	29	949	0.8055	3.1790

Note: I,II: pure forests; III: mixed forest

Table 33. Species diversity index analysis of 3 populations in *D. punctatus tabulaeformis*

Community	Sample plot	$FeSO_4$ volume of using (mL)	Organic matter content (g·kg^{-1})	Average
	1	12	20.12	
I	2	1	47.79	32.70
	3	8	30.18	
	1	13	17.61	
II	2	8	30.18	22.64
	3	12	20.12	
	1	3	42.76	
III	2	7	32.70	32.28
	3	11.5	21.38	

Note: I, II: pure forests; III: mixed forest

Table 34. Soil organic matter content of 3 populations in *D. punctatus tabulaeformis*

Community	Sample plot	Acid standard solution volume of using (ml)	Soil total nitrogen content (g·kg^{-1})	Average
	1	1.1	0.011	
III	2	1.6	0.018	0.0107
	3	0.6	0.003	
	1	0.5	0.002	
II	2	1	0.009	0.0051
	3	0.7	0.005	
	1	1.2	0.012	
I	2	0.5	0.002	0.0049
	3	0.45	0.001	

Table 35. Soil total nitrogen content of 3 populations in *D. punctatus tabulaeformis*

Population numbers	II	I	III	Total
Polymorphic bands	107	99	91	124
Polymorphism(%)	73.29	67.81	62.33	84.93
Observed number of alleles	1.7329	1.6781	1.6233	1.8493
Effective number of alleles	1.4804	1.3694	1.3441	1.4371
Nei's gene diversity	0.2683	0.2127	0.2024	0.2561
Shannon's Information index	0.3935	0.3193	0.3046	0.3877

Table 36. The genetic diversity of 3 populations in *D. punctatus tabulaeformis*

Normalizing characteristic vector	Factor 1	Factor 2	Factor 3
Shannon-weaver index of diversity	0.2403	0.3895	-0.1527
Simpson diversity index	0.2218	0.3852	0.1777
Gradient	0.1933	0.0608	-0.4748
Exposure	0.4173	-0.1938	0.1688
Crown density	-0.1848	0.4354	-0.3303
Organic matter	-0.3535	-0.3378	-0.1414
Nitrogen	-0.3636	-0.1911	0.3595
Tree height	-0.0288	0.4434	0.4204
Diameter at breast height	-0.2028	0.216	0.4462
Incidence extent	0.4173	-0.1938	0.1688
Stand type	0.4173	-0.1938	0.1688

Table 37. Principle component analysis of 3 populations in *D. punctatus tabulaeformis*

3.2 The influence of ecological factors on the genetic diversity of *D. punctatus spectabilis*

Correlation analysis by SPSS software of 4 populations of *D. punctatus spectabilis* in different geographical areas shows a significant negative relation between genetic diversity and elevation, and the genetic diversity within populations was positively related with the annual temperature and moisture. In opposite, the genetic diversity was weakly related with latitude. Besides, there was a positive relation between the genetic distance and the distance of elevation, which shows that geographic isolation has obstruct effect on gene flow (table38,39). Generally, the genetic distance between groups in the same population is certainly related with the geographical distance and host($p<0.01$,high notable correlativity).

	QC~TC	TC~JC	JC~SC	SC~QC	TC~SC	QC~JC
Altitude distance (m)	247	40	520	717	480	197
Genetic distances	0.0720	0.0521	0.0500	0.0808	0.0733	0.0813
Relationship quotiety			0.2780			

Table 38. Correlation of the genetic diversitys within 4 populations of *D. punctatus spectabilis* and the ecological factors

Ecologocal variables	Genetic diversity estimated by Shannon index	Genetic diversity estimated by Nei index
Latitude	0.0920	0.0730
Altitude	-0.6250	-0.6380
Annual temperature	0.4800	0.4650
Annual moisture	0.5610	0.5640

Table 39. Relationship of genetic distances and geographical distances (altitude distances) between 4 populations of *D. punctatus* spectabilis

3.3 The influence of ecological factors on the genetic diversity of *D. punctatus* Walker

Three populations of *D. punctatus* Walker and three populations of geographic subspecies of *D. punctatus* Walker—*D. punctatus spectabilis* Bulter were examined and compared by RAPD. The influence of ecological factors on the genetic diversity is also discussed by the correlation analysis. The correlations between the genetic index and ecological and geographic factors are significant (table40,41).

Code	Location	Longitude and latitude	Altitude (m)	Annual Precipitation (mm)	Annual mean temperature (°C)	Annual mean moisture (%)
HTM	Huitong , Hunan	26° 40'N 109°26'E	260	1304.2	16.3	75
TDM	Tongdao, Hunan	26° 07'N 109°46'E	280	1480.7	16.6	75
YJM	Yujiang , Jiangxi	28° 11'N 116°54'E	310	1788.8	17.6	76
SYC	Shenyang, Liaoni	41° 11' N 119°23'E	900	500	8.3	37
JZC	Jinzhou, Liaoning	40° 27' N 119°51'E	380	640	9.1	55
QDC	Qingdao, Shandong	35°30' N 119°37'E	183	800	12.2	73

Table 40. Origin and the ecological factors of the tested *D. punctatus* Walker materials

Ecological variables	*A*	*Ae*	*h*	*I*	*P*
Longitude	-0.6658	-0.7367	-0.7506	-0.7470	-0.6686
Latitude	-0.3489	-0.4182	-0.4048	-0.4151	-0.3499
Altitude	-0.3564	-0.3312	-0.3487	-0.3357	-0.3601
Annual precipitation	0.6324	0.9161*	0.8632*	0.8927*	0.6370
Annual temperature	0.7246	0.8507*	0.8554	0.8570*	0.7282
Annual moisture	0.6265	0.6034	0.6423	0.6212	0.6302

Note: * p<0.05, notable correlativity; ** p<0.01, high notable correlativity; *A:* Number of alleles per locus; *Ae:* Effective number of alleles per locus; *I:* The Shannon information index; *h:* Nei's genetic index *P:* Percentage of polyporphic loci

Table 41. Correlation of the genetic diversity and the ecological factors of 6 populations of *D. punctatus* Walker

3.4 The influence of ecological factors on the genetic diversity of 5 sub-populations of *D. punctatus tabulaeformis*

The population genetic variability and genetic structure of 5 sub-populations of *D. punctatus tabulaeformis* in Pingquan city, Hebei province were analyzed using SSR. The influence of ecological factors on the genetic diversity is also discussed by the correlation analysis. The genetic index has positive relation with ecological factors weakly. The genetic index has positive relationship with ecological factors but not significant(table42). Stand type, slope, aspect and altitude were the main traits account for the genetic diversity according to the Principal Component Analysis (PCA).

Ecological variables	*A*	*Ae*	*I*	*Ho*	*He*	*h*	*P*
Altitude (m)	0.8238	0.8552	0.8158	-0.1849	0.6753	0.6717	0.6023
Aspect	-0.7505	-0.3947	-0.3622	0.7219	-0.1426	-0.1371	-0.3294
Slope	0.8251	0.8348	0.8591	0.2665	0.8227	0.8216	0.5535

Note:*P<0.05, **P<0.01, ***P<0.001 *A*: Number of alleles per locus; *Ae*: Effective number of alleles per locus; *I*: The Shannon information index; *Ho*: Observed heterozygosity; *He*: Expected heterozygosity; *P*: Percentage of polyporphic loci

Table 42. Correlation of genetic diversity and the ecological factors of five sub-populations of *D. punctatus tabulaeformis*

4. Conclusions

For *Dendrolimus*, beside of population control, there were a few evidences for genetic structure and population differentiation of them. For such aspect, this paper may be a more effective approach, and morphological diversities, allozyme, RAPD, AFLP, ISSR, mitochondrial DNA and SSR should be the best probe tools.

The results shows that a very high level of genetic differentiation and a very low gene flow among the species of populations of *D. punctatus* When $Nm < 1$, the mutational pressure is not strong enough to prevent that allele from reaching high frequency, gene flow can't counterbalance genetic drift, and genetic drift will result in substantial local differentiation. There are great differences on genetic diversity of *Dendrolimus* populations in different forest stands and site conditions. The correlations between the genetic index and ecological and geographic factors are significant.

5. References

[1] Brower A V Z, Desalle R. Patterns of mitochondrial versus nuclear DNA sequence divergence among Nymphalidae butterflies: the utility of wingless as a source of characters for phylogenetic inference. Insect Molecular Biology, 1998, 7:73-82.

[2] Chen N, Zhu G P, Hao J S, Zhang X P , Su C Y , Pan H C, Wu D X. Molecular phylogenetic analysis of the main lineages of Nymphalidae (Lepidoptera, Rhopalocera) based on mitochondrial 16S rDNA sequences. Acta Zoologica Sinica, 2007, 53(1): 106-115.

[3] Dai J X, Zhen Z M. Phylogenetic relationships of eleven species of Pentatominae based on sequences of cytochrome b gene. Chinese Bulletin of Entomology, 2005, 42(4):395-399.

[4] Dai J X, Zhang D Z. Phylogenetic analysis of eight species of Pimeliiae based on COⅡ and Cyt b gene sequences. Chinese Bulletin of Entomology, 2008, 45(4):554-558.

[5] Fu J Y, Han B Y. A RAPD analysis on genetic diversity of populations of *Aleurocanthus spiniferus* from seven tea gardens in eastern China. *Acta Ecologica Sinica.* 2007, 27(5): 1887–1894.

[6] Gao B J,Gao L J, Hou J H, *et al*. Genetic diversity of *Dendrolimus* (Lepidoptera) population from different geographic area. *Acta Ecologica Sinica.* 2008, 28(2): 842–848.

[7] Huang Y C, Li W F, Lu W, Chen Y J, Zhang Y P. Mitochondrial DNA ND4 sequence variation and phylogeny of five species of Bostrychidae (Coleoptera). Acta Entomologica Sinica, 2001, 44(4):494-500.

[8] Ji Y J, Hua Y P, Liu Y D. Ten polymorphic microsatellite markers developed in the mason pine moth *Dedrolimus punctatus* Walker(Lepidoptera:Lasiocampidae). Molecular Ecology Notes, 2005, 5:5911-5913.

[9] Jane K H , Clare L H, Calvin D *et al*. Genetic diversity in butterflies: interactive effects of habitat fragmentation and climate-driven range expansion. *Biology Letter*. 2006, 2: 152–154.

[10] Li A L, Xu A Y, Shen X J, Tang S M, Zhang Z F, Pan S Y. Analysis of segment sequences and molecular evolution between *Bombyx mori* and *Bombyx mandarina* using mitochondrial Cyt b gene. Canye Kexue, 2004, 30(1)80-84.

[11] Luo C, Yao Y, Wang R J, Yan F M , Hu D X , Zhang Z L. The use of mitochondrial cytochrome oxidase I (mt CO I)gene sequences for the identification of biotypes of *Bemisia tabaci* (Gennadius)in China. Acta Entomologica Sinica, 2002, 45(6):759-763.

[12] Martel C, Re´jasse A, Rousset F, *et a* . Host-plant-associated genetic differentiation in Northern French populations of the European corn borer. *Heredity,* 2003, 90: 141–149.

[13] Nangong Z Y, Gao B J, Liu J X, Yang J. Genetic diversity of geographical populations of four *Dendrolimus* species (Lepidoptera: Lasiocampidae) in China based on allozyme analysis. Acta Entomologica Sinica, 2008, 51(4):417-423.

[14] Pichon A, Arvanitakis L, Roux O, *et al*. Genetic differentiation among various populations of the diamondback moth, *Plutella xylostella* (Lepidoptera: Yponomeutidae). *Bull Entomology Res.* 2006, 96(2): 137–144.

[15] Ren Z M, Ma E B, Guo Y P. Genetic relationships among *Oxya agavisa* and other relative species revealed by cyt b sequences. Acta Genetica Sinica, 2002, 29(6):507-513.
[16] Ren Z M, Ma E B, Guo Y P. Mitochondrial DNA sequences and interrelationships of *Oxya japonica* from different Areas in China. Acta Entomologica Sinica , 2003,46(1):51-57.
[17] Sun Q X, Zhang Y L. Analyses of DNA sequence polymorphism in the mitochondrial 16S rRNA gene of Pentatominae (Hemiptera:Pentatomidae). Entomotaxonomia, 2004, 26(2):107-113.
[18] Wang R J, Wan H, Long Y, Lei G C, Li S W. Phylogenetic analysis of *Polyura* in China inferred from mitochondrial COII sequences(Lepidoptera: Nymphalidae). Acta Entomologica Sinica, 2004, 47(2): 243-247.
[19] Yuan Y Y, Gao B J, Li M, Yuan S L, Zhou G N. The genetic diversity of *Dendrolimus tabulaeformis* Tsai et Liu in forests of different stand types. Acta ecologica Sinica, 2008,28(5):2099-2106.
[20] Yuan Y Y, Gao B J, Li M, *et al*. The genetic diversity of *Dendrolimus tabulaeformis* Tsai et Liu in forests of different stand types. *Acta ecologica sinica*. 2008, 28(5): 2099–2106.
[21] Zhang A B, Kong X B, Li DM, *et al*. DNA fingerprinting evidence for the phylogenetic relationship of eight species and subspecies of *Dendrolimus* (Lepidoptera: Lasiocampidae) in China. *Acta Entomolqogica Sinica*, 2004, 47 (2): 236–242.
[22] Zeng W M, Jiang G F, Zhang D Y, Hong F. Evolutionary relationships among six Chinese grasshoppers of two genera of Catantopidae(Orthoptera: Acridoidea) inferred from mitochondrial 12S rRNA gene sequences. Acta Entomologica Sinica, 2004, 47 (2): 282-252.
[23] Zhang A B, Kong X B, Li D M, Liu Y Q. DNA fingerprinting evidence for the phylogenetic relationship of eight species and subspecies of *Dendrolimus* (Lepidoptera: Lasiocampidae) in China. Acta Entomologica Sinica, 2004, 47(2): 236-242.

15

Biodiversity in a Rapidly Changing World: How to Manage and Use Information?

Tereza C. Giannini, Tiago M. Francoy,
Antonio M. Saraiva and Vera L. Imperatriz-Fonseca
University of São Paulo
Brazil

1. Introduction

Biodiversity science has been evolving quickly and moved from a focus on systematics and taxonomy in the 1970–80s, to a more dynamic view of biodiversity's role in ecosystem functioning throughout the 1990s. The early 2000s have placed biodiversity within the context of ecosystem services and human well-being, and some efforts are currently focusing on putting this concept into practice, and on valuing and mapping ecosystem services in order to shed light on economic and environmental consequences of decisions (Larigauderie and Mooney, 2010a).

Ecosystem services are defined as the benefits that humans obtain from ecosystems (Seppelt et al., 2011). The Millennium Ecosystem Assessment (MA, 2005) contributed substantially to pose the ecosystem services concept as a policy tool to achieve the sustainable use of natural resources bringing a broad research approach, where ecological, economic and institutional perspectives are integrated to produce insights into human impacts on ecosystems and the welfare effects of management policies.

In December 2010, the United Nations Environment Programme (UNEP) was asked to convene a meeting to determine modalities and institutional arrangements of a new assessment body to track causes and consequences of anthropogenic ecosystem change (Perrings et al., 2011). This was an important step to the foundation of the Intergovernmental Platform on Biodiversity and Ecosystem Services (IPBES) that works closely with UNESCO, FAO, UNDP and other relevant organizations (Larigauderie and Mooney, 2010b). The establishment of IPBES provides an important link with international policy, proposing a relationship between key scientific organizations, environmental policy bodies, and research funding organizations, which is a critical feature to address both scientific capacity and the policy relevance of research aiming to build capacity for and strengthen the use of science in policy making.

As pointed out by Mooney et al. (2009), the capacity of ecosystems to deliver essential services to society is already under stress and it is urgent to track the changing status of ecosystems, deepen the understanding of the biological underpinnings for ecosystem service delivery and develop new tools and techniques for maintaining and restoring resilient biological and social systems. Additionally, solving problems posed by global change requires coordinated international research, and as much attention to social science as it does to natural science (Carpenter et al. 2009).

Pollination is considered as a key element of ecosystem services (Daily, 1997). Ollerton et al. (2011) estimated that the proportion of animal-pollinated species is near 78% in temperate-zone communities and 94% in tropical communities and that the global number and proportion of animal pollinated angiosperms is 308 006, which means 87.5% of the estimated species-level diversity of flowering plants.

The decline of pollinators has received attention since the 1990 decade (Buchmann e Nabhan, 1996; Kearnes et al., 1998). Recently, multiple drivers were suggested as the main causes to this decline (Schweiger et al., 2010; Potts et al., 2010) such as loss and fragmentation of habitat, aggressive agricultural practices, pathogens, invasive species and climate changes.

In order to achieve a better understanding about pollinator species threats it is necessary new research approaches, especially considering the necessity to build useful public policies to protect them. Here, we discuss new approaches to research on pollinators, especially bees, based mostly on Information Technology tools, such as, Biodiversity Databases, DNA Barcode, Morphometric Analysis and Species Distribution Modeling. At the end, a study case is presented, considering some Brazilian bee species and the potential impact of climate change on their distribution.

2. Digitization and sharing of biological collection data

According to Chapman (2009) the Earth's biodiversity is estimated to comprise approximately 11.3 million species, from which less than 2 million have been formally described by science. These figures reveal the limited knowledge we have which is a key issue for the preservation and sustainable use of biodiversity and ecosystem services. In order to fill that gap more field data is necessary to discover and map the biodiversity before it is gone. Nonetheless a lot can be done with the existing data, if it becomes more available and if other techniques are applied to analyze the data.

Traditionally biodiversity primary data are hosted in biological collections distributed around the world. They vary broadly in relation to size and organization, ranging from large, structured, well-documented and maintained museum collections to small sets of specimens kept by individual researchers with limited resources. Both data sources are important as they may cover different gaps - taxonomic and geographic - in our quest to know life on Earth.

The most traditional users of biological collections have been taxonomists and systematists that use them for identifying, naming and classifying species, for studying the diversity of species and the relationships among them through time (Baird, 2010). However, while these studies are essential for the development of other disciplines, such as ecology, biological collections are also essential data sources to help answer questions that interest and may involve many more individuals and knowledge areas including basic biology, human economics, and public health.

Typically they help address questions on natural resource inventories, on the occurrence and distribution of species over space and time; on the reasons for changes that may have occurred; the effect of environmental change – including climate change - on biodiversity (Scoble, 2010). This applies to native (wild) species, to economically important species, to infectious disease vectors, and to invasive species for which distribution prediction can be very helpful.

Despite the broad use already in place, biological collection data still has a great potential to be used in research, on natural and agricultural resources management, on education and on sustainability science (Scoble, 2010).

A broader, more open and easier access to specimen data is vital to distribute information and in turn create knowledge (Canhos et al., 1994; Baird, 2010). However for this to become effective it is necessary to digitize data and make it available on the web. Only then we will be able to make plain use of the wealth of data and information which is hardly accessible in many cases in collections throughout the world and which, in many cases, only integrated can provide a better picture of a species scenario.

The digitization of collection data is in itself a challenge. It implies an important effort in terms of cost and time, which sometimes competes with other demands on those who digitize. The cost-effectiveness of data digitization is not easy to prove, especially when resources are scarce, although its scientific value can be agreed upon. In cases where an economically important question can directly benefit from the data, this can be less of a problem. Since both volume and quality of data are essential, digitization in a larger scale demands the effort to be prioritized, focused and sustained, according to Scoble (2010). The author also mentions the difference in digitization efforts that is required for different taxa, such as plants and insects, as a result of the methods used for mounting the specimens and the labels that contain the data to be digitized.

Currently biological data digitization is a global effort which is led by institutions such as GBIF and TDWG. The Global Biodiversity Information Facility (GBIF, www.gbif.org) was created in 2001 after a recommendation from a working group of the Megascience Forum of the Organization for Economic Cooperation and Development (OECD), and is open to participation of any country or international organization that agrees with its purpose of making scientific biodiversity information freely available. Its three core services and products are: "1. an information infrastructure - an Internet-based index of a globally distributed network of interoperable databases that contain primary biodiversity data; 2- Community-developed tools, standards and protocols - the tools data providers need to format and share their data; and 3 - Capacity-building - the training, access to international experts and mentoring programs that national and regional institutions need to become part of a decentralized network of biodiversity information facilities". Besides developing tools to be used by itself and by others, such as a data portal, GBIF provides access to more than 276 million occurrence registers (including specimens and observations) integrating in a single access point data of data providers from all over the world.

Other regional and national initiatives have collaborated and participated actively on the global effort towards digitizing and standardizing biological data: in Europe (ENHSIN - European Natural History Specimen Information Network, and EDIT - European Distributed Institute of Taxonomy), in America (IABIN - InterAmerican Biodiversity Information Network, with data from many countries in the continent - www.iabin.net).

The Biodiversity Information Standards (TDWG - www.tdwg.org), also known as the Taxonomic Databases Working Group, was originally formed to establish international collaboration among biological database projects. It now focuses on the development of standards for the exchange of biological data, having as mission also the promotion of the standards. Maybe the most important existing standard is Darwin Core (DwC), a standard for exchange of biological information. It is primarily based on taxa and their occurrence in nature as documented by observations, specimens, and samples, and related information. Other important standard is a protocol for data exchange, TAPIR - TDWG Access Protocol for Information Retrieval (www.tdwg.org/standards), which allows data to be exchanged among different systems, using agreed upon standards, such as DwC.

Currently one trend within the community of biodiversity informatics is to develop new standards for other contents, expanding from the current specimen/observation focus to other aspects of biodiversity, such as genomic data, interaction data (Saraiva et al., 2009) and species data, and multimedia data, such as images. The new contents will broaden the scope of the data networks and offer new possibilities for data analysis hopefully allowing address issues that are even closer to societal needs, cross-cutting different disciplines.

Specimen and observation data are fundamental to develop distribution models using ecological niche concepts. Molecular data and images are key to identify a specimen and to study the relationship between individuals and populations.

3. DNA barcode

One of the biggest problems faced today by researcher is the lack of specialized personnel to identify biodiversity. In times of a rapidly changing world and fast loss of habitats and biodiversity, it is almost impossible to measure the existing species in the ecosystems. There is a lack of identification keys and genera revisions. The taxonomists are few and normally have more complexes problems to focus the simple species identification for the general public or to scientists from other science areas. The development of alternative tools to assess biodiversity other than traditional taxonomy is an urgent need.

At this point, the astonishing development of molecular biology in the last years has indicated new alternatives that can be used to identify species. Since the final of the last century, mitochondrial DNA (mtDNA) has been used as a very interesting and powerful alternative tool. This molecule has an enormous potential due to extremely peculiar and unique characteristics, like being a small circular genome, with high evolutionary rates but well conserved in animals (Arias et al., 2003). In neotropical stingless bees, it has been largely used in populational studies and in the evaluation of genetic diversity (Francisco et al., 2001; 2008; Brito & Arias, 2005; May-Itzá et al., 2010).

Although very controversial, the use of this molecule to identify species was proposed in 2003. Based on the principle that differences on the sequence of the genes are greater between species than within species, the proposition consists in sequencing approximately 650 base pair from the beginning of the Cytochrome Oxidase I (CO-I) gene and comparing it among the species (Herbert et al., 2003). According to recent revisions, the studies show the efficiency of these genic regions to discriminate species and it is working well in the vast majority of the animal cases studied until that moment (more than 95%) (Vogler and Monaghan, 2007; Waugh, 2007).

In fact, the use of a mtDNA sequence to identify cryptic species is constantly reported in literature. However, the standardization of the procedures and the establishment of some guidelines are the novelty in the DNA barcode proposition (Brown et al., 1999; Mitchell and Samways, 2005). Briefly, the following sequence is proposed in this procedure:

- the standardization of the region to be used;
- the possibility of an operation to be largely used, since the sequencing methodologies are becoming more and more accessible;
- the obligatory deposition of vouchers in entomological collections, in order to facilitate future studies of combined morphological + molecular studies;

- an accurate organization of large data bank available to the general public at the CBOL (Consortium for the Barcode of Life) website (Mitchell, 2008).

In bees, some studies are corroborating the effectiveness of the technique in species identification. The complete bee fauna of a taxonomically well resolved region was tested and the 150 species were correctly identified. Together with these results, they also identified some cryptic species and joined individuals from different sexes in the same species. In this last case, most of these species description was based in individuals from only one gender (Sheffield et al., 2009).

Another example is the use this approach, combined with traditional morphological analysis, in a study of a taxonomically extremely difficult group of bees, the subgenus *Dialictus* (family Halictidae; genus *Lasioglossum*). In this case, DNA barcoding proved essential for the delimitation of numerous species that were morphologically almost indistinguishable. The main conclusion of these studies is that DNA barcoding is efficient at the detection of cryptic species, associating the sexes of dimorphic species, associating the castes of species with strong queen-worker dimorphism and as a generally useful tool for basic identification (Gibbs, 2009). (For a revision of successful cases see Packer et al., 2009).

A global campaign to barcode the bees of the world has been initiated (see the website at: www.bee-bol.org).

4. Morphometric analysis

The first attempts to classify bee subspecies of *Apis mellifera* was based mainly in differences in color and body size. However, since there is a great superimposition in these parameters, most of the classification systems based in these characteristics failed in correctly identify the individuals (Ruttner, 1988). In 1940, Goetze proposed a large number of measures, to be taken from several parts of the bee body in order to better differentiate the geographical ecotypes present all over its wide geographical distribution that encompasses Africa, Europe and parts of Asia.

However, all the analysis used until the moment were based in uni-variate statistics, which takes into consideration only one measure at a time and the range of the measures often overlap and turn more difficult to achieve a precise identification. It was only after the works of DuPraw (1964, 1965a; b) that the usage of multivariate statistics was proposed and, with help of Principal Component Analysis and Discriminant Analysis that the identifications became more precise. An important advance is also proposed in this series of works, where DuPraw (1965a) indicated the use of measures that are independent of size, like angles between vein junctions in the wings, avoiding the environmental effects, like food availability, parasites and others.

After these propositions and a series of small studies, it is published a guide to discriminate the subspecies of *A. mellifera* (Ruttner et al., 1978). In this work, the authors propose approximately 40 measures to be taken from several parts of at least 20 bees per colony to achieve a good confidence in the classification. It was based on morphometric results that the existence of evolutionary branches in *A. mellifera* (Ruttner et al., 1978) that were later confirmed by mitochondrial DNA (Franck et al., 2000), microsatellites (Estoup et al., 1995) and SNPs (Withfield et al., 2006). In spite of being very informative and confident, this kind of analysis is often very time consuming.

More recently, allied to the development of computational methods, the analysis became faster and some of them completely automated as ABIS (Automated Bee Identification

System) (Schröder et al., 2001). The first step in this process was the construction of a semi-automated system based on features extracted from the images of the forewings, in which the user had to plot landmarks in the wing vein intersections (Schröder et al., 1995). A full automated version of the software was developed with some modifications, like the automated identification of the landmarks and the implementation of a non-linear discriminant analysis, which improved the identification rates of the individuals (Schröder et al., 2001). In this process, ABIS extract more than 300 features related to create a "fingerprint" of each species. This pattern is stored in a databank and each new wing loaded in the software is compared to the databank in order to identify the species. It was very efficient in discriminating the European species of the genus *Andrena, Colletes* and *Bombus* (Schröder et al., 2001), Africanized honey bees (Drauschke et al., 2007; Francoy et al., 2008) and also *Euglossa* species (Francoy et al., unpublished data).

Another morphometric technique that is presenting very interesting results concerning shape variation is geometric morphometrics (Bookstein, 1991). While studies with standard morphometrics analyze shape variation using co-variation of pairs of linear measures, geometric morphometrics is based on the variation of the relative positions of the landmarks and therefore, is able to describe more clearly any changes in shape and also to graphically reconstruct these differences (for a detailed description of the method, see Rohlf and Marcus, 1993).

The first attempt to use this methodology in the patterns of wing venation to differentiate bee groups was done in Africanized honey bees and in the subspecies that formed this hybrid. The relative warps analysis of the landmark positions in the wing was able to correctly classify 85% of the individuals in the correct group and the higher error rates were found in two subspecies that belong to the same evolutionary branch (Francoy et al., 2008). It is important to state that these bees are not easily distinguishable even for the well established standard morphometric methods. Another interesting result from this work is the possibility of correctly identify 99.2% of the Africanized sample, which is always a concern in areas newly occupied for these bees. Additionally, the usage of these methods allows a quicker identification than the traditional methods, once it can be done in a few minutes while the identification through standard morphometrics is more time consuming, around a few hours per colony. Still in honey bees, it was already demonstrated that other European subspecies (Tofilsky, 2008) can be identified using this methodology as well as different species from the genus *Apis* (Rattanawannee et al., 2010).

In stingless bees, the first work using this methodology demonstrated the power of the technique to identify cryptic biodiversity (Francisco et al., 2008). Colonies from two distinct populations of *Plebeia remota* kept in the same apiary for more than 10 years do not presented any gene flow. Until that moment, the populations were considered as the same species. This result was reinforced by other molecular markers, like mitochondrial DNA and cuticular hydrocarbons, which pointed in the same direction of the morphological data. It was also very informative to discriminate species with very little or no external morphological differences, like species from the genus *Eubazus* (Villemant et al., 2007), *Bombus* (Aytekin et al., 2007) and *Euglossa* (Francoy et al., unpublished results). Other works also indicated the efficiency of the technique in stingless bees. When working with bees from the same genus, studies showed 93.4% of success in the discrimination of 6 species of *Plebeia* (Francoy et al., unpublished results). It was also demonstrated the discrimination of sub-populations of *Nannotrigona testaceicornis* (Mendes et al., 2007), differences between the wings of males and workers in stingless bees (Francoy et al., 2009) and in honey bees

(Rattanawannee et al., 2010), and also the efficiency in discriminating the species when large datasets are used at the same time, as the correct identification of 93% of the individuals into the respective group for 34 different species of stingless bees (Francoy et al., unpublished results).

Together with the efficiency in identifying species based only in the patterns of wing venation, perhaps one of the most important applications of this kind of analysis is the possibility of mapping intra-specific variability within a species and consequently, tracking the geographical origin of samples. It allows the researchers to evaluate this variability in several geographic scales. The stingless bee *Melipona beecheii* has a geographical distribution that ranges from Mexico to Costa Rica, where it inhabits the most varied environments. An analysis of the patterns of wing venation of bees from Mexico, Nicaragua, El Salvador, Guatemala and Costa Rica indicated marked differences among the populations, correctly re-assigning 87% of the individuals to the respective group (Francoy et al., 2011). It was also demonstrated that it is valid for other examples, like bees from the genus *Peponapis* in North America (Bischoff et al., 2009), *Apis florea* (Kandemir et al., 2009) and *Apis mellifera* populations (Özkhan and Kandemir, 2010) and *Nannotrigona testaceicornis* sub-populations (Mendes et al., 2007).

Another geometric approach that is very promising is the outline of wing cells. It has already been demonstrated that features extracted from a single wing cell can discriminate *Apis mellifera* subspecies (Francoy et al., 2006). Based on this principle, it was proposed that people already knew how to manage colonies to transportation around 3000 years ago. In an archaeological excavation in the middle of the Jordan valley in Northern Israel it was found what appears to be a well-organized apiary. Two of the hives contained charred honey comb remains with many honey bee body remains. Although most remains were damaged, two wings with clear cells were of sufficient quality to perform morphometric measurements comparable to those available for present-day subspecies over the entire distribution range of *A. mellifera*. However, as only small parts were available, only feature extracted from single wing cells were compared. It was determined that the wings belonged to the subspecies currently living in parts of Turkey, instead of the one living in Israel. Since the climatic data indicated no extreme climate and vegetation change in the last 3000 years, the authors concluded that the beekeepers already knew how to transport colonies across long distances and kept importing bees from Turkey because of the more suitable behaviour of the *A. m. anatoliaca* bees rather than the original *A. m. siriaca* (Bloch et al., 2010).

Despite effective species discrimination from application of landmark or outline-based methods used independently, the combined results of these two methods is only now being investigated. In an exploratory study, five species of *Euglossa* were analyzed using landmark and outline based methods in order to compare the efficiency of both. Regarding the landmark analysis, 18 landmarks were used and achieved 84% of correct identifications. In the outline based analysis, a complete exploratory characterization of all wing cells was made and the wing cells that better discriminated the five species correctly re-assigned 77% of the individuals to the respective group. However, when using the features extracted in both analysis in a combined matrix, the correct classification rates achieved 91% (Francoy et al., unpublished results).

In order to improve the process and to make the analysis faster and more precise, new tools are being developed for a complete automation of the system like algorithms to automatic identification of the landmarks (Bueno, 2010) and new processes of features extraction that make the entire process more reliable and efficient (Buani, 2010). The automated identification system uses two computational algorithms to complete the recognition and classification of bee species. The first algorithm, named kNN (k-Nearest Neighbour), is used to select and extract morphometric features related to the distances between the landmarks plotted in the wing veins intersections from the pictures. The second algorithm, named FkNN (Fuzzy kNN) implements a variation of the Fuzzy Logic for species classification. For an optimized result, the features selection involves a statistical analysis which evaluates the better landmarks for the classification process and only the most informative are used in the species characterization in the Fuzzy Logic (Buani, 2010).

The morphometric analysis of forewing is a very powerful tool to describe species variation and also to identify species based on landmark and outline morphometric methods. Allied to that, morphometric analysis is a fast, inexpensive and informative method to be used in the characterization of species and its variation.

5. Modeling species distribution

Innovative techniques are urgent to understand species geographic distribution, especially considering the impact of global changes. A computation technique was developed to model species distribution. This tool received different names, which will be considered here as synonymous, e. g., species distribution modeling, ecological niche modeling, and recently, habitat suitability modeling.

Species distribution modeling (SDM) can combine georeferenced occurrence data points (latitude x longitude) with different data sets that characterize the environment where the focal species occur. These sets are combined and analyzed aiming to build a representation of ecological requirements of the focal species or, in other words, a representation of their ecological niche. The final result can be projected in the geographic space, indicating the areas that are suitable to the focal species and can be potentially occupied by it.

Usually, these data sets are comprised by abiotic features, such as temperature, precipitation and altitude, which describe the environment where the species occur. But SDM can include data about occurrences of interacting species (biotic features) that are also responsible for shaping geographical distribution, such as other species involved with the focal species on mutualism, competition or parasitism. And finally, it can include data about the species dispersal capacities, in order to estimate their capability of occupying new environments. These features are the base of modelling and its conceptual framework can be found mainly in Soberon (2010) (but also see Elith and Leathwick, 2009). Nowadays, abiotic, biotic and dispersion capacities can be integrated in BIOMOD (Thuiller, 2003) a computational system developed to R (The R Foundation for Statistical Computing) that can be used to model species distribution.

Considering the relationship between bees and plants, interactions are key aspects to include in SDM and have been considered its main challenge (Elith and Leathwick, 2009). Interactions are also of special concern when considering scenarios of global change (Schweiger et al., 2010). Mismatches between the correspondences of geographic areas of obligate interacting species due to climate change were already suggested (Stralberg et al., 2009). Besides, the relationship between pollinators and their host plants includes a temporal correspondence, in which plants synchronize the flowering period with their pollinators' activity, and this correspondence can also be changed due to climate changes (Hulme, 2011)

Interacting species are closely related to their geographic areas of distribution (Giannini et al., 2010; Giannini et al., 2011). For example, when characterizing the flower visitors' composition at the whole range of their host plant distribution, Espíndola et al. (2011) found geographically structured variability of the prevailing visitor. They suggested that climate is driving the specificity of this interaction, by potentially affecting the phenology of one or both interacting species, providing an example of the direct effect that the abiotic environment can have on the plant–insect interaction.

This is in accordance with Thompson (2005) who suggested a geographic mosaic theory of coevolution stating that interspecific interactions commonly exhibit geographic selection mosaics and trait remixing among populations. From this view, the form and trajectory of coevolutionary selection vary across landscapes. In addition, gene flow and metapopulation dynamics continually shift traits among populations, thereby continually altering the structure of local selection.

Laine (2009) reviewed 29 studies that support this theory, concluding that natural coevolutionary selection produces genetic differentiation among populations and may be an important mechanism promoting diversity in nature given how different types of interactions show divergence, and how variable the causes promoting such divergence are. One of the remarkable results of this review is the spatial scale over which it is possible to find divergent coevolutionary trajectories. Variation was detected in populations separated by some hundreds of kilometers highlighting the potential for the environment to create geographically variable selection trajectories. For example, analyzing a rare and endangered solitary bee (*Colletes floralis*), Davis et al. (2010) found an extremely high genetic differentiation among populations at the extreme edges of the species range. Also, Pellissier et al. (2010) analyzed how the traits of different pollination syndromes influence the distributions of plant species in interaction with pollinators. They used a combination of environmental descriptors and found a potential effect of the pollinator on the spatial distribution of plant species. Also, analyses of a system involving the Japanese camellia and its obligate seed predator, found that the sizes of the plant defensive apparatus (pericarp thickness) and the weevil offensive apparatus (rostrum length) clearly correlated with each other across geographically structured populations (Toju and Sota, 2006).

Therefore, intermingled with environmental (abiotic) and interaction (biotic) features, geographical distribution is also related to species evolutionary trends, determining patterns of genetic diversity and trait variation across space. New approaches are necessary to analyze the importance of these complex features. Recently, Pavoine et al. (2011) suggested a framework based on a mathematical method of ordination to analyze phylogeny, traits, abiotic variables and space in a plant community. Another example can be found in Diniz Filho et al. (2009) proposing an integrated framework to study spatial patterns in genetic diversity within local populations, coupling genetic data, SDM and landscape genetics. Also, Kuparinen and Schurr (2007) developed a framework to link the spatio-temporal dynamics of plant populations and genotypes, and a similar approach was suggested for modeling the variation of geographical distribution of animals in a climate change scenario (Kearney and Porter, 2009).

To analyze the multiple drivers shaping the species geographical distribution is a challenge that will be met by integrating different fields of research. In order to attain the objective of predicting impacts on species distribution due to global changes, it is necessary to consider that species are genetically heterogeneous entities and, in order to protect them, it is necessary to protect its genetic diversity. As species diversity might act as insurance against environmental changes, genetic diversity should also have the potential to protect communities from environmental variability (Lavergne et al., 2010). Taubmann et al. (2011)

analyzed the genetic population structure of the endangered mayfly (*Ameletus inopinatus*) in its European range genotyping hundreds of individuals from different populations. They found variations in genetic diversity and also projected the distribution of species through SDM for the year 2080 finding some areas of regional habitat loss. By relating these range shifts to the population genetic results, they were able to identify conservation units that, if preserved, would maintain high levels of the present-day genetic diversity and continue to provide long-term suitable habitat under future climate change scenarios.

Most ecological forecasting of future species ranges is based on models that generally ignore evolution and assume that the mechanistic relationship between species abundance and environmental characteristics is unchanged at the timescale of the projection (Lavergne et al., 2010). But there is accumulating evidence that evolution can proceed fast (Hairston et al., 2005) and genetic variation for adaptation - and more generally for traits defining species ecological niches - is common both between and within populations, suggesting a high level of local adaptation to climate at a fine scale (Pearman et al., 2008). Adaptation and dispersal are often presented as alternative mechanisms whereby a population can respond to changing environmental conditions playing a crucial role in tracking favorable environmental conditions through space (Pease et al., 1989). Thus migration of different genotypes could have important consequences for the evolution of geographical distribution limits (Davis et al., 2005).

Addressing the main aspects discussed here about distribution of species, it was suggested that the new trends on SDM, regarding the impacts of global changes on species diversity, are niche evolution, phylogeographic and phylogenetic research (Zimmermann et al., 2010). As pointed out by Gilman et al. (2010), the key question is not the effects resulting from global change on individual species, but the stability of the system as a whole. Integrated fields of research will allow novel analysis of both historical and contemporary drivers of species ranges, and will likely provide new possibilities to understand present day species distributions and project them to the future.

Species distribution modeling presents some steps and requires expertise knowledge about the focal species and also ecology, geography and clime. The following summarized steps are suggested.

5.1 Occurrence points

It is necessary to prepare a database with presence and absence points of the focal species. This step can include the georeferencing of points (latitude x longitude) and the exclusion of doubtful or inaccurate information. Occurrence data can be obtained from biodiversity data providers such as Global Biodiversity Information Facility (GBIF) and The Inter-American Biodiversity Information Network (IABIN). Also, to search occurrences in the literature or perform new local surveys can provide additional information.

5.2 Environmental data sets

This step aims to obtain the environmental layers to be used. Usually, they must be on 'raster' format, in which a matrix of cells is used to build the image. Generally, a Geographic Information System (GIS) software is necessary, such as ArcGIS (ESRI Inc.) or DIVA GIS (LizardTech, Inc. and University of California) to prepare them for SDM. Climate data sets can be found in the WordClim (Hijmans et al., 2005) and in the International Center of Tropical Agriculture (CIAT) websites. Categorical data sets can be found in World Wild Life

Fund (WWF) (Terrestrial Ecorregion - Olson et al., 2001) and Global Land Cover database (Bicheron et al., 2008).

5.3 Algorithm

Algorithms are finite sequences of instructions for calculating a function. There are lots of algorithms for SDM. Two important ones are Maxent (Maximum Entropy - Phillips et al., 2006) and Genetic Algorithm for Rule set Production (GARP - Stockwell and Peters, 1999) that have been successfully applied to small data sets with presence-only occurrence points (Wisz et al., 2008). GARP and other algorithms can be found in openModeller (Santana et al., 2008), a computational system to perform SDM. Another system available to SDM is BIOMOD (Biodiversity Modelling – Thuiller, 2003) developed to R (The R Foundation for Statistical Computing) that presents nine algorithms.

5.4 Model evaluation

Generating an independent data set is necessary to evaluate the model accuracy. This data set can be obtained through new surveys on supplementary areas or dividing the original data set in two. One of the data sets will be used to generate the models (train data) and the other to test it (test data). Usually, it is suggested to divide the original data set randomly and without reposition in 70% of the data to generate the model, and 30% to test it (Fielding and Bell, 1997; Hirzel and Guisan, 2002). It is also possible to divide the data considering their spatial pattern (Peterson et al., 2008). The area under the receiver operating characteristic curve (AUC) is usually used to evaluate the models. Values of AUC range from 0.5 for models with no predictive ability, to 1.0 for models giving perfect predictions (Swets, 1988). Some authors discussed the use of AUC to evaluate the model accuracy (Peterson et al., 2008) but nowadays, it is the most important method to this end (but see other alternatives on Thuiller, 2003). It is also possible to check the model accuracy conducting new surveys on the areas suggested as potential areas of occurrence by modelling.

6. Example of tools' integration using pollinators

Preliminary analyses regarding some Brazilian pollinators and species distribution modeling were recently done. We chose some species of *Melipona* and *Centris* bees (Apidae; Hymenoptera) to forecast the impact of future climate changes (Saraiva et al., in press; Giannini et al., in press).

Melipona genus is comprised by eusocial bees that are mainly associated to Atlantic Forest, an endangered moist forest of Brazil. This genus was suggested as an important pollinator to this ecosystem by Ramalho (2004). On the other hand, *Centris* is comprised by solitary species that are especially important to plants that produce floral oil, on which they depend to build their nest and feed their offspring (Simpson 1989).

The analyzed species of *Melipona* were also reported as pollinators of some fruit crops, as açai (*Euterpe oleracea*), avocado (*Persea americana*) and guava (*Psidium guajava*) (Castro, 2002; Venturieri *et al.*, 2008). The *Centris* species were reported as pollinators of acerola (*Malphighia punicifolia*), murici (*Byrsonima crassifolia*), cashew (*Anacardium occidentale*) and tamarind (*Tamarindus indica*) (Castro, 2002; Freitas *et al.*, 2002; Vilhena and Augusto, 2007; Rego *et al.*, 2006; Ribeiro *et al.*, 2008; Siqueira, 2010). Recently, Nunes-Silva et al. (2010) also demonstrated the importance of *M. fasciculate* on performing buzz pollination on tomatoes.

Some of these crops are cultivated regionally, and in this case, the loss of native pollinators can cause a potentially higher impact on local economy.
To forecast the future distribution of some species pertaining to these two genera, we used a moderate scenario of climate change to the years 2050 and 2080 and we find an alarming decreasing of potential areas of occurrence to all species. *Melipona* species presented the highest decrease, but *Centris* species presented a remarkable fragmentation on future suitable habitats. For genetic reasons, bees are particularly sensitive to the effects of small population size through the production of sterile males (Zayed and Packer, 2005; Alves *et al.*, 2011). Therefore, small fragmented areas could not be able to maintain viable population.
Nevertheless, some areas were highlighted as possible areas of conservation to these bees, because they are suitable now and will remain like this in the future. These areas can be considered as important areas in a large scale study about pollinators' conservation. Another interesting approach to conserve genetic diversity of these organisms is to associate the genetic variability of the populations with the future distribution maps and check if the populations with larger genetic variability are found in the possible remaining refuge areas. As discussed previously, this genetic variability can be measured by morphometric or molecular methods. The comparison of the current distribution of the genetic variability with the future scenarios for species distribution is of fundamental importance to verify if this variability will be naturally preserved or if other public actions will be necessary to a more effective conservation.

7. Acknowledgment

To FAPESP (04/15801-0), CNPq (575069/2008-2) and Research Center on Biodiversity and Computing (BioComp).

8. References

Alves, D. A.; Imperatriz-Fonseca, V. L.; Francoy, T. M.; Santos-Filho, P. S.; Billen, J.; Wenseleers, T.(2011). Successful maintenance of a stingless bee population despite a severe genetic bottleneck. *Conservation Genetics*, Vol. 12, No. 3, pp. 647-658, 1566-0621.

Arias, M. C.; Francisco, F. O.; Silvestre, D. (2003). O DNA mitocondrial em estudos populacionais e evolutivos de meliponíneos. In: Melo, G. A. R. and Alves-dos-Santos, I. (eds) *Apoidea Neotropica: homenagem aos 90 anos de Jesus Santiago Moure*, p. 305-309, UNESC, Criciúma.

Aytekin, A.M.; Terzo, M.; Rasmont, P.; Çagatay, N. (2007). Landmark based geometric morphometric analysis of wing shape in *Sibiricobombus* Vogt (Hymenoptera: Apidae: *Bombus* Latreille). *Annales de la Société Entomologique de France*, Vol 43, pp. 95-102, 0037-9271.

Baird, R. (2010). Leveraging the fullest potential of scientific collections through digitization. *Biodiversity Informatics*, Vol. 7, No. 2, pp. 130 - 136, 1546-9735.

Bicheron, P.; Defourny, P.; Brockmann, C.; Schouten, L.; Vancutsem, C.; Huc, M.; Bontemps, S.; Leroy, M.; Archard, F; Herold, M; Ranera, F; Arino, O. (2008). *GLOBCOVER - Products description and validation report.* Medias France, Paris.

Bischoff, I.; Schröder, S.; Misof, B. (2009). Differentiation and range expansion of North American squash bees *Peponapis pruinosa* (Apidae: Apiformes) populations assessed by geometric wing morphometry. *Annals of the Entomological Society of America*, Vol 102, pp. 60–69, 0013-8746.

Bloch, G.; Francoy, T. M.; Wachtel, I.; Paniz-Cohen, N.; Fuchs, S.; Mazar, A. (2010). Industrial apiculture in the Jordan valley during Biblical times with Anatolian honeybees. *PNAS*, Vol 107, pp. 11240 - 11244, 1091-6490.

Bookstein, F.L. (1991). *Morphometric tools for landmark data.* Cambridge University Press, 0521383854, Cambridge.

Brito, R.M. & Arias, M.C. (2005). Mitochondrial DNA characterization of two *Partamona* species (Hymenoptera, Apidae, Meliponini) by PCR+RFLP and sequencing. *Apidologie,* Vol 36, pp. 431- 438, 1297-9678.

Brown, B.; Emberson, R.M.; Paterson, A.M. (1999). Mitochondrial CO I and CO II provide useful markers for *Weiseana* (Lepidoptera, Hepialidae) species identification. *Bulletin of Entomological Research*, Vol 89, pp. 287 - 294, 0007-4853.

Buani, B.E. (2010). Aplicação da Lógica Fuzzy KNN e Análises Estatísticas para Seleção de Características e Classificação de Abelhas. *Master Thesis.* Escola Politécnica da Universidade de São Paulo.

Buchmann, S.; Nabhan, G. (1996). *The forgotten pollinators.* Island Press, 1559633530, Washington.

Bueno, J. F. (2010). Modelo de Arquitetura de Referência para o Sistema Automatizado de Classificação de Abelhas baseado em Reconhecimento de Padrões. *PhD Thesis.* Escola Politécnica da Universidade de São Paulo.

Canhos, V. P.; Souza, S.; Giovanni, R.; Canhos, D. A. L. (2004). Global biodiversity informatics: setting the scene for a "new world" of ecological modeling. *Biodiversity Informatics*, Vol. 1, No. 1, pp. 1-13, 1546-9735.

Carpenter, S. R.; Mooney, H. A.; Agard, J.; Capistrano, D.; DeFriese, R. S.; Díaz, S.; Dietz, T.; Duraiappah, A. K.; Oteng-Yeboahi, A.; Pereira, H. M.; Perrings, C.; Reid, W. V.; Sarukhan, J.; Scholes, R. J.; Whyte, A. (2009). Science for managing ecosystem services: beyond the Millennium Ecosystem Assessment. *PNAS*, Vol. 106, No. 5, pp. 1305-1312, 1091-6490.

Castro, M.S. (2002) Bee fauna of some tropical and exotic fruits: potential pollinators and their conservation. In: Kevan P., Imperatriz Fonseca V.L. (eds). *Pollinating Bees.* Ministry of Environment, Brasília, p. 275-288.

Chapman, A.D. (2009). *Numbers of Living Species in Australia and the World.* Report for the Australian Biological Resources Study, 2nd ed., 80 pp. Canberra, Australia.

Daily, G.C. (1997) *Nature's Services: Societal Dependence on Natural Ecosystems.* Island Press, 1559634766, Washington.

Davis, M. B.; Shaw, R. G.; Etterson, J. R. (2005). Evolutionary responses to changing climate. *Ecology*, Vol. 86, No. 7, pp. 1704–14, 0962-1083.

Davis, E. S.; Murray, T. E.; Fitzpatrick, U.; Brown, M. J. F.; Paxton, R. J. (2010). Landscape effects on extremely fragmented populations of a rare solitary bee, *Colletes floralis. Molecular Ecology*, Vol. 19, No. 22, pp. 4922–4935, 0962-1083.

Diniz-Filho, J. A. F.; Nabout, J. C.; Bini, L. M.; Soares, T. N.; Campos Telles, M. P.; Marco, P.; Collevatti, R. G. (2009). Niche modelling and landscape genetics of *Caryocar brasiliense* (Pequi tree, Caryocaraceae) in Brazilian Cerrado. *Tree Genetics & Genomes,* Vol. 5, No. 4, pp. 617-627, 1614-2942.

Drauschke, M.; Steinhage, V.; Pogoda, A.; Müller S.; Francoy T.M.; Wittmann, D. (2007). Reliable Biometrical Analysis in Biodiversity Information Systems. In: *Proceedings of the 7th International Workshop on Pattern Recognition in Information Systems*. Funchal, Portugal pp. 25-36.

Dupraw, E. J. (1964). Non-Linnean Taxonomy. *Nature,* Vol 202, pp. 849 - 852, 0028-0836.

Dupraw, E. J. (1965a). Non-Linnean taxonomy and the systematics of honeybees. *Systematic Zoology,* Vol 14, pp. 1 - 24, 1365-3113.

Dupraw, E. J. (1965b). The recognition and handling of honeybee specimens in non-Linnean taxonomy. *Journal of Apicultural Research,* Vol 4, pp. 71 - 84, 2078-6913.

Elith, J.; Leathwick, J. R. (2009). Species distribution models: ecological explanation and prediction across space and time. *Annual Review of Ecology, Evolution and Systematics,* Vol. 40, pp. 677–97, 1543-592X.

Espíndola, A.; Pellissier, L.; Alvarez, N. (2011). Variation in the proportion of flower visitors of *Arum maculatum* along its distributional range in relation with community-based climatic niche analyses. *Oikos,* Vol. 120, No. 5, pp. 728–734, 0030-1299.

Estoup, A.; Garnery, L.; Solignac, M.; Cornuet, J. M. (1995). Microsatellite variation in honey-bee (*Apis mellifera* L.) populations - Hierarchical genetic structure and test of the infinite allele and stepwise mutation models. *Genetics,* Vol 140, pp. 679-695, 0016-6731.

Fielding, A. H.; Bell, J. F. (1997). A review of methods for the assessment of prediction errors in conservation presence/absence models. *Environmental Conservation,* Vol. 24, No. 1, pp. 38–49, 0376-8929.

Francisco, F.O.; Nunes-Silva, P.; Francoy, T.M.; Wittmann, D.; Imperatriz-Fonseca, V.L.; Arias, M.C.; Morgan, E.D. (2008). Morphometrical, biochemical and molecular tools for assessing biodiversity. An example in *Plebeia remota* (Holmberg, 1903) (Apidae, Meliponini). *Insectes Sociaux,* Vol 55, pp. 231-237, 0020-1812.

Francisco, F.O.; Silvestre, D.; Arias, M.C. (2001). Mitochondrial DNA characterization of five species of *Plebeia* (Apidae: Meliponinae): RFLP and restriction maps. *Apidologie,* Vol 32, pp. 323–332, 1297-9678.

Franck, P.; Garnery, L.; Solignac, M.; Cornuet, J. M. (2000). Molecular confirmation of a fourth lineage in honeybees from the Near East. *Apidologie,* Vol 31, pp. 167-180, 1297-9678.

Francoy, T.M.; Prado, P.P.R.; Gonçalves, L.S.; Costa, L.D.; De Jong, D. (2006). Morphometric differences in a single wing cell can discriminate *Apis mellifera* racial types. *Apidologie,* Vol 37, pp. 91-97, 1297-9678.

Francoy, T.M.; Wittmann, D.; Draushcke M.; Müller S.; Steinhage V.; Bezerra-Laure, M.A.F.; De Jong, D.; Gonçalves, L.S. (2008). Identification of Africanized honey bees through wing morphometrics: two fast and efficient procedures, *Apidologie,* Vol 39, pp. 488-494, 1297-9678.

Francoy, T.M.; Silva, R.A.O.; Nunes-Silva, P.; Menezes, C.; Imperatriz-Fonseca, V.L. (2009). Gender identification in five genera of stingless bees (Apidae, Meliponini) based on wing morphology. *Genetics and Molecular Research,* Vol 8, pp. 207-214, 1676-5680.

Francoy, T.M.; Grassi, M.L.; Imperatriz-Fonseca, V.L.; May-Itzá, W.J.; Quezada-Euán, J.J.G. (2011). Geometric morphometrics of the wing as a tool for assigning genetic lineages and geographic origin to *Melipona beecheii* (Hymenoptera: Meliponini). *Apidologie,* doi:10.1007/s13592-011-0013-0, 1297-9678.

Freitas, B. M.; Paxton, R. J.; Holanda-Neto, J. P. (2002) Identifying pollinators among an array of flower visitors, and the case of inadequate cashew pollination in NE Brazil. In: Kevan, P., Imperatriz Fonseca, V.L. (eds). *Pollinating Bees.* Ministry of Environment, Brasília, p.229-244.

Giannini, T. C.; Saraiva, A. M.; Alves dos Santos, I. (2010). Ecological niche modeling and geographical distribution of pollinator and plants: a case study of *Peponapis fervens* (Smith, 1879) (Eucerini: Apidae) and *Cucurbita* species (Cucurbitaceae). *Ecological Informatics,* Vol. 5, No. 1, pp. 59-66, 1574-9541.

Giannini, T. C.; Lira-Saade, R.; Ayala, R.; Saraiva, A. M.; Alves dos Santos, I. (2011). Ecological niche similarities of *Peponapis* bees and non-domesticated *Cucurbita* species. *Ecological Modelling,* Vol. 222, No. 12, pp. 2011-2018, 0304-3800.

Giannini, T. C.; Acosta, A. L.; Saraiva, A. M.; Alves dos Santos, I.; Garófalo, C. (in press). A. Impacto de mudanças climáticas em abelhas solitárias: um estudo de caso envolvendo duas espécies de *Centris.* In: Imperatriz-Fonseca, V. L.; Canhos, D. A. L.; Saraiva, A. M. (editors). *Polinizadores no Brasil.* São Paulo, Instituto de Estudos Avançados.

Gibbs, J.J. (2009). New Species in the *Lasioglossum petrellum* Species Group Identified through an Integrative Taxonomic Approach. *The Canadian Entomologist,* Vol 141, pp. 371 - 396, 0008-347X.

Gilman, S. E.; Urban, M. C. Tewksbury, J.; Gilchrist, G. W.; Holt, R. D. (2010). A framework for community interactions under climate change. *Trends in Ecology and Evolution,* Vol. 25, No. 6, pp. 325-331, 0169-5347.

Goetze, G. (1940). *Die beste Biene.* Liedlof Loth Michaelis, Leipzig.

Hairston, N. H.; Ellner, S. P.; Geber, M. A.; Yoshida, T.; Fox, J. A. (2005). Rapid evolution and the convergence of ecological and evolutionary time. *Ecology Letters,* Vol. 8, No. 10, pp. 1114–27, 1461-0248.

Herbert, P.D.N.; Ratnasingham, S.; De Waard J.R. (2003). Barcoding animal life: citocrome c oxidase sbunit 1 divergences among closely related species. *Proceedings of the Royal Society of London. Series B, Biological Sciences,* Vol 270, pp. 313-321, 1800-1905.

Hijmans, J. R.; Cameron, S. E.; Parra, J. L.; Jones, P. G.; Jarvis, A. (2005). Very high resolution interpolated climate surfaces for global land areas. *International Journal of Climatology,* Vol. 25, No. 15, pp. 1965–1978, 0899-8418.

Hirzel, A.; Guisan, A. (2002). Which is the optimal sampling strategy for habitat suitability modelling. *Ecological Modelling,* Vol. 157, Nos. 2-3, pp. 331–341, 0304-3800.

Hulme, P. (2011). Contrasting impacts of climate driven flowering phenology on changes in alien and native plant species distributions. *New Phytologist,* Vol. 189, No. 1, pp. 272-281, 1469-8137.

Jiménez-Valverde, A.; Lobo, J. M.; Hortal, J. (2008). Not as good as they seem: the importance of concepts in species distribution modeling. *Diversity and Distributions,* Vol. 14, No. 6, pp. 885-890, 1472-4642.

Kandemir, I., Moradi, M.G., Özden, B., Özkan, A. (2009). Wing geometry as a tool for studying the population structure of dwarf honey bees (*Apis florea* Fabricius 1876) in Iran. *Journal of Apicultural Research.* Vol 48, pp. 238–246, 2078-6913.

Kearney, M.; Porter, W. (2009). Mechanistic niche modelling: combining physiological and spatial data to predict species' ranges. *Ecology Letters,* Vol. 12, No. 4, pp.334–50, 1461-0248.

Kearns, C. A.; Inouye, D. W.; Waser, N. M. (1998). Endangered mutualisms: the conservation of plant-pollinator interactions. *Annual Review of Ecology, Evolution, and Systematics,* Vol. 29, pp. 83-112, 1543-592X.

Kuparinen, A.; Schurr, F. M. (2007). A flexible modelling framework linking the spatio-temporal dynamics of plant genotypes and populations: application to gene flow from transgenic forests. *Ecological Modeling,* Vol. 202, Nos. 3-4, pp. 476–86, 0304-3800.

Laine, A. L. (2009). Role of coevolution in generating biological diversity: spatially divergent selection trajectories. *Journal of Experimental Botany,* Vol. 60, No. 11, pp. 2957–2970, 0022-0957.

Larigauderie, A.; Mooney, H. A. (2010a). The International Year of Biodiversity: an opportunity to strengthen the science–policy interface for biodiversity and ecosystem services. *Current Opinion in Environmental Sustainability,* Vol. 2, No. 1, pp. 1-2, 1877-3435.

Larigauderie, A.; Mooney, H. A. (2010b). The Intergovernmental science-policy Platform on Biodiversity and Ecosystem Services: moving a step closer to an IPCC-like mechanism for biodiversity. *Current Opinion in Environmental Sustainability,* Vol. 2, No. 1, pp. 9-14, 1877-3435.

Lavergne, S.; Mouquet, N.; Thuiller, W.; Ronce, O. (2010). Biodiversity and climate change: integrating evolutionary and ecological responses of species and communities. *Annual Review of Ecology, Evolution and Systematics,* Vol. 41, pp.321–50, 1543-592X.

May-Itzá, W.J.; Quezada-Euán, J.J.G.; Medina, L.A.M.; Enriquez, E.; De la Rua, P. (2010). Morphometric and genetic differentiation in isolated populations of the endangered Mesoamerican stingless bee *Melipona yucatanica* (Hymenoptera: Apoidea) suggest the existence of a two species complex. *Conservation Genetics,* vol 5, pp. 2079 - 2084, 1566-0621.

Mendes, M.F.M.; Francoy, T.M.; Nunes-Silva, P.; Menezes, C.; Imperatriz-Fonseca, V.L. (2007). Intra-populational variability of *Nannotrigona testaceicornis* Lepeletier 1836 (Hymenoptera, Meliponini) using relative warp analysis. *Bioscience Journal,* Vol 23, pp. 147 - 152, 1981-3163.

Millennium Ecosystem Assessment. (2005). *Ecosystems and Human Well-Being: General Synthesis.* Island Press, 9781597260404, Washington.

Mitchell, A. (2008). DNA barcoding demystified. *Australian Journal of Entomology.* Vol 47, pp. 169 - 173, 1326-6756.

Mitchell, A.; Samways, M.J. (2005). DNA evidence that the morphological 'forms' of *Palpopleura lucia* (Drury) are separate species (Odonata: Libellulidae). *Odonatologica,* Vol 4, pp. 173 - 178, 0375-0183.

Mooney, H. A.; Larigauderie, A.; Cesario, M.; Elmquist, T.; Hoegh-Guldberg, O.; Lavorel, S.; Mace, G. M.; Palmer, M.; Scholes, R.; Yahara, T. (2009). Biodiversity, climate change, and ecosystem services. *Current Opinion in Environmental Sustainability,* Vol. 1, No. 1, pp. 46-54, 1877-3435.

Nunes-Silva, P.; Hnrcir, M.; Imperatriz-Fonseca, V. L. (2010) A polinização por vibração. *Oecologia Australis,* Vol. 14, No. 1, pp. 140-151, 2177-6199.

Ollerton, J.; Winfree, R.; Tarrant, S. (2011). How many flowering plants are pollinated by animals? *Oikos,* Vol. 120, No. 3, pp. 321–326, 0030-1299.

Olson, D.M.; Dinerstein, E.; Wikramanayake, E.D.; Burgess, N.D.; Powell, G.V.N.; Underwood, E.C.; D'Amico, J.A.; Itoua, I.; Strand, H.E.; Morrison, J.C.; Loucks, C.J.; Allnutt, T.F.; Ricketts, T.H.; Kura, Y.; Lamoreux, J.F.; Wettengel, W.W.; Hedao, P.; Kassem, K.R. (2001). Terrestrial ecoregions of the world: a new map of life on earth. *BioScience,* Vol. 51, No. 11, pp. 933-938, 1981-3163.

Öskan, A.; Kandemir, I. (2010). Discrimination of western honey bee populations in Turkey using geometric morphometric methods. *Journal of Apicultural Research,* Vol 87, pp. 24 - 26, 2078-6913.

Packer, L.; Gibbs, J.; Sheffield, C.; Hanner, R. (2009). DNA barcoding and the mediocrity of morphology. *Molecular Ecology Resources,* Vol 9, pp. 42-50, 1755-098X.

Pavoine, S.; Vela, E. Gachet, S.; Belair, G.; Bonsall, M. B. (2011). Linking patterns in phylogeny, traits, abiotic variables and space: a novel approach to linking environmental filtering and plant community assembly. *Journal of Ecology,* Vol. 99, No. 1, pp. 165–175, 1365-2745.

Pease, C. P.; Lande, R.; Bull, J. J. (1989). A model of population growth, dispersal and evolution in a changing environment. *Ecology,* Vol. 70, pp. 1657–64, 0012-9658.

Pearman, P. B.; Guisan, A.; Broennimann, O.; Randin, C. F. (2008). Niche dynamics in space and time. *Trends Ecology Evolution,* Vol. 23, No. 11, pp. 149–58, 0169-5347.

Pellissier , L.; Pottier , J.; Vittoz , P.; Dubuis, A.; Guisan, A. (2010). Spatial pattern of floral morphology: possible insight into the effects of pollinators on plant distributions. *Oikos,* Vol. 119, No. 11, pp. 1805–1813, 0030-1299.

Perrings, C.; Duraiappah, A.; Larigauderie, A.; Mooney, H. (2011). The Biodiversity and Ecosystem Services Science-Policy Interface. *Science,* Vol. 331, No. 6021, pp. 1139-1140, 0036-8075.

Peterson, A. T.; Papes, M.; Soberon, J. (2008). Rethinking receiver operating characteristic analysis applications in ecological niche modeling. *Ecological Modelling,* Vol. 213, No. 1, pp. 63-72, 0304-3800.

Potts, S.G.; Biesmeijer, J.C.; Kremen, C.; Neumann, P.; Schweiger, O.; Kunin, W.E. (2010) Global pollinator declines: trends, impacts and drivers. *Trends in Ecology and Evolution* Vol. 25, No. 6, pp. 345-353, 0169-5347.

Phillips, S. J.; Anderson, R. P.; Schapire, R. E. (2006). Maximum entropy modeling of species geographic distributions. *Ecological Modelling,* Vol. 190, Nos. 3-4, pp. 231–259, 0304-3800.

Ramalho, M. (2004). Stingless bees and mass flowering trees in the canopy of Atlantic Forest: a tight relationship. *Acta Botânica Brasilica,* Vol. 18, No. 1, pp. 37-47, 0102-3306.

Rattanawannee, A.; Chanchao, C.; Wongsiri, S. (2010). Gender and species identification of four native honey bees (Apidae:*Apis*) in Thailand based on wing morphometic analysis. *Annals of the Entomological Society of America,* Vol 106, pp. 965 - 970, 0013-8746.

Rego, M. M. C.; Albuquerque, P. M. C.; Ramos, M. C.; Carreira, L. M. (2006) Aspectos da biologia de nidificação de *Centris flavifrons* (Friese) (Hymenoptera: Apidae, Centridini), um dos principais polinizadores do murici (*Byrsonima crassifolia* L. Kunth, Malpighiaceae), no Maranhão. *Neotropical Entomology,* Vol. 35, pp. 579-587, 1519-566X.

Ribeiro, E. K. M.; Rêgo, M. M. C.; Machado, I. C. S. (2008). Cargas polínicas de abelhas polinizadoras de *Byrsonima chrysophylla* Kunth. (Malpighiaceae): fidelidade e fontes alternativas de recursos florais. *Acta Botanica Brasilica,* Vol. 22, pp. 165-171, 0102-3306.

Rohlf, F. J.; Marcus, L. F. (1993). A Revolution in Morphometrics. *Trends in Ecology & Evolution,* Vol 8, pp. 129-132, 0169-5347.

Ruttner F. (1988). *Biogeography and taxonomy of honeybees.* Springer Verlag, 0387177817, Berlin.

Ruttner, F.; Tassencourt, L.; Louveaux, J. (1978). Biometrical-statistical analysis of the geographic variability of Apis mellifera L.. *Apidologie* Vol. 9, pp. 363–381, 1297-9678.

Santana, F. S; Siqueira, M. F.; Saraiva, A. M.; Correa, P. L. P. (2008). A reference business process for ecological niche modeling. *Ecological Informatics,* Vol. 3, No. 1, pp. 75-86, 1574-9541.

Saraiva, A. M.; Cartolano Junior, E. A.; Giovanni, R.; Giannini, T. C.; Correa, P. L. P. (2009). Exchanging specimen interaction data using Darwin Core. In: *Proceedings of Annual Conference of Biodiversity Information Standards* (TDWG), Montpellier.. http://www.tdwg.org/proceedings/article/view/552.

Saraiva, A. M.; Acosta, A. L.; Giannini, T. C.; Carvalho, C. A. L.; Alves, R. M. O.; Drummond, M. S.; Blochtein, B.; Witter, S.; Alves dos Santos, I.; Imperatriz-Fonseca, V. L. (*in press*). Influência das alterações climáticas sobre a distribuição de algumas espécies de *Melipona* no Brasil. In: Imperatriz-Fonseca, V. L.; Canhos, D. A. L.; Saraiva, A. M. (editors). *Polinizadores no Brasil.* Instituto de Estudos Avançados, São Paulo.

Schröder, S.; Drescher, W.; Steinhage, V.; Kastenholtz, B. (1995). An Automated Method for the Identification of Bee Species (Hymenoptera: Apoidea). *Proc. Intern. Symp. on Conserving Europe's Bees*: International Bee Research Association & Linnean Society. 6-7 p.

Schröder, S.; Wittmann, D.; Drescher, W.; Roth, V.; Steinhage, V.; Cremers, A. B. (2001). The new key to bees: Automated Identification by image analysis of wings. In: Kevan P., Imperatriz Fonseca V.L. (eds). *Pollinating Bees.* Ministry of Environment, Brasília, p. 229-236

Schweiger, O.; Biesmeijer, J. C.; Bommarco, R.; Hickler, T.; Hulme, F. E.; Klotz, S.; Kuhn, I.; Moora, M.; Nielsen, A.; Ohlemuller, R.; Petanidou, T.; Potts, S. G.; Pysek, P.; Stout, J. C.; Sykes, M. T.; Tscheulin, T.; Vila, M.; Walther, G. R.; Westphal, C.; Winter, M.; Zobel, M.; Settele, J. (2010). Multiple stressors on biotic interactions: how climate change and alien species interact to affect pollination. *Biological Review,* Vol. 85, No. 4, pp. 777–795, 1464-7931.

Scoble, M. J. (2010), Natural history collections digitization: rationale and value. *Biodiversity Informatics,* Vol. 7, No. 2, pp. 77 – 80, 1546-9735.

Seppelt, R.; Dormann, C. F.; Eppink, F. V.; Lautenbach, S.; Schmidt, S. (2011). A quantitative review of ecosystem service studies: approaches, shortcomings and the road ahead. Journal of Applied Ecology, Vol. 48, No. 3, pp. 630-636, 1365-2664.

Sheffield, C.S.; Herbert, P.D.N.; Kevan, P.G; Packer, L. (2009). DNA barcoding a regional bee (Hymenoptera: Apoidea) fauna and its potential for ecological studies. *Molecular Ecology Resources,* Vol 9, pp. 196 - 207, 1755-098X.

Simpson, B. B. (1989). Krameriaceae. *Flora Neotropica,* Vol. 49, pp. 1-109, 0071-5794.

Siqueira, K. M. M. (2010). Polinização da aceroleira (*Malpighia emarginata*). In: Ribeiro, M. F. (ed). *Segunda semana de polinizadores.* Embrapa Semiárido, Petrolina, p. 24-35.

Soberón, J. (2010). Niche and area of distribution modeling: a population ecology perspective. *Ecography,* Vol. 33, No. 1, pp. 159-167, 0906-7590.

Stockwell, D.; Peters, D. (1999). The GARP modelling system: problems and solutions to automated spatial prediction. *International Journal of Geographical Information Science,* Vol. 13, No. 2, pp. 143-158, 1365-8816.

Stralberg, D.; Jongsomjit, D.; Howell, C.; Snyder, M.; Alexander, J.; Wiens, J.; Root, T. (2009). Re-shuffling of species with climate disruption: a no-analog future for California birds. *PLoS One,* Vol. 4, No. 9, pp. 1-8, 1932-6203.

Swets, K. A. (1988). Measuring the accuracy of diagnostic systems. *Science,* Vol. 240, No. 4857, pp. 1285–1293, 0036-8075.

Taubmann, J.; Theissinger, K.; Feldheim, K.; Laube, I.; Graf, W.; Haase, P.; Johannesen, J.; Pauls, S. (2011). Modelling range shifts and assessing genetic diversity distribution of the montane aquatic mayfly *Ameletus inopinatus* in Europe under climate change scenarios. *Conservation Genetics,* Vol. 12, No. 12, pp. 503-515, 1566-0621.

Thompson, J. N. (2005). *The geographic mosaic of coevolution.* University of Chicago Press, 9780226797618, Chicago.

Thuiller, W. (2003). BIOMOD - optimizing predictions of species distributions and projecting potential future shifts under global change. *Global Change Biology,* Vol. 9, N. 10, pp. 1353–1362, 1354-1013.

Tofilski, A. (2008). Using geometric morphometrics and standard morphometry to discriminate three honeybee subspecies. *Apidologie,* Vol 39, pp. 558-563, 1297-9678.

Toju, H.; Sota, T. (2006). Imbalance of predator and prey armament: geographic clines in phenotypic interface and natural selection. *American Naturalist,* Vol. 167, No. 1, pp. 105–117, 0003-0147.

Venturieri, G. C.; Souza, M. S.; Pereira, C. A. B.; Rodrigues, S. T. (2008) Potencial nectarífero do açaizeiro (*Euterpe oleracea* Mart. - Arecaceae) na Amazônia Oriental. In: *Encontro Sobre Abelhas.* Ribeirão Preto, p. 154-158.

Vilhena, A. M. G. F.; Augusto, S. C. (2007) Polinizadores da aceroleira *Malpighia emarginata* DC (Malpighiaceae) em área de cerrado no triângulo mineiro. B*ioscience Journal,* Vol. 23: No. 0, pp. 14-23 1981-3163, 1981-3163..

Villemant, C.; Simbolotti, G.; Kenis, M. (2007). Discrimination of *Eubazus* (Hymenoptera, Braconidae) sibling species using geometric morphometrics analysis of wing venation. *Systematic Entomology,* Vol 32, pp. 625 - 634, 1365-3113.

Vogler, A.P.; Monagham, M.T. (2007). Recent advances in DNA taxonomy. *Journal of Zoological Systematics and Evolutionary Research,* Vol 45, pp. 1 - 10, 1439-0469.

Waugh, J. (2007). DNA barcoding in animal species: progress, potential and pitfalls. *Bioessays.* Vol. 29, pp. 188 - 197, 1521-1878.

Wisz, M. S.; Hijmans, R. J.; Li, J.; Peterson, A. T.; Graham, C. H.; Guisan, A.; NCEAS Predicting Species Distributions Working Group. (2008). Effects of sample size on the performance of species distribution models. *Diversity and Distributions,* Vol. 14, No. 5, pp. 763-773, 1366-9516.

Whitfield, C. W.; Behura, S. K.; Berlocher, S. H.; Clarck, A. G.; Johnston, J. S.; Sheppard, W. S.; Smith, D. R.; Suarez, A. V.; Weaver, D.; Tsutsui, N. D. (2006). Thrice out of Africa: Ancient and recent expansions of the honey bee, *Apis mellifera. Science,* Vol 314, pp. 642-645, 0036-8075.

Zayed, A.; Packer, L. (2005) Complementary sex determination substantially increases extinction proneness of haplodiploid populations. *Proceedings of the National Academy of Sciences*, Vol. 102, No. 30, pp. 10742–10746, 7949-7952.

Zimmermann, H. E.; Edwards Jr., T. C.; Graham, C. H.; Pearman, P. B.; Svenning, J. C. (2010). New trends in species distribution modelling. *Ecography*, Vol. 33, No. 6, pp. 985-989, 0906-7590.